Annals of Mathematics Studies

Number 40

ANNALS OF MATHEMATICS STUDIES

Edited by Emil Artin, Robert C. Gunning, and Marston Morse

1. Algebraic Theory of Numbers, *by* Hermann Weyl
3. Consistency of the Continuum Hypothesis, *by* Kurt Gödel
6. The Calculi of Lambda-Conversion, *by* Alonzo Church
10. Topics in Topology, *by* Solomon Lefschetz
11. Introduction to Nonlinear Mechanics, *by* N. Kryloff *and* N. Bogoliuboff
15. Topological Methods in the Theory of Functions of a Complex Variable, *by* Marston Morse
16. Transcendental Numbers, *by* Carl Ludwig Siegel
17. Problème Général de la Stabilité du Mouvement, *by* M. A. Liapounoff
19. Fourier Transforms, *by* S. Bochner *and* K. Chandrasekharan
20. Contributions to the Theory of Nonlinear Oscillations, Vol. I, *edited by* S. Lefschetz
21. Functional Operators, Vol. I, *by* John von Neumann
22. Functional Operators, Vol. II, *by* John von Neumann
24. Contributions to the Theory of Games, Vol. I, *edited by* H. W. Kuhn *and* A. W. Tucker
25. Contributions to Fourier Analysis, *edited by* A. Zygmund, W. Transue, M. Morse, A. P. Calderon, *and* S. Bochner
26. A Theory of Cross-Spaces, *by* Robert Schatten
27. Isoperimetric Inequalities in Mathematical Physics, *by* G. Polya *and* G. Szego
28. Contributions to the Theory of Games, Vol. II, *edited by* H. W. Kuhn *and* A. W. Tucker
29. Contributions to the Theory of Nonlinear Oscillations, Vol. II, *edited by* S. Lefschetz
30. Contributions to the Theory of Riemann Surfaces, *edited by* L. Ahlfors *et al.*
31. Order-Preserving Maps and Integration Processes, *by* Edward J. McShane
32. Curvature and Betti Numbers, *by* K. Yano *and* S. Bochner
33. Contributions to the Theory of Partial Differential Equations, *edited by* L. Bers, S. Bochner, *and* F. John
34. Automata Studies, *edited by* C. E. Shannon *and* J. McCarthy
35. Surface Area, *by* Lamberto Cesari
36. Contributions to the Theory of Nonlinear Oscillations, Vol. III, *edited by* S. Lefschetz
37. Lectures on the Theory of Games, *by* Harold W. Kuhn. In press
38. Linear Inequalities and Related Systems, *edited by* H. W. Kuhn *and* A. W. Tucker
39. Contributions to the Theory of Games, Vol. III, *edited by* M. Dresher, A. W. Tucker *and* P. Wolfe
40. Contributions to the Theory of Games, Vol. IV, *edited by* R. Duncan Luce *and* A. W. Tucker
41. Contributions to the Theory of Nonlinear Oscillations, Vol. IV, *edited by* S. Lefschetz
42. Lectures on Fourier Integrals, *by* S. Bochner. In press
43. Ramification Theoretic Methods in Algebraic Geometry, *by* S. Abhyankar. In press

CONTRIBUTIONS TO THE THEORY OF GAMES

VOLUME IV

R. J. AUMANN
B. R. GELBAUM
D. B. GILLIES
J. H. GRIESMER
H. M. GURK
J. C. HARSANYI
J. R. ISBELL
G. K. KALISCH
J. G. KEMENY
R. D. LUCE
W. H. MILLS
E. D. NERING
J. VON NEUMANN
L. S. SHAPLEY
M. SHUBIK
W. VICKREY

Edited by

A. W. Tucker and R. D. Luce

PRINCETON, NEW JERSEY
PRINCETON UNIVERSITY PRESS
1959

L. C. Card A51-8925

This research was supported in part by the Office of Naval Research. Reproduction, translation, publication, use and disposal in whole or in part by or for the United States Government is permitted.

Printed in the United States of America

To the Memory

of

JOHN VON NEUMANN

(1903-1957)

PREFACE

This fourth (and final) volume of CONTRIBUTIONS TO THE THEORY OF GAMES is devoted to recent work in n-person games, the part of his theory which probably most interested the late John von Neumann. It seems especially fitting, therefore, that this volume be dedicated to the memory of John von Neumann, creator of the Theory of Games.

The preparation of this Study and much of the refereeing and editing have been done at Princeton University in the Department of Mathematics as part of the work of a Logistics Project sponsored by the Office of Naval Research. Project members who have participated in the task have been M. L. Balinski, J. H. Griesmer, H. W. Kuhn, H. D. Mills, M. Richardson, G. L. Thompson, P. Wolfe. Others who have generously assisted in the refereeing have been H. M. Gurk, G. K. Kalisch, W. H. Mills, J. W. Milnor, O. Morgenstern, J. F. Nash, E. D. Nering, H. Raiffa, L. S. Shapley, R. M. Solow, R. M. Thrall. The typing of the master copy has been expertly done by Mrs. Euthie Anthony (the bibliography by Mrs. Virginia Nonziato). Mr. John Boles, Science Editor of the Princeton University Press, has watched over publication details with sympathetic interest. To all these individuals, for their friendly cooperation, and to the contributors for their patience, we express our sincere thanks.

R. D. Luce

July 1958 A. W. Tucker

CONTENTS

CONTRIBUTIONS TO THE THEORY OF GAMES, IV

INTRODUCTION

More than two-thirds of von Neumann and Morgenstern's THEORY OF GAMES AND ECONOMIC BEHAVIOR is concerned with a theory of non-strictly competitive games. Because of its special assumptions — uninhibited coalition formation and disruption, and freedom to make side payments among the players — the theory cannot be considered complete. Couple both this emphasis in their extremely influential volume and the restrictions of the theory with the fact that such games are potentially far more important in the social sciences than two-person zero-sum games, and one might have predicted a surge of research papers in the area. It is common knowledge that this has not happened, at least not during the period soon after the publication of their book. Yet a definite trend can be seen: in Study 24, the first volume of CONTRIBUTIONS TO THE THEORY OF GAMES, there were only two papers that could be said to fall in this area; however, three years later in the second volume, Study 28, one of four sections had to be allotted to n-person theory; and now with this Study we have a whole volume devoted to the topic. The prediction may yet come true if enough time is allowed.

In broad outline one can see what has happened in the past decade and a half. Mathematically, the minimax concept for two-person zero-sum games was, and is, the most elegant and unassailable feature of game theory, and it has attracted many mathematicians to explore its ramifications. New methods of proof of the classical result, extensions to infinite games and modifications for stochastic games, applications to statistical decision problems, and inter-relationships with linear programming have filled the literature. Only Nash, in his important paper on non-cooperative games, has seriously considered generalizing the minimax concept to n-person games. Another focus of activity has been two-person non-strictly competitive games, especially the bargaining problem where the players reapportion their common set of goods. Here some, though not all, of the difficulties of the general game problem exhibit themselves. One might expect that success in this area, and there has been some, would lead to natural generalizations for more than two players, but little work of this type has appeared (the few suggestions in Luce and Raiffa, GAMES AND DECISIONS, and Harsanyi's much more detailed exploration in this Study are all).

Those who have worked on n-person games have largely confined their attention to the problems posed by von Neumann and Morgenstern, even though few have been really satisfied with this formulation of general conflicts. Let us explore why the dissatisfaction. First, the central construct, the characteristic function, is obtained by considering the "two-person" games that arise when a coalition of players is opposed by its complement, and the minimax value from two-person theory is assigned to the coalition. The reduction of the conflict to two opposing coalitions seems to prejudge the problem. In addition, the assumption of unlimited and perfectly coordinated cooperation among the players, the lack of direct threats within and among the coalitions, and the assumed conservation of utility under side payments have seemed to many workers to stretch the mathematical idealization too far from any conceivable reality. There can be no doubt that some such simplification is needed if we are to have a tractable theory, but examples exist (see p. 351 of McKinsey, INTRODUCTION TO THE THEORY OF GAMES) which suggest that, at least in some cases, the real problem is lost in passing to the characteristic function.

Second, the intuitive basis and empirical meaning of the von Neumann-Morgenstern solution notion has remained tantalizingly elusive. One defines the domination of one distribution of payments — an imputation — by another imputation relative to a particular coalition, and a solution is any set of imputations with the properties (i) that every imputation not in the set is dominated by at least one in the set and (ii) that no imputation in the set is dominated by any other in the set. These "stable", and almost always overwhelmingly rich, sets of imputations seem neither to prescribe rational behavior nor to predict behavior with sufficient precision to be of empirical value.

Third, although neither of the above points need be telling criticisms for the mathematician, surely he will demand that the resulting structure be both mathematically deep and elegant if it is to hold his attention. Over the years doubt has increased about the elegance of solution theory. The mathematical character of solutions has been extremely complex from the first, becoming more so with almost every paper on the subject, and now conclusively demonstrated by Shapley (Paper 4).

With this as background, it is not difficult to see a natural organization of the papers into two broad categories: those that accept the characteristic function formulation and the von Neumann-Morgenstern definition of a solution, and those that do not. Each of these categories subdivides into three divisions. Among those on solutions, some papers are concerned with general propositions about solutions, others with the solutions of a restricted class of games, and still others with games having solutions of a particular type. To be sure, few papers fall precisely into just one division, but the general orientation of most of the papers seems clear. Within the other general class, there are papers that attempt to formulate alternative equilibrium concepts within the characteristic function framework, others that abandon both the solution and characteristic function concepts but attempt to generalize some notion from two-person theory, and finally those that seem to begin almost totally from scratch.

In discussing the papers, we have grouped them (with one exception) according to this scheme, even though the distribution of papers is far from uniform. It may serve to suggest which aspects of the theory are receiving less than their just due. In the following paragraphs we take up each paper separately, attempting to indicate its salient points.

PAPER 1, a translation of von Neumann's classic 1928 paper, falls outside the above classification; indeed, to a large degree, it sets the stage for the other papers. The historical importance of this paper would be sufficient reason to republish it in its original form, but in addition we felt that its usefulness would be augmented if it were in English. There is no need to attempt a summary of this paper, so we turn to the others.

General Theorems about Solutions (PAPERS 2-4)

A paper that we would have particularly liked to include, had it existed, would have settled the question whether or not all finite games have von Neumann-Morgenstern solutions. On a number of occasions, the late Professor von Neumann indicated that he felt this to be the most important open problem in n-person theory. Although others have placed less importance on this problem, every one working in the area is vexed that the question is still unresolved, especially since there rarely seems to be a shortage of solutions in games that have been investigated.

We do have, however, two papers that are definitely related to this problem. Kalisch and Nering extend the concept of a characteristic function to countably infinite sets of players in PAPER 2, and they exhibit a class of these games for which no solutions exist. Although this result does not in any sense settle the question for a finite number of players, it is of mathematical interest in its own right and it appears to provide some insight into the finite problem.

Gillies' PAPER 3, attacks the problem of solutions in finite games directly, and he obtains a variety of results that bear upon the question of existence of solutions. Two of the more interesting are, first, a proof that a positive fraction of all games have solutions and that a positive fraction of all non-zero-sum games have a unique solution, and second, a possible method for reducing the question of existence to a special class of games. An (n+1)-person pyramid game can be created from any given n-person game by adjoining to the original game what may be termed an indispensible player and defining the characteristic function as follows: a coalition not including the new player has value 0, and one including him has the value that the coalition of the other players has in the original game. Domination in pyramid games is acyclic, and so the existence of solutions should be easier to show. Gillies conjectures that every solution in the original game is contained in a solution of the associated pyramid game.

A number of other issues are explored in this paper that are pertinent to an understanding of the general notion of a solution. Among these are results on the relationships among characteristic functions and more general set functions that leave the domination relation invariant and a generalization of the important notion of a decomposable game defined in terms of solutions rather than directly in terms of the characteristic function.

In addition to their existence, another fact of the problem of solutions is their regularity, i.e., the ease with which they may be described mathematically. The profusion of solutions that have been uncovered in earlier work (also see PAPERS 5 through 13 of this Study) and the difficulty that one has in knowing whether all the solutions of a game have been found has suggested that they may be none too regular, that many pathologies may exist. In PAPER 4 Shapley has given a sharp, and disheartening, form to this intuition. A class of n-person games is exhibited for which an arbitrary closed subset of a certain (n-3)-dimensional region the simplex of imputations together with another, disjoint, closed subset forms a solution. The latter set depends upon the arbitrary one, but the important point is that the two components are disjoint. The fact that the arbitrary component need only be closed makes it clear that almost any naive conjecture about the regularity of solutions is bound to fail; as Shapley says, "there is practically no limit to the possible complexity of solutions."

Solutions in Restricted Classes of Games (PAPERS 5-11)

Given Shapley's result, it is clear that to describe solutions of games one must confine his attention to restricted classes of games or to restricted classes of solutions or both. Actually, most of the following papers do both, but usually one restriction seems the more dominant and so we have classified them accordingly.

In PAPER 5 attention is confined to symmetric n-person zero-sum games. Let v_q denote the characteristic function value of a coalition having q members and let $a = \max \{v_q/q \mid 1 \leq q \leq n\}$ and $p = \min \{q \mid v_q/q = a\}$. Gelbaum calls any coalition of p players a minimax coalition. Intuitively, each member of a minimax coalition should receive a payoff of a. Using this as a lead, he constructs symmetric (and sometimes finite) solutions for large classes of symmetric n-person zero-sum games. The development rests upon certain special assumptions concerning the characteristic function, but games meeting these conditions are shown to exist.

In PAPER 6, Nering provides symmetric solutions to all general-sum symmetric four-person games. The solutions given include as special cases the symmetric solutions to zero-sum symmetric four-person games found by von Neumann and Morgenstern. Although the supply of solutions provided seems to be quite generous, they are actually all modifications of a relatively simple basic type. Analytic proofs that these proposed solutions are in fact solutions are very tedious, so only a sample proof is included. Much more rewarding is the geometric viewpoint that led to their discovery.

It is well known that the essential zero-sum four-person games can be identified in a natural way with the points of a cube. Mills' PAPER 7 studies the games on the faces of the cube, and he determines their finite solutions. Although these games have a bewildering variety of infinite solutions, none of them has more than one finite solution, and most have none whatsoever. Indeed, only those games corresponding to the vertices and to the diagonal connecting the vertices representing three-person games with a dummy player have finite solutions. Each interior point of this diagonal has a seven point solution.

In PAPER 8, Shapley studies the solution set of a certain class of n-person games which represent a type of market. The principal feature of the market, as reflected in the characteristic function

$$v(S) = \min(|S \cap M|, |S \cap N|) \quad ,$$

where $M \cup N$ is the whole set of players divided into a group of "buyers" and a group of "sellers", is that players of opposite types can enter into profitable arrangements, whereas those of the same type cannot. The solutions to these games form one parameter sets. The parameter can be interpreted as the average net market price, or, when transformed, as the total profit of the sellers. These solutions are closely related to the "bargaining curves" that have been observed in many other game solutions. Although a complete list of solutions is not obtained, a bound on their location is given and they are characterized as monotonic arcs in the simplex of imputations. A major subclass of solutions is determined explicitly, and in certain cases this turns out to include all of the solutions.

In Study 28, Shapley defined and investigated the important quota games, which are games with a characteristic function v such that a vector ω exists having the properties

$$v(\{i, j\}) = \omega_i + \omega_j$$

$$v(\{1, 2, \ldots, n\}) = \sum_{i=1}^{n} \omega_i$$

A class of "quota" solutions based upon dominations over two element sets was exhibited. In PAPER 9, Kalisch generalizes these games in two ways. First, he replaces the equalities in the above conditions by inequalities. Necessary and sufficient conditions are presented for a characteristic function to satisfy these relaxed conditions and for a characteristic function to have analogues of Shapley's quota solutions. Second, he replaces the first of the two conditions by its analogue for m-element sets instead of two-element ones. The m-quota solutions, analoguous to the two-quota ones, are based upon dominations over the m-element sets. A necessary and sufficient condition is given that ensures the existence of m-quota solutions.

It will be recalled that von Neumann and Morgenstern pointed out that the set of all n-person constant-sum games can be considered as a convex polyhedron of dimension $2^{n-1} - n - 1$. Simple games, i.e., ones with but two values, are extreme points (vertices) of these polyhedra; however, for $n \geq 5$, not all extreme points correspond to simple games. Extreme games are just barely essential; coalitions tend to form only to an extent quite restricted by the characteristic function. Gurk investigates the non-simple extreme games with smallest n, namely 5, in PAPER 10. He gives a characterization which can be used to determine them explicitly and to count them. They all have exactly three values. A general solution to these games is presented. Finally, Gurk shows that his results cannot be extended directly to games with more than five players.

Taking his lead from Gurk's work, Griesmer studies the general class of extreme games with three values in PAPER 11. As a preliminary, a method is given for deciding whether an arbitrary game is extreme or not. A set of homogenous linear equations is associated with the game; the game is extreme if and only if the solution set of this system of equations is $\{0\}$. Next, the set of extreme games with three values is characterized in two steps. First, conditions are imposed upon the partition of coalitions into winning, losing, and intermediate types. Second, a "chain" condition employed by Gurk is imposed to guarantee extremeness. One may interpret this condition as a "reluctance" on the part of intermediate coalitions to consider coalition changes.

To any game with three values, a linear graph may be associated by treating the intermediate coalitions as nodes and introducing edges between these nodes that correspond to disjoint coalitions. Griesmer shows that a game with three values is extreme if and only if every node of the associated graph is part of a cycle with an odd number of edges, repetition of edges being allowed. The coefficient matrix of the system of inequalities defining the set of all n-person constant-sum games is next studied and properties of the system relating to the set of extreme games with two and three values are developed. Finally, discriminatory and finite solutions for these games are discussed; they exhibit the property that the minimal winning and minimal intermediate coalitions are most likely to form.

Restricted Classes of Solutions (PAPERS 12-13)

Although, as we have classified them, only two papers fall into this division, as against seven devoted to solutions for special classes of games, one cannot help but feel that Shapley's result on the irregularity of solutions underscores the importance of work in this direction. The first of these papers attacks the problem as a conceptual issue; the second explores a special class of solutions that have arisen from the study of particular games.

In PAPER 12, Vickrey raises the question whether some imputation sets, including some solutions, may not be "self-policing" in the sense that deviations from the set are naturally inhibited. Specifically, A is such a set if for each imputation $a \in A$ and imputation $b \notin A$ such that b dominates a with the effective set H, there is a member of H who receives less in c than in a for every $c \in A$ that dominates b. That is to say, some person required for each deviation from the self-policing set is bound to suffer in a return to that set. Solutions that are self-policing are called <u>strong</u>. Vickrey shows that the main solution of any constant-sum homogeneous majority game is the unique strong solution. Some constant-sum games, e.g., a 10-person homogeneous non-majority game described in the paper, have several strong solutions; others, e.g., the constant-sum non-homogeneous majority games, have no strong solutions.

A series of weaker notions are also discussed; however, it really seems that somewhat stronger conditions are of greater interest because strong solutions appear to be extremely numerous in variable-sum non-simple games. Vickrey suggests the following idea: Suppose that one postulates an average expectation, such as the Shapley value, for each player. In a given imputation, a player is considered privileged if he is receiving more than this expectation, and an imputation is called firm, relative to the given value, if there does not exist a set of non-privileged players that is effective against it. A solution is called firm if each of its imputations is firm. He shows that the strong solutions of the simple constant-sum games are all firm with respect to the Shapley value. What other relations hold between strong and firm solutions is not known.

In PAPER 13 Gurk and Isbell investigate those solutions, called simple solutions, in which each player i has only two possible payoffs, $x_i \geq 0$ or 0. They give necessary and sufficient conditions for a vector of payoffs to be in a simple solution of a given game, and the solution, which is uniquely determined by the vector, is described. All simple solutions of symmetric games and of decomposable games are determined. An algorithm to find all simple solutions of a simple game, together with some internal results about these solutions, is given. For example, they show that any solution in which each player has only two positions, x_i and y_i, must in fact be a simple solution. In general, even a strong simple game may have many simple solutions; however, some criteria for uniqueness are found which show, in particular, that the five-person simple games and the six-person strong simple games have unique simple solutions. The paper concludes with a list of the strong simple six-person games, extending von Neumann and Morgenstern's enumeration for $n \leq 5$, and a proof that every strong simple game with n non-dummy players has at least n minimal winning sets.

Other Concepts Resting on the Characteristic Function (PAPERS 14-15)

In PAPER 14, which is related to PAPER 8, Shubik considers a set of bargainers partitioned into two sets M and N. Initially, each member of M has a units of a certain commodity and 0 of another and each member of N has 0 of the first and b of the second. All players are assumed to have the same utility function $\psi(x, y)$, where x denotes units of the first commodity and y units of the second. It is supposed that the characteristic function is given by

$$v(S) = |S|\psi\left(a\frac{|S \cap M|}{|S|}, \quad b\frac{|S \cap N|}{|S|} \right) .$$

These are n-person extensions of a two-person trading game first investigated by Edgeworth. He treated the undominated imputations, i.e., the core, as the solution, and he conjectured that as the number of bargainers on each side is increased the core would become increasingly more determinate until in the limit a single imputation, corresponding to the market solution in a model of pure competition, would result. Shubik shows that this is the case provided that the ratio of the number of persons in M to that in N is fixed in the passage to the limit. Conditions are also established for the core to be vacuous and for it to be one-dimensional in the limit.

Elsewhere (see e.g., Chapter 10 of GAMES AND DECISIONS) an equilibrium notion for games in characteristic function form called ψ-stability has been suggested. It has the feature that "outcomes" are pairs consisting of an imputation and a partition of the players into coalitions. The central idea is that not all one-step changes from a given coalition partition are possible, but only those that are in a sense "near" it. In PAPER 15 Luce argues that a specific restriction on coalition changes may hold for the two four-person constant-sum games in an experiment reported by Kalisch, Milnor, Nash, and Nering (DECISION PROCESSES), and the data are reanalyzed in terms of ψ-stability theory. For each observed coalition partition, the set of admissible imputations is determined, and for 29 of the 32 cases the observed imputations lie within the limits of the predictions, which vary in their specificity. Whether or not the particular analysis is relevant, the regrouping of the data by coalition partitions results in considerable regularity in the observed imputations, suggesting that theories which characterize equilibria only in terms of imputations are missing an important aspect of cooperative games.

Generalization of Concepts from Two-Person Theory (PAPERS 16-17)

Aumann treats in PAPER 16 the problem of cooperative n-person games in which side payments are prohibited. The players are assumed to be free to form coalitions and to correlate their strategy choices, as in characteristic function theory, but, since they are not permitted to alter their payoffs by side payments, the analysis is more akin to Nash's non-cooperative theory than to the usual cooperative theories. In selecting "equilibrium" strategies, the guiding idea is to ask what is reasonable when the game is repeated a large, unspecified number of times. Let c be an n-tuple of strategies, one for each player in the game. If, by forming a coalition and choosing strategies different from those in c, each player of a set B can expect a larger payoff than from c, then it hardly seems reasonable to expect c to be in equilibrium. If, however, as a result of this change at least one member of B is exposed to the possibility of getting less than in c, then it is not so clear that the defection from c can be arranged. An n-tuple c is said to be an <u>acceptable point</u> if it has the property that for every coalition B and every defection from c, the complimentary coalition has at least one correlated strategy that will inflict a loss on at least one member of B. (Note the close conceptional relationship between this idea and Vickrey's strong solutions.) In a two-person zero-sum game, the payoff from an acceptable point equals the value of the game. Not all games have acceptable points: a three-person zero-sum game is presented that does not.

Another approach to repeated plays is to treat the infinite sequence of games as a "supergame", defining strategies and average payoffs in some suitable manner. Aumann gives such a definition. He then defines a <u>strong equilibrium point</u> — generalizing Nash's definition of an equilibrium point — to be an n-tuple c of strategies in the supergame for which no coalition profits by defecting from c when the remaining players do not alter their choices. Since strategy choices are one-shot affairs in the supergame, it is unnecessary to worry about retaliation by the other players. Aumann's central theorem establishes that the set of payoff n-tuples corresponding to acceptable points is identical to the set corresponding to the strong equilibrium points.

One of the more interesting studies of a class of two-person non-zero-sum games is Nash's paper on the bargaining problem. In PAPER 17, Harsanyi presents a generalization for n-person games. Assuming that every proper subset is less effective than the total set of players, Harsanyi's model postulates that each subset — called a <u>syndicate</u> — publicly guarantees certain payoffs — called <u>dividends</u> — to its members. A player's final payoff is the sum of dividends guaranteed to him by the various syndicates of which he is a member. In addition, each syndicate is assumed to announce a joint strategy — a <u>threat</u> — that its members will follow against the rest of the players should no acceptable distribution of payoffs be reached. The members of each of the syndicates must, in turn, agree upon the dividends and threat strategy of their syndicate by bargaining. The ultimate outcome of the game, therefore, results from these systems of agreements, and Harsanyi derives certain equilibrium conditions that the agreements must satisfy. These imply that the cooperative n-person bargain is equivalent to a (2^n-1)-person non-cooperative game among the syndicates.

Sometimes, e.g., when utility is transferable, the equilibrium conditions uniquely determine the payoffs. When this is not so, another round of bargaining is introduced which leads to a unique equilibrium point. The unique payoffs resulting from transferable utility are shown to form a Shapley value based not on the ordinary characteristic function, but on the characteristic function arising from the assumption that each coalition plays the zero-sum game of differences in total payoff to itself and to its complement.

New Concepts (PAPERS 18-19)

In PAPER 18, Isbell develops a theory for cooperative games that admit continuous, but not necessarily linear, transfers of utility. His approach stems from two criticisms of previous procedures: the contention that the values of a utility function for a "rational" person should form a bounded set, and a criticism, due to McKinsey, that the characteristic function is an inadequate description of certain situations. Isbell's theory rests upon an analogue of the characteristic function which he calls a <u>projective form</u>. It is a function on sets of players whose values form curved surfaces in utility spaces; when the utility is linearly transferable, the surfaces are linear subspaces. This form is obtained, as in the von Neumann-Morgenstern theory, by considering two-player "quotient" games, but, unlike their theory, Isbell's depends upon an arbitration scheme that is an extension of Raiffa's theory fitted into the context of bounded utility. One particular scheme, called the <u>ratio scheme</u>, is adopted which assumes there is an absolute zero of utility. If a two-person game has a "threat point" (x, y), the arbitrated solution is the largest possible proportional imputation (tx, ty). Given the projective form, an analogue of the von Neumann-Morgenstern solution theory is developed.

Aside from criticisms of the characteristic function representation of a game, one of the more telling criticisms of solution theory is the multiplicity (often a continuum) of solutions and the number of imputations within each. Although in part this is a result of the loss of information in passing from the normal to the characteristic function form, it is widely felt that such indeterminateness is inherent in classical game theory because so little is specified about the players. The difficult question, then, is what more to assume. In PAPER 19, Kemeny suggests that a numerical function, not unlike a utility function, be associated with each player. It is to serve as a measure of his bargaining effectiveness, subsuming both the ideas of bargaining ability and power. (No axiomatization in terms of more primitive concepts, similar to that of utility in terms of choices among gambles, is given.) Kemeny assumes that these functions have been so determined that a coalition will split the profits equally, relative to the utility functions of its members. Not only is such a division always possible, but it is also unique. Each individual can, therefore, rank the coalitions according to preference, and this ranking can be utilized to determine which coalition will occur. Normally, but not always, a unique coalition is selected, in which case the division of profits is uniquely determined. Several examples are worked out in detail.

At the end of this Study there is a reasonably complete BIBLIOGRAPHY of game theory prepared by D. M. and G. L. Thompson at the instigation of Professor Oskar Morgenstern. It has seemed desirable to permit overlapping with the bibliographies in Studies 24 and 28, in order to have a single self-contained bibliography.

ON THE THEORY OF GAMES OF STRATEGY[1]

John von Neumann

[A translation by Mrs. Sonya Bargmann of "Zur Theorie der Gesellschaftsspiele," Mathematische Annalen 100 (1928), pp. 295-320.]

INTRODUCTION

1. The present paper is concerned with the following question:

> n players $S_1, S_2, \ldots, S_n$ are playing a given game of strategy, $\mathfrak{G}$. How must one of the participants, S_m, play in order to achieve a most advantageous result?

The problem is well known, and there is hardly a situation in daily life into which this problem does not enter. Yet, the meaning of this question is not unambiguous. For, as soon as $n > 1$ (i.e., $\mathfrak{G}$ is a game of strategy in the proper sense), the fate of each player depends not only on his own actions but also on those of the others, and their behavior is motivated by the same selfish interests as the behavior of the first player. We feel that the situation is inherently circular.

Hence we must first endeavor to find a clear formulation of the question. What, exactly, is a game of strategy? A great many different things come under this heading, anything from roulette to chess, from baccarat to bridge. And after all, any event - given the external conditions and the participants in the situation (provided the latter are acting of their own free will) - may be regarded as a game of strategy if one looks at the effect it has on the participants.[2] What element do all these things have in common?

---

[1] A shortened version of this paper has been presented to the Goettingen Mathematical Society on December 7, 1926.

[2] This is the principal problem of classical economics: how is the absolutely selfish "homo economicus" going to act under given external circumstances?

We may assume that it is the following:

> A game of strategy consists of a certain series of events each of which may have a finite number of distinct results. In some cases, the outcome depends on chance, i.e., the probabilities with which each of the possible results will occur are known, but nobody can influence them. All other events depend on the free decision of the players S_1, $S_2, \ldots, S_n$. In other words, for each of these events it is known which player, S_m, determines its outcome and what is his state of information with respect to the results of other ("earlier") events at the time when he makes his decision. Eventually, after the outcome of all events is known, one can calculate according to a fixed rule what payments the players $S_1, S_2, \ldots, S_n$ must make to each other.

It is easy to bring this somewhat qualitative explanation into a precise form. The definition of a game of strategy would then be the following:

> For a complete description of a game $\mathfrak{G}$, the following data are necessary, which in their entirety are the "rules of the game."
>
> α) The number of events or "draws" depending on chance, and the number of events or "steps" depending on the free decision of the individual players must be specified. Let these numbers be z and s respectively, and let us denote the "draws" by $E_1, E_2, \ldots, E_z$, the "steps" by $F_1, F_2, \ldots, F_s$.
>
> β) The number of possible results of each "draw", E_μ, and of each "step", F_ν, must be specified. Let these numbers be M_μ and N_ν respectively. ($\mu = 1, 2, \ldots, z$, $\nu = 1, 2, \ldots, s$.) We shall denote the results, for short, by their numbers $1, 2, \ldots, M_\mu$ and $1, 2, \ldots, N_\nu$ respectively.
>
> γ) For each "draw" E_μ the probabilities $\alpha_\mu^{(1)}$, $\alpha_\mu^{(2)}, \ldots, \alpha_\mu^{(M_\mu)}$ of the different results $1, 2, \ldots, M_\mu$ must be given. Obviously, we have
>
> $$\alpha_\mu^{(1)} \geq 0,\ \alpha_\mu^{(2)} \geq 0,\ \ldots,\ \alpha_\mu^{(M_\mu)} \geq 0$$
>
> $$\alpha_\mu^{(1)} + \alpha_\mu^{(2)} + \ldots + \alpha_\mu^{(M_\mu)} = 1\ .$$

δ) For every "step" F_ν, the player S_m who determines the outcome of this "step" ("whose step" F_ν is) must be specified. In addition, the numbers of all "draws and steps" must be specified of whose outcome the player is informed at the time he makes his decision concerning F_ν. (These "draws" and "steps" we shall call "earlier" than F_ν.)

In order that this whole scheme be consistent and permit of a temporal-causal interpretation, there must be no cycles $F_{\nu_1}, F_{\nu_2}, \ldots, F_{\nu_p}, F_{\nu_{p+1}} = F_{\nu_1}$, and F_{ν_q} must always be "earlier" than $F_{\nu_{q+1}}$ $(q = 1, 2, \ldots, p)$.

ε) Finally, n functions $f_1, f_2, \ldots, f_n$ must be given. Each of them depends on $z + s$ variables which take on the values

$$1, 2, \ldots, M_1; \quad 1, 2, \ldots, M_2; \quad \ldots; \quad 1, 2, \ldots, M_z;$$
$$1, 2, \ldots, N_1; \quad 1, 2, \ldots, N_2; \quad \ldots; \quad 1, 2, \ldots, N_s$$

respectively. These functions are real-valued and

$$f_1 + f_2 + \cdots + f_n \equiv 0$$

holds identically. If in the course of a play which has been completed the results of the z "draws" and s "steps" were $x_1, x_2, \ldots, x_z, y_1, y_2, \ldots, y_s$ respectively, $(x_\mu = 1, 2, \ldots, M_\mu, y_\nu = 1, 2, \ldots, N_\nu;$ $\mu = 1, 2, \ldots, z, \nu = 1, 2, \ldots, s)$, the players S_1, $S_2, \ldots, S_n$ obtain from each other[3] the amounts

$$f_1(x_1, \ldots, x_z, y_1, \ldots, y_s), f_2(x_1, \ldots, x_z, y_1, \ldots, y_s), \ldots, f_n(x_1, \ldots, x_z, y_1, \ldots, y_s) .$$

(On closer inspection it can be seen that in spite of this somewhat lengthy description we are dealing here with quite simple matters. Actually, in several respects our definition might have been somewhat more

---

[3] The identity

$$f_1 + f_2 + \cdots + f_n \equiv 0$$

expresses that the players make payments to each other only, but collectively they neither gain nor lose.

general. We could, e.g., have included the case that the M_μ, N_ν and $\alpha_1^{(\mu)}, \alpha_2^{(\mu)}, \ldots, \alpha_{M_\mu}^{(\mu)}$ depend on the results of the "earlier" "draws" and "steps", and the like. It is easy to see, however, that such generalizations would be inessential.)

2. With this definition the concept of a game of strategy is precisely defined. But it also becomes clear that, as we indicated at the beginning of 1., the expression "S_m tries to achieve a result as advantageous as possible" is rather obscure. What constitutes the most advantageous result for the player S_m is obviously the largest possible value of f_m, but how can any value of f_m be "achieved" by S_m? By himself, S_m is in no position to fix the value of f_m! The value of f_m depends on the variables $x_1, \ldots, x_z, y_1, \ldots, y_s$, only part of which are determined by S_m's decision (viz. those y_ν for which S_m has the "step" F_ν, i.e., those for which $S_{(F_\nu)} = S_m$). All other variables y_ν depend on the decisions of the participants and all variables x_μ depend on chance.

In our case, the "unforeseeable" chance event is actually the factor which it is easiest to deal with. In fact, assume that a particular f_m depends only on those y_ν which are decided on by S_m ($S_{(F_\nu)} = S_m$), and in addition on the x_μ (which are determined by chance). In that case, S_m can at least anticipate this much: If I make certain moves, I can expect such and such results (i.e., values of f_m) with such and such probabilities (since the probabilities $\alpha_1^{(\mu)}, \alpha_2^{(\mu)}, \ldots, \alpha_{M_\mu}^{(\mu)}$ are given) - regardless of how the other players act! If we now assume that "a most advantageous result" is the highest possible value of the expectation (and this assumption or a similar one has to be made in order to apply the methods of the theory of probability)[4] we have, in principle, solved our problem. For, we have here a simple maximum problem: The values of those variables y_ν which S_m has to determine must be so chosen by him that the expected value of f_m (which depends only on these variables y_ν) becomes as large as possible.

It is this type of game which in the theory of probability is treated in the so-called "theory of games of chance." A typical example is roulette: Let the number of players be $k + 1$ ($S_1, \ldots, S_k$ are the "pointers", S_{k+1} is the "banker"), S_{k+1} has no influence whatsoever on the game,[5] and the result achieved by S_ℓ, f_ℓ, ($\ell = 1, 2, \ldots, k$) only

---

[4] We shall not enter on the well-known objections to the use of the expected value (and the ensuing attempts to replace the latter by the so-called moral expectation or similar concepts). The difficulties that form the subject of our considerations are of a different nature.

[5] Anyway, he has no need to, since according to the rules of the game his gain after each play is 2.70% of the turn-over.

depends on chance and his own actions.[6]

The name alone, "game of chance", indicates that the main emphasis is put on the variables x_μ, which are dependent on chance, and not on the variables y_ν, which are subject to the decisions of the players. But this is exactly what we are interested in. We shall try to investigate the effects which the players have on each other, the consequences of the fact (so typical of all social happenings!) that each player influences the results of all other players, even though he is only interested in his own.

§1. GENERAL SIMPLIFICATIONS

1. The definition of a game of strategy given in the Introduction is rather complicated, which may appear justified in view of the fact that games of strategy may be arbitrarily complex. Nevertheless it is possible to bring all games falling under this definition into a much simpler normal form, in a way, into the simplest form that is at all conceivable. We contend that it is sufficient to consider games of the following kind:

$z = 1$ (i.e., only one "draw" takes place).

$s = n$, the ν-th "step" being that of the player S_ν ($S_{(F_\nu)} = S_\nu$).

The relation "earlier" is never realized (i.e., each player must make his dispositions without knowing anything about the other participants or about the "draw").

The play thus takes the following course: Each player S_m ($m = 1, 2, \ldots, n$) chooses a number $1, 2, \ldots, N_m$ without knowing the choices of the others. Now a "draw" takes place in which the numbers $1, 2, \ldots, M$ will appear with the probabilities $\alpha_1, \alpha_2, \ldots, \alpha_M$. The results achieved by the players are (if the "draw" and the n "steps" have resulted in $x, y_1, y_2, \ldots, y_n$)

$$f_1(x, y_1, \ldots, y_n),\ f_2(x, y_1, \ldots, y_n),\ \ldots,\ f_n(x, y_1, \ldots, y_n) \ .$$

This apparently far-reaching restriction is actually not essential, for the following reason:

Let the "steps" of the players S_m ($S_{(F_\nu)} = S_m$) be those with the numbers $\nu_1^{(m)}, \nu_2^{(m)}, \ldots, \nu_{\sigma_m}^{(m)}$. Obviously, it is inadmissible to make the assumption that S_m might be able to tell, before the start of the

[6] As may be guessed, on the basis of the preceding footnote, the unambiguous result in this case for the behavior of the pointers is rather trivial: if possible they should have the turn-over zero; the closer they approximate it, so much the better!

game, what his choices for all these steps are going to be. It would mean a restriction of his free will and change his chances (for the worse). For, S_m's decision in each of these "steps" will generally be significantly influenced by the results of the "draws" and "steps" known to him at the moment of his decision.

On the other hand, it may well be assumed that before the play has started he knows how to answer the following question: What will be the outcome of the $\nu_k^{(m)}$-th "step" ($k = 1, 2, \ldots, \sigma_m$) provided the results of all "draws" and "steps" "earlier" than $\nu_k^{(m)}$ are available? In other words, the player knows beforehand how he is going to act in a precisely defined situation: he enters the play with a theory worked out in detail. Even if this may not be the case for a particular player, it is clear that such an assumption will certainly not spoil his chances.

Accordingly, we define the "strategy" of a player S_m as follows:

> In order to describe completely the "strategy" of a player S_m ($m = 1, 2, \ldots, n$) the following specifications are necessary:
>
> As before, let S_m have the "steps" with the numbers $\nu_1^{(m)}, \nu_2^{(m)}, \ldots, \nu_{\sigma_m}^{(m)}$ and assume that at the moment when he decides on the $\nu_k^{(m)}$-th "step" ($k = 1, 2, \ldots, \sigma_m$) the results of the "draws" and "steps" with the numbers $\bar{\mu}_1^{(m,k)}, \bar{\mu}_2^{(m,k)}, \ldots, \bar{\mu}_{\alpha_{m,k}}^{(m,k)}$ and $\bar{\nu}_1^{(m,k)}, \bar{\nu}_2^{(m,k)}, \ldots, \bar{\nu}_{\beta_{m,k}}^{(m,k)}$ respectively are available to him, that is, they are "earlier" than $\nu_k^{(m)}$. For each possible combination of results of the "draws" and "steps" mentioned above (obviously, there is only a finite number of such combinations) it must be specified what S_m's decision with respect to the $\nu_k^{(m)}$-th "step" is going to be (i.e., what will be the outcome of this step).

One sees immediately that only a finite number of strategies is available to S_m, which we shall call $S_1^{(m)}, S_2^{(m)}, \ldots, S_{\Sigma_m}^{(m)}$.

It can now easily be shown (using the assumption on the absence of cycles in the Introduction, 1., definition of a game of strategy, (δ)) that the course of a play is described in a permissible and unambiguous manner if we specify

1. which strategies $\bar{S}^{(1)}, \bar{S}^{(2)}, \ldots, \bar{S}^{(n)}$ are being used by the players $S_1, S_2, \ldots, S_n$ respectively,

2. what are the results of the "draws" $E_1, E_2, \ldots, E_z$.

Two points should be noted here. In the first place, it is inherent in the concept of "strategy" that all the information about the actions of the participants and the outcome of "draws" a player is able to obtain or to infer is already incorporated in the "strategy." Consequently, each player must choose his strategy in complete ignorance of the choices of the rest of the players and of the results of the "draws."

Secondly, it has become entirely immaterial that the "draws" $E_1, E_2, \ldots, E_n$ are <u>separate</u> events (where for E_μ, $\mu = 1, 2, \ldots, z$, the numbers $1, 2, \ldots, M_\mu$ will occur with the respective probabilities $\alpha_\mu^{(1)}, \alpha_\mu^{(2)}, \ldots, \alpha_\mu^{(M_\mu)}$) since the players must act, i.e., choose their "strategies" without knowing the outcome of the "draws." But if this is the case nothing prevents us from combining all z "draws" into a single "draw", H, the outcome of which will be the aggregates of numbers

$$x_1, x_2, \ldots, x_z \quad (x_\mu = 1, 2, \ldots, M_\mu, \ \mu = 1, 2, \ldots, z)$$

with their respective probabilities $\alpha_1^{(x_1)} \cdot \alpha_2^{(x_2)} \cdot \ldots \cdot \alpha_z^{(x_z)}$, or, what amounts to the same thing, the numbers $1, 2, \ldots, M$ ($M = M_1 \cdot M_2 \cdot \ldots \cdot M_z$) with their respective probabilities, which we shall call $\beta_1, \beta_2, \ldots, \beta_M$.

Thus we can modify 2. in the following way:

2'. The result of the "draw" H must be specified. (H may have the results $1, 2, \ldots, M$ with the respective probabilities $\beta_1, \beta_2, \ldots, \beta_M$.)

The choices (interpreted as "steps") of the players $S_1, S_2, \ldots, S_n$ in 1. together with the "draw" in 2'. are fully equivalent to the original game $\mathfrak{G}$ (if one takes the fact into account that each "step" is taken in complete ignorance of all other circumstances), and they evidently constitute a game $\mathfrak{G}'$ which, indeed, is of the simple form mentioned at the beginning of this secton.

2. The last element to be eliminated from the game, since from our point of view it is inessential, is the "draw". This is done by replacing the actual results for the individual players by their expected values. To be exact:

If the players $S_1, S_2, \ldots, S_n$ have chosen the "strategies"

$$S_{u_1}^{(1)}, S_{u_2}^{(1)}, \ldots, S_{u_n}^{(1)} \qquad (u_m = 1, 2, \ldots, \Sigma_m, m = 1, 2, \ldots, n)$$

and if the outcome of the "draw" H has been the number v (= 1, 2, ..., M), then let the results for the players $S_1, S_2, \ldots, S_n$ be

$$f_1(v, u_1, \ldots, u_n), f_2(v, u_1, \ldots, u_n), \ldots, f_n(v, u_1, \ldots, u_n) .$$

(We may disregard the fact that we are dealing with "strategies" and not with actual "steps" and simply speak of the choices $u_1, u_2, \ldots, u_n$.)

If only the choices of $u_1, u_2, \ldots, u_n$ are known, but not yet the "draw" v, the expected values of $f_1, f_2, \ldots, f_n$ will be

$$g_m(u_1, \ldots, u_n) = \sum_{v=1}^{M} \beta_v f_m(v, u_1, \ldots, u_n) \qquad (m = 1,2, \ldots, n)$$

($f_1 + f_2 + \ldots + f_n \equiv 0$ implies $g_1 + g_2 + \ldots + g_n \equiv 0$). It is entirely in the spirit of the probabilistic method to discount the "draw" altogether and to deal exclusively with the expected values $g_1, g_2, \ldots, g_n$. In doing so we obtain the following basic type of a game of strategy which is even more schematized and simplified.

> Each of the players $S_1, S_2, \ldots, S_n$ chooses a number, S_m choosing one of the numbers $1, 2, \ldots, \Sigma_m$[7] (m = 1, 2, ..., n). Each player must make his decision without being informed about the choices of the other participants. After having made their choices $x_1, x_2, \ldots, x_n$ ($x_m = 1, 2, \ldots, \Sigma_m$, m = 1, 2, ..., n) the players receive the following amounts respectively:
>
> $$g_1(x_1, \ldots, x_n), g_2(x_1, \ldots, x_n), \ldots, g_n(x_1, \ldots, x_n)$$
>
> (where identically $g_1 + g_2 + \ldots + g_n \equiv 0$).

The rules of the game have thus been obtained in a form which retains only those characteristics of a game of strategy which are essential to our consideration - and as we have just shown, essentially without loss

[7] In addition, we could also make all Σ_m equal to each other by assuming a Σ which is not smaller than any Σ_m and subdividing each Σ_m-th case in $\Sigma - \Sigma_m + 1$ subcases, each of which would have the same effect as the original. But this simplification is inessential.

of generality. Nothing is left of a "game of chance". The actions of the players determine the result completely (since everything takes place as if each of the players has his eye on the expected value only). As a result, the feature which was emphasized at the end of the Introduction emerges in a particularly clear form: each g_m depends on all $x_1, x_2, \ldots, x_n$.

The standard case of probability theory that g_m depends on x_m only (which, of course, cannot hold for all m) now appears to be entirely trivial.

§2. THE CASE $n = 2$

1. Since we cannot proceed any further with the same generality, it is now appropriate to consider the simplest case for n. The case $n = 0$ is meaningless, and so is the case $n = 1$ (since $g_1 + g_2 + \ldots + g_n \equiv 0$); neither involves an actual game of strategy. So we shall now investigate the case $n = 2$.

Since $g_1 + g_2 \equiv 0$, we can put $g_1 = g$, $g_2 = - g$. The description of a general two-person game is then as follows:

> The players S_1, S_2 choose arbitrary numbers among the numbers $1, 2, \ldots, \Sigma_1$ and $1, 2, \ldots, \Sigma_2$ respectively, each one without knowing the choice of the other. After having chosen the numbers x and y respectively, they receive the sums $g(x, y)$ and $- g(x, y)$ respectively.

$g(x, y)$ may be any function (defined for $x = 1, 2, \ldots, \Sigma_1$, $y = 1, 2, \ldots, \Sigma_2$!).

It is easy to picture the forces struggling with each other in such a two-person game. The value of $g(x, y)$ is being tugged at from two sides, by S_1 who wants to maximize it, and by S_2 who wants to minimize it. S_1 controls the variable x, S_2 the variable y. What will happen?

2. After S_1 has chosen the number x ($x = 1, 2, \ldots, \Sigma_1$), his result $g(x, y)$ still depends on the choice y of S_2, but in any event $g(x, y) \geq \mathrm{Min}_y\, g(x, y)$. And by an appropriate choice of x this lower limit can be made equal to $\mathrm{Max}_x\, \mathrm{Min}_y\, g(x, y)$ (and not any larger!). I.e., if S_1 so wishes, he certainly can make $g(x, y)$

$$\geq \mathrm{Max}_x \mathrm{Min}_y g(x, y)$$

(irrespective of what S_2 does!). The same argument holds for S_2. If

S_2 so wishes, he certainly can make $g(x, y)$

$$\leq \text{Min}_y \text{Max}_x g(x, y)$$

(irrespective of what S_1 does!).

If now

$$\text{Max}_x \text{Min}_y g(x, y) = \text{Min}_y \text{Max}_x g(x, y) = M$$

it follows from the above, as well as from the fact that S_1 wants to maximize $g(x, y)$ and S_2 wants to minimize it, that $g(x, y)$ will have the value M. For, S_1 is interested in making it large and can keep it from becoming smaller than M. S_2, on the other hand, is interested in making it small and can keep it from becoming larger than M. Hence it will have the value M.

Though, in general,

$$\text{Max}_x \text{Min}_y g(x, y) \leq \text{Min}_y \text{Max}_x g(x, y)$$

it is not at all true that the $=$ sign always holds. Actually, it is easy to exhibit such $g(x, y)$ for which the $<$ sign holds, that is, for which the above consideration breaks down. The simplest example of this kind is the following:

$$\Sigma_1 = \Sigma_2 = 2, \quad g(1, 1) = 1, \quad g(1, 2) = -1,$$
$$g(2, 1) = -1, \quad g(2, 2) = 1 .$$

(Evidently, Max Min $= -1$ and Min Max $= 1$.)

Another example is the so-called game of "Morra":[8]

$$\Sigma_1 = \Sigma_2 = 3, \quad g(1, 1) = 0, \quad g(1, 2) = 1, \quad g(1, 3) = -1,$$
$$g(2, 1) = -1, \quad g(2, 2) = 0, \quad g(2, 3) = 1,$$
$$g(3, 1) = 1, \quad g(3, 2) = -1, \quad g(3, 3) = 0 .$$

(Here, too, Max Min $= -1$ and Min Max $= 1$.)

The fact that this difficulty comes up can also be realized in the following way:

$\text{Max}_x \text{Min}_y g(x, y)$ is the best result that S_1 can achieve if

[8] Also called "gangster baccarat." In the usual formulation, 1, 2, 3 are called "Paper", "Stone", "Scissors" ("Paper covers the stone, scissors cut the paper, stone grinds the scissors").

he is "found out" by S_2; if whenever S_1 plays x, S_2 plays a y such that $g(x, y) = \text{Min}_y\, g(x, y)$. (According to the rules of the game S_2 was not supposed to know how S_1 was going to play, he would have to infer it in some other way. This is what we mean to indicate by the expression "finding out". In the same way, the best result that S_2 can achieve if S_1 has found him out is $\text{Min}_y \text{Max}_x\, g(x, y)$. If the two numbers are equal this means: it makes no difference which of the two players is the better psychologist, the game is so insensitive that the result is always the same. It is obvious that this is not the case for the two games just mentioned: here, everything depends on finding the adversary out, on guessing whether he is going to choose 1 or 2 (or 1 or 2 or 3).

The fact that the two quantities Max Min and Min Max are different means that it is impossible for each of the two players, S_1 and S_2, to be cleverer than the other.

3. Still, it is possible, by use of an artifice, to force the equality of the two above-mentioned expressions.

To this purpose, the possibilities of action for the two players S_1 and S_2 are extended as follows: At the beginning of the game, S_1 is not asked to choose one of the numbers $1, 2, \ldots, \Sigma_1$. He only has to specify Σ_1 probabilities

$$\xi_1, \xi_2, \ldots, \xi_{\Sigma_1} \quad (\xi_1 \geq 0, \xi_2 \geq 0, \ldots, \xi_{\Sigma_1} \geq 0, \xi_1 + \xi_2 + \ldots + \xi_{\Sigma_1} = 1)$$

and then draw the numbers $1, 2, \ldots, \Sigma_1$ from an urn containing these numbers with the probabilities $\xi_1, \xi_2, \ldots, \xi_{\Sigma_1}$. He then chooses the number drawn. This may look like a restriction of his free will: it is not he who determines x. But this is not so. Because if he really wants to get a particular x, he can specify $\xi_x = 1$, $\xi_u = 0$ (for $u \neq x$). On the other hand, he is protected against his adversary "finding him out"; for, if, e.g., $\xi_1 = \xi_2 = 1/2$, nobody (not even he himself!) can predict whether he is going to choose 1 or 2!

S_2 is supposed to act in the same way. He also chooses Σ_2 probabilities $\eta_1, \eta_2, \ldots, \eta_{\Sigma_2}$ and proceeds accordingly.

Let us denote the sequence $\xi_1, \xi_2, \ldots, \xi_{\Sigma_1}$ by ξ and the sequence $\eta_1, \eta_2, \ldots, \eta_{\Sigma_2}$ by η. If S_1 chooses ξ and S_2 chooses η, S_1 has the expected value

$$h(\xi, \eta) = \sum_{p=1}^{\Sigma_1} \sum_{q=1}^{\Sigma_2} g(p, q)\, \xi_p \eta_q$$

and S_2 has the expected value $-h(\xi, \eta)$. The new function $h(\xi, \eta)$ includes the old one, $g(x, y)$, in the following sense: if $\xi_x = \eta_y = 1$ and $\xi_u = \eta_v = 0$ (for $u \neq x$, $v \neq y$), then $h(\xi, \eta) = g(x, y)$.

We can now apply the same consideration to $h(\xi, \eta)$ we applied to $g(x, y)$. If S_1 has made the choice ξ, his expected value is at least $\text{Min}_\eta h(\xi, \eta)$. Hence, he is in a position to obtain the minimal expected value $\text{Max}_\xi \text{Min}_\eta h(\xi, \eta)$ (irrespective of what S_2 does!). In the same way, S_2 can keep the expected value of S_1 from exceeding the maximal value $\text{Min}_\eta \text{Max}_\xi h(\xi, \eta)$. Again we have

$$\text{Max}_\xi \text{Min}_\eta h(\xi, \eta) \leq \text{Min}_\eta \text{Max}_\xi h(\xi, \eta)$$

and the question is whether the equality sign always holds.

Evidently, in this case our chances are better than they were for $g(x, y)$; for, $g(x, y)$ could be any function, whereas $h(\xi, \eta)$ is a bilinear form! Even though $h(\xi, \eta)$ is essentially a generalization of $g(x, y)$, yet it is a function of a much simpler type than $g(x, y)$. In fact, we shall prove in Section 3 that the relation

$$\text{Max}_\xi \text{Min}_\eta h(\xi, \eta) \equiv \text{Min}_\eta \text{Max}_\xi h(\xi, \eta)$$

holds for all bilinear forms $h(\xi, \eta)$ (where Max_ξ is taken for all ξ fo which $\xi_1 \geq 0, \ldots, \xi_{\Sigma_1} \geq 0, \xi_1 + \cdots + \xi_{\Sigma_1} = 1$, and Min_η is taken for all η for which $\eta_1 \geq 0, \ldots, \eta_{\Sigma_2} \geq 0, \eta_1 + \cdots + \eta_{\Sigma_2} = 1$).

4. Anticipating the result we put

$$\text{Max}_\xi \text{Min}_\eta h(\xi, \eta) = \text{Min}_\eta \text{Max}_\xi h(\xi, \eta) = M \quad .$$

Let $\mathfrak{A}$ be the set of all ξ for which $\text{Min}_\eta h(\xi, \eta)$ assumes its maximal value M, and let $\mathfrak{B}$ be the set of all η for which $\text{Max}_\xi h(\xi, \eta)$ assumes its minimal value M. From these definitions the relations below follow immediately.

(1) If ξ belongs to $\mathfrak{A}$, then always $h(\xi, \eta) \geq M$

(2) If η belongs to $\mathfrak{B}$, then always $h(\xi, \eta) \leq M$

(3) If ξ does not belong to $\mathfrak{A}$, there exists an η for which $h(\xi, \eta) < M$

(4) If η does not belong to $\mathfrak{B}$, there exists a ξ for which $h(\xi, \eta) > M$

(5) If ξ belongs to $\mathfrak{A}$ and η belongs to $\mathfrak{B}$, then $h(\xi, \eta) = M$.

On the basis of the relations (1) to (5) the following statement seems justified:

> Clearly, S_1 must choose a complex ξ belonging to $\mathfrak{A}$, and S_2 must choose a complex η belonging to $\mathfrak{B}$. For any such choice, a play has the value M or $-M$ for S_1 and S_2 respectively.

Evidently, a two-person game can be called "fair" if $M = 0$; and it can be called "symmetric" if the players S_1, S_2 have the same roles. I.e., if on interchanging ξ and η (which presupposes that $\Sigma_1 = \Sigma_2$) $h(\xi, \eta)$ and $-h(\xi, \eta)$ are also interchanged, in other words, if

$$h(\xi, \eta) = -h(\eta, \xi)$$

or, equivalently,

$$g(x, y) = -g(y, x)$$

i.e., if the bilinear form $h(\xi, \eta)$, or else the matrix $g(x, y)$ is skew-symmetric. In this case, the game is, of course, also "fair", as can be seen as follows:

$$-\text{Max}_\xi \text{Min}_\eta h(\xi, \eta) = \text{Min}_\xi \text{Max}_\eta - h(\xi, \eta) = \text{Min}_\xi \text{Max}_\eta h(\eta, \xi)$$
$$= \text{Min}_\eta \text{Max}_\xi h(\xi, \eta)$$

i.e.,

$$-M = M, \quad M = 0\ .$$ [9]

One can easily see that in our two examples (in 2.) $M = 0$ since $\mathfrak{A}$ contains only $\xi_1 = \xi_2 = 1/2$ and $\xi_1 = \xi_2 = \xi_3 = 1/3$ respectively, and $\mathfrak{B}$ contains only $\eta_1 = \eta_2 = 1/2$ and $\eta_1 = \eta_2 = \eta_3 = 1/3$ respectively. I.e., both games are "fair" ("Paper, Scissors, Stone" is even symmetric), and in both examples each player must choose all numbers at random, all of

---

[9] Use is made of the fact that Max Min = Min Max, i.e., we have applied our rather deep theorem on bilinear forms. Trivially -- i.e., from Max Min $\leq$ Min Max -- it would only follow that

$$\text{Max Min} \leq 0, \quad \text{Min Max} \geq 0\ .$$

While this paper was put into its final form, I learned of the note of E. Borel in the Comptes Rendus of Jan. 10, 1927 ("Sur les systèmes de formes linéaires...et la théorie du jeu," pp. 52-55). Borel formulates the question of bilinear forms for a symmetric two-person game and states that no examples for Max Min < Min Max are known.

Our result above answers his question.

them with the same probability.

The following point should be emphasized: Although in Section 1 chance was eliminated from the games of strategy under consideration (by introducing expected values and eliminating "draws"), it has now made a spontaneous reappearance. Even if the rules of the game do not contain any elements of "hazard" (i.e., no draws from urns) - as e.g., the two examples in 2. - in specifying the rules of behavior for the players it becomes imperative to reconsider the element of "hazard". The dependence on chance (the "statistical" element) is such an intrinsic part of the game itself (if not of the world) that there is no need to introduce it artificially by way of the rules of the game: even if the formal rules contain no trace of it, it still will assert itself.

§3. PROOF OF THE THEOREM "Max Min = Min Max"

1. Let us slightly change our notation by replacing Σ_1 by $M + 1$, Σ_2 by $N + 1$ and $g(p, q)$ by α_{pq}. We then have

$$h(\xi, \eta) = \sum_{p=1}^{M+1} \sum_{q=1}^{N+1} \alpha_{pq} \xi_p \eta_q \quad .$$

Because of the conditions

$$\xi_1 + \cdots + \xi_M + \xi_{M+1} = 1, \quad \eta_1 + \cdots + \eta_N + \eta_{N+1} = 1$$

the complex ξ is already determined by $\xi_1, \ldots, \xi_M$ and so is the complex η by $\eta_1, \ldots, \eta_N$. Thus

$$h(\xi, \eta) = \sum_{p=1}^{M} \sum_{q=1}^{N} u_{pq} \xi_p \eta_q + \sum_{p=1}^{M} v_p \xi_p + \sum_{q=1}^{N} w_q \eta_q + r \quad .$$

(There is no need to specify explicitly the coefficients in terms of α_{pq}.) We shall make use of only some of the properties of $h(\xi, \eta)$ and investigate continuous functions of two sets of variables $f(\xi, \eta)$ which satisfy the following conditions:

(K). If $f(\xi', \eta) \geq A$, $f(\xi'', \eta) \geq A$, then $f(\xi, \eta) \geq A$ for every $0 \leq \vartheta \leq 1$, $\xi = \vartheta\xi' + (1 - \vartheta)\xi''$ (i.e., $\xi_p = \vartheta\xi_p' + (1 - \vartheta)\xi_p''$, $p = 1, 2, \ldots, M$). If $f(\xi, \eta') \leq A$, $f(\xi, \eta'') \leq A$, then $f(\xi, \eta) \leq A$ for every $0 \leq \vartheta \leq 1$, $\eta = \vartheta\eta' + (1 - \vartheta)\eta''$ (i.e., $\eta_q = \vartheta\eta_q' + (1 - \vartheta)\eta_q''$, $q = 1, 2, \ldots, N$).

(It is clear that $h(\xi, \eta)$, being linear in the ξ as well as in the η, has the property (K).) For these functions $f(\xi, \eta)$ we are going to prove that

$$\mathrm{Max}_\xi \mathrm{Min}_\eta (f(\xi, \eta) = \mathrm{Min}_\eta \mathrm{Max}_\xi f(\xi, \eta)$$

where Max_ξ is taken over the range $\xi_1 \geq 0, \ldots, \xi_M \geq 0, \xi_1 + \cdots + \xi_M \leq 1$ and Min_η is taken over the range $\eta_1 \geq 0, \ldots, \eta_N \geq 0, \eta_1 + \cdots + \eta_N \leq 1$. We can also write

$$\begin{aligned}
&\underset{\substack{\xi_1\geq 0\\ \xi_1\leq 1}}{\mathrm{Max}_{\xi_1}} \underset{\substack{\xi_2\geq 0\\ \xi_1+\xi_2\leq 1}}{\mathrm{Max}_{\xi_2}} \cdots \underset{\substack{\xi_M\geq 0\\ \xi_1+\cdots+\xi_M\leq 1}}{\mathrm{Max}_{\xi_M}} \underset{\substack{\eta_1\geq 0\\ \eta_1\leq 1}}{\mathrm{Min}_{\eta_1}} \underset{\substack{\eta_2\geq 0\\ \eta_1+\eta_2\leq 1}}{\mathrm{Min}_{\eta_2}} \cdots \underset{\substack{\eta_N\geq 0\\ \eta_1+\cdots+\eta_N\leq 1}}{\mathrm{Min}_{\eta_N}} f(\xi, \eta)\\
= &\underset{\substack{\eta_1\geq 0\\ \eta_1\leq 1}}{\mathrm{Min}_{\eta_1}} \underset{\substack{\eta_2\geq 0\\ \eta_1+\eta_2\leq 1}}{\mathrm{Min}_{\eta_2}} \cdots \underset{\substack{\eta_N\geq 0\\ \eta_1+\cdots+\eta_N\leq 1}}{\mathrm{Min}_{\eta_N}} \underset{\substack{\xi_1\geq 0\\ \xi_1\leq 1}}{\mathrm{Max}_{\xi_1}} \underset{\substack{\xi_2\geq 0\\ \xi_1+\xi_2\leq 1}}{\mathrm{Max}_{\xi_2}} \cdots \underset{\substack{\xi_M\geq 0\\ \xi_1+\cdots+\xi_M\leq 1}}{\mathrm{Max}_{\xi_M}} f(\xi, \eta) .
\end{aligned}$$

2. We introduce the following notation:

$$M^{\xi_r} f(\xi_1, \ldots, \xi_r, \eta_1, \ldots, \eta_s) = \underset{\substack{\xi_r\geq 0\\ \xi_1+\cdots+\xi_r\leq 1}}{\mathrm{Max}_{\xi_r}} f(\xi_1, \ldots, \xi_r, \eta_1, \ldots, \eta_s)$$

$$M^{\eta_s} f(\xi_1, \ldots, \xi_r, \eta_1, \ldots, \eta_s) = \underset{\substack{\eta_s\geq 0\\ \eta_1+\cdots+\eta_s\leq 1}}{\mathrm{Min}_{\eta_s}} f(\xi_1, \ldots, \xi_r, \eta_1, \ldots, \eta_s)$$

Clearly, M^{ξ_r} and M^{η_s} eliminate the dependence of f on ξ_r and η_s respectively. We are going to prove: if f satisfies the condition (K) (in 1.), then

$$\begin{aligned}
&M^{\xi_1} M^{\xi_2} \cdots M^{\xi_p} M^{\eta_1} M^{\eta_2} \cdots M^{\eta_q} f\\
&= M^{\eta_1} M^{\eta_2} \cdots M^{\eta_q} M^{\xi_1} M^{\xi_2} \cdots M^{\xi_p} f .
\end{aligned}$$

Evidently, we need only prove the following two assertions:

α) If $f = f(\xi_1, \ldots, \xi_r, \eta_1, \ldots, \eta_s)$ is continuous and has the property (K), the same holds for $M^{\xi_r} f$ and $M^{\eta_s} f$.

β) If $f = f(\xi_1, \ldots, \xi_r, \eta_1, \ldots, \eta_s)$ is continuous and has the property (K), then

$$M^{\xi_r} M^{\eta_s} f = M^{\eta_s} M^{\xi_r} f \quad .$$

We first prove (α). It is sufficient to consider $M^{\xi_r} f$; the same considerations apply to $M^{\eta_s} f$.

We have

$$\begin{aligned} M^{\xi_r} f(\xi_1, \ldots, \xi_r, \eta_1, \ldots, \eta_s) &= f^*(\xi_1, \ldots, \xi_{r-1}, \eta_1, \ldots, \eta_s) \\ &= \operatorname{Max}_{\substack{\xi_r \\ \xi_r \geq 0 \\ \xi_1 + \cdots + \xi_r \leq 1}} f(\xi_1, \ldots, \xi_r, \eta_1, \ldots, \eta_s) \end{aligned}$$

It is obvious that the continuity of f implies that of f^*. We still have to examine the two properties in (K).

First, let

$$f^*(\xi_1', \ldots, \xi_{r-1}', \eta_1, \ldots, \eta_s) \geq A, \quad f^*(\xi_1'', \ldots, \xi_{r-1}'', \eta_1, \ldots, \eta_s) \geq A \; .$$

The f^* represent maximal values of f on finite intervals. Since f is continuous, they are actually assumed, say, for ξ_r' and ξ_r'' respectively.

Then

$$f(\xi_1', \ldots, \xi_r', \eta_1, \ldots, \eta_s) \geq A, \quad f(\xi_1'', \ldots, \xi_r'', \eta_1, \ldots, \eta_s) \geq A$$

and since f satisfies (K) (we put $\xi_1 = \vartheta\xi_1' + (1 - \vartheta)\xi_1'', \ldots, \xi_{r-1} = \vartheta\xi_{r-1}' + (1 - \vartheta)\xi_{r-1}''$ and $\xi_r = \vartheta\xi_r' + (1 - \vartheta)\xi_r''$)

$$f(\xi_1, \ldots, \xi_r, \eta_1, \ldots, \eta_s) \geq A \quad .$$

The inequalities

$$\xi_r' \geq 0, \quad \xi_1' + \cdots + \xi_r' \leq 1, \quad \xi_r'' \geq 0, \quad \xi_1'' + \cdots + \xi_r'' \leq 1$$

imply directly

$$\xi_r \geq 0, \quad \xi_1 + \cdots + \xi_r \leq 1 \quad .$$

Hence, a fortiori, for the maximum f^*

$$f^*(\xi_1, \ldots, \xi_{r-1}, \eta_1, \ldots, \eta_s) \geq A \quad .$$

Second, let

$$f^*(\xi_1, \ldots, \xi_{r-1}, \eta_1', \ldots, \eta_s') \leq A, \quad f^*(\xi_1, \ldots, \xi_{r-1}, \eta_1'', \ldots, \eta_s'') \leq A \ .$$

In virtue of the maximum property of f^* we have

$$f(\xi_1, \ldots, \xi_r, \eta_1', \ldots, \eta_s') \leq A, \quad f(\xi_1, \ldots, \xi_r, \eta_1'', \ldots, \eta_s'') \leq A$$

for all ξ_r for which

$$\xi_r \geq 0, \quad \xi_1 + \ldots + \xi_r \leq 1 \quad .$$

Since f satisfies (K), this again implies

$$f(\xi_1, \ldots, \xi_r, \eta_1, \ldots, \eta_s) \leq A$$

$(\eta_1 = \vartheta\eta_1' + (1 - \vartheta)\eta_1'', \ldots, \eta_s = \vartheta\eta_s' + (1 - \vartheta)\eta_s'')$. And since this holds for all ξ_r mentioned above, we have

$$f^*(\xi_1, \ldots, \xi_{r-1}, \eta_1, \ldots, \eta_s) \leq A \quad .$$

This completes the proof of our assertion (α).

3. We will now show that always (i.e., for all $\xi_1, \ldots, \xi_{r-1}, \eta_1, \ldots, \eta_{s-1}$) $M^{\xi_r} M^{\eta_s} f = M^{\eta_s} M^{\xi_r} f$. If in $f(\xi_1, \ldots, \xi_r, \eta_1, \ldots, \eta_s)$ we keep the variables $\xi_1, \ldots, \xi_{r-1}, \eta_1, \ldots, \eta_{s-1}$ fixed, then f, as a function of ξ_r, η_s alone, evidently still satisfies the condition (K). It remains for us to prove (writing ξ, η for ξ_r, η_s):

> If $f(\xi, \eta)$ is a continuous function and if $f(\xi', \eta) \geq A$, $f(\xi'', \eta) \geq A$ for $\xi' \leq \xi \leq \xi''$ implies that $f(\xi, \eta) \geq A$, and if $f(\xi, \eta') \leq A$, $f(\xi, \eta'') \leq A$ for $\eta' \leq \eta \leq \eta''$ implies that $f(\xi, \eta) \leq A$, then
>
> $$\underset{0\leq\xi\leq a}{\mathrm{Max}_\xi} \ \underset{0\leq\eta\leq b}{\mathrm{Min}_\eta} f(\xi, \eta) = \underset{0\leq\eta\leq b}{\mathrm{Min}_\eta} \ \underset{0\leq\xi\leq a}{\mathrm{Max}_\xi} f(\xi, \eta)$$

(We write a and b for $1 - \xi_1 - \ldots - \xi_{r-1}$ and $1 - \eta_1 - \ldots - \eta_{s-1}$ respectively.)

The assertion can also be formulated in the following way: There exists a "saddle point" ξ_0, η_0 $(0 \leq \xi_0 \leq a,\ 0 \leq \eta_0 \leq b)$, i.e., $f(\xi_0, \eta)$ assumes its minimum for $\eta = \eta_0$ in $0 \leq \eta \leq b$ and $f(\xi, \eta_0)$ assumes its maximum for $\xi = \xi_0$ in $0 \leq \xi \leq a$.

First of all, we evidently have

$$\mathrm{Max}_\xi\ \mathrm{Min}_\eta\ f(\xi, \eta) \leq \mathrm{Min}_\eta\ \mathrm{Max}_\xi\ f(\xi, \eta)$$

and secondly, the assertion just formulated implies that

$$\mathrm{Max}_\xi\ \mathrm{Min}_\eta\ f(\xi, \eta) \geq \mathrm{Min}_\eta\ f(\xi_0, \eta) = f(\xi_0, \eta_0)$$

$$\mathrm{Min}_\eta\ \mathrm{Max}_\xi\ f(\xi, \eta) \leq \mathrm{Max}_\xi\ f(\xi, \eta_0) = f(\xi_0, \eta_0)$$

hence

$$\mathrm{Max}_\xi\ \mathrm{Min}_\eta\ f(\xi, \eta) = \mathrm{Min}_\eta\ \mathrm{Max}_\xi\ f(\xi, \eta) = f(\xi_0, \eta_0) \quad .$$

Our task is now to find a pair ξ_0, η_0 with the desired properties.

Let ξ be fixed. For which values of η, $0 \leq \eta \leq b$, does $f(\xi, \eta)$ assume its minimum? The answer is easy. Since f is continuous, this set is closed, and because of the second assumption about f ($f(\xi, \eta') \leq A$, $f(\xi, \eta'') \leq A$ implies $f(\xi, \eta) \leq A$ for all $\eta' \leq \eta \leq \eta''$) it is convex. But the only closed and convex sets of real numbers are closed intervals. Therefore, this set is a subinterval of the interval $0, b$; we call it $K'(\xi), K''(\xi)$.

If η is fixed, we conclude in the same way that those ξ, $0 \leq \xi \leq a$, for which $f(\xi, \eta)$ assumes its maximum form a closed subinterval of $0, a$; we call it $L'(\eta), L''(\eta)$.

Evidently, always $K'(\xi) \leq K''(\xi)$ and $L'(\eta) \leq L''(\eta)$. Furthermore, the continuity of $f(\xi, \eta)$ implies that $K'(\xi)$, $L'(\eta)$ and $K''(\xi)$, $L''(\eta)$ are lower and upper semi-continuous functions respectively.[10]

Let now ξ^* be fixed. We form the set of all ξ^{**} with the following property: There exists an η^* such that $f(\xi^*, \eta)$ assumes its

---

[10] Let us indicate the proof for $K'(\xi)$. It will be the same for the other three functions.

If $K'(\xi) = 0$, the assertion is trivial since always $K'(\zeta) \geq 0$. Let $K'(\xi) > 0$. For $0 \leq \eta \leq K'(\xi) - \epsilon\ (\epsilon > 0)$, there exists a $\delta > 0$ such that $f(\xi, \eta) \geq \mathrm{Min}_\eta\ f(\xi, \eta) + \delta$, by the definition of $K'(\xi)$. Hence, if ζ is sufficiently close to ξ, we still have $f(\zeta, \eta) \geq \mathrm{Min}_\eta\ f(\zeta, \eta) + \delta/2$ (because both $f(\xi, \eta)$ and $\mathrm{Min}_\eta\ f(\xi, \eta)$ are continuous); i.e., $f(\zeta, \eta)$ does not assume its minimum (in η, for $0 \leq \eta \leq b$) in $0 \leq \eta \leq K'(\xi) - \epsilon$. Therefore $K'(\zeta) > K'(\xi) - \epsilon$, and $K'(\zeta)$ is lower semi-continuous as we have asserted.

minimal value (in $0 \leq \eta \leq b$) at $\eta = \eta^*$ and $f(\xi, \eta^*)$ assumes its maximal value (in $0 \leq \xi \leq a$) at $\xi = \xi^{**}$. I.e., we form the union of all intervals $L'(\eta^*) \leq \xi^{**} \leq L''(\eta^*)$ where η^* assumes all values in the interval $K'(\xi^*) \leq \eta^* \leq K''(\xi^*)$.

In the interval $K'(\xi^*) \leq \eta^* \leq K''(\xi^*)$ the lower semi-continuous function $L'(\eta^*)$ assumes its minimum and the upper semi-continuous function $L''(\eta^*)$ assumes its maximum. Hence the set of ξ^{**} contains a minimal as well as a maximal element. It also contains every intermediate element ξ', which can be demonstrated in the following way: If it were not so, every interval $L'(\eta^*)$, $L''(\eta^*)$ would lie either entirely to the left or entirely to the right of ξ', and both kinds would exist (those belonging to the smallest as well as to the largest ξ^{**}). Since η^* runs over an interval, both kinds of η^* would have a common limit-point. Since both $L'(\eta^*) \leq \xi'$ and $L''(\eta^*) \geq \xi'$ occur arbitrarily close to η' (and L', L'' are lower and upper semi-continuous respectively), it follows that $L'(\eta') \leq \xi'$, $L''(\eta'') \geq \xi'$; i.e., ξ' belongs indeed to one of the intervals, namely, to $L'(\eta')$, $L''(\eta')$.

Our ξ^{**} thus form a closed subinterval of $0, a$, which we shall call $H'(\xi^*)$, $H''(\xi^*)$. $H'(\xi^*)$ is the minimum of the $L'(\eta^*)$, $H''(\xi^*)$ is the maximum of the $L''(\eta^*)$, for $K'(\xi^*) \leq \eta^* \leq K''(\xi^*)$. It is easy to see that $H'(\xi^*)$ and $H''(\xi^*)$ are again lower and upper semi-continuous respectively (this is implied by the corresponding properties of $K'(\xi^*)$, $K''(\xi^*)$, and $L'(\eta^*)$, $L''(\eta^*)$).

It remains to find a $\xi^*(0 \leq \xi^* \leq a)$ which at the same time is a ξ^{**}, i.e., a ξ^* for which $H'(\xi^*) \leq \xi^* \leq H''(\xi^*)$.

If no such ξ^* existed, every interval $H'(\xi^*)$, $H''(\xi^*)$ would lie either entirely to the left or entirely to the right of ξ^*, and both kinds would exist ($\xi^* = a$ and $\xi^* = 0$). Since ξ^* runs over an interval, both kinds of ξ^* would have a common limit-point ξ'. Since arbitrarily close to ξ' both $H'(\xi^*) \leq \xi^*$ and $H''(\xi^*) \geq \xi^*$ occur (and H', H'' are lower and upper semi-continuous respectively), it follows that $H'(\xi') \leq \xi'$, $H''(\xi') \geq \xi'$, i.e., ξ' belongs indeed to the interval $H'(\xi')$, $H''(\xi')$.

We have now proved our last assertion (and hence the assertion (β)), which concludes the proof of our theorem.

§4. THE CASE n = 3

Having dealt in Sections 2 and 3 with the case $n = 2$ we now proceed to the next case, $n = 3$.

Consider a three-person game characterized, according to the description at the end of Section 1, by three functions g_1, g_2, g_3 of three variables x, y, z ($x = 1, 2, \ldots, \Sigma_1$, $y = 1, 2, \ldots, \Sigma_2$, $z = 1, 2, \ldots, \Sigma_3$), where identically

$$g_1 + g_2 + g_3 \equiv 0 \quad .$$

In the case $n = 2$ it was possible to strictly determine the value of a play for each of the players S_1 and S_2 with the results

$$\text{value for } S_1 = \text{Max}_\xi \text{Min}_\eta \sum_{p=1}^{\Sigma_1} \sum_{q=1}^{\Sigma_2} g(p, q)\xi_p\eta_q = \text{Max}_\xi \text{Min}_\eta \sum_{p=1}^{\Sigma_1} \sum_{q=1}^{\Sigma_2} g_1(p, q)\xi_p\eta_q$$

$$\text{value for } S_2 = -\text{Min}_\eta \text{Max}_\xi \sum_{p=1}^{\Sigma_1} \sum_{q=1}^{\Sigma_2} g(p, q)\xi_p\eta_q = \text{Max}_\eta \text{Min}_\xi \sum_{p=1}^{\Sigma_1} \sum_{q=1}^{\Sigma_2} g_2(p, q)\xi_p\eta_q$$

where

$$\text{value for } S_1 + \text{value for } S_2 = 0 \quad .$$

Let us now try in the case $n = 3$ to compute the values of a play for the three players S_1, S_2, S_3. Assume these values to be w_1, w_2, w_3 respectively. For these values to be satisfactory under any conditions and without further discussion they clearly should have the following property: No two players must be able, by forming a coalition, to achieve an expected value exceeding the sum of the "values of a play" assigned to them. Furthermore, $w_1 + w_2 + w_3 = 0$ must hold, since the players make payments only to each other.

By putting

$$\text{Max}_\xi \text{Min}_\eta \sum_{p=1}^{\Sigma_1} \sum_{q=1}^{\Sigma_2} \sum_{r=1}^{\Sigma_3} (g_1(pqr) + g_2(pqr))\xi_{pq}\eta_r = M_{1,2}$$

$$\text{Max}_\xi \text{Min}_\eta \sum_{p=1}^{\Sigma_1} \sum_{q=1}^{\Sigma_2} \sum_{r=1}^{\Sigma_3} (g_1(pqr) + g_3(pqr))\xi_{pr}\eta_q = M_{1,3}$$

$$\text{Max}_\xi \text{Min}_\eta \sum_{p=1}^{\Sigma_1} \sum_{q=1}^{\Sigma_2} \sum_{r=1}^{\Sigma_3} (g_2(pqr) + g_3(pqr))\xi_{qr}\eta_p = M_{2,3}$$

(the ξ_{pq} form a system of probabilities, as do the η_r; similarly ξ_{pr}, η_q and ξ_{qr}, η_p), S_1 and S_2 by forming a coalition can play an ordinary two-person game against S_3, thereby procuring for themselves (in accordance with the above) the expected value $M_{1,2}$. The same holds for S_1 and S_3, and for S_2 and S_3 regarding the expected values $M_{1,3}$ and $M_{2,3}$ respectively. Hence we must have

$$w_1 + w_2 \geq M_{1,2} \quad w_1 + w_3 \geq M_{1,3} \quad w_2 + w_3 \geq M_{2,3}$$

$$w_1 + w_2 + w_3 = 0 \quad .$$

Clearly, this is possible if and only if

$$M_{1,2} + M_{1,3} + M_{2,3} \leq 0 \quad .$$

As we are going to show in 2., it is always true that

$$M_{1,2} + M_{1,3} + M_{2,3} \geq 0$$

and it is easy to give examples in which the $>$ sign holds. Such an example is provided by the following three-person game:

> $\Sigma_1 = \Sigma_2 = \Sigma_3 = 3$. If among the x_1, x_2, x_3 (i.e., among the choices of S_1, S_2, S_3, formerly also denoted by x, y, z) there are two for which $x_\mu = \nu$, $x_\nu = \mu$, then μ, ν form a "true couple". Clearly, there will be precisely one true couple, or none at all.
>
> If no "true couple" exists, let $g_1 = g_2 = g_3 = 0$. If there is a "true couple", let it be μ, ν and let the third of the numbers 1, 2, 3 be λ. Let $g_\mu = g_\nu = 1$, $g_\lambda = -2$.

In this game, evidently $M_{1,2} = M_{1,3} = M_{2,3} = 2$ (any two S_μ, S_ν in coalition can choose ν and μ respectively to form a "true couple" and thus take from the third player, S_λ, the amount 2!). Hence $M_{1,2} + M_{1,3} + M_{2,3} = 6 > 0$.

The reason that in this game any attempt at valuation is bound to fail is the following: In order to gain the amount 2, it is only necessary for any two of the three players to get together. They are then in a position to rob the third one without any ado, in spite of the fact that the rules of the game are strictly symmetrical, i.e., the game is formally fair.[11] The symmetry would imply the value for each player to be zero, but this is obviously wrong. If two players only want to, they can procure for themselves a gain of 2. How is this contradiction to be resolved?

2. Let us proceed systematically.

$$M_{1,2} + M_{1,3} + M_{2,3} \geq 0$$ [12]

always holds. For, evidently,

$$M_{1,2} = \text{Max}_\xi \text{Min}_\eta \sum_{p=1}^{\Sigma_1} \sum_{q=1}^{\Sigma_2} \sum_{r=1}^{\Sigma_3} (g_1(pqr) + g_2(pqr))\xi_{pq}\eta_r$$

(according to our theorem on two-person games)

$$= \text{Min}_\eta \text{Max}_\xi \sum_{p=1}^{\Sigma_1} \sum_{q=1}^{\Sigma_2} \sum_{r=1}^{\Sigma_3} (g_1(pqr) + g_2(pqr))\xi_{pq}\eta_r$$

$$= - \text{Max}_\eta \text{Min}_\xi \sum_{p=1}^{\Sigma_1} \sum_{q=1}^{\Sigma_2} \sum_{r=1}^{\Sigma_3} g_3(pqr)\xi_{pq}\eta_r \quad .$$

We have to prove, therefore, that

$$\text{Max}_{\eta'} \text{Min}_{\xi'} \sum_{p,q,r} g_3(pqr)\xi'_{pq}\eta'_r + \text{Max}_{\eta''} \text{Min}_{\xi''} \sum_{p,q,r} g_2(pqr)\xi''_{pr}\eta''_q$$

$$+ \text{Max}_{\eta'''} \text{Min}_{\xi'''} \sum_{p,q,r} g_1(pqr)\xi'''_{qr}\eta'''_p \leq 0$$

---

[11] This shows that our example is anything but a "pathological" game. Actually, it is a fairly frequent and typical case. Accordingly, we shall see in Section 4, 3 and Section 5, 1 that it is even the general case of a three-person game.

[12] Intuitively, this is immediately clear. S_2 and S_3 in coalition against S_1 can at best obtain $M_{2,3}$, hence S_1 for himself (against all others) can at best obtain $- M_{2,3}$ (because of our theorem on two-person games). Likewise, S_2 for himself can at best obtain $- M_{1,3}$. S_1 and S_2 in coalition can at best obtain $M_{1,2}$. "Unity is strength," i.e.,

$$- M_{2,3} - M_{1,3} \leq M_{1,2} \qquad M_{1,2} + M_{1,3} + M_{2,3} \geq 0 \quad .$$

i.e., for all systems $\bar{\eta}'$, $\bar{\eta}''$, $\bar{\eta}'''$

$$\text{Min}_{\xi'} \sum_{p,q,r} g_3(pqr)\xi'_{pq}\bar{\eta}'_r + \text{Min}_{\xi''} \sum_{p,q,r} g_2(pqr)\xi''_{pr}\bar{\eta}''_q$$

$$+ \text{Min}_{\xi'''} \sum_{p,q,r} g_1(pqr)\xi'''_{qr}\bar{\eta}'''_p \leq 0 \quad .$$

This is actually the case. For, if we put

$$\xi'_{pq} = \bar{\eta}'''_q\bar{\eta}''_q, \quad \xi''_{pr} = \bar{\eta}'''_r\bar{\eta}'_r, \quad \xi'''_{qr} = \bar{\eta}''_q\bar{\eta}'_r$$

we obtain (because of $g_1 + g_2 + g_3 \equiv 0$)

$$\sum_{p,q,r} g_3(pqr)\bar{\eta}'_r\bar{\eta}''_q\bar{\eta}'''_p + \sum_{p,q,r} g_2(pqr)\bar{\eta}'_r\bar{\eta}''_q\bar{\eta}'''_p + \sum_{p,q,r} g_1(pqr)\bar{\eta}'_r\bar{\eta}''_p\bar{\eta}'''_q = 0 \ .$$

We know already that the $>$ sign actually occurs. The case of equality must, therefore, be regarded as a degenerate limiting case.

Let us assume that the player S_1 makes a claim on a gain of w_1 per play. How can he enforce it? Evidently, in two ways.

First of all, he can try to play alone. Essentially, this amounts to setting up a two-person game in which he is on one side and S_2, S_3 (in coalition) on the other. The value of a play is for him $- M_{2,3}$. This solution is acceptable only if $w_1 \leq - M_{2,3}$. Let us, therefore, assume $w_1 > - M_{2,3}$.

Then only the second possibility remains. S_1 must try to get S_2 or S_3 as an ally. In coalition with S_2 or S_3 he can win the sum $M_{1,2}$ or $M_{1,3}$ per play. Since he wants to keep w_1 for himself, he can offer the sum $M_{1,2} - w_1$ to S_2, or $M_{1,3} - w_1$ to S_3 as the price of the alliance. It is out of the question, however, that S_2 or S_3 will accept this offer if in coalition with each other they can win more than $(M_{1,2} - w_1) + (M_{1,3} - w_1)$ per play, i.e., if

$$(M_{1,2} - w_1) + (M_{1,3} - w_1) < M_{2,3}, \quad w_1 > 1/2\ (M_{1,2} + M_{1,3} - M_{2,3}) \quad .$$

Hence we can say that S_1 has no hope at all of satisfying a claim w_1 which is

$$> - M_{2,3}, \quad > 1/2\ (M_{1,2} + M_{1,3} - M_{2,3}) \quad .$$

The second number is $\geq$ the first (because $M_{1,2} + M_{1,3} + M_{2,3} \geq 0$), hence

$$w_1 \leq 1/2\ (M_{1,2} + M_{1,3} - M_{2,3}) = \bar{w}_1$$

must hold. In the same way, it can be shown that

$$w_2 \leq 1/2\ (M_{1,2} + M_{2,3} - M_{1,3}) = \bar{w}_2$$

$$w_3 \leq 1/2\ (M_{1,3} + M_{2,3} - M_{1,2}) = \bar{w}_3$$

must hold.

These upper limits for the claims $\bar{w}_1$, $\bar{w}_2$, $\bar{w}_3$ of the three players can easily be obtained. If, say, S_1, S_2 enter into a coalition, they can achieve the gain $M_{1,2} = \bar{w}_1 + \bar{w}_2$ (against S_3). In the same way, S_1, S_3 or S_2, S_3 can, by entering an alliance, make sure of the gain $M_{1,3} = \bar{w}_1 + \bar{w}_3$ and $M_{2,3} = \bar{w}_2 + \bar{w}_3$ respectively. Hence, the highest possible and still completely justified claims of the three players S_1, S_2, S_3 in a play are $\bar{w}_1$, $\bar{w}_2$, $\bar{w}_3$ respectively.

3. How is this valuation compatible with the impossibility we found in 1. to make a general valuation? If $M_{1,2} + M_{1,3} + M_{2,3} = 0$, no difficulty arises. In this case

$$\bar{w}_1 = - M_{2,3}, \quad \bar{w}_2 = - M_{1,3}, \quad \bar{w}_3 = M_{1,2}$$

i.e., each player can push his claims all by himself, without the help of another (and in the face of a possible coalition of his opponents). All three players can satisfy their claims simultaneously, and accordingly

$$\bar{w}_1 + \bar{w}_2 + \bar{w}_3 = 0 \quad .$$

The situation is different for $M_{1,2} + M_{1,3} + M_{2,3} > 0$. Since

$$\bar{w}_1 > - M_{2,3}, \quad \bar{w}_2 > - M_{1,3}, \quad \bar{w}_3 > - M_{1,2}$$

no player alone can satisfy his claim, and since

$$\bar{w}_1 + \bar{w}_2 + \bar{w}_3 = 1/2\ (M_{1,2} + M_{1,3} + M_{2,3}) > 0$$

it is impossible for all three of them to obtain their desired gains at the same time. But because

$$\bar{w}_1 + \bar{w}_2 = M_{1,2} \qquad \bar{w}_1 + \bar{w}_3 = M_{1,3} \qquad \bar{w}_2 + \bar{w}_3 = M_{2,3}$$

each pair of players having entered an alliance (to rob the third) is assured of its success. Both players can completely satisfy their claims, while the third player will receive only $-M_{2,3}$, $-M_{1,3}$, $-M_{1,2}$ respectively per play, and therefore, his gain will fall short of his justified claim by the amount of $1/2\ (M_{1,2}+M_{1,3}+M_{2,3})$.

This can be formulated as follows: Each of the three players S_1, S_2, S_3 must endeavor to ally himself with another player. If he succeeds, he receives per play

$$1/2\ (M_{1,2}+M_{1,3}-M_{2,3}), \quad 1/2\ (M_{1,2}+M_{2,3}-M_{1,3}),$$
$$1/2\ (M_{1,3}+M_{2,3}-M_{1,2})$$

respectively. If he does not succeed (i.e., if the two others form a coalition), he receives only

$$-M_{2,3}, \quad -M_{1,3}, \quad -M_{1,2}$$

respectively. Still another way of describing the situation, and possibly the most concise one, would be the following:

α) A play has for the players S_1, S_2, S_3 the respective "basic values"

$$v_1 = 1/3\ (M_{1,2}+M_{1,3}-2M_{2,3}), \quad v_2 = 1/3\ (M_{1,2}+M_{2,3}-2M_{1,3}),$$
$$v_3 = 1/3\ (M_{1,3}+M_{2,3}-2M_{1,2}) \ .$$

Since $v_1+v_2+v_3=0$, this is a proper valuation.

β) But for each of any two players entering an alliance against the third there exists the possibility to win $1/6\ D$ in excess of the above "basic values", while the third one sustains a loss of $1/3\ D$ (also in excess of his "basic value"). We have

$$D = M_{1,2}+M_{1,3}+M_{2,3} > 0 \ .$$ [13]

(The first case, $D = M_{1,2}+M_{1,3}+M_{2,3} = 0$, can also be included in this formulation. Here, (α) is the result, and -- since $D=0$ -- (β) is vacuous.)

---

[13] Incidentally

$$v_1 = -M_{2,3}+1/3\ D, \quad v_2 = -M_{1,3}+1/3\ D, \quad v_3 = -M_{1,2}+1/3\ D \ .$$

This solution shows immediately that the three-person game is essentially different from a game between two persons. The actual game strategy of the individual player recedes into the background. It does not offer anything new since the formation of coalitions (which is bound to take place) makes the play a two-person game. But the value of a play for the player does not only depend on the rules of the game. Rather, it is a question -- at least as soon as $D > 0$ -- of which of the three equally possible coalitions $S_1, S_2; S_1, S_3; S_2, S_3$ has been formed. A new element enters, which is entirely foreign to the stereotyped and well-balanced two-person game: struggle.

§5. PRELIMINARY REMARKS ON GAMES FOR $n > 3$

1. For $n > 3$ it has not yet been possible to obtain results of general validity. It may be that the best way to proceed will be in analogy to the cases $n = 2, 3$ which we have already dealt with. Let us recapitulate.

$n = 2$. We define

$$M = \text{Max}_\xi \text{Min}_\eta \sum_{p,q} g_1(pq)\xi_p\eta_q .$$ [14]

An individual play has for the players S_1, S_2 the values $M, -M$ respectively.

$n = 3$. We define

$$M_{1,2} = \text{Max}_\xi \text{Min}_\eta \sum_{p,q,r} (g_1(pqr) + g_2(pqr))\xi_{pq}\eta_r$$

$$M_{1,3} = \text{Max}_\xi \text{Min}_\eta \sum_{p,q,r} (g_1(pqr) + g_3(pqr))\xi_{pr}\eta_q$$

$$M_{2,3} = \text{Max}_\xi \text{Min}_\eta \sum_{p,q,r} (g_2(pqr) + g_3(pqr))\xi_{qr}\eta_p$$ [14]

$$D = M_{1,2} + M_{1,3} + M_{2,3}$$

[14] The Max_ξ and Min_η are to be taken for all systems of probabilities, i.e., we require that

$$\text{all } \xi_p \geq 0, \quad \Sigma_p\xi_p = 1; \quad \text{all } \xi_{pq} \geq 0, \quad \Sigma_{pq}\xi_{pq} = 1; \quad \text{etc.}$$

and similarly,

$$\text{all } \eta_p \geq 0, \quad \Sigma_p\eta_p = 1; \quad \text{etc.}$$

$D \geq 0$, and we distinguish two cases, viz. $D = 0$ and $D > 0$.

$D = 0$. In this case, a play has for the players S_1, S_2, S_3 the values $-M_{2,3}$, $-M_{1,3}$, $-M_{1,2}$ respectively.

$D > 0$. In this case, a play has for the players S_1, S_2, S_3 the "basic values" $-M_{2,3} + 1/3\ D$, $-M_{1,3} + 1/3\ D$, $-M_{1,2} + 1/3\ D$ respectively. In order to obtain the correct values, however, another term has to be added to the "basic values". This is due to the fact that each of any two players who form a coalition against the third (no matter which two) can procure an additional gain of $1/6\ D$, while the third player sustains a loss of $1/3\ D$ per play (in excess of his basic value).

From this summary it becomes clear that the cases $n = 2$ and $n = 3$ with $D = 0$ are of the same type. The case $n = 3$, $D > 0$, however, as was already established at the end of Section 4, represents a new type. We shall denote these two types by the terms strictly-determined and symmetric non-strictly determined respectively, which are self-explanatory.

Is there any chance of reducing all games of strategy to these same two types, even if $n > 3$? Or are new complications to be anticipated? In particular, the possibility of asymmetrical non-strictly determined types has to be contemplated, i.e., types for which the significant possibilities of coalition-formation are not symmetrically distributed among all players. For $n = 3$ this possibility cannot arise. Any possible asymmetries of the rules of the game are entirely absorbed by the "basic values" of the three players. But all players are equally capable of forming coalitions, all three coalitions S_1, S_2; S_1, S_3; S_2, S_3 are equally possible. Let us investigate this question somewhat more closely.

2. In order to characterize a general n-person game we introduce the following constants:

$$M_{\{\mu_1, \mu_2, \ldots, \mu_k\}} = \mathrm{Max}_{\xi}\ \mathrm{Min}_{\eta} \sum_{p_1=1}^{\Sigma_1} \sum_{p_2=1}^{\Sigma_2} \cdots$$

$$\sum_{p_n=1}^{\Sigma_n} (g_{\mu_1}(p_1, \ldots, p_n) + \ldots + g_{\mu_k}(p_1, \ldots, p_n))\xi_{p_{\mu_1}, \ldots p_{\mu_k}} \eta_{p_{\nu_1}, \ldots, p_{\nu_{n-k}}}$$

where $\mu_1, \mu_2, \ldots, \mu_k$ are any k distinct numbers among the numbers $1, 2, \ldots, n$, and $\nu_1, \nu_2, \ldots, \nu_{n-k}$ are the remaining numbers (Max_ξ is to be taken for all $\xi_{p_{\mu_1},\ldots,p_{\mu_k}} \geq 0$, $\Sigma\xi_{p_{\mu_1},\ldots,p_{\mu_k}} = 1$, and Min_η for all $\eta_{p_{\nu_1},\ldots,p_{\nu_{n-k}}} \geq 0$, $\Sigma\eta_{p_{\nu_1},\ldots,p_{\nu_{n-k}}} = 1$). Clearly, $M_{\{\mu_1,\mu_2,\ldots,\mu_k\}}$ is the sum per play which the coalition of the players $S_{\mu_1}, \ldots, S_{\mu_k}$ is able to obtain from the coalition of the players $S_{\nu_1}, \ldots, S_{\nu_{n-k}}$ (since, in fact, the game is a two-person game).

Evidently, $M_{\{\ \}} = 0$. Our theorem on two-person games further implies that $M_{\{\mu_1,\ldots,\mu_k\}} = -M_{\{\nu_1,\ldots,\nu_{n-k}\}}$. Finally, let $\mu_1, \ldots, \mu_k$; $\nu_1, \ldots, \nu_\ell$; $\rho_1, \ldots, \rho_{n-k-\ell}$ be three subsets of $1, 2, \ldots, n$, complementary to each other. If the players $S_{\mu_1}, \ldots, S_{\mu_k}$ as well as $S_{\nu_1}, \ldots, S_{\nu_\ell}$ and $S_{\rho_1}, \ldots, S_{\rho_{n-k-\ell}}$ form fixed coalitions, this is a three-person game, and we have (putting primes on the quantities involved in this kind of game

$$M'_{1,2} = M_{\{\mu_1,\ldots,\mu_k,\nu_1,\ldots,\nu_\ell\}}$$

$$M'_{1,3} = M_{\{\mu_1,\ldots,\mu_k,\rho_1,\ldots,\rho_{n-k-\ell}\}} = -M_{\{\nu_1,\ldots,\nu_\ell\}}$$

$$M'_{2,3} = M_{\{\nu_1,\ldots,\nu\ ,\rho_1,\ldots,\rho_{n-k-\ell}\}} = -M_{\{\mu_1,\ \ldots,\ \mu_k\}} .$$

But according to our results for three-person games we have

$$M'_{1,2} + M'_{1,3} + M'_{2,3} \geq 0$$

i.e.,

$$M_{\{\mu_1,\ldots,\mu_k,\nu_1,\ldots,\nu_\ell\}} \geq M_{\{\mu_1,\ldots,\mu_k\}} + M_{\{\nu_1,\ldots,\nu_\ell\}} .$$

We recapitulate:

> A given n-person game assigns to each subset $\mu_1, \mu_2, \ldots, \mu_k$ of $1, 2, \ldots, n$ a constant $M_{\{\mu_1,\ldots,\mu_k\}}$ (that is the sum per play which the coalition of the players $S_{\mu_1}, \ldots, S_{\mu_k}$ is able to obtain from the coalition of the other players). The system of the constants $M_{\{\mu_1,\ldots,\mu_k\}}$ always satisfies

the following three conditions:

1. $M_{\{\ \}} = 0$

2. $M_{\{\mu_1,\ldots,\mu_k\}} + M_{\{\nu_1,\ldots,\nu_{n-k}\}} = 0$ if $\mu_1, \ldots, \mu_k$ and $\nu_1, \ldots, \nu_{n-k}$ are complementary subsets of $1, 2, \ldots, n$.

3. $M_{\{\mu_1,\ldots,\mu_k,\nu_1,\ldots,\nu_\ell\}} \geq M_{\{\mu_1,\ldots,\mu_k\}} + M_{\{\nu_1,\ldots,\nu_\ell\}}$ if $\mu_1, \ldots, \mu_k$ and $\nu_1, \ldots, \nu_\ell$ are disjoint subsets of $1, 2, \ldots, n$.[15]

It is not difficult to prove the converse, i.e., to specify, for each system of numbers $M_{\{\mathfrak{M}\}}$ ($\mathfrak{M}$ ranging over all 2^n subsets of $1, 2, \ldots, n$) satisfying the conditions 1. to 3., a game of strategy for which the above constants have precisely these values $M_{\{\mathfrak{M}\}}$. We refrain from discussing here such an example, which is not deep at all.

3. I venture the conjecture that the complex of valuations and coalitions in a game of strategy is determined by these 2^n constants alone. We have seen that this is true for $n = 2, 3$;[16] for $n > 3$ a general proof has yet to be found. While in the case $n = 2$ no coalition at all is possible and for $n = 3$ only one type of coalition is conceivable (i.e., "two against one"), the number of possibilities increases rapidly for $n > 3$. If $n = 4$ one must already decide whether a coalition "three against one" or "two against two" is going to be formed, i.e., which alliance offers the best chances to the participants. If $n = 4$ it is still possible to discuss the principal cases (on the basis of the $M_{\{\mathfrak{M}\}}$ alone!), but a satisfactory general theory is as yet lacking.

If our conjecture is correct, we have brought all games of strategy into a natural and final normal form. Each system of 2^n constants $M_{\{\mathfrak{M}\}}$ satisfying the conditions 1. to 3. represents a class of

---

[15] Intuitively, this assertion is as clear as the one considered in the footnote 12, p. 34.

[16] For $n = 2$

$$M_{\{\ \}} = 0, \quad M_{\{1\}} = M, \quad M_{\{2\}} = -M, \quad M_{\{1,2\}} = 0$$

and for $n = 3$

$$M_{\{\ \}} = 0, \ M_{\{1\}} = -M_{2,3}, \ M_{\{2\}} = -M_{1,3}, \ M_{\{3\}} = -M_{1,2}, \ M_{\{1,2\}} = M_{1,2},$$

$$M_{\{1,3\}} = M_{1,3}, \ M_{\{2,3\}} = M_{2,3}, \ M_{\{1,2,3\}} = 0 \ .$$

"tactically equivalent" games of strategy.[17]

In conclusion I would like to add that a later publication will contain numerical calculations of some well-known two-person games (Poker, though with certain schematical simplifications, Baccarat). The agreement of the results with the well-known rules of thumb of the games (e.g., proof of the necessity to "bluff" in poker) may be regarded as an empirical corroboration of the results of our theory.

---

[17] Another possibility of normalizing the $M_{\{\mathfrak{M}\}}$ consists of introducing "basic values" $v_1, v_2, \dots, v_n$ for the players $S_1, S_2, \dots, S_n$ -- in analogy to the "values" (of a play for $n = 2$ and the "basic values" for $n = 3$. For the values exceeding the v_1 one obtains, of course, the new constants

$$M^*_{\{\mathfrak{M}\}} = M_{\{\mathfrak{M}\}} - \sum_{p \text{ in } \mathfrak{M}} v_p$$

Appropriately, the v_p are chosen such that

$$M^*_{\{1\}} = M^*_{\{2\}} = \dots = M^*_{\{n\}}, \quad v_1 + v_2 + \dots + v_n = 0$$

i.e., all players playing for themselves are equally strong, and differences are due only to the various possibilities of coalitions.

(From 1. - 3. it follows easily that the common value of

$$M^*_{\{1\}}, M^*_{\{2\}}, \dots, M^*_{\{n\}} \leq 0 \quad .$$

If it is zero, all $M^*_{\mathfrak{M}} = 0$, i.e., after payment of the "basic values" the play is strictly determined. Hence this common value represents a kind of measure of how non-strictly determined the game is, i.e., a measure of the tactical possibilities the game offers.)

COUNTABLY INFINITELY MANY PERSON GAMES[1]

G. K. Kalisch and E. D. Nering

We consider a set function $v(S)$ defined for all subsets $S \subset I$ (where I is a countable set) with the usual properties of a characteristic function of an essential normalized zero sum n-person game; see [1] for notations and technical vocabulary. We shall speak of a game Γ with characteristic function v (see [1], p. 241) without tracing the connection in our case; the procedure is essentially the same as in the finite case (see [1], p. 243).

We shall make the following definition: an imputation is a vector with countably many components, $\alpha = \{\alpha_i\}$, such that $\alpha_i \geq -1$, $\Sigma|\alpha_i| < \infty$, $\Sigma\alpha_i = 0$. The concepts of domination and solution are defined as usual (see [1], p. 264).

THEOREM. If $v(S)$ has the property that $v(S) \leq v(S - (i))$ for all $i \in S$ and all $S \subset I$, then the corresponding game Γ has no solution.

PROOF. Suppose that the set V of imputations is a solution. Then, as in the finite case, $V \neq \emptyset$ (see [1], p. 278). Observe next that there must exist in V an imputation with a positive component since V cannot consist of the zero vector alone which fails to dominate $\{1, -1, 0, \ldots\}$. Consider now $\alpha \in V$ such that $\alpha_i > 0$. To simplify the exposition, we let $i = 1$. Then $\beta = \{\alpha_1 - \epsilon, \alpha_2 + \epsilon/2, \ldots, \alpha_n + \epsilon/2^{n-1}, \ldots\}$ for $0 < \epsilon < \alpha_1$ is such that $\beta \succ \alpha$ and hence $\beta \notin V$. Therefore there exists $\gamma \in V$ such that $\gamma \succ \beta$ on a set T. We observe that $1 \in T$ since otherwise $\gamma \succ \alpha$, and therefore, $0 < \alpha_1 - \epsilon < \gamma_1$; $S \subset T - (1)$ is not effective for γ, while T is effective for γ:

$$\gamma_1 + \sum_{i \in T-(1)} \gamma_i \leq v(T), \quad \sum_{i \in T-(1)} \gamma_i > v(T - (1)) \quad ;$$

[1] The preparation of this paper was sponsored (in part) by the Office of Naval Research.

hence

$$0 < \gamma_1 < v(T) - v(T - (1)) \quad ,$$

which is an impossibility.

Examples of games satisfying the hypothesis of the theorem are the following: if S is finite and $|S|$ is the number of elements in S, then let $v(S) = -|S|$; if both S and $I - S$ are infinite sets, let $v(S) = 0$; if $I - S$ is finite, let $v(S) = |I - S|$. Another possible assignment is $-1, 0, 1$, respectively, for these three kinds of sets. Note, incidentally, that no finite game satisfies the hypothesis of our theorem.

It is interesting to remark that both the above games do have solutions if we restrict ourselves to imputations almost all of whose components are 0. In the first case the set of all imputations constitutes the (unique) solution since domination must take place on a finite set and all finite sets are flat (see [1], p. 275); in the second case there is a solution consisting of the set V of all imputations α such that $\Sigma_{i \epsilon T} \alpha_i \geq -1$ for all finite sets T. The consistency of V is a consequence of the fact that no $\alpha \epsilon V$ can be dominated by any of our present "almost everywhere 0" imputations, since domination must take place on finite sets and these are not effective for imputations which try to dominate imputations $\alpha \epsilon V$. We now must show that if $\alpha \notin V$ there exists $\beta \epsilon V$ such that $\beta \succ \alpha$. We note that if $\alpha \notin V$ there exists a minimal nonempty finite set F such that $\Sigma_{i \epsilon F} \alpha_i < -1$; we then construct β such that $\beta_i > \alpha_i$ for all $i \epsilon F$, $\Sigma_{i \epsilon F} \beta_i = -1$, $\beta_j = 1$ for some $j \notin F$, $\beta_k = 0$ otherwise; this β clearly dominates α and is in V. This solution is unique as may be seen from the proof of the consistency of V.

If v does not satisfy the hypothesis of the theorem then we may have solutions even in the case where the imputations have infinitely many nonzero components. Consider, for example, the following function: $v(S) = -1$ for $S = (i)$, $v(S) = 1$ for $S = I - (i)$, $v(S) = 0$ otherwise. We get the following family of solutions (one solution for each positive integer t): V_t consists of all imputations α such that, for some integer s $(0 \leq s < \alpha)$ and some real number x $(0 \leq x \leq 1)$,

$$\alpha_i = \begin{cases} -x & \text{if} \quad i = s + 1 \quad , \\ x & \text{if} \quad i = s + t \quad , \\ 0 & \text{otherwise} \quad . \end{cases}$$

The main conclusion to be drawn from the theorem is that it is impossible to prove the existence of solutions for all finite n-person games by means of the ordering relation of domination alone (see [1], p. 266 and following); in some way the compactness of the set of admissible imputations may come in. But the examples of countably many person games with solutions (as well as the case of imputations with almost all components equal to 0) show that the complete story is undoubtedly very complicated.

BIBLIOGRAPHY

[1] von NEUMANN, J., and MORGENSTERN, O., Theory of Games and Economic Behavior, Princeton, 2nd ed. 1947.

G. K. Kalisch
E. D. Nering

University of Minnesota

SOLUTIONS TO GENERAL NON-ZERO-SUM GAMES[1]

Donald B. Gillies

The main problem at present in n-person game theory is to determine whether each game ([1] pp. 504-533: General Non-Zero-Sum Games) has at least one solution in the von Neumann-Morgenstern sense. While a considerable amount of work has been done studying special classes of games, no general systematic theory has yet appeared. It is the purpose of this paper to give the beginnings of such a theory.

The principal results are these:

(1) A positive fraction of all games have solutions, and a positive fraction of all non-zero-sum games have unique solutions.

(2) In all but a lower-dimensional set of games, open regions about each vertex of the simplex of imputations are excluded from any solution, because some player would receive too much in such a region. Discriminatory solutions exist for every game in that lower dimensional set. Furthermore, the restricted set of imputations has the desirable property that any set stable for them is a solution. The excluded imputations just don't matter.

(3) In games for which the value is sufficiently small, a set of imputations, called the core, is in every solution. The core is closed and convex.

(4) A regular feature of games is that the relations, "dominates" and "does not dominate," are preserved if certain player sets are disregarded. If $v(S)$ is any set function, there exists a set function $v^*(S)$ which is superadditive and defines the same dominations. $v^*(S) \geq v(S)$ always, and if $v^*(S) > v(S)$ then S is certainly unnecessary.

(5) Games are considered equivalent if there exists a 1-1 domination-preserving mapping of their imputation spaces. If a game is defined by an arbitrary set-function, this game is equivalent to a game

---

1 The preparation of this paper was sponsored (in part) by the Office of Naval Research.

defined by a superadditive set-function. Furthermore, two games can have different superadditive set-functions and still be equivalent.

(6) The definitions of a decomposable game and a dummy will be generalized by being treated from the point of view of solutions. A larger class of games will be shown to be decomposable (or, as a special case, to have dummies).

(7) An approach will be sketched, which might lead to a general existence theorem for solutions. An (n+1)-person "pyramid game" is obtained by adjoining one indispensible player. To obtain its old-game value in the pyramid game, a coalition must also contain the extra player; otherwise it is utterly defeated. In a pyramid game, domination is an acyclic relation, so it should be easier to show the existence of solutions. It seems to be true that every solution to the original game is contained in a solution to the pyramid game.

§1. STABLE SETS FOR GENERAL REGIONS

Let E^n be a real n-dimensional Euclidean space with fixed coordinates indexed by $1, 2, \ldots, n$. Let $I_n = (1, 2, \ldots, n)$ be the full index set. Subsets of I_n are written $S, S_1, \ldots, T, T_1, \ldots$.

DEFINITION. Subsets of I_n are called <u>coalitions</u>. The number of indices in a set S is written $|S|$. If $|S| = k$, S is called a <u>k-person set</u>, or <u>k-person coalition</u>. A one-person coalition is called a <u>player</u>.

Points in E^n are written $\alpha, \beta, \gamma, \ldots$ etc. and are defined once and for all by

$$\alpha = (a_1, a_2, \ldots, a_n),$$
$$\beta = (b_1, b_2, \ldots, b_n),$$
$$\gamma = (c_1, c_2, \ldots, c_n), \text{ etc.}$$

DEFINITION. For any given α in E^n, the coordinate a_i is interpreted as the <u>payoff to player (i)</u>. The payoff to a coalition is the sum of the payoffs to its several players. The payoff to coalition S will be written $\Sigma_S a_i$ meaning $\Sigma_{i \epsilon S} a_i$.

Let v be a real-valued set-function defined on all sets $S \subseteq I_n$, satisfying $v(\emptyset) = 0$, where $\emptyset$ is the empty set.

DEFINITION. With respect to v, a set S is called <u>effective for</u> α if the payoff of S via α is not greater than $v(S)$, that is if $\Sigma_S a_i \leq v(S)$. S is called <u>strictly effective</u> (for α) if the inequality is strict, and <u>exactly effective</u> in the case of equality. If $\Sigma_S a_i > v(S)$,

S is called <u>ineffective</u> for α.

DEFINITION. The <u>S-dominion of</u> α, written $\text{dom}_S\alpha$, is the empty set if S is ineffective, and if S is effective, $\text{dom}_S\alpha = \{\beta \mid b_i < a_i \text{ for all } i \text{ in } S\}$, which we will write $\{\beta \mid b_i < a_i \text{ for } S\}$. If $\beta \subseteq \text{dom}_S\alpha$, α is said to <u>dominate</u> β <u>via</u> S, written $\alpha \underset{S}{\succ} \beta$. The dominion of a point is the union of its S-dominions, and the dominion of a point-set is the union of the dominions of its various points.

THEOREM 1. dom_S and dom are monotone non-decreasing functions of point sets into open point-sets.

PROOF. $\text{dom}_S\alpha$ is either the null set or $\bigcap_S \{\beta \mid b_i < a_i\}$, a finite intersection of open sets. Thus $\text{dom}_S\alpha$ is an open set. If Q is any point set, $\text{dom}_S Q = \bigcup_{\alpha \in Q} \text{dom}_S\alpha$, which is a union of open sets and therefore open. Similarly dom (Q) is open. If R is any point set, $\text{dom}_S(Q \cup R) = \text{dom}_S Q \cup \text{dom}_S R \supseteq \text{dom}_S Q$. Hence dom_S is a monotone non-decreasing function. Similarly dom is a monotone non-decreasing function.

THEOREM 2. $\underset{S}{\succ}$ is a partial ordering.

PROOF. (a) If $\alpha \underset{S}{\succ} \beta$ then $a_i > b_i$ for S which contradicts $a_i = b_i$ or $b_i > a_i$ for S. Hence at most one of $\alpha \underset{S}{\succ} \beta$, $\alpha = \beta$, $\beta \underset{S}{\succ} \alpha$ can hold between α and β.

(b) If $\alpha \underset{S}{\succ} \beta$, $\beta \underset{S}{\succ} \gamma$, then

$$a_i > b_i \text{ for } S,$$
$$b_i > c_i \text{ for } S,$$

so

$$a_i > c_i \text{ for } S.$$

But since S is effective for α, $\alpha \underset{S}{\succ} \gamma$.

DEFINITION. If K, P are point sets in E^n, and if K = P - dom K, then K is called <u>P-stable</u>. Thus K consists of those points of P which are not in the dominion of K, so

(a) K is contained in P
(b) K and dom K are disjunct
(c) $K \cup \{P \cap \text{dom } K\} = P$.

The pair (v, P) may be considered a generalized game, where v is a value without the restriction of superadditivity, P is a set of generalized extended imputations without necessarily the restriction

$a_i \geq v((i))$ for α in P, and K is a solution. Since various imputation spaces have been proposed for defining a solution, and since one deals with irregular subsets of imputations while studying solutions, it is preferable to keep P as general as possible, even though P-stable sets do not always exist.

DEFINITION. The set P - dom P is called the <u>P-core</u> or <u>core</u>, which may be empty.

THEOREM 3. If K is P-stable and $K \subseteq Q \subseteq P$ then

(3.1) K is closed in the relative topology of P

(3.2) K contains the core

(3.3) K is Q-stable.

PROOF. (3.1) dom K is open, and $P \cap \text{dom } K$ is open in the relative topology of P. But K and $P \cap \text{dom } K$ are disjunct sets whose union is P, so K is closed in the relative topology.

(3.2) $P \supseteq K$ and since dom is a monotone non-decreasing mapping,

$$\text{dom } P \supseteq \text{dom } K,$$

hence

$$P - \text{dom } P \subseteq P - \text{dom } K = K\ ,$$

so

$$P - \text{dom } P \subseteq K\ .$$

Hence K contains the core.

(3.3) Since $P \supseteq Q \supseteq K$,

$$P - \text{dom } K \supseteq Q - \text{dom } K \supseteq K - \text{dom } K$$

but

$$P - \text{dom } K = K \quad \text{since } K \text{ is P-stable},$$

and

$$K - \text{dom } K = K \quad \text{since } K \text{ and } \text{dom } K \text{ are disjunct},$$

so

$$K \supseteq Q - \text{dom } K \supseteq K$$

hence

$$K = Q - \text{dom } K, \quad \text{and } K \text{ is Q-stable.}$$

We have found a set, possibly empty, called the core, which is in every stable set. The points in the core are those points undominated by any point in P, that is, the set of relative maxima in P with respect to the relation of domination. Hence all points of P which are not in the core have the property that they are dominated by points of P. For some points, such as those in the dominion of the core, the relation of being dominated is so strong that we can guarantee that such a point is in no stable set.

> LEMMA 1. For a given α, if there exists a β in P such that $\beta \in\!\!- \alpha$ and if for every $\gamma \in\!\!- \beta$ it follows that $\gamma \in\!\!- \alpha$, then α cannot be in any stable set.

PROOF. If K is any stable set, then β is either in K or in dom K. If β is in K, α is in dom β, so α is in dom K. If β is in dom K, then K contains γ with $\gamma \in\!\!- \beta$. Hence $\gamma \in\!\!- \alpha$, and α is in dom K. Hence α is always in dom K, and never in K.

DEFINITION. If P contains α and β with $\beta \in\!\!- \alpha$, and if, for every γ in P for which $\gamma \in\!\!- \beta$ it follows that $\gamma \in\!\!- \alpha$, then β is said to <u>majorize</u> α, written $\beta \longrightarrow \alpha$.

Lemma 1 can now be re-stated: A sufficient condition for α to be in no P-stable set is the existence of β with $\beta \longrightarrow \alpha$. Note that no assumptions have been made regarding the sets effective for the dominations in question: if $\beta \longrightarrow \alpha$ and $\gamma \in\!\!- \beta$ we admit the possibility that $\gamma \in\!\!- \beta$ via S_1 and $\gamma \in\!\!- \alpha$ via S_2, where $S_1 \neq S_2$. Majorization will be shown to be very important in the study of stable sets. We show that it is a partial ordering with the property that every linearly ordered increasing subset has at least one limit point, and that the relation $\longrightarrow$ is preserved when taking limits. Then it will be shown that if P is closed, for any partial ordering of this type there exists a set of maxima. This permits the operation of simultaneously eliminating all majorized points -- any stable set is contained in the set of maxima, and any set stable over the maxima is stable over P.

> LEMMA 2. $\longrightarrow$ is a partial ordering, that is,
>
> (2.1) At most one of $\beta \longrightarrow \alpha$, $\beta = \alpha$, $\alpha \longrightarrow \beta$ can hold, and
>
> (2.2) If $\gamma \longrightarrow \beta$ and $\beta \longrightarrow \alpha$ then $\gamma \longrightarrow \alpha$.

PROOF. (2.1) If $\beta \longrightarrow \alpha$, then since $\beta \in\!\!- \alpha$, it follows $\beta \neq \alpha$. If $\beta \longrightarrow \alpha$ and $\alpha \longrightarrow \beta$ then: $\alpha \in\!\!- \beta$ from $\alpha \longrightarrow \beta$, but since $\alpha \in\!\!- \beta$ and $\beta \longrightarrow \alpha$, then $\alpha \in\!\!- \alpha$, a contradiction. Hence at most one of the three possibilities holds.

(2.2) Since $\gamma \longrightarrow \beta$, any point dominating γ must dominate β, and also γ dominates β. Since $\beta \longrightarrow \alpha$, any point dominating β must dominate α. Hence any point dominating γ, and γ itself, dominates α. Hence $\gamma \longrightarrow \alpha$ which was to be shown.

LEMMA 3. If $\{\alpha^i\}$ is an infinite sequence of points in E^n with at least one limit point, there exists a Cauchy subsequence in which each coordinate is monotone (constant, increasing, or decreasing), which converges to any given limit point of the original sequence.

PROOF. Let α be a given limit point of $\{\alpha^i\}$. Choose a subsequence $\{\beta^i\}$ of $\{\alpha^i\}$ as follows:

$$\beta^1 = \alpha^1$$

$$\beta^{i+1}$$

is a point α^j in the sequence $\{\alpha^i\}$ which occurs after the point β^i, and has the property

$$|b_k^{i+1} - a_k| < 2^{-i} \text{ for all } k, \quad i \geq 1.$$

Since α is a limit point of $\{\alpha^i\}$ this construction can always be made, and it can easily be shown that $\{\beta^i\}$ is a Cauchy subsequence with limit α.

For any coordinate x_j, the sequence of coordinates $\{b_j^i\}$ satisfies at least one of the following conditions:

(3.1) There is an infinity of points $\{\beta^i\}$ for which $b_j^i = a_j$.

(3.2) For every $\epsilon > 0$, there is an infinity of points $\{\beta^i\}$ for which $a_j - \epsilon < b_j^i < a_j$.

(3.3) For every $\epsilon > 0$, there is an infinity of points $\{\beta^i\}$ for which $a_j < b_j^i < a_j + \epsilon$.

Select one of these conditions that holds. If (3.1) is selected, choose a subsequence of $\{\beta^i\}$ satisfying $b_j^i = a_j$ for every point β^i of the subsequence. If (3.2) is selected, choose i_1 to be any index i for which $b_j^{i_1} < a_j$, i_2 any index greater than i_1 for which $b_j^{i_1} < b_j^{i_2} < a_j$. (The existence of such an index is shown by setting $\epsilon = a_j - b_j^{i_1} > 0$.) Continuing in this way an infinite subsequence can be found monotone increasing in the coordinate x_j. Similarly if (3.3) is selected a subsequence can be found monotone decreasing in the coordinate x_j. Since a subsequence of a Cauchy subsequence is itself a Cauchy subsequence with the same limit point, we have constructed a Cauchy subsequence monotone in the coordinate x_j.

Perform this construction first to $\{\beta^i\}$ with $j = 1$, then to the resulting subsequence with $j = 2$, then to the resulting subsequence with $j = 3$, etc. Continuing in this way, after n successive constructions, a Cauchy subsequence is obtained, monotone in all n coordinates, with limit α.

LEMMA 4. The relation $\longrightarrow$ is preserved in the limit, that is, if $\{\alpha^i\}$ is a sequence for which $\alpha^{i+1} \longrightarrow \alpha^i$ holds for every i, and if α is any limit point of this sequence, then $\alpha \longrightarrow \alpha^i$ for every i.

PROOF. From Lemma 3, select $\{\beta^i\}$, a Cauchy subsequence monotone in each coordinate, with limit α. To prove the lemma, it is sufficient to prove the following:

(4.1) If the lemma holds for $\{\beta^i\}$ it holds for $\{\alpha^i\}$.

(4.2) If $\gamma \in\!\!- \alpha$ then $\gamma \in\!\!- \beta^i$.

(4.3) $\alpha \in\!\!- \beta^i$.

(4.1): Since $\longrightarrow$ is a transitive relation, it can be shown by induction that $\alpha^j \longrightarrow \alpha^i$ for every $j > i$, and that $\beta^j \longrightarrow \beta^i$ for every $j > i$. Since $\{\beta^i\}$ is a subsequence, β^{i+1} coincides with some α^j for which $j \geq i + 1 > i$, so $\beta^{i+1} \longrightarrow \alpha^i$ always. But if $\alpha \longrightarrow \beta^{i+1} \longrightarrow \alpha^i$ then $\alpha \longrightarrow \alpha^i$, so it is sufficient to prove the lemma for $\{\beta^i\}$.

(4.2): If $\gamma \in\!\!- \alpha$ via S, then $c_j > a_j$ for S. Since S is a finite set, there exists a positive ϵ so small that

$$c_j > a_j + \epsilon \text{ for } S.$$

Since α is the Cauchy limit of $\{\beta^i\}$,

$$a_j + \epsilon > b_j^k \text{ for } j \text{ in } S,$$

holds for all sufficiently large k, and in particular for some $k > i$. Hence $c_j > b_j^k$ for S holds for this k, and

$$\gamma \in\!\!- \beta^k \longrightarrow \alpha^i ,$$

so

$$\gamma \in\!\!- \beta^i .$$

(4.3): Let S be the index set for which the coordinates of $\{\beta^i\}$ are monotone increasing functions of i. Since $\beta^{i+1} \longrightarrow \beta^i$,

$\beta^{i+1} \in\!\!- \beta^i$ via some set T_i defined for each i.

Since $b_j^{i+1} > b_j^i$ for j in T_i, $T_i \subseteq S$, and S is not empty. Now S contains a finite number of subsets, and each T_i coincides with one of these subsets. Since there is an infinity of T_i, then there is an infinity of T_i equal to some subset T of S. It will be shown that T is effective for α. For any β^i, there exists $j > i$ with $\beta^{j+1} \in\!\!- \beta^j$ via T, so T is strictly effective for β^j. But $b_k^j > b_k^i$ for k in $T \subseteq S$, since the coordinates are monotone increasing in S. Hence $\Sigma_T b_k^j > \Sigma_T b_k^i$, so T is strictly effective for β^i. Hence T is strictly effective for all β^i. Now $\Sigma_T a_k = \lim \Sigma_T b_k^i \leq v(T)$ since $\Sigma_T b_k^i < v(T)$ always. Hence T is effective for α. But $a_k > b_k^i$ for k in T so $\alpha \in\!\!- \beta^i$ via T. This proves (4.3) and completes the proof of the lemma.

LEMMA 5. If $P \supseteq Q \ni \alpha$ and if $\mathcal{R}$ is a partial ordering of P, define $Q(\alpha) = \{\beta | \alpha \mathcal{R} \beta,\ \beta \in Q\}$. Then Q has a maximum in P if and only if $Q - Q(\alpha)$ has a maximum in P.

PROOF. If γ is a maximum in P, it is sufficient to show that

(5.1) $\gamma \mathcal{R}(Q - \gamma)$ implies $\gamma \mathcal{R}(Q - Q(\alpha) - \gamma)$, and

(5.2) $\gamma \mathcal{R}(Q - Q(\alpha) - \gamma)$ implies $\gamma \mathcal{R}(Q - \gamma)$.

(5.1): By definition $Q \supseteq Q(\alpha)$, so $\gamma \mathcal{R}(Q - \gamma)$ implies a fortiori $\gamma \mathcal{R}(Q - Q(\alpha) - \gamma)$.

(5.2): Since $\alpha \notin Q(\alpha)$, and $\gamma \mathcal{R}(Q - Q(\alpha) - \gamma)$, either $\gamma = \alpha$ or $\gamma \mathcal{R} \alpha$. In either case, since $\mathcal{R}$ is transitive, $\gamma \mathcal{R} Q(\alpha)$, so $\gamma \mathcal{R}(Q - \gamma)$. This completes the proof of the lemma.

LEMMA 6. If every point α of P satisfies $\Sigma a_i \leq M$ for a fixed M, and if Q is a subset of P linearly ordered with respect to $\longrightarrow$, there exists Q^*, a bounded (linearly ordered) subset of Q, such that Q has a maximum in P if and only if Q^* has a maximum in P.

PROOF. Since $\Sigma a_i \leq M$, Q^* is bounded provided $a_i \geq k_i$ for $i = 1, 2, \ldots, n$, and k_i are constants. Suppose first that $a_i \geq v((i))$ for every i, and every α. Set $k_i = v((i))$ and $Q^* = Q$ and the lemma is trivially true. Otherwise choose α, i so $a_i < v((i))$, and define $Q_1 = Q - Q(\alpha)$. By Lemma 5, Q_1 is equivalent to Q as far as maxima are concerned. Suppose, per absurdum, that Q_1 contains β with $b_i < a_i$. Then $\beta \longrightarrow \alpha$ since β is in Q_1 and not in $Q(\alpha)$, and $\alpha \leftarrow\!\!\in \beta$ via (i). Hence $\alpha \leftarrow\!\!\in \alpha$, a contradiction. Hence Q_1 contains only points β for which $b_i \geq a_i$.

If Q_1, although bounded in x_i is not bounded in some coordinate x_j, a set Q_2 may be defined in a similar fashion, equivalent to Q_1 and to Q. Continuing in this fashion, a set Q_k is obtained, in at most n steps, bounded in every coordinate, and equivalent to Q. Set $Q^* = Q_k$. Then Q^* satisfies the lemma.

LEMMA 7. If

(7.1) $\mathcal{R}$ is a partial ordering of $P \subset E^n$.

(7.2) The limit α of a Cauchy sequence $\{\alpha^i\}$ satisfying $\alpha^{i+1} \mathcal{R} \alpha^i$ is maximum contained in P.

(7.3) Q is a bounded linearly ordered subset of P,

then Q has a maximum in P.

PROOF. If Q is finite, Q contains a maximum. Suppose then that Q contains an infinity of points but Q contains no maximum.

A point-set L is said to have property A, if, for any point β in Q there is an infinity of points $\{\alpha^i\}$ in $Q \cap L$ with $\alpha^i \mathcal{R} \beta$. Since Q has an infinity of points, is linearly ordered, and has no maximum, Q and any superset of Q have property A. Furthermore, if a set with property A be decomposed into sets V, W, not necessarily disjunct, then at least one of V and W has property A. For suppose not -- then there exist points β^1 and β^2 in Q such that (a) and (b) hold:

(a) Only a finite number of points $\{\alpha^i\}$ in $V \cap Q$ satisfy $\alpha^i \mathcal{R} \beta^1$.

(b) Only a finite number of points $\{\alpha^i\}$ in $W \cap Q$ satisfy $\alpha^i \mathcal{R} \beta^2$.

Since Q is linearly ordered, one of $\beta^1 \mathcal{R} \beta^2$, $\beta^2 \mathcal{R} \beta^1$ holds, say $\beta^1 \mathcal{R} \beta^2$. Since $V \cup W$ has property A, there is an infinity of points $\{\alpha^i\}$ in $(V \cup W) \cap Q$ for which $\alpha^i \mathcal{R} \beta^1$. Since only a finite number of these are in V (from (a)), W must contain an infinity of $\{\alpha^i\}$ with $\alpha^i \mathcal{R} \beta^1$. But $\beta^1 \mathcal{R} \beta^2$ so $\alpha^i \mathcal{R} \beta^2$ for each of these α^i, contradicting (b). Hence at least one of V, W has property A.

By repeated application of this argument, it can be shown that if a set having property A is decomposed into any finite number of subsets, not necessarily disjunct, then at least one of these subsets will have property A.

Since Q is bounded, it can be enclosed in a cube K_0, having property A since K_0 is a superset of Q. Bisect K_0 into 2^n closed subcubes by means of planes parallel to the coordinate axes. One of these, called K_1, will have property A. Similarly construct $K_2, K_3, \ldots$. By construction, K_i has property A for each i, and $\cap K_i$ is a point, called α. For any point β in Q, choose a point β^1 in $K_1 \cap Q$ with $\beta^1 \mathcal{R} \beta$, and in general, choose β^{i+1} in $K_{i+1} \cap Q$ with $\beta^{i+1} \mathcal{R} \beta^i$. Then $\{\beta^i\}$ is a Cauchy sequence with limit α, so $\alpha \mathcal{R} \beta$, for any point β in Q. Hence α is a maximum. This contradicts the assumption that Q has no maximum. This contradiction proves the lemma.

THEOREM 4. If

(4.1) P contains the limit of each Cauchy sequence $\{\alpha^i\}$ for which $\alpha^{i+1} \longrightarrow \alpha^i$ (for example if P is closed), and if

(4.2) there exists a constant M, such that $\Sigma a_i \leq M$ holds for every point α of P,

then the set P* of points of P maximal with respect to $\longrightarrow$, has the property that P*-stability is equivalent to P-stability.

PROOF. By Lemma 1, all stable sets are contained in P*, so <u>P-stability implies P*-stability</u>.

Let Q be any subset of P which is linearly ordered with respect to $\longrightarrow$. By Lemmata 3, 4, 5, 6 a bounded set Q* can be chosen such that Q has a maximum in P if and only if Q* has a maximum in P. By Lemma 7, Q* has a maximum in P. Hence, by Zorn's lemma there exists a set $P^* \subseteq P$ such that:

(a) If α, β are in P*, never $\alpha \longrightarrow \beta$ over P

(b) If α is in P - P*, there exists a β in P* with $\beta \longrightarrow \alpha$.

Let K be P*-stable. Then for any α in P - P*, there is a β in P* with $\beta \longrightarrow \alpha$. If β is in K, dom K contains α. If K contains γ such that $\gamma \in\!\!- \beta$, since $\beta \longrightarrow \alpha$ it follows that $\gamma \longrightarrow \alpha$. Hence P - P* is contained in dom K, and K is P-stable. Hence <u>P*-stability implies P-stability</u>.

COROLLARY 5. If P, P* are defined as in Theorem 4, and $P \supseteq Q \supseteq P^*$, then by (3.3) of Theorem 3, P-stability is equivalent to Q-stability. Hence any relation implying majorization but not necessarily implied by it, may be used to construct a set $Q \supseteq P^*$, to which the theorem applies.

COROLLARY 6. If P, P^* are defined as in Theorem 4, and C is the core, then P^* and dom C are disjunct, and a necessary and sufficient condition for the core to be the unique stable set is $P^* = C$.

PROOF. $C = P - \text{dom } P$, by definition. If α is a point in C, then α is not dominated by any point of P. Hence, if $\alpha \in\!\!- \beta$, $\alpha \longrightarrow \beta$ too, so β must be in $P - P^*$. Hence P^* and dom C are disjunct. If $P^* \neq C$, there are points of P^* undominated by C, so $P^* = C$ is a necessary condition for C to be the unique stable set. It is also a sufficient condition, since any stable set K satisfies $P^* \supseteq K \supseteq C$.

Since P^* and dom C are disjunct, in passing from P to P^* we have eliminated the effect of the core: the two regions C, $P^* - C$ are independent in the sense that no point in one region can dominate a point in the other. The core is in every stable set, and stable sets differ only in which points of $P^* - C$ they contain.

COROLLARY 7. The relation of majorization may be redefined for points in P^* alone, (a weaker condition), and a set P^{**} constructed such that $P^{**} \subseteq P^*$, and P-stability is equivalent to P^{**}-stability.

PROOF. The existence of $P^{**} \subseteq P$ is guaranteed by Theorem 4, and it is only necessary to verify that P^{**} and $P - P^*$ are disjunct. Suppose not, and let α be a point in both sets. Since α is in $P - P^*$, P^* contains $\beta \longrightarrow \alpha$ where $\longrightarrow$ is defined in terms of P. A fortiori $\beta \longrightarrow \alpha$, when $\longrightarrow$ is defined in terms of P^*. Hence α is not a maximum of P^*, so α is not in P^{**}, the required contradiction.

It is therefore possible, in theory, to construct successively smaller sets P^{**}, P^{***}, ..., each of which contains all P-stable sets, and such that the stability of any set need only be verified for the smallest of the sequence. Examples have been found for which $P^{**} \neq P^*$, but it is not known whether this can happen when P is the usual simplex of imputations.

§2. SPECIAL REGIONS AND EQUIVALENCE OF SET-FUNCTIONS

For a fixed set-function $v(S)$ defined on the subsets of I_n, we have considered a general region $P \subset E^n$, restricted, at most, by the two conditions:

For α in P, $\Sigma a_i \leq M$
P contains the limits of
Cauchy majorization sequences.

(A sufficient condition would be that P is closed in any bounded region.)

To obtain sharper results, we now consider four regions of E^n:

$$E = \{\alpha \mid \Sigma a_i = M\}$$
$$\bar{E} = \{\alpha \mid \Sigma a_i \leq M\}$$
$$A = \{\alpha \mid \Sigma a_i = M \text{ and } a_i \geq v((i)) \text{ for all } i\}$$
$$\bar{A} = \{\alpha \mid \Sigma a_i \leq M \text{ and } a_i \geq v((i)) \text{ for all } i\}$$

Two lemmata will be proved, leading to a theorem that all $\bar{E}$-stable sets ($\bar{A}$-stable sets) are obtained from the E-stable sets (A-stable sets) by adjoining the part of the $\bar{E}$-core ($\bar{A}$-core) lying outside E (or A). This reduces the problem to the study of E and A.

LEMMA 1. If $\bar{P} = \bar{E}$ or $\bar{A}$ has core $\bar{C}$ and $P = E$ or A respectively, and has core C; then $\bar{P}^* - \bar{C} \subseteq P$.

PROOF. It is sufficient to prove that if α is in $\bar{P} - P - \bar{C}$ ($\Sigma a_i < M$ and some point dominates α), then α is majorized.

Since some point dominates α, there is some set S strictly effective for α: $\Sigma_S a_i < v(S)$. Choose $\epsilon > 0$ such that

$$\epsilon \leq \frac{1}{|S|} \operatorname{Min}\,(M - \Sigma a_i,\ v(S) - \Sigma_S a_i)$$

and define

$$\beta : b_i = a_i + \epsilon \text{ for } i \text{ in } S$$
$$= a_i \text{ otherwise.}$$

Then

$$\Sigma_S b_i = \Sigma_S a_i + |S|\epsilon \leq \Sigma_S a_i + v(S) - \Sigma_S a_i = v(S),$$

so S is effective for β, and $\Sigma b_i = \Sigma a_i + |S|\epsilon \leq \Sigma a_i + M - \Sigma a_i = M$, so β is in $\bar{P}$. Now $b_i > a_i$ for i in S so $\bar{P} \ni \beta \in\!\!- \alpha$ and $b_i \geq a_i$ for all i, so any point which dominates β also dominates α. Hence $\bar{P} \ni \beta \longrightarrow \alpha$. This proves the lemma.

LEMMA 2. If α is a point of E^n, and β is a point of $\bar{P}$ such that $\beta \underset{S}{\in\!\!-} \alpha$, $S \neq I_n$, then a point γ can be found in P, such that $\gamma \underset{S}{\in\!\!-} \alpha$, and any point which dominates γ also dominates β. Therefore

$$\underset{S}{\text{dom}}\ P = \underset{S}{\text{dom}}\ \bar{P}$$

and

$$\underset{S}{\text{dom}}\ C = \underset{S}{\text{dom}}\ \bar{C}$$

and

$$C = \bar{C} \cap P \quad .$$

PROOF. Define γ by

$$\begin{aligned} c_i &= b_i \quad \text{for } i \text{ in } S \\ &= b_i + \epsilon \quad \text{otherwise,} \end{aligned}$$

where

$$\epsilon = \frac{1}{n - |S|}(M - \Sigma b_i) \geq 0 \quad .$$

Then $\Sigma c_i = \Sigma b_i + (n - |S|)\epsilon = M$, so γ is in P. Since $c_i = b_i$ for i in S and $\beta \underset{S}{\in\!\!-} \alpha$, $\gamma \underset{S}{\in\!\!-} \alpha$. Since $c_i \geq b_i$ always, any point that dominates γ, also dominates β.

(1) If β is a general point of $\bar{P}$ then

$$\underset{S}{\text{dom}}\ \beta = \underset{S}{\text{dom}}\ \gamma \quad ,$$

so

$$\underset{S}{\text{dom}}\ P \supseteq \underset{S}{\text{dom}}\ \bar{P} \quad .$$

Since $P \subseteq \bar{P}$

$$\underset{S}{\text{dom}}\ P \subseteq \underset{S}{\text{dom}}\ \bar{P} \quad .$$

Hence

$$\underset{S}{\text{dom}}\, P = \underset{S}{\text{dom}}\, \bar{P} \ .$$

(2) Again, if β is a general point of $\bar{C}$, γ is in $\bar{C}$ also, so

$$\underset{S}{\text{dom}}\, \bar{C} = \underset{S}{\text{dom}}\, (\bar{C} \cap P).$$

Since $P \subseteq \bar{P}$,

$$\text{dom}\, P \subseteq \text{dom}\, \bar{P} \ ,$$

$$P - \text{dom}\, P \supseteq P - \text{dom}\, \bar{P} \ ,$$

$$C \supseteq \bar{C} \cap P \ .$$

(3) Suppose α is a point in $C - \bar{C} \cap P$, which is therefore undominated by any point in P, but dominated by a point β in $\bar{P} - P$. Then $\Sigma a_i > \Sigma b_i$ so $\beta \in\!\!- \alpha$ via $S \neq I_n$. Then P contains $\gamma \in\!\!- \alpha$ via S, a contradiction. Hence

$$C - \bar{C} \cap P$$

is empty, and

$$C = \bar{C} \cap P$$

and

$$\underset{S}{\text{dom}}\, C = \underset{S}{\text{dom}}\, (\bar{C} \cap P) = \underset{S}{\text{dom}}\, \bar{C} \ .$$

THEOREM 8. If $Q = P - C - \text{dom}\, C$, then K is P-stable if and only if $K - C$ is Q-stable, and K is $\bar{P}$-stable if and only if $K - \bar{C}$ is Q-stable.

PROOF. It is sufficient, by Corollary 6, to show

(8.1) $Q \supseteq P^* - C$

(8.2) $Q \supseteq \bar{P}^* - \bar{C}$.

(8.1) Since $P^* \supset C$ and $P^* \cap \text{dom}\, C = \emptyset$, and $P^* \subseteq P$, then

$$P^* - C = P^* - C - \text{dom}\, C \subseteq P - C - \text{dom}\, C = Q \ .$$

Hence

$$Q \supseteq P^* - C \ .$$

(8.2) By Lemma 1, $\bar{P}^* - \bar{C} \subseteq P$. Now

$$\begin{aligned}\bar{P}^* - \bar{C} &= \bar{P}^* - \bar{C} - \operatorname{dom} \bar{C} \\ &\subseteq P - \bar{C} - \operatorname{dom} \bar{C} \\ &= P - \bar{C} \cap P - [\cup_{S \neq I_n} \operatorname{dom} \bar{C}] \cup \operatorname{dom}_{I_n} \bar{C} \\ &= P - C - \operatorname{dom} C \cup \operatorname{dom}_{I_n} \bar{C} \\ &= P - C - \operatorname{dom} C - \operatorname{dom}_{I_n} \bar{C} \; .\end{aligned}$$

But if $\alpha \in\!\!- \beta$ via I_n, $\Sigma a_i > \Sigma b_i$, so $M > \Sigma b_i$, and β cannot be in P. Hence

$$P - \operatorname{dom}_{I_n} \bar{C} = P,$$

and

$$\bar{P}^* - \bar{C} \subseteq P - C - \operatorname{dom} C = Q \; .$$

We have shown that there is a 1-1 correspondence between A-stable sets and $\bar{A}$-stable sets, and between E-stable sets and $\bar{E}$-stable sets, namely to obtain an A-stable set from an $\bar{A}$-stable set, delete $\bar{C} - C$; to obtain an $\bar{A}$-stable set from an A-stable set, adjoin $\bar{C} - C$; etc. Stability for one implies stability for the other. We may therefore confine ourselves to the sets A, E.

THEOREM 9. The E-core is contained in the A-core, and is bounded.

PROOF. It will be shown first that the E-core is contained in A. Suppose not. Then there exists a point α in E - dom E and an index i such that $a_i < v((i))$. By hypothesis α is dominated by no point of E. Define β by

$$b_i = a_i + (n - 1)\epsilon$$
$$b_j = a_j - \epsilon \quad \text{for} \quad j \neq i$$

where

$$0 < \epsilon \leq \frac{1}{n - 1} (v((i)) - a_i) \; .$$

Then $\Sigma b_i = \Sigma a_i + (n - 1)\epsilon - (n - 1)\epsilon = \Sigma a_i = M$. Hence β is in E. But

$$a_i < b_i = a_i + (n - 1)\epsilon \leq v((i))$$

so $\beta \underset{(i)}{\in\!\!-} \alpha$, a contradiction. Hence the E-core is contained in A. Thus

$$E - \text{dom } E = A - \text{dom } E$$

but

$$\text{dom } E \supseteq \text{dom } A \quad \text{since} \quad E \supset A$$

so

$$A - \text{dom } E \subseteq A - \text{dom } A \ .$$

Hence the E-core is contained in the A-core. Since the E-core is contained in A it is bounded.

THEOREM 10. For a given v, there exist n quantities $k_1, k_2, \ldots, k_n$, such that if α satisfies $a_i > k_i$ for some i, then either α is in the core or α is majorized.

PROOF. Define

$$k_i = \max_{S \supseteq i} (v(S) - v(S - (i))) \geq v((i))$$

since $v(\emptyset) = 0$, and suppose α satisfies $a_i > k_i$. Either α is in the core, or else α is dominated by some point, so that there is at least one set strictly effective for α. In the latter case we shall show that α is majorized. Choose $\epsilon > 0$ so small that β, defined by

$$b_i = a_i - (n - 1)\epsilon$$
$$b_j = a_j + \epsilon \quad \text{for} \quad j \neq i$$

has the same strictly effective sets, and $b_i \geq k_i$. Then $a_i > b_i$, and $a_j < b_j$ for $j \neq i$. To show $\beta \longrightarrow \alpha$ it is necessary to show

(10.1) $\beta \in\!\!- \alpha$

(10.2) If $\gamma \in\!\!- \beta$ then $\gamma \in\!\!- \alpha$.

(10.1) Let S be a set strictly effective for β. If $S \not\ni i$, then $b_j > a_j$ for S so $\beta \underset{S}{\in\!\!-} \alpha$. If $S \ni i$, $v(S - (i)) \geq v(S) - k_i \geq \Sigma_S b_j - b_i = \Sigma_{S-(i)} b_j$ so $S - (i)$ is effective for β and $\beta \underset{S-(i)}{\in\!\!-} \alpha$.

(10.2) If $\gamma \in\!\!- \beta$ via S then $c_j > b_j$. If $S \not\ni i$, then $b_j > a_j$ for S so $c_j > a_j$ for S and $\gamma \underset{S}{\in\!\!-} \alpha$. If $S \ni i$, $v(S - (i)) \geq v(S) - k_i \geq \Sigma_S c_j - c_i = \Sigma_{S-(i)} c_j$ since $c_i > b_i \geq k_i$, so $S - (i)$ is effective for γ. Now $c_j > b_j > a_j$ for $S - (i)$ so $\gamma \underset{S-(i)}{\in\!\!-} \alpha$.

COROLLARY 11. E-stable sets are bounded.

PROOF. If $R = \{\alpha \mid a_i \leq k_i$ for all $i\}$ where the quantities k_i are defined as in the proof of Theorem 10, then the core is contained in A, and any point outside of the core but in some stable set must be contained in R. Hence $R \cup A$ contains every stable set. Since both A and R are bounded, every stable set is bounded.

THEOREM 12.

(12.1). A necessary and sufficient condition for an A-stable set to be E-stable is that it touch each face of the simplex A, that is, for each j in I_n, the stable set contains a point $\beta^{(j)}$ such that $b_j^{(j)} = v((j))$.

(12.2). A necessary and sufficient condition for and E-stable set to be A-stable is that it be contained in A.

PROOF. (12.1). Sufficiency. Suppose K is A-stable and satisfies (12.1). Define

$$Q_j = \{\alpha \mid \Sigma a_i = M \text{ and } a_j < v((j))\}$$

$$\begin{aligned} E &= \{\alpha \mid \Sigma a_i = M\} \\ &= \{\alpha \mid \Sigma a_i = M \text{ and } a_j \geq v((j)) \text{ for all } j\} \cup \textstyle\bigcup_{j \in I_n} \{\alpha \mid \Sigma a_i = M \text{ and } a_j < v((j))\} \\ &= A \cup_{I_n} Q_j \ . \end{aligned}$$

For every j, K contains $\beta^{(j)}$ such that

$$b_j^{(j)} = v((j)) \ .$$

But if α satisfies

$$a_j < v((j))$$

then

$$v((j)) = b_j^{(j)} > a_j$$

so

$$\beta^{(j)} \underset{(j)}{\succ} \alpha \ .$$

Hence

$$\operatorname{dom}_{(j)} \beta^{(j)} \supseteq Q_j$$

so

$$\operatorname{dom} K \supseteq \cup_{I_n} Q_j \quad .$$

Now

$$K = A - \operatorname{dom} K$$

so

$$K = A \cup_{I_n} Q_j - \operatorname{dom} k$$

$$= E - \operatorname{dom} K,$$

so K is E-stable.

Necessity. Suppose K is A-stable, but there exists some index j such that, for every β in K, $b_j > v((j))$. Since K is a bounded set, a quantity N can be chosen so large that, if $\alpha = \alpha(N)$ is defined by

$$a_i = \frac{M}{n} + N \quad \text{if} \quad i \neq j$$

$$a_j = \frac{M}{n} - (n - 1)N \quad ,$$

then for each β in K,

$$a_i > b_i \quad \text{for} \quad i \neq j$$

and

$$a_j < v((j)) < b_j \quad .$$

Suppose K is E-stable. Then K must contain some point β such that $\beta \in\!\!- A$ via some set S. Hence $b_i > a_i$ for S and S is effective for β. But $b_j > a_j$ only, so S = (j), and, by hypothesis, (j) is not effective for β. Hence K is not E-stable. This proves that the condition was necessary.

(12.2). If K = E - dom K = A - dom K, then $K \subseteq A$, so the condition is necessary. If $K \subseteq A$ and K = E - dom K, then $K \cap A = E \cap A - \operatorname{dom} K$ so K = A - dom K. This proves the sufficiency of $K \subseteq A$.

THEOREM 13. A necessary condition for the core to be A-stable or E-stable is that it touches each face of A.

PROOF. By Theorem 12, it is sufficient to show that if $a_i > v((i))$ for some i and every α in C, then C is not A-stable. Suppose, therefore, that C is A-stable, so C is closed. Define

$$\bar{a}_i = \inf_{\alpha \epsilon C} (a_i) > v((i)).$$

Then $\bar{a}_i$ is attained by some point $\bar{\alpha}$ in C. So $a_i \geq \bar{a}_i$ for every α in C, but for $\bar{\alpha}$ this is an equality. Define

$$\beta : b_i = v((i))$$

$$b_j = \bar{a}_j + \frac{\bar{a}_i - v((i))}{n - 1} \quad \text{for } j \neq i \ .$$

Then β is in A, $b_j > \bar{a}_j$ for $j \neq i$ and $b_i < \bar{a}_i$. Since $b_i < \bar{a}_i$, β is not in C. Hence, by hypothesis, β must be in dom C. But if $\bar{\alpha} \in\!\!- \beta$, this domination must be via (i) which is impossible since $\bar{a}_i > v((i))$. Hence C contains $\gamma \neq \bar{\alpha}$ such that $\gamma \in\!\!- \beta$.

If $\gamma \in\!\!- \beta$ via S then S contains i and $c_i = \bar{a}_i$, for if

$$S \not\ni i, \ c_j > b_j > \bar{a}_j \quad \text{for } S \text{ so } \gamma \in\!\!- \bar{\alpha}$$

and if

$$c_i > \bar{a}_i, \ c_j > \bar{a}_j \quad \text{for } S \text{ so } \gamma \in\!\!- \bar{\alpha}$$

since it is not true that $c_i > b_i > \bar{a}_i$. Now

$$v(S) \geq \Sigma_S c_j = \bar{a}_i + \Sigma_{S-(i)} c_j > \bar{a}_i + \Sigma_{S-(i)} b_j > \bar{a}_i + \Sigma_{S-(i)} \bar{a}_j$$

so

$$v(S) \geq \Sigma_S c_j > \Sigma_S \bar{a}_j \ .$$

Since γ, $\bar{\alpha}$ are in A, $S \neq I_n$ so $I_n - S$ is not empty and

$$\Sigma_{I_n - S} \bar{a}_j > \Sigma_{I_n - S} c_j \ .$$

We can therefore choose a point δ such that

$$d_j = c_j \quad \text{for } j \text{ in } I_n - S$$

$$d_j = \frac{\bar{a}_j + c_j}{2} \quad \text{for } j \text{ in } S - (i)$$

and

$$d_i = \bar{a}_i + \Sigma_{S-(i)} \frac{c_j - \bar{a}_j}{2} \quad .$$

Now $\Sigma d_j = \Sigma c_j$ so δ is in E. Since

$$c_j > \bar{a}_j \text{ for } j \text{ in } S - (i) \, ,$$

$$d_j > \bar{a}_j \text{ for } j \text{ in } S - (i)$$

and

$$d_i > \bar{a}_i \quad .$$

Hence δ is in A and $d_j > \bar{a}_j$ for S. But

$$\Sigma_S d_i = \Sigma_S c_j \leq v(S)$$

so S is effective for δ. Hence $\delta \succ \bar{\alpha}$, contradicting the original assumption that $\bar{\alpha}$ was in C. Hence C is not A-stable if it does not touch each face of A.

§3. EQUIVALENCE OF GAMES

In Section 1, an arbitrary real set-function v, satisfying only $v(\emptyset) = 0$, was used to define domination and stability. On the basis of this, two sets, the core C and the set of unmajorized points P* were defined, having the properties that if K is P-stable, then

$C \subseteq K \subseteq P^*$ and also

P*-stability implies P-stability.

In Section 2, the region P was limited to one of four possible regions, and the stability problem was reduced to the stability problem for two of these regions: A and E. General theorems were obtained without the usual restrictions: $M = v(I_n)$, or v is superadditive. These restrictions appear to have originated in the economic interpretation of games, and imposing them does not seem to simplify the search for stable sets. Indeed, the most convenient way to approach the solutions to decomposable games as well as game equivalence is to reject these conditions.

For simplicity we shall now confine ourselves to A-stability, although, a parallel theory could be constructed for E-stability.

DEFINITION. A _game_ is a real valued set function v, satisfying

$v(\emptyset) = 0$, defined on the subsets of I_n, together with a constant M from which A is defined. A solution to (M, v) is an A-stable set. A point in A is called an imputation.

DEFINITION. Two games (M, v) and (M', v') are called equivalent if there exists a $1 - 1$ dominion-preserving transformation between their imputation-spaces A and A'.

If T is such a transformation, it is sufficient that T preserve dominions of single imputations, since the dominion of a set is the union of the dominions of its members. Since T carries complements of dominions into complements of dominions, it carries solutions into solutions. The relation of game equivalence is reflexive, symmetric and transitive, so it is an equivalence relation, and if $A = A'$, the class of dominion-preserving transformations forms a group. T need not carry S-dominions into S-dominions, and there are important cases when it does not.

THEOREM 14. (M, v) is equivalent to a game in which

(14.1) Players are re-numbered (a player permutation) or

(14.2) $v'(S) = \lambda v(S)$ for λ some positive constant, and $a_i' = \lambda a_i$ for $\alpha \sim \alpha'$, $M' = \lambda M$ or

(14.3) $v'(S) = v(S) + \Sigma_S \ell_i$ where $\ell_1, \ell_2, \ldots, \ell_n$ are fixed constants, $a_i' = a_i + \ell_i$ for $\alpha \sim \alpha'$, $M' = M + \Sigma \ell_i$

or a combination of these transformations.

Since this theorem is a trivial extension of the results in Theory of Games [1], the proof will be omitted.

DEFINITION. A game (M, v) is said to be in $0 - 1$ reduced form if $v((i)) = 0$ for $i = 1, 2, \ldots, n$ and $M =$ either 1, 0 or -1.

A reduced form (M', v') of any game (M, v) is obtained by setting

$$M' = \frac{M - \Sigma_{I_n} v((i))}{|M - \Sigma_{I_n} v((i))|} \quad \text{if this is not } \frac{0}{0}$$

and

$$M' = 0 \quad \text{if} \quad M - \Sigma_{I_n} v((i)) = 0 ,$$

and setting

$$v'(S) = v(S) - \Sigma_S v((i)) \quad \text{if} \quad M' = 0,$$

$$= \frac{v(S) - \Sigma_S v((i))}{|M - \Sigma_{I_n} v((i))|} \quad \text{if} \quad M' \neq 0 .$$

It may be readily verified that (M', v') is equivalent to (M, v), that $v'((i)) = 0$ for $i = 1, 2, \ldots, n$, and that $M' = 1, 0$ or -1. If $M' = 0$, A consists of the point $(0, 0, \ldots, 0)$ which is the unique solution. In this case, (M, v) is called an inessential game. If $M' = -1$, A is empty so there is no problem. There remains only the case $M' = 1$.

THEOREM 15. Any game is equivalent to a game with superadditive set function.

PROOF. It is sufficient to consider (M, v) already in $0 - 1$ reduced form. If $M = -1$ or 0, there are, respectively, no imputations or one imputation in A. In neither case does A contain a point dominated by another point, so (M, v) is equivalent to (M, v') for any set function v' satisfying $v'((i)) = 0$ for $i = 1, 2, \ldots, n$. If we define v' by $v'(S) = 0$ for every S, then v' is superadditive, and (M, v') is equivalent to (M, v).

There remains only the case $M = 1$. For a set $S \subseteq I_n$, let $\mathcal{S}$ be the class of partitions of S into non-empty subsets:

$$\mathcal{S} = \left\{(S_1, S_2, \ldots,) \mid S_i \neq \emptyset, \quad S_i \cap S_j = \emptyset, \quad \cup S_i = S\right\} .$$

Define

$$v'(S) = \operatorname*{Max}_{\mathcal{S}} \Sigma v(S_i).$$

Since $\mathcal{S}$ contains the partition $S_1 = S$, $v'(S) \geq v(S)$. We show first that v' is superadditive, that is, if S and T are disjunct,

$$v'(S) + v'(T) \leq v'(S \cup T) .$$

Let $\mathcal{T}$ be the class of partitions of T, and $\mathcal{U}$ the class of partitions of $S + T$. Denote by $\mathcal{S} \times \mathcal{T}$ the topological product of $\mathcal{S}$ and $\mathcal{T}$, that is, the class of partitions of $\mathcal{U}$ comprising every combination of one partition of S and one partition of T. Then $\mathcal{U} \supseteq \mathcal{S} \times \mathcal{T}$, and

$$\begin{aligned} v'(S \cup T) &= \max_{\mathcal{U}} \Sigma v(S_i) \\ &\geq \max_{\mathcal{S}\times\mathcal{T}} \Sigma v(S_i) \\ &= \max_{\mathcal{S}} \Sigma v(S_i) + \max_{\mathcal{T}} \Sigma v(S_i) \\ &= v'(S) + v'(T) . \end{aligned}$$

Hence v' is superadditive.

To show that (M, v) is equivalent to (M, v'), given that $v'(S) \geq v(S)$, it is sufficient to show that $\alpha \succ \beta$ via S for v' implies $\alpha \succ \beta$ for v. Now $a_i > b_i$ for S, and

$$\begin{aligned} \Sigma_S a_i &\leq v'(S) \\ &= \max_{\mathcal{S}} \Sigma v(S_i) . \end{aligned}$$

Suppose that this maximum is attained by the partition

$$(S_1, S_2, \ldots, S_P), \quad P \geq 1 .$$

Then

$$v'(S) = v(S_1) + v(S_2) + \ldots + v(S_P) .$$

Then one of $S_1, S_2, \ldots, S_P$ is effective for α, for v. Suppose not. Then

$$\begin{aligned} &\Sigma_{S_1} a_i > v(S_1) \\ &\vdots \\ &\Sigma_{S_P} a_i > v(S_P) \end{aligned}$$

so

$$\Sigma_{S_1} a_i + \ldots + \Sigma_{S_P} a_i > v(S_1) + \ldots + v(S_P) = v'(S)$$

so

$$\Sigma_S a_i > v'(S),$$

a contradiction. Hence some S_j is effective for α, for v, and $a_i > b_i$ for S_j so $\alpha \succ \beta$ via S_j, for v. This completes the proof of the theorem.

The preceding proof illustrates the fact that if a subset of S is effective, it does not matter whether S is effective or not - any domination via S can be achieved via the effective subset instead. For such sets, the value of the set-function can be altered, provided it is not made too large, for S is never a minimal winning set. For A, the value $v(I_n)$ plays no role, since domination in A is never via I_n.

The quantity $v(I_n) - M$ is analogous to the excess ([1], p. 364), and, as will be shown, if M is allowed to vary, the critical question is when A-stability and $\bar{A}$-stability coincide. This happens provided $\bar{C} = C$ which means that either C is empty or $v(I_n) \geq M$ so $\bar{C} - C$ is empty.

DEFINITION. A set S is called vital for (M, v) if there exists imputations α, β in A for which $\alpha \underset{S}{\in} \beta$, and this domination is not achieved via any subset of S.

EXAMPLE. In the 0 - 1 reduced form, if $v(12) = v(23) = v(31) = \frac{1}{2}$ and $v(123) \leq \frac{3}{4}$ then (123) is not vital, since α in A implies $a_1 \geq 0$, $a_2 \geq 0$, $a_3 \geq 0$, and therefore if $a_1 + a_2 + a_3 \leq 3/4$, one of: $a_1 + a_2 \leq \frac{1}{2}$, $a_2 + a_3 \leq \frac{1}{2}$, $a_3 + a_1 \leq \frac{1}{2}$ must hold. It is evident that for a set to be vital, it is not sufficient that its value merely exceed the sum of the values of the sets in any non-trivial partition.

THEOREM 16. Domination can always be achieved via a vital set.

PROOF. Suppose $\alpha \in \beta$ via S, and S is not vital. Then S properly contains a subset S_1 such that $\alpha \in \beta$ via S_1. If S_1 is vital, we are done. Otherwise S_1 properly contains a subset S_2 such that $\alpha \in \beta$ via S_2. Continuing in this way, after at most a finite number of steps, a set S_k is obtained, such that $|S_k| \geq 2$ (since players are not effective for any A-domination, $|S_k| \neq 1$), $\alpha \underset{S_k}{\in} \beta$, and via no subset of S_k does $\alpha \in \beta$. Hence S_k is vital.

COROLLARY 17. If (M, v) is in the 0 - 1 reduced form, and v' is defined by

$$v'(S) = 0 \text{ if } S \text{ is not vital,}$$
$$v'(S) = \text{Min } (v(S), M) \text{ if } S \text{ is vital,}$$

then (M, v) is equivalent to (M, v').

PROOF. If, for v, $\alpha \in \beta$ then this domination can be achieved via some vital set S. Hence $\Sigma_S a_i \leq v(S)$. But since $a_i \geq v((i)) = 0$ for all i,

$$\Sigma_{I_n - S} a_i \geq 0$$

Now

$$M = \Sigma a_i = \Sigma_{I_n - S} a_i + \Sigma_S a_i \geq \Sigma_S a_i \ .$$

Hence

$$\Sigma_S a_i \leq M$$

and

$$\Sigma_S a_i \leq v(S)$$

so

$$\Sigma_S a_i \leq \min\ (v(S), M) = v'(S) \ ,$$

so S is effective for α when v' is the set function. Hence $\alpha \in\!\!- \beta$ via S for v'. Similarly, if $\alpha \in\!\!- \beta$ via S for v',

$$\Sigma_S a_i \leq \min\ (v(S), M)$$

so, a fortiori,

$$\Sigma_S a_i \leq v(S)$$

and $\alpha \in\!\!- \beta$ via S for v. This proves the corollary.

(M, v') as defined in the corollary will be called the <u>normal form</u>. The set function v is adjusted so that non-vital sets have value 0 and are effective nowhere, and vital sets have their original value, provided that this value does not exceed M. If the value exceeds M, it is replaced by M, since such sets are still effective everywhere.

A set S is vital if and only if the equalities and inequalities

$$a_i \geq 0, \ \Sigma a_i = 1, \ \Sigma_S a_i \leq v(S), \ \Sigma_T a_i > v(T) \ \text{ for each } \ T \subset S$$

have at least one solution, so the enumeration of vital sets is a linear programming problem.

THEOREM 18. The core is the intersection with A of closed half-spaces of the form

$$\{\alpha \mid \Sigma_S a_i \geq v(S) \ \text{ where } \ S \ \text{ is vital}\},$$

where v is the normalized value.

PROOF. Define

$$H_S = \{\alpha \mid \Sigma_S a_i \geq v(S)\} .$$

It is required to prove

$$C = A \bigcap_{\substack{\text{vital}\\ \text{sets}\\ S}} H_S .$$

Suppose first that α is a point in C and that there exists a vital set S such that α is not in H_S. Then

$$M = 1 \geq v(S) > \Sigma_S a_i$$

$$1 = \Sigma_{I_n} a_i .$$

so

$$\Sigma_{I_n - S} a_i > 0 .$$

Choose $\epsilon : 1 > \epsilon > 0$ so small that

$$\Sigma_S a_i + \epsilon \Sigma_{I_n - S} a_i \leq v(S)$$

and define β by

$$b_i = (1 - \epsilon) a_i \quad \text{for } i \text{ in } I_n - S$$

$$= a_i + \frac{\epsilon}{|S|} \Sigma_{I_n - S} a_j \quad \text{for } i \text{ in } S.$$

Now β is in A, since

$$\Sigma b_i = \Sigma_S a_i + \epsilon \Sigma_{I_n - S} a_i + (1 - \epsilon) \Sigma_{I_n - S} a_i = \Sigma a_i = 1 ,$$

and $a_i \geq 0$ for i in $I_n - S$ implies $b_i \geq 0$ for i in $I_n - S$, and $b_i > a_i$ for i in S. But

$$\Sigma_S b_i = \Sigma_S a_i + \epsilon \Sigma_{I_n - S} a_i \leq v(S),$$

so S is effective for β. Hence $\beta \succ \alpha$, a contradiction, since α was supposed to be contained in the core. Hence $C \subseteq A \cap H_S$, where the intersection is taken over vital sets.

Suppose next that α is a point in $A \cap H_S - C$. Hence A contains $\beta \succ \alpha$, and this domination can be achieved via some vital set S. Now $\Sigma_S b_i \leq v(S)$, and $a_i < b_i$ for S, so $\Sigma_S a_i < \Sigma_S b_i \leq v(S)$, and α is not in H_S. This contradicts the assumption that α is in $A \cap H_S$.

This contradiction proves the theorem.

COROLLARY 19. The core is closed and convex; its boundary consists of points having at least one exactly effective vital set or player; and its interior consists of points having no vital set effective.

PROOF. H_S and A are closed and convex, so C is a finite intersection of closed convex sets. Therefore C is closed and convex. The boundary of $\cap H_S$ consists of points

$$\{\alpha \mid \Sigma_S a_i \geq v(S) \text{ for every vital set, } \Sigma_S a_i = v(S) \text{ for some vital set}\},$$

so the boundary of $\cap H_S$ consists of points having at least one exactly effective vital set. The boundary of A consists of points

$$\{\alpha \mid \Sigma a_i = 1, a_i \geq 0, \text{ and } a_j = 0 \text{ for some } j\}$$

so it consists of points having at least one player exactly effective. The boundary of $A \cap H_S$ consists of boundary points either of A or of $\cap H_S$, so it consists of points having at least one exactly effective vital set or player. Similarly it may be shown that the interior of C consists of points having no vital set effective.

DEFINITION. A game (M, v) in normalized form is called <u>semi-simple</u> if there exists some vital set S, having no vital subsets, satisfying $v(S) = 1 = M$.

THEOREM 20. Every semi-simple game has a solution.

PROOF. Let (M, v) be semi-simple and let S be a minimal vital set satisfying $v(S) = 1$. Since S is vital, S is neither I_n nor a player. Define

$$V = \{\alpha \mid \Sigma_S a_i = 1, \text{ and } \alpha \text{ is in } A\}.$$

It will be shown that V is a solution, that is:

(20.1) $$\text{dom } V \supseteq A - V$$

(20.2) $$V \cap \text{dom } V = \emptyset .$$

(20.1): For β in A, $\Sigma b_i = 1$, so

$$A - V = \{\beta \mid b_j > 0 \text{ for some } j \text{ in } I_n - S, \beta \text{ in } A\}.$$

For any β in $A - V$, define

$$a_i = 0 \text{ for } i \text{ in } I_n - S$$

$$a_i = b_i + \frac{1}{|S|} \Sigma_{I_n - S} b_j \text{ for } i \text{ in } S.$$

It may be verified that

$$a_i \geq 0 \text{ always}$$

$$\Sigma a_i = 1$$

$$\Sigma_S a_i = 1 = v(S)$$

so α is in V, and S is effective for α. But since

$$b_i \geq 0 \text{ always}$$

and

$$b_j > 0 \text{ for some } j \text{ in } I_n - S$$

therefore

$$\frac{1}{|S|} \Sigma_{I_n - S} b_i > 0$$

so

$$a_i > b_i \text{ for } S .$$

Hence $\alpha \underset{S}{\in\!\!-} \beta$ and $\text{dom } V \supseteq A - V$.

(20.2): Suppose $V \cap \text{dom } V \neq \emptyset$, that is, there exist α, β in V such that $\alpha \in\!\!- \beta$ via some set T. So

$$a_i > b_i \text{ for } T$$

but

$$a_i = b_i = 0 \text{ for } I_n - S$$

so

$$T \subseteq S .$$

Since T may be chosen to be a vital set, and S has no proper subsets, $T = S$. Hence $\Sigma_S a_i > \Sigma_S b_i$ or $1 > 1$, a contradiction. Hence $V \cap \text{dom } V = \emptyset$, and V is a solution.

For any point α in A, define two classes $\mathcal{S}$, $\mathcal{S}_1$, of vital sets effective for α as follows:

If S is in $\mathcal{S}$, and $T \subset S$, never $v(T) - \Sigma_T a_i \geq v(S) - \Sigma_S a_i$

If S is in $\mathcal{S}_1$, and $T \subset S$, never $v(T) - \Sigma_T a_i > v(S) - \Sigma_S a_i$.

Then $\mathcal{S}_1 \supseteq \mathcal{S}$. Define

$$R = \cup_{S \in \mathcal{S}} S, \quad R_1 = \cup_{S \in \mathcal{S}_1} S, \quad \text{so} \quad R_1 \supseteq R.$$

THEOREM 21.

(21.1) If R is not empty, and β is such that $b_i < a_i$ for R, then $\alpha \longrightarrow \beta$.

(21.2) If R_1 is not empty, and there exists an i in $I_n - R_1$ such that $a_i > 0$, then α is majorized unless α is in the core.

PROOF. (21.1): Since R is not empty, $\mathcal{S}$ contains some effective vital set S, and $a_i > b_i$ for $S \subseteq R$. Hence $\alpha \in\!\!-\, \beta$ via S. Suppose next that $\gamma \in\!\!-\, \alpha$ via some set S not in $\mathcal{S}$. Then, by the definition of $\mathcal{S}$, there is a $T \subset S$ such that

$$v(T) - \Sigma_T a_i \geq v(S) - \Sigma_S a_i$$

(21.1a)

$$v(T) \geq v(S) - \Sigma_S a_i + \Sigma_T a_i.$$

We show that this set T can be chosen to be an element of $\mathcal{S}$. First, this set T is effective for γ (shown in the present proof) so $\gamma \underset{T}{\in\!\!-} \alpha$. Hence, T can be chosen to be vital. Now if $T \notin \mathcal{S}$, we can again choose a subset, by the definition of $\mathcal{S}$. This process will eventually terminate, yielding a $T \in \mathcal{S}$, since each time a genuinely smaller subset is chosen, and since $|T| = 2$ trivially belongs to $\mathcal{S}$, since its only subsets are players. Since $\gamma \in\!\!-\, \alpha$ via S, $c_i > a_i$ for S and

$$\Sigma_S c_i \leq v(S) .$$

Substituting this in (21.1a), we have

$$\begin{aligned} v(T) &\geq \Sigma_S c_i - \Sigma_S a_i + \Sigma_T a_i \\ &= \Sigma_T c_i + \Sigma_{S-T}(c_i - a_i) \\ &> \Sigma_T c_i \quad \text{since} \quad c_i > a_i \quad \text{for} \quad S - T . \end{aligned}$$

Hence T is effective for γ and $\gamma \in\!\!-\, \alpha$ via $T \subset R$. Hence $c_i > a_i > b_i$ for T and $\gamma \in\!\!-\, \beta$ via T. Hence $\alpha \longrightarrow \beta$.

(21.2): define

$$\beta : b_j = a_j + \epsilon \text{ for } R_1$$
$$= a_j \text{ for } I_n - R_1 - (i)$$
$$b_j = a_1 - |R_1|\,\epsilon$$

where ϵ is chosen so small that

(1) $\Sigma_S a_j < v(S)$ implies $\Sigma_S b_j < v(S)$
(2) $\Sigma_S a_j > v(S)$ implies $\Sigma_S b_j > v(S)$
(3) for $S \supset T$ and T in $\mathcal{S}$
$v(T) - \Sigma_T a_j > v(S) - \Sigma_S a_j$ implies
$v(T) - \Sigma_T b_j > v(S) - \Sigma_S b_j$.

There is a finite number of inequalities, each satisfied by any sufficiently small positive ϵ, so there exists $\epsilon > 0$ satisfying them all. For β, define $\mathcal{S}(\beta)$ by analogy with $\mathcal{S}$.

Since α is not in the core, there is some set strictly effective for α in R_1 and therefore strictly effective for β, so $\mathcal{S}(\beta)$ is not empty. If $\mathcal{S}(\beta) \subseteq \mathcal{S}_1$ then $R(\beta) \subseteq R_1$ and $b_i > a_i$ for R_1 so $\beta \longrightarrow \alpha$. It is therefore sufficient to show that $\mathcal{S}(\beta) \subseteq \mathcal{S}_1$. Suppose S is a set in $\mathcal{S}(\beta)$ but not in $\mathcal{S}_1$. Then since $\Sigma_S a_j > v(S)$ implies $\Sigma_S b_j > v(S)$, S must be effective for α. Hence it may be shown, by reasoning similar to that of the proof to (21.1), that there exists $T \subset R_1$ such that

$$v(T) - \Sigma_T a_j > v(S) - \Sigma_S a_j$$

so

$$v(T) - \Sigma_T b_j > v(S) - \Sigma_S b_j \quad .$$

Hence S is not in $\mathcal{S}(\beta)$, a contradiction which proves that $\mathcal{S}(\beta) \subseteq \mathcal{S}_1$. Hence $\beta \longrightarrow \alpha$.

Theorem 21 is a generalization of Theorem 10. Suppose the core is empty. Theorem 10 states that there is a definite maximum amount which a player can hope to receive in any imputation of a solution, namely the amount by which he increases the value of some coalition by joining it. Theorem 16 states that the strategic possibilities of an imputation depend only on the vital coalitions whose payoff does not exceed the value of the game. Theorem 21 states that certain imputations are excluded because the vital coalitions are penalized in an inequitable way. A vital set S, instead of receiving $v(S)$, is offered $\Sigma_S a_i$ and is asked to accept a deficit $v(S) - \Sigma_S a_i$. The most injured coalitions are those for which no subset is offered as great a deficit. If any player is not a member of a most injured coalition, but would receive more than the absolute minimum (zero), then the imputation is untenable.

Theorem 21 is best possible for the 3-person non-zero sum games - the union of all solutions to each game just fills the region not eliminated by Theorem 21.

DEFINITION. A player in no vital set is a <u>dummy</u>.

DEFINITION. A game for which the core is empty is called a strong game.

DEFINITION. The game (M, v) is said to be <u>decomposable</u> if there exists a partition of I_n into two or more disjunct sets such that any vital set is contained in some one of the sets of the partition.

Let $S_1, S_2, \ldots, S_p$ be a decomposition of I_n minimal in the sense that if any set S_i is split into two or more sets, or if one or more players are removed from S_i, then the resulting partition is not a decomposition partition. Since each player in US_i is in at least one vital set, $I_n - US_i$ consists of the dummies, if any. Therefore, "minimal decomposition" is used in this paper, not in the usual sense, since single element sets (dummies) are excluded from the minimal decomposition.

THEOREM 22. If $S_1, S_2, \ldots, S_p$ is a minimal decomposition of a strong game (M, v), and α is a point in a solution,

(22.1) For each S_i there is at least one vital set contained in S_i effective for α,

(22.2) If β is a point in the same solution, then $\Sigma_{S_i} a_j = \Sigma_{S_i} b_j$ holds for each S_i of the partition.

PROOF. (22.1): A useful, but weaker, form of Theorem 21 is this: If an imputation is in $A^* - C$, the effective vital sets cover each non-zero component. Since α is in a solution, α is in A^*, and C is empty since (M, v) is strong. If $a_j > 0$ for some j in S_i, there is a vital set containing j effective for α. Since (M, v) is decomposable, this set is contained in S_i. But if $a_j = 0$ for every j in S_i, since S_i does not consist of dummies, there exists a set $S \subseteq S_i$ such that $v(S) > 0$. Hence S is effective for α.

(22.2): Suppose $\Sigma_{S_i} a_j \neq \Sigma_{S_i} b_j$ for some S_i of the partition. Define γ by

$c_j = a_j$ if j is in a set S_k of the partition for which $\Sigma_{S_k} b_j \geq \Sigma_{S_k} a_j$.

$c_j = b_j$ if j is in a set S_k of the partition for which $\Sigma_{S_k} b_j < \Sigma_{S_k} a_j$.

Then $c_j \geq \min(a_j, b_j)$ and $\Sigma c_j < M$, so γ is in $\bar{A} - A$. Since the core is empty, $\bar{A} - A$ is in the dominion of any A-stable set, so the solution contains $\delta \succ \gamma$ via S. But S may be chosen to be a vital set contained in some S_k, so either

$$c_i = a_i \quad \text{for} \quad S \subseteq S_k$$

or

$$c_i = b_i \quad \text{for} \quad S \subseteq S_k.$$

Hence either $\delta \underset{S}{\succ} \alpha$ or $\delta \underset{S}{\succ} \beta$, a contradiction. Hence $\Sigma_{S_i} a_j = \Sigma_{S_i} b_j$.

THEOREM 23. If V is a solution to the strong decomposable game (M, v), and α is any point in V, and $M_i = \Sigma_{S_i} a_j$, then

(23.1) $\Sigma M_i = M$ since each dummy receives 0,

(23.2) For β in V, $\Sigma_{S_i} b_j = M_i$

(23.3) The point

$$\gamma : c_j = a_j \quad \text{for} \quad j \text{ in } S_i$$
$$c_j = b_j \quad \text{otherwise,}$$

is in V.

(23.4) If γ is a point such that $\Sigma_{S_i} c_j \leq M_i$, and for no point α of V does $c_j = a_j$ for S_i, then γ is dominated via a subset of S_i by some point of V.

PROOF. (23.1): Suppose i belongs to no vital set, and there exists an $\alpha \in V$ with $a_i > 0$. Construct β such that

$$b_i = a_i - (n - 1)\epsilon$$
$$b_j = a_j + \epsilon \qquad j \neq i .$$

Either $\beta \succ \alpha$ via some set $S \subseteq S_i$, or else all $S \subseteq S_i$ are either ineffective or exactly effective for α so that α is in the core. In either case we have a contradiction.

(23.2): This is a re-statement of (22.2).

(23.3): γ is either in V or dom V. If γ is in dom V, there exists δ in V with $\delta \in\!\!- \gamma$ via S, and S may be taken to be a vital set contained in some S_k. If $k = i$, $\delta \in\!\!- \alpha$; if $k \neq i$, $\delta \in\!\!- \beta$, a contradiction in either case. Hence γ is not in dom V, but in V.

(23.4): γ satisfies $\Sigma_{S_i} c_j \leq M_i$, and for no α of V does $c_j = a_j$ for S_i. Let α be any point of V, and define

$$\beta : b_j = a_j \text{ for } j \text{ not in } S_i ,$$

$$b_j = c_j + \frac{1}{|S_i|} (M_i - \Sigma_{S_i} c_j) \text{ for } j \text{ in } S_i .$$

If β is in V, then $M_i - \Sigma_{S_i} c_j > 0$ since $\beta \neq \gamma$ so $b_j > c_j$ for S_i. Since β must have some effective vital set S contained in S_i, $\beta \in\!\!- \gamma$ via S. If β is not in V, then V contains some point $\delta \in\!\!- \beta$ via some set S. If S is contained in $S_k \neq S_i$ then $\delta \in\!\!- \alpha$, a contradiction. Hence S is contained in S_i and $d_i > b_i \geq c_i$ so $\delta \in\!\!- \gamma$ via $S \subseteq S_i$, which was to be shown.

A subgame or component game (M_i, v_i) is a game defined on the $|S_i|$-dimensional Euclidean space E^{S_i} whose coordinates are indexed by the integers in S_i, and

$$A_i = \left\{\alpha \mid \Sigma a_i + M_i,\ a_i \geq 0 \text{ and } \alpha \text{ is in } E^{S_i}\right\}$$

$$\bar{A}_i = \left\{\alpha \mid \Sigma a_i \leq M_i,\ a_i \geq 0 \text{ and } \alpha \text{ is in } E^{S_i}\right\} .$$

v_i is the restriction of v to S_i and its subsets. Let V_i be the projection of V on A_i, and call C_i, $\bar{C}_i$ the A_i-core and $\bar{A}_i$-core, respectively. The results obtained so far may be re-stated:

1. Any point α of A which is the composition of one point from each of $V_1, V_2, \ldots, V_p$, is in V.
2. V_i is $\bar{A}_i$-stable, which means that V_i is A-stable and $\bar{C}_i = C_i$.

These conditions have been shown to be necessary.

THEOREM 24. Necessary and sufficient conditions that V be a solution to the strong decomposable game (M, v) are

(24.1) There exist p constants $M_1, M_2, \ldots, M_p$ such that $\Sigma M_i = M$, and for any α in V, $\Sigma_{S_i} a_i = M_i$

(24.2) For each (M_i, v_i), $\bar{C}_i = C_i$ and V_i is A_i-stable.

(24.3) Any point composed of a point each from $V_1, V_2, \ldots, V_p$ is in V.

PROOF. Necessity has been shown. For sufficiency, suppose β is a point in $A - V$ but not in dom V. Since $\Sigma b_j = 1$, it is not possible that $\Sigma_{S_i} b_j > M_i$ for all i. Hence either

$$\Sigma_{S_i} b_j < M_i \text{ for some } i,$$

or

$$\Sigma_{S_i} b_j = M_i \text{ for } i = 1, 2, \ldots, p.$$

If $\Sigma_{S_i} b_j < M_i$ for some i, the point $\beta^{(i)}$ formed by projecting β on E^{S_i}, is in $\bar{A}_i - A_i$, so V_i contains some point $\alpha^{(i)} \succ \beta^{(i)}$. Hence V contains some point $\alpha \succ \beta$. Similarly, if $\Sigma_{S_i} b_j = M_i$ for $i = 1, 2, \ldots, p$ define $\beta^{(1)}, \beta^{(2)}, \ldots, \beta^{(p)}$ to be the projections of β on $E^{S_1}, E^{S_2}, \ldots, E^{S_p}$. If $\beta^{(i)} \in V_i$ for $i = 1, 2, \ldots, p$ then β is in V. Otherwise there exists an i for which $\beta^{(i)} \notin V_i$. Hence $\beta^{(i)}$ is in $A_i - V_i$ so there exists $\alpha^{(i)}$ in V_i with $\alpha^{(i)} \succ \beta^{(i)}$. Hence there exists α in V with $\alpha \succ \beta$. This proves the sufficiency of the conditions.

The possible range of each individual M_i may be computed by using the condition $\bar{C}_i = C_i$. Let μ_i be the lower bound of the numbers $\{\mu\}$ for which the equations and inequalities

$$a_j \geq 0 \quad \text{for } S_i$$

$$\Sigma_S a_j > v(S) \quad \text{for every } S \subseteq S_i$$

$$\Sigma_{S_i} a_j = \mu,$$

have a solution. Then μ_i is not attained, and the range of M_i is

$$0 \leq M_i \leq \mu_i \, .$$

THEOREM 25. A positive fraction of all n-person games have a unique solution consisting of the core.

PROOF. The value v is defined by the $2^n - n - 2$ numbers $\{v(S)\}$ for $S \neq \emptyset, I_n$ or a player. Imposing the restriction $v(S) < \frac{1}{n}$ for every S selects a full-dimensional class of values $\{v\} \cdot \{v\}$

contains, incidentally, a full dimensional class of values for which every $S \neq \emptyset$, I_n or a player, is a vital set. But $\max_i\{v(S \cup \{i\}) - v(S)\} < \frac{1}{n}$, and for any point α in A, since $\Sigma a_i = 1$, there is some component $a_j \geq \frac{1}{n}$. Hence by Theorem 10, α is majorized unless it is in the core. Since the core contains the point $(\frac{1}{n}, \frac{1}{n}, \ldots, \frac{1}{n})$ it is not empty, and $A^* = C$ since A is closed and bounded. Since C is C-stable (A^*-stable) it is A-stable. Hence the core is the unique solution.

COROLLARY 26. A positive fraction of all (n+1)-person constant-sum games have discriminatory solutions.

PROOF. If v is super-additive and satisfies $v(S) < \frac{1}{n}$ for $S \neq I_n$ and $v(I_n) = 1$, an (n+1)-person constant-sum game can be defined over $I_n \cup (n + 1)$ by setting

$$v^*(S) = v(S) \quad \text{for} \quad S \subseteq I_n$$

$$v^*(S) = 1 - v(I_n - S) \quad \text{otherwise.}$$

It can be readily verified that the set of points consisting of points in the core of the original game with an (n+1)st coordinate, identically 0, adjoined, is a solution to the (n+1)-person constant-sum game. Since the value of an (n+1)-person constant-sum game is given by $2^n - n - 2$ numbers, the class of constant-sum (n+1)-person games having discriminatory solutions of this form is full-dimensional.

§4. AN APPROACH TO A GENERAL EXISTENCE THEOREM FOR SOLUTIONS

The results obtained so far - that every semi-simple game has a solution, and that every game for which the value is sufficiently small (and hence the core sufficiently large) has a unique solution consisting of the core alone - have depended on somehow setting up a partial ordering which implies domination but need not be implied by it. Another approach is to define, for the game (M, v) a related game called a pyramid game, for which domination is acyclic. It is believed that the task of finding solutions to the pyramid game is easier than for the original game, and it is also believed that a solution to the one problem would also solve the other.

DEFINITION. An <u>essential player</u> is a player contained in every vital set. A <u>pyramid game</u> is a game with one or more essential players.

Given a game (M, w), where w is defined for subsets of I_n, a game (M, v) can be defined, where v is defined on the subsets of I_{n+1} as follows:

$$v(S) = 0 \text{ if } S \subseteq I_n$$
$$v(S + (n + 1)) = w(S).$$

If the original simplex A_0 for (M, w) is embedded in the new simplex A for (M, v), it consists of the face $x_{n+1} = 0$. If this face is visualized as the base of the simplex A, then S-dominions or "domination cones" open downward toward this base because of the necessary condition

$$\alpha \in\!\!- \beta \text{ implies } a_{n+1} > b_{n+1} .$$

Thus one can visualize the passage from (M, w) to (M, v) as erecting a "pyramid" over the original simplex. The point $(0, 0, \ldots, 0, 1)$ has no strictly effective set and therefore is contained in the core. For any number m in $0 \leq m \leq M$, the simplex

$$A_m = \{\alpha \mid a_{n+1} \geq m, \alpha \text{ in } A\}$$

may be defined. A_m consists of those points of A for which $x_{n+1} \geq m$. Note that if the essential player is discarded and the original game value restored for I_n, the base of A_m which is $\{\alpha \mid \Sigma_{I_n} a_i = M - m\}$ corresponds to the situation in decomposable games in which a set is given a constant payoff. In this sense we are stacking or pyramiding all of these various games together.

THEOREM 26. If V is a solution to the pyramid game (M, v) then $V \cap A_m$ is a solution to (M, v) defined over A_m.

PROOF. If α is a point in V but not in A_m, then $a_{n+1} < m$. If $\alpha \in\!\!- \beta$, then

$$a_{n+1} > b_{n+1} .$$

Hence

$$\text{dom } \alpha \cap A_m = \emptyset .$$

Hence

$$A_m - \text{dom } V = A_m - A_m \cap \text{dom } V = A_m \cap V .$$

THEOREM 27. If Q is a set in A composed of sets $\{P_i\}$ such that $P_i \subseteq \{\alpha \mid a_{n+1} = k_i\}$, a plane parallel to the base of the pyramid, where $\{k_i\}$

are constants, and either

(27.1) The planes are finite in number, or

(27.2) The planes are denumerably infinite in number, and such that each limit plane (which need not contain points of Q) is a finite distance from the next lower plane.

Then there exists a unique Q-stable set.

(This is an instance of a "regressively finite" relation, for which solutions have been shown to exist by Richardson [4].)

PROOF. (27.1): By a permutation of indices, $\{k_i\}$ satisfy

$$1 \geq k_1 \geq k_2 \geq \cdots \geq k_r \geq 0 .$$

Define

$$Q_1 = P_1$$

$$Q_{i+1} = P_{i+1} - \text{dom} \bigcup_{j \leq i} Q_j \qquad 1 \leq i \leq r - 1 .$$

Any Q-stable set must contain Q_1, and hence $Q_2, \ldots,$ and hence Q_r. Since $Q \cap \text{dom}\ Q_r = \emptyset$

$$\cup Q_i = \cup P_i - \text{dom}\ \cup Q_i$$

so

$$\cup Q_i \text{ is Q-stable} .$$

(27.2): This follows from (27.1) by transfinite induction since solutions are closed in the relative topology of Q.

In particular, if Q is any finite set of imputations, planes can be found containing every point of Q and satisfying (27.1), so any finite set Q has a unique Q-stable set. There is the possibility of approximating to an A-stable set by inserting more and more planes $\{P_i\}$. But since the values $\{k_i\}$ are nowhere dense in [0, 1] this task seems formidable.

THEOREM 28. If m is sufficiently close to 1, an A_m-stable set exists.

PROOF. Let

$$\epsilon = \max_{v(S)<1} (v(S)) .$$

Let $1 > m > \epsilon$, and let α be any point of A_m. Then for S to be effective for α,

$$\Sigma_S a_i \leq v(S)$$

$$a_{n+1} + \sum_{S-(n+1)} a_i \leq v(S) .$$

But

$$a_{n+1} \geq m > \epsilon$$

so

$$v(S) - \epsilon > \sum_{S-(n+1)} a_i \quad \text{and} \quad v(S) > \epsilon .$$

Hence $v(S) = 1$, and S is effective throughout A_m. If no set S satisfies $v(S) = 1$, then the core coincides with A_m and is the unique stable-set. Otherwise the game is semi-simple in A_m and a stable set exists.

DEFINITION. If Q satisfies $Q \cap \text{dom}\, Q = \emptyset$, and no point in $A - \text{dom}\, Q$ is distant more than ϵ from Q, Q is called an <u>ϵ-solution</u>.

Intuitively this means that no change greater than ϵ in the payoff can be proposed because of the ϵ-stability of Q.

THEOREM 29. For any $\epsilon > 0$, a pyramid game has a finite ϵ-solution.

PROOF. A rule for adding points, one at a time, on a descending sequence of planes parallel to the base will be given. Assume inductively that the process has been continued down to the plane $x_{n+1} = t$ $(1 \geq t \geq 0)$, that is, a sequence $\alpha^1, \alpha^2, \ldots, \alpha^k$ of points has been found such that

$$1 \geq a^1_{n+1} \geq a^2_{n+1} \geq a^k_{n+1} = t ,$$

and never

$$\alpha^i \in\!\!- \alpha^j ,$$

and if β is not in $\text{dom}(\alpha^1, \alpha^2, \ldots, \alpha^k)$, either $b_{n+1} \leq t$ or there exists some j such that α^j is distant not more than ϵ from β.

Let $\mathfrak{A}^k$ be the point-set in A distant at least ϵ from every point of $(\alpha^1, \alpha^2, \ldots, \alpha^k)$ and undominated by any of these. $\mathfrak{A}^k$ is closed since it is the intersection of two closed sets. Define

$$\bar{t} = \sup_{\beta \in \mathcal{U}^k} (b_{n+1}) \quad .$$

Then $\bar{t}$ is attained for some β called α^{k+1} and $\bar{t} \leq t$. α^{k+1} is the next point added, and satisfies the inductive assumptions. Since A is compact, this process terminates after a finite number of steps since no two points are nearer than ϵ from each other.

If ϵ tends to zero, it can be shown that limit sets exist, however they may not be A-stable because

(1) the limit of points having a set S ineffective may have S exactly effective

(2) the limit of points, each dominating a fixed point, need not dominate the fixed point, since one or more components may become equal, in the limit, to components of the fixed point.

The importance of pyramid games lies in the possibility that a solution, under suitable conditions (for example, connectedness) when intersected with the base, always solves the subgame. In the case of a three-person non-zero-sum game, a pyramid game can be constructed over it, and the connected solutions, intersected with any plane parallel to the base, solve the corresponding games. Thus solutions to games with excess are unified if suitable properties can be established generally.

Also, for all but a lower dimensional set of games, $v(S) < 1$ for all S, and the core is the unique solution to the pyramid game in a neighbourhood of $(0, 0, \ldots, 0, 1)$. We can pass from any solution of the game on the base to any other solution by going up the pyramid, possibly as high as the core, and taking another continuation down again. This would unify all solutions to a particular game -- a perplexing problem at present.

BIBLIOGRAPHY

[1] von NEUMANN, J. and MORGENSTERN, O., Theory of Games and Economic Behavior, Princeton 1944, 2nd edition, 1947.

[2] SHAPLEY, L. S., "Notes on the n-person game III: some variants of the von Neumann-Morgenstern definition of solution", RAND Corporation report RM-817 (unpublished).

[3] McKINSEY, J. C. C., "Isomorphism of games and strategic equivalence", Contributions to the Theory of Games, Vol. 1. Annals of Mathematics Study No. 24 (Princeton, 1950), pp. 117-130.

[4] RICHARDSON, Moses, "Solutions of irreflexive relations", Annals of Mathematics 58 (1953), pp. 573-590.

Donald B. Gillies

University of Illinois

A SOLUTION CONTAINING AN ARBITRARY CLOSED COMPONENT

L. S. Shapley

A class of n-person game solutions is described in this note which consist of an arbitrary closed subset, J, of a certain $(n-3)$-dimensional region of the fundamental simplex, together with one other closed set, $K - H$, depending on but disjoint from J. The arbitrariness of J illustrates the great variety of solution sets that are possible in the von Neumann-Morgenstern theory, and provides a ready source of examples, pleasant and pathological, for the testing of conjectures about the topological properties of solutions in general.

Let $N = \{1, \ldots, n\}$ be the set of players, $n \geq 4$. We shall consider the simple game with winning coalitions $N - \{1\}$, $N - \{2\}$, $N - \{3\}$, and N.[1] In the $0 - 1$ normalization, the imputation space A is the simplex of nonnegative n-tuples $\alpha = \langle \alpha_1, \ldots, \alpha_n \rangle$ with sum 1. Let $\text{dom}_k\, \beta$ denote the set of $\alpha \in A$ with $\alpha_i < \beta_i$, all $i \neq k$. Then the set of imputations dominated by β is

$$\text{dom}\, \beta = \text{dom}_1\, \beta \cup \text{dom}_2\, \beta \cup \text{dom}_3\, \beta \quad ,$$

and the defining properties of a solution V can be expressed

$$V \cap \text{dom}\, V = \emptyset \quad \text{and} \quad V \cup \text{dom}\, V = A \quad ,$$

where, of course, $\text{dom}\, V$ denotes

$$\sup_{\beta \in V} \text{dom}\, \beta \quad .$$

We shall now define the sets J, K, and H referred to above. Let U be the simplex of nonnegative $(n-3)$-tuples $u = (u_4, \ldots, u_n)$ with sum ≤ 1. Let u_0 denote the difference $1 - \Sigma_4^n u_j$, and let U_0 be the

[1] It can be described as a direct product of the simple majority game on $\{1, 2, 3\}$ and the pure bargaining (or "unanimous") game on $\{4, \ldots, n\}$.

face of U on which $u_o = 0$. Let C be an arbitrary, closed set in $U - U_o$, and let J be the image of C under the one-one map:

$$(u_4, \ldots, u_n) \longleftrightarrow \langle 0, \frac{u_o}{2}, \frac{u_o}{2}, u_4, \ldots, u_n \rangle$$

of U into A. This is the "arbitrary" part of the solution.

Let ρ be the distance function in U:

$$\rho(u, w) = \sum_4^n |u_j - w_j|$$

and let

$$\rho(u, W) = \inf_{w \in W} \rho(u, w)$$

for sets W in U. Note that

$$\rho(u, W) \leq \rho(u, v) + \rho(v, W) \quad ,$$

and that $\rho(u, U_o) = u_o$ for all $u \in U$.

The set K can now be defined; it consists of all imputations of the form:

$$\langle \frac{u_o - \rho(u, \bar{C})}{2}, x, y, u_4, \ldots, u_n \rangle, \quad u \in U,$$

where $\bar{C} = C \cup U_o$. Observe that the first component is always nonnegative, and that the sum of the second and third is determined by u; they run between 0 and $[u_o + \rho(u, \bar{C})]/2$, sweeping out a line segment for each fixed u in $U - U_o$. Thus K is a continuous hypersurface of dimension $n - 2$, separating A into two parts. It is disjoint from J since any imputation in $J \cap K$ would have $u \in C$ and $u_o - \rho(u, \bar{C}) = 0$, implying both $u_o \neq 0$ and $u_o = 0$.

Finally, H is defined as $K \cap \text{dom}\, J$, the part of K dominated by J. Since $\alpha_1 = 0$ in J, we have $H = K \cap \text{dom}_1 J$. We shall show that $J \cup (K - H)$ is a solution.

Our construction is illustrated in Figure 1 for $n = 4$. J consists of two dots and a dash, so spaced as to make the "holes" H clearly visible. An inspection of the horizontal cross-sections shows a relationship between our solutions and the familiar symmetric and discriminatory solutions of the 3-person simple majority game. (The dotted lines are medians.)

The proof that $J \cup (K - H)$ is a solution is most conveniently presented as a series of "corrections" to the near-solution K. Thus,

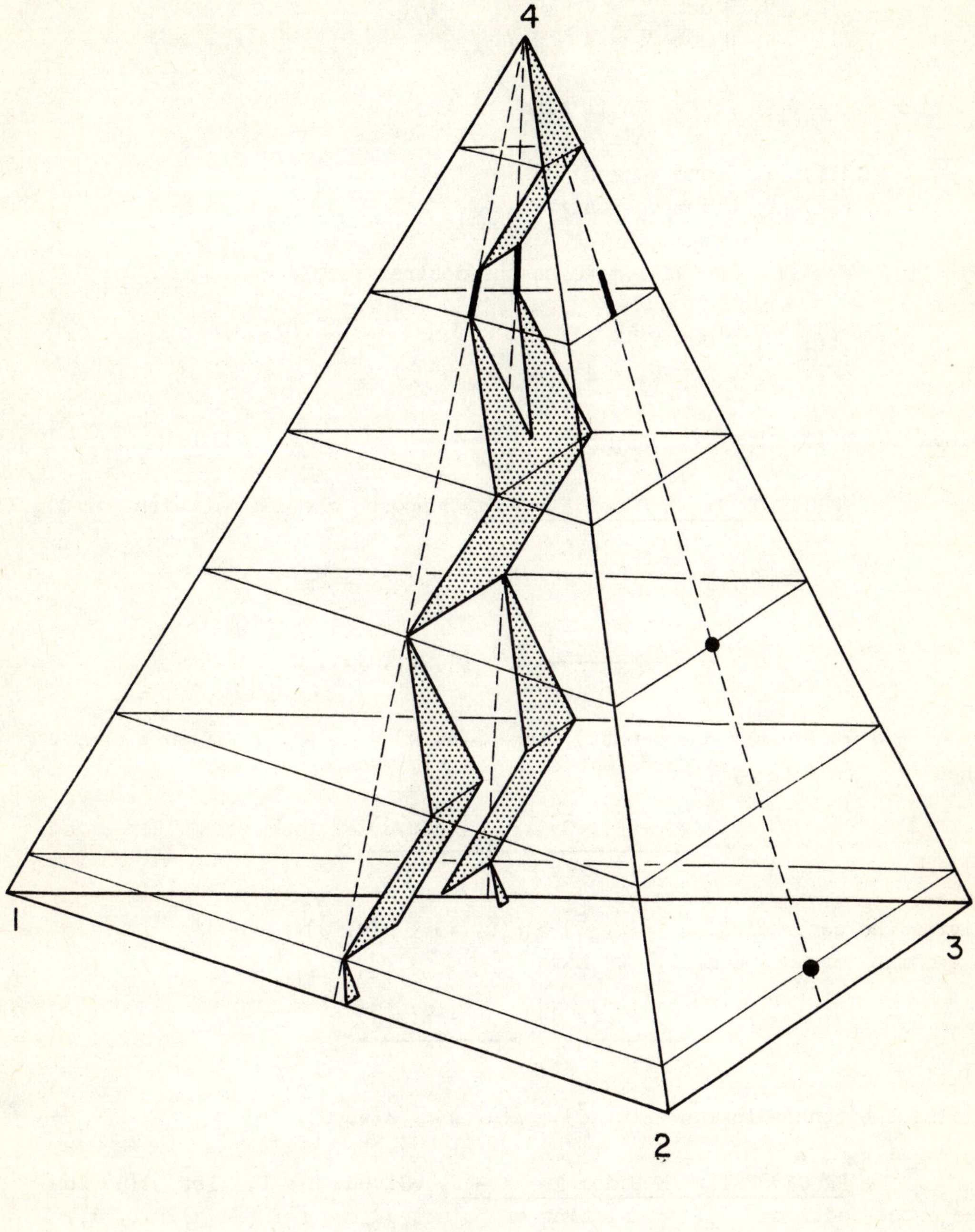

FIGURE 1

putting $V_1 = K$, we shall show

$$\text{(I)}\quad V_1 \cap \text{dom } V_1 = \emptyset$$
$$\text{(II)}\quad V_1 \cup \text{dom } V_1 = A - J .$$

Putting $V_2 = V_1 \cup J$, we shall find

$$\text{(III)}\quad V_2 \cap \text{dom } V_2 = H$$
$$\text{(IV)}\quad V_2 \cup \text{dom } V_2 = A .$$

Finally, $V_3 = V_2 - H$ will give us the desired result

$$\text{(V)}\quad V_3 \cap \text{dom } V_3 = \emptyset$$
$$\text{(VI)}\quad V_3 \cup \text{dom } V_3 = A ,$$

since $(K \cup J) - H = J \cup (K - H)$.

PROOF OF I: $K \cap \text{dom } K = \emptyset$. Suppose $\alpha \in \text{dom } \beta$, with $\alpha, \beta \in K$:

$$\alpha = < \frac{u_0 - \rho(u,\bar{C})}{2}, x, y, u_4, \ldots, u_n > ,$$

$$\beta = < \frac{w_0 - \rho(w,\bar{C})}{2}, x', y', w_4, \ldots, w_n > .$$

Then $u < w$ (in all components), and $\rho(u, w) = u_0 - w_0$. If $\alpha \in \text{dom}_1 \beta$ then $x + y < x' + y'$, or

$$\frac{w_0 + \rho(w,\bar{C})}{2} > \frac{u_0 + \rho(u,\bar{C})}{2} ,$$

giving the contradiction $\rho(w, \bar{C}) > \rho(u, w) + \rho(u, \bar{C})$. But if $\alpha \in \text{dom}_2 \beta$ or $\alpha \in \text{dom}_3 \beta$, we have

$$\frac{w_0 - \rho(w,\bar{C})}{2} > \frac{u_0 - \rho(u,\bar{C})}{2}$$

giving the contradiction $\rho(u, \bar{C}) > \rho(u, w) + \rho(w, \bar{C})$.

PROOF OF II: $K \cup \text{dom } K = A - J$. Given $u \in U$, let $A(u)$ be the cross-section of A consisting of points of the form $< \alpha_1, \alpha_2, \alpha_3, u_4, \ldots, u_n >$. Define $K(u) = K \cap A(u)$; it is a single point if $u \in U_0$, otherwise it is a line segment parallel to the $\alpha_1 = 0$ face of $A(u)$, not more than halfway from that face to the opposite vertex (see Figure 2). Define $L(u) = A(u) \cap \overline{\text{dom}}\, K(u)$, where "$\overline{\text{dom}}$" is taken with respect to the

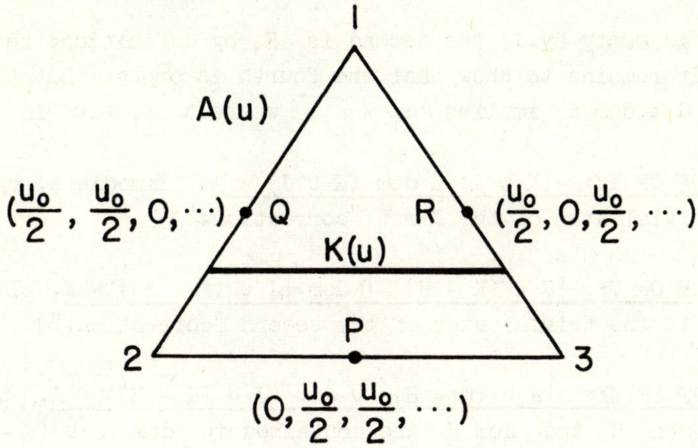

FIGURE 2

two-person sets $\{1, 2\}$, $\{1, 3\}$, and $\{2, 3\}$. (The "$\overline{\text{dom}}$" pattern in $A(u)$ is the same as the domination pattern of the 3-person simple majority game.) It is clear that $K(u)$ and $L(u)$ are disjoint and exhaust $A(u)$, except in the extreme case $K(u) = \overline{QR}$ (see the figure), when the point P is not covered. But the extreme case occurs only when $\rho(u, \bar{C}) = 0$ and $u \notin U_o$ — that is, precisely when $u \in C$ — and we see that the uncovered point is an element of J. To sum up, if α is an arbitrary point of A, then it is either in J, in K, or in $L(u)$ for some $u \notin U_o$. If the latter, we have in fact $\alpha \in \text{dom}\, K$, since we can pick $u' > u$ and have $\alpha \in \text{dom}\, K(u')$ by making the differences $u_j' - u_j$ sufficiently small, by the uniform continuity of ρ. Thus $J \cup K \cup \text{dom}\, K = A$, and half of our result, namely: $K \cup \text{dom}\, K \supseteq A - J$, has been established.

To prove the reverse inequality: $K \cup \text{dom}\, K \subseteq A - J$, it suffices to show that J and $\text{dom}\, K$ are disjoint. Suppose therefore that $\alpha \in \text{dom}\, \beta$ where $\alpha \in J \cap A(u)$, $\beta \in K \cap A(w)$ for some $u, w \in U$. If $\alpha \in \text{dom}_1\, \beta$, then we would have $\alpha_1 > \beta_1$, impossible since $\alpha_1 = 0$. If $\alpha \in \text{dom}_2\, \beta$, then $\alpha_3 < \beta_3$ and $u < w$, by definition. But $\alpha_3 = u_o/2$ and $\beta_3 \leq [w_o + \rho(w, \bar{C})]/2$, hence $\rho(w, \bar{C}) > u_o - w_o = \rho(w, u)$. This is also impossible, since $u \in C$. The case $\alpha \in \text{dom}_3\, \beta$ is similar. This completes the proof.

<u>PROOF OF III: $(K \cup J) \cap \text{dom}\,(K \cup J) = H$</u>. Expanding the left member of III, we have

$$(K \cap \text{dom}\,K) \cup (K \cap \text{dom}\,J) \cup (J \cap \text{dom}\,K) \cup (J \cap \text{dom}\,J) \quad .$$

The first term is empty by I; the second is H by definition; the third is empty by II. It remains to show that the fourth is empty. But this is trivial, since $\alpha \in \text{dom}\,\beta$ implies $\alpha_1 \neq \beta_1$, whereas $\alpha_1 \equiv 0$ in J.

<u>PROOF OF IV: $(K \cup J) \cup \text{dom}\,(K \cup J) = A$</u>. Immediate, by II. (This is the trivial step of the first "correction.")

<u>PROOF OF V: $[J \cup (K - H)] \cap \text{dom}\,[J \cup (K - H)] = \emptyset$</u>. Immediate, by III. (This is the trivial step of the second "correction.")

<u>PROOF OF VI: $[J \cup (K - H)] \cup \text{dom}\,[J \cup (K - H)] = A$</u>. We intend to prove that both H and $\text{dom}\,H$ are contained in $\text{dom}\,[J \cup (K - H)]$. Then the left member of VI can be written:

$$[J \cup (K - H)] \cup H \cup \text{dom}\,[J \cup (K - H)] \cup \text{dom}\,H \quad ,$$

which is equal to A, by IV. We shall show in particular that H and $\text{dom}_1\,H$ are contained in $\text{dom}\,J$, and that $\text{dom}_2\,H$ and $\text{dom}_3\,H$ are contained in $\text{dom}\,(K - H)$. First, by definition, we have $H \subseteq \text{dom}\,J$; also:

$$\text{dom}_1\,H \subseteq \text{dom}_1\,(\text{dom}\,J) = \text{dom}_1\,(\text{dom}_1\,J) \subseteq \text{dom}_1\,J \subseteq \text{dom}\,J \quad ,$$

using the transitivity of "dom_k". (The last two inclusions are actually equalities.) This leaves $\text{dom}_2\,H$ and $\text{dom}_3\,H$. Suppose that $\alpha \in \text{dom}_2\,\beta$, $\beta \in H$. Define β' by

$$\begin{cases} \beta'_1 = \beta_1, \ \beta'_2 = 0, \quad \beta'_3 = \beta_2 + \beta_3 \quad , \\ \beta'_j = \beta_j, \quad j = 4, \ldots, n \quad . \end{cases}$$

Then β', like β, is in K, and clearly $\alpha \in \text{dom}_2\,\beta'$. However, β' is <u>not</u> in H. Indeed, if it were, there would be $\gamma \in J$ with $\beta' \in \text{dom}\,\gamma = \text{dom}_1\,\gamma$ and we would have:

$$\beta'_1 = \beta'_1 + \beta'_2 = 1 - \sum_3^n \beta'_j > 1 - \sum_3^n \gamma_j = \gamma_1 + \gamma_2$$
$$= \gamma_2 = \gamma_3 > \beta'_3 = \beta'_2 + \beta'_3 \quad ;$$

which is an impossible inequality for $\beta' \in K$. Hence $\text{dom}_2 H \subseteq \text{dom}_2(K - H)$. Similarly $\text{dom}_3 H \subseteq \text{dom}_3(K - H)$. This completes the proof.

REMARKS

(1) It is easy to see that $K - H$ is a connected set, by considering imputations like the β' of the last proof.

(2) Our particular games are not constant-sum, but their constant-sum extensions ($n + 1$ players) will possess discriminatory solutions that are essentially the same as those we have constructed. There is no reason to believe that solutions with arbitrary closed components cannot be found in a much wider class of games — perhaps in almost all games.

(3) It would be interesting to know of more general situations in which the method of successive corrections:

$$V_{n+1} = (V_n - X) \cup Y$$

where $V_n \cap \text{dom } V_n = X$, $V_n \cup \text{dom } V_n = A - Y$, can be shown to lead to a solution.

(4) Two interesting questions regarding cardinality are not answered by our present example, since $K - H$ is always uncountable, namely: Do countably infinite solutions ever occur? Is there an upper bound (say 2^n) to the number of imputations in a finite solution?

L. S. Shapley

The RAND Corporation

SYMMETRIC ZERO-SUM n-PERSON GAMES[1]

B. R. Gelbaum

§1. INTRODUCTION

In a symmetric, zero-sum, n-person game, those coalitions, which can provide the largest pro rata gain for their members, seem to provide an intuitive and heuristic starting point for the construction of solutions. Among the coalitions which provide maximum pro rata gain are those which are smallest in size and these are what may be called the minimax coalitions, (minimal size, maximum pro rata gain).

The following results provide solutions based on the concept of minimax coalitions for symmetric, zero-sum, n-person games for large classes of integers n.

§2. NOTATIONS

$v(S)$ is the characteristic function of the symmetric, zero-sum, n-person game Γ. Since $v(S)$ depends only on the number, $q(S) = q$, of persons in S, we shall set v_q to be the value of $v(S)$ for any S for which $q(S) = q$, $q = 0, 1, \ldots, n$. Finally the letters a, p, m, r will be used in the senses given by:

$$a = \sup \{(v_q)/q \mid q = 1, 2, \ldots, n\}$$

$$p = \inf \{q \mid (v_q)/q = a\}$$

$$n = mp + r, \quad 0 \leq r < p.$$

§3. SOLUTIONS UNDER CERTAIN CONDITIONS

LEMMA 1. $a > 0$.

[1] The preparation of this paper was supported (in part) by the Office of Naval Research.

PROOF. $$\frac{v_{n-1}}{n-1} = \frac{1}{n-1} > 0 \quad .$$

LEMMA 2. $r > 0$.

PROOF. If $r = 0$, then $n = pm$, and $0 = v_n \geq mv_p = mpa$. Thus $a \leq 0$, and Lemma 1 is contradicted.

LEMMA 3. If $q = kp$, k an integer and $1 \leq q \leq n$, then $v_q = qa$.

PROOF. Since $q = kp$, we see $v_q \geq kv_p = kpa = qa$. Thus $(v_q)/q \geq a$, whereas $(v_q)/q \leq a$, by definition of a. Thus $v_q = qa$.

NOTE. The converse of Lemma 3 is false. For example, let $n = 5$, $v_1 = -1$, $v_2 = -\frac{3}{4}$, $v_3 = \frac{3}{4}$, $v_4 = 1$, $v_5 = 0$. Then $p = 3$, $a = \frac{1}{4}$, and yet $v_4 = 4a$ while $3 (= p)$ fails to divide $4 (= q)$.

LEMMA 4. For any non-empty set S, if S is flat, i.e., if $v(S) = -q(S) = -q$, then $q \leq r$.

PROOF. If $q > r$, then $v_r = -r$ and $0 < q = v_{n-q} > r = v_{n-r}$. Thus

$$\frac{v_{n-q}}{n-q} > \frac{v_{n-r}}{n-q} > \frac{v_{n-r}}{n-r} = \frac{v_{mp}}{mp} = \frac{mpa}{mp} = a$$

(Lemma 3). This contradicts the definition of a.

LEMMA 5. $1/(n-1) \leq a \leq r/pm < 1/m$.

PROOF. $1/(n-1) = (v_{n-1})/(n-1) \leq a$; $v_{pm} = pma = -v_r \leq r$; $r < p$.

NOTE. If $n = 3$, then $p = 2$, $r = 1$, $m = 1$, $a = \frac{1}{2}$ and thus the result above is the best possible.

THEOREM 3.1. If $r = p - 1$ and $v_{p-1} = -(p-1)$, then the set V of all permutations of the vector

$$\alpha(r) = \left\{ a, a, \ldots, a, \frac{-pma}{r}, \ldots, \frac{-pma}{r} \right\} \tag{1}$$

(pm components equal to a, r components equal to $-pma/r$) is a (finite, symmetric) solution of Γ.

PROOF. $\alpha(r)$ is an imputation since $a > 0 > -1$ and $-pma/r = -v_{pm}/r = v_r/r = -r/r = -1$, and the components have sum zero. Note that the set of players $(pm + 1, pm + 2, \ldots, n)$ is effective for α, and that the fact that α is an imputation does not depend on the hypothesis $v_{p-1} = -(p - 1)$.

Let dom V (dominion of V) be the set of imputations which are dominated by some imputation in V. We show

(a) $V \cap \text{dom } V = \emptyset$ (the empty set), i.e., that V is consistent, and

(b) $V \cup \text{dom } V$ = set of all imputations, i.e., that V is extradominative.

Ad(a): Since (p-1)-element sets are flat, at least p-element sets are required for domination. However, if α_1 and α_2 are two imputations of V, then at most $p - 1$ components of α_1 are greater than the corresponding components of α_2, whence α_1 cannot dominate α_2 and $V \cap \text{dom } V = \emptyset$.

Ad(b): Let $\beta \notin V$. If β has at least p components less than a, then $\beta \in \text{dom } V$, with domination of β arising via inequalities of the form $a > \beta_i$, where β_i are p components, all less than a. The α which has components equal to a for the corresponding set, S, of indices is one for which S is effective, since $v(S) = v_p = pa = \Sigma_{i \in S} \alpha_i$.

Let the number of components of β which are less than a be $p - k$, $k > 0$. Then $k < p$, since $\Sigma_1^n \beta_i = 0$. Let these components be $\beta_{i_1}, \beta_{i_2}, \ldots, \beta_{i_{p-k}}$, and let $\beta_{j_1}, \ldots, \beta_{j_{n-p+k}} \geq a$.

Then $-1 \leq \beta_{i_e}$ and

$$-(p - k) \leq \sum_e \beta_{i_e} = -\sum_t \beta_{j_t} \leq -(n - p + k)\,a \leq -pma = -(p - 1);$$

and so $p - k \geq p - 1$, $k \leq 1$. Thus $k = 1$ and $n - p + k = mp$.

If any $\beta_{i_e} > -1$, then $\Sigma_e \beta_{i_e} > -(p - 1)$ and

$$\sum_1^n \beta_i > -(p - 1) + pma = -(p - 1) + v_{pm} = v_{p-1} + v_{pm} = 0 \ ,$$

contradicting the fact that $\Sigma_1^n \beta_i = 0$. Thus all $\beta_{i_e} = -1$.

If any $\beta_{j_t} > a$, then $\Sigma_1^n \beta_i > -(p - 1) + pma = 0$, yielding again a contradiction. We could then conclude that $\beta \in V$, in contradiction with the original assumption: $\beta \notin V$.

NOTE. The set V, regardless of hypotheses on r or v_r, is always consistent. For the only inequalities via which intradomination in V can occur are of the form $a > -pma/r$. Since the number, k, of these is not greater than $r < p$, and since for such k, $v_k < ka$, effectivity and hence domination as well are absent. We observe also that if $n = 3$, the hypotheses of Theorem 3.1 are satisfied and our formula for α reduces to the familiar imputation $(1/2, 1/2, -1)$ which, with its two other permutations, make up the unique symmetric solution for the normalized 3-person game.

LEMMA 6. If $v_{r-1} = -(r-1)$, then for any <u>imputation</u>

$$\gamma = \{a, a, \ldots, a, x_1, x_2, \ldots, x_r\} \tag{2}$$

(first pm components equal to a), the inequalities $x_i \leq a$ obtain.

PROOF. $\Sigma_1^r x_i = -pma = v_r$. Assume $x_{i_0} > a$. Then $a < x_{i_0} = v_r - \Sigma_{i \neq i_0} x_i \leq v_r + r - 1 = -pma + r - 1$, or $a < (r-1)/(pm+1)$. But

$$a \geq \frac{v_{pm+1}}{pm+1} = \frac{v_{n-(r-1)}}{pm+1} = \frac{-v_{r-1}}{pm+1} = \frac{r-1}{pm+1} .$$

This contradiction implies the assertion of the lemma.

THEOREM 3.2. If $r = p - 1$ and if $v_{p-2} = -(p-2)$, then W, the set of permutations of all γ (Lemma 6), is a (symmetric) solution of Γ.

PROOF. Ad consistency of W: From Lemma 6 we see that the only inequalities which can obtain between corresponding components of two different imputations $\gamma^{(1)}, \gamma^{(2)}$ of W are of the form $a > x_i^{(1)}$ or $x_i^{(1)} > x_i^{(2)}$, and there can be no more than $r = p - 1$ such inequalities. For the sake of effectivity, (since $v_{p-2} = -(p-2)$), precisely $p - 1$ inequalities must obtain. Since $\Sigma_1^{p-1} x_i^{(2)} = v_{p-1}$, we see that domination of $\gamma^{(2)}$ by $\gamma^{(1)}$ is impossible because of the absence of effectivity for $\gamma^{(1)}$.

Ad extradomination by W: If $\beta \notin W$ and if β has at least p components less than a, then $\beta \in \operatorname{dom} W$ by virtue of p element domination via inequalities of the form $a > \beta_i$.

If the number of components of β which are less than a is

$p - k$, $0 < k$, then, modulo a permutation, $\beta = \{a + \epsilon_1, \ldots, a + \epsilon_e, \beta_{e+1}, \ldots, \beta_n\}$ where $\epsilon_i \geq 0$, $i = 1, 2, \ldots, e$, $e \geq pm$, $\epsilon_i \geq \epsilon_j$ if $i < j$. Let $\gamma = \{a, a, \ldots, a, x_{pm+1}, \ldots, x_n\}$, (first pm components equal to a) where

$$x_i = \beta_i + \frac{\Sigma_{j=1}^{pm} \epsilon_j}{p - 1}, \quad i = pm + 1, \ldots, n \ .$$

Since $\beta \notin W$, we see that $\Sigma_1^{pm} \epsilon_j > 0$ and thus $x_i > \beta_i$. Since β is an imputation,

$$ea + \sum_{j=1}^{e} \epsilon_j + \sum_{i=e+1}^{n} \beta_i = 0 \ .$$

Thus γ is an imputation, since

$$pma + \sum_{i=pm+1}^{n} x_i = pma + \sum_{j=1}^{pm} \epsilon_j + \sum_{i=e+1}^{n} \beta_i + \sum_{i=pm+1}^{e} (a + \epsilon_i)$$

$$= pma + (e - pm)a + \sum_{i=pm+1}^{e} \epsilon_i + \sum_{j=1}^{pm} \epsilon_j + \sum_{i=e+1}^{n} \beta_i$$

$$= ea + \sum_{j=1}^{e} \epsilon_j + \sum_{i=e+1}^{n} \beta_i = 0 ,$$

and clearly $\gamma_i \geq -1$.

Thus $\gamma \in W$ and $\gamma \succ \beta$ via inequalities on the last $p - 1$ components of each, since

$$\sum_{i=pm+1}^{n} x_i = - pma = v_{p-1}$$

insures effectivity. This completes the proof of Theorem 3.2.

Let W_r denote the set of all permutations of the imputation $\alpha(r)$ in (1). Let W_{p-1} denote the set of all permutations of the imputation

$$\alpha(p - 1) = \left\{ a, a, \ldots, a, - [n - (p - 1)] \frac{a}{p - 1}, \ldots, - [n - (p - 1)] \frac{a}{p - 1} \right\}$$

where the first $n - (p-1)$ components are equal to a and the remaining

$(p - 1)$ components are equal to $-[n - (p - 1)]\frac{a}{p - 1}$. (Note that $-[n - (p - 1)]\frac{a}{p - 1} \geq - 1$.)

Then we have the following result.

THEOREM 3.3. For arbitrary admissible values of r, if $v_r = - r$ and $v_{r+1} \geq (r + 1 - \frac{n}{(p-1)})a$, then $W_r \cup W_{p-1}$ is a (finite, symmetric) solution for Γ.

PROOF. We observe first that W_r is the same as V of Theorem 3.1.. Thus W_r is internally consistent. W_{p-1} is internally consistent, since among the components of two different imputations $\alpha^{(1)}, \alpha^{(2)}$, of W_{p-1}, at most $p - 1$ inequalities of the form $\alpha_i^{(1)} > \alpha_i^{(2)}$ can obtain, and all are of the form $a > b = - (n - (p - 1))\frac{a}{(p-1)}$. However, if $0 < k \leq p - 1$, then $v_k < ka$ and thus effectivity for $\alpha^{(1)}$ is absent and domination of $\alpha^{(2)}$ by $\alpha^{(1)}$ is impossible.

Domination from W_r to W_{p-1} cannot obtain (i.e., $W_{p-1} \cap \text{dom } W_r = \emptyset$) for the same reasons.

Finally, $W_r \cap \text{dom } W_{p-1} = \emptyset$. For if $\alpha(p - 1) \succ \alpha(r)$, the inequalities, via which this domination occurs, must obtain for the indices of a flat set and must be of the form $(\alpha(p - 1))_i > - pma/r$. Thus $W_r \cup W_{p-1}$ is consistent.

We show $W_r \cup W_{p-1}$ is extradominative. Let B_k denote the set of imputations having precisely k components less than a, $k = 1, 2, \ldots, n$. Then clearly $B_r = W_r$, $B_{p-1} \supset W_{p-1}$, $B_s = \emptyset$ if $s < r$, $B_k \cap B_e = \emptyset$ if $k \neq e$ and $B_k \subset \text{dom } W_r$ if $k \geq p$. Thus we need only show that if $B = B_{r+1} \cup B_{r+2} \cup \ldots \cup B_{p-1}$, then $D = B - W_{p-1} \subset \text{dom } (W_r \cup W_{p-1})$. Thus let $\beta \in D$, say $\beta \in B_{r+j}$, $r + 1 \leq r + j \leq p - 1$. Then for this β we have

$$\sum{}' \beta_i = - \sum{}'' \beta_i \leq - (n - (r + j))a \leq - (n - (p - 1))a ,$$

where Σ' denotes summation over those indices i for which $\beta_i < a$, and Σ'' denotes summation over the remaining indices. Hence for some i_o for which $\beta_{i_o} < a$ it must be true that

$$\beta_{i_o} \leq - \frac{(n-(r+j))a}{r + j} \leq - \frac{(n-(p-1))a}{r + j} \leq - \frac{(n-(p-1))a}{p - 1} .$$

Equality in these relations obtains throughout if and only if $\beta \in W_r \cup W_{p-1}$. This last is excluded by our assumption concerning β. Thus $\beta_{i_o} < - (n - (p - 1)) \cdot \frac{a}{(p-1)}$. Thus we can choose $\alpha \in W_{p-1}$ such that for r indices $i_1, i_2, \ldots, i_r$ $\alpha_{i_r} = a > \beta_{i_r}$ and $\alpha_{i_o} = - (n - (p - 1))\frac{a}{(p-1)} > \beta_{i_o}$.

The assumption concerning v_{r+1} now implies that for this α, $\alpha \in \beta$.

§4. EXISTENCE THEOREMS

We now consider questions of existence of games which fall into the categories delineated in the preceding theorems. We shall denote the conditions given below by the symbols preceding them.

(A) $r = p - 1$

(B) $v_r = - r$

(C) $v_{r-1} = - (r - 1)$

(D) $v_{r+1} \geq \left(r + 1 - \frac{n}{p - 1}\right)a, \; a = \frac{r}{pm}$.

NOTE. (D) implies (B); (B) implies (C). The latter is clear. Moreover, $a = r/pm$ implies $v_{pm} = r$, whence $0 = u_{pm} + u_r = r + u_r$, whence (D) implies (B). First we show a class of games for which (A) and (B) obtain, i.e., games which fall in the domain of Theorem 3.1.

THEOREM 4.1. The class of games for which (A) and (B) obtain is not empty.

PROOF. Let $a = \frac{(p-1)}{pm}$, $n = pm + (p - 1)$ and let $v_t = pka - e$ if $t = pk + e$, $0 \leq e < p$, $0 \leq k \leq m$. Direct computation by exhaustion, confirms the fact that the function v_t, so defined, is the characteristic function of a game Γ and that for this v_t, (A) and (B) obtain.

THEOREM 4.2. If there is a q such that $2 \leq q \leq p - 2$, $q(r + 1) \leq n$ and $r - q(r + 1) \equiv 0(p)$, then condition (D) cannot obtain.

PROOF. If (D) obtains, then $q(r + 1 - \frac{n}{(p-1)})a \leq q\, v_{r+1} \leq v_{qr+q} = - v_{pm+r-qr-q}$, since $q(r + 1) \leq n$. But if $r - q(r + 1) \equiv 0(p)$, then $v_{pm+r-qr-q} = (pm + r - qr - q)a$. Thus

$$q\left(r + 1 - \frac{n}{(p-1)}\right) \leq - (n - q(r + 1)) \quad ,$$

$$- \frac{nq}{(p-1)} \leq - n \quad ,$$

$$q \geq p - 1 \quad ,$$

contradicting $q \leq p - 2$.

Theorem 4.2 is not vacuous as the following example reveals:

Let $p = 4$, $q = 2$, $r = 2$, $n = 6$, $m = 1$.

Despite this result, however, there are games for which (D) does obtain, e.g., $p = 3$, $n = 4$, $r = 1$, $a = \frac{1}{3}$, $m = 1$, $v_4 = 0$, $v_3 = 1 = - v_1$, $v_2 = 0 = (1 + 1 - \frac{4}{2})a$.

THEOREM 4.3. If $2r + 1 \leq p \leq 2r + 2$, then there is an integer $M(r)$ such that if $m \geq M(r)$, then there is a symmetric zero-sum, n-person game Γ, $n = pm + r$, whose characteristic function satisfies condition (D), and, a fortiori, conditions (B) and (C).

PROOF. The proof is divided into two sections corresponding to the cases $r > 1$ and $r = 1$.

CASE (a): $r > 1$. If $r > 1$, then for each m let v_t be defined as follows:

$$v_t = \begin{cases} - t & , 0 \leq t \leq r \\ [r + 1 - n/(p - 1)]a & , t = r + 1 \\ r(1 - m)/m - v_{r+1} & , t = p - 1 \\ pa & , t = p \\ [(s - 1)/(p - r - 2)][v_{p-1} - v_{r+1}] + v_{r+1} , & \\ \qquad r + 1 \leq t = r + s \leq p - 1 \neq 2 & \\ pka + v_e & , t = pk + e, 0 \leq k \leq m, 0 \leq e \leq p - 1 \end{cases}$$

where $a = r/pm$ and $n = pm + r$.

We show now that if m is large enough then v_t is a true characteristic function of a game.

Note first that if m is large, v_t satisfies the condition $v_t \geq - t$. For if m is large then the following inequalities obtain:

$$v_t > \begin{cases} - t & , 0 \leq t \leq r \\ - \frac{1}{2} > - (r + 1) & , t = r + 1 \\ - r > - 2r \geq - (p - 1) & , t = p - 1 \\ 0 > - p & , t = p \\ - s > - r - s & , r + 1 \leq t = r + s \leq p - 1 \neq 2 \\ v_e \geq - e \geq - pk - e & , t = pk + e, 0 \leq k \leq m, \\ & \quad 0 \leq e \leq p - 1 \quad . \end{cases}$$

Observe also that asymptotically, if $r + 1 \leq e \leq p - 1$,

$$v_e = -\frac{r}{p-1} + \frac{u}{w}\left[-r + \frac{2r}{p-1}\right] = -r\left[\frac{1}{p-1} + \frac{u}{w}\frac{(p-3)}{(p-1)}\right] < 0 \quad ,$$

where $u = e - r - 1$, $w = p - r - 2$. Obviously $v_e \leq 0$ if $0 \leq e \leq r$. Thus

$$\frac{v_t}{t} = \frac{(kr/m) + v_e}{pk + e} \leq \frac{(kr/m)}{pk + e} \leq \frac{(kr/m)}{pk} = \frac{r}{pm} \quad .$$

The motivation for the formula for v_t, in the range $r + 1 \leq t \leq r + s$ is the following one: We must insure the relation $v_h + v_{n-h} = 0$, $0 \leq h \leq n$. But if $h = pk + e$, $0 \leq k \leq m$, $0 \leq e \leq p - 1$, then $n - h = p(m - k) + r - e = p(m - k - 1) + p + r - e$. If $r - e \geq 0$, then

$$v_h = \frac{kr}{m} - e \quad ,$$

$$v_{n-h} = \frac{(m-k)}{m} r - (r - e) \quad ,$$

$$v_h + v_{n-h} = 0 \quad .$$

If $r - e < 0$, then

$$v_h = \frac{kr}{m} + v_e \quad ,$$

$$v_{n-h} = \left(\frac{m - k - 1}{m}\right) r + v_p + r - e \quad .$$

In order to make $v_h + v_{n-h} = 0$, we must have

$$v_e + v_{p+r-e} = \frac{(1-m)r}{m} \quad .$$

Our formulation is a linear interpolation which produces this equality.

Clearly $v_o = v_{pm+r} = 0$. The remainder of the proof will consist in verifying superadditivity for v_t for large m. Thus let $t_i = pk_i + e_i$, $0 \leq k_i \leq m$, $0 \leq e_i \leq p - 1$. We treat the problem by examining the following mutually exclusive and (because of the symmetry) exhaustive cases:

(i) $e_1 = e_2 = 0$

(ii) $e_1 = 0$, $e_2 > 0$

(iii) $0 < e_1, e_2, e_1 + e_2 \leq r$

(iv) $0 < e_1, e_2 \leq r$, $r < e_1 + e_2 \leq p - 1$

$$\text{(v)}\quad 0 < e_1 \le r,\ \ r < e_2,\ \ e_1 + e_2 \le p - 1$$

$$\text{(vi)}\quad 0 < e_1 \le r,\ \ r < e_2;\ \ e_1 + e_2 = p + q,\quad 0 \le q \le r - 1$$

$$\text{(vii)}\quad r < e_1,\ e_2;\ \ e_1 + e_2 = p + q \qquad\qquad 0 \le q \le p - 2\ .$$

The analysis proceeds as follows:

Ad (i):
$$\begin{aligned} v_{t_1} &= pk_1a, \\ v_{t_2} &= pk_2a, \\ v_{t_1+t_2} &= p(k_1 + k_2)a = v_{t_1} + v_{t_2}\ . \end{aligned}$$

Ad (ii):
$$\begin{aligned} v_{t_1} &= pk_1a, \\ v_{t_2} &= pk_2a + v_{e_2}, \\ v_{t_1+t_2} &= p(k_1 + k_2)a + v_{e_2} = v_{t_1} + v_{t_2}\ . \end{aligned}$$

Ad (iii):
$$\begin{aligned} v_{t_1} &= pk_1a - e_1, \\ v_{t_2} &= pk_2a - e_2, \\ v_{t_1+t_2} &= p(k_1 + k_2)a - (e_1 + e_2) = v_{t_1} + v_{t_2}\ . \end{aligned}$$

Ad (iv):
$$\begin{aligned} v_{t_1} &= pk_1a - e_1, \\ v_{t_2} &= pk_2a - e_2, \\ v_{t_1+t_2} &= p(k_1 + k_2)a + v_{e_1+e_2} \ge p(k_1 + k_2)a - e_1 - e_2 \\ &= v_{t_1} + v_{t_2}\ . \end{aligned}$$

Ad (v):
$$\begin{aligned} v_{t_1} &= pk_1a - e_1, \\ v_{t_2} &= pk_2a + v_{e_2}, \\ v_{t_1+t_2} &= p(k_1 + k_2)a + v_{e_1+e_2}\ . \end{aligned}$$

Let $e_1 + e_2 = r + s$, $e_2 = r + \sigma$. Then

$$v_{t_1+t_2} - v_{t_1} - v_{t_2} = v_{e_1+e_2} + e_1 - v_{e_2}$$

$$= \frac{(s-1)}{(p-r-2)} [v_{p-1} - v_{r+1}] + v_{r+1} + e_1$$

$$- \frac{(\sigma-1)}{(p-r-2)} [v_{p-1} - v_{r+1}] - v_{r+1}$$

$$= \frac{(s-\sigma)}{(p-r-2)} \left\{ \frac{t(1-m)}{m} - 2 \left[r + 1 - \frac{n}{(p-1)} \right] \frac{r}{pm} \right\} + e_1 .$$

But $s - \sigma = e_1$. Thus if $p = 2r + 1$, the above becomes

$$e_1 \left\{ 1 + \frac{1}{r-1} \left\{ \frac{r}{m} - r - \frac{2(r+1)r}{(2r+1)m} + 2 \frac{r}{2r} + \frac{2r \cdot r}{2r \cdot (2r+1)m} \right\} \right\}$$

$$= e_1 \left\{ 1 + \frac{1}{r-1} [1 - r] \right\} = 0 .$$

If $p = 2r + 2$ the expression reduces to

$$e_1 \left\{ 1 + \frac{1}{r} \left\{ \frac{r}{m} - r - \frac{2(r+1)r}{2(r+1)m} + \frac{2r}{2r+1} + \frac{2r}{2r+1} \cdot \frac{r}{2(r+1)m} \right\} \right\}$$

$$= e_1 \left\{ \frac{r}{m(2r+1)(r+1)} + \frac{2}{2r+1} \right\} > 0 .$$

Ad (vi):

$$v_{t_1} = pk_1 a - e_1 ,$$

$$v_{t_2} = pk_2 a + v_{e_2} ,$$

$$v_{t_1+t_2} = p(k_1 + k_2 + 1)a + v_q = p(k_1 + k_2)a + pa - q .$$

Let Δ denote the difference $v_{t_1+t_2} - (v_{t_1} + v_{t_2})$, then

$$\Delta = pa - q + e_1 - v_{e_2} .$$

Let $e_2 = r + \sigma$. Then

$$\Delta = pa - q + e_1 - \frac{(\sigma-1)}{(p-r-2)} \left\{ r \frac{(1-1)}{m} - 2 \left[(r+1) - \frac{n}{p-1} \right] \frac{r}{pm} \right\} - \left[r - 1 - \frac{n}{p-1} \right] \frac{r}{pm}$$

$$= \frac{r}{m} \left[1 - \frac{\sigma - 1}{p - r - 2} \right] + 2 \left(\frac{\sigma - 1}{p - r - 2} \right) \left[\frac{(r+1)r}{pm} - \frac{r^2}{(p-1)pm} \right] - \frac{r(r+1)}{pm} + \frac{r^2}{(p-1)pm}$$

$$- q + e_1 - \left(\frac{\sigma - 1}{p - r - 2} \right) \left\{ - r + \frac{2r}{p-1} \right\} + \frac{r}{p-1} .$$

As $m \longrightarrow \infty$,

$$\Delta \longrightarrow - q + e_1 + r \frac{(\sigma-1)}{(p-r-2)}\left\{\frac{p-3}{p-1}\right\} + \frac{r}{p-1} .$$

However, since $e_1 + e_2 = p + q$, we have $- q + e_1 = p - e_2 = p - r - \sigma > 0$ and so we get

$$\lim \Delta = p - r - \sigma + r \frac{(\sigma-1)(p-3)}{(p-r-2)(p-1)} + r/_{p-1} > 0, \quad \text{as} \quad m \longrightarrow \infty .$$

Ad (vii):
$$v_{t_1} = pk_1 a + v_{e_1},$$
$$v_{t_2} = pk_2 a + v_{e_2},$$
$$v_{t_1+t_2} = p(k_1 + k_2)a + pa + v_q, \quad 0 \le q \le p - 2 .$$

Since $\Delta = v_{t_1+t_2} - (v_{t_1} + v_{t_2})$ we have

$$\Delta = pa + v_q - v_{e_1} - v_{e_2} .$$

Let $e_1 = r + s_1$, $e_2 = r + s_2$, and assume first that $q > r$, $q = r + s$, $s > 0$. Then

$$\begin{aligned}\Delta = pa &+ \frac{(s-1)}{(p-r-2)}\left[r/_m - r - 2\left[r + 1 - \left(n/_{(p-1)}\right)\right]r/_{pm}\right] + v_{r+1}\\ &- \frac{(s_1-1)}{(p-r-2)}\left[r/_m - r - 2\left[r + 1 - \left(n/_{(p-1)}\right)\right]r/_{pm}\right] - v_{r+1}\\ &- \frac{(s_2-1)}{(p-r-2)}\left[r/_m - r - 2\left[r + 1 - \left(n/_{(p-1)}\right)\right]r/_{pm}\right] - v_{r+1}\\ = pa &+ \frac{(s-s_1-s_2+1)}{(p-r-2)}\left[\frac{r}{m} - r - 2\left[r + 1 - \frac{n}{p-1}\right]r/_{pm}\right] - v_{r+1} .\end{aligned}$$

As $m \longrightarrow \infty$ this becomes

$$\begin{aligned}&\frac{(s-s_1-s_2+1)}{(p-r-2)}\left[r\frac{(3-p)}{(p-1)}\right] + \frac{3}{p-1}\\ &= \frac{r}{p-1}\left[\frac{(s-s_1-s_2+1)(3-p)}{p-r-2} + 1\right]\\ &= \frac{r}{p-1}\left[\frac{(r-p+1)(3-p)}{p-r-2} + 1\right],\end{aligned}$$

since $p + q = p + r + s = 2r + s_1 + s_2$.

Since $2r + 1 \le p$, we see that $r + 1 \le p - r$ and $r + 1 - p \le - r < 0$.

Thus $(r - p + 1)(r - p) > 0$ and thus for large m, $\Delta > 0$.

Suppose next that $q \leq r$, and let $p = 2r + 1$. Then $q = s_1 + s_2 - 1$, and

$$\begin{aligned}\Delta &= pa - q - v_{e_1} - v_{e_2} \\ &= pa - (s_1 + s_2 - 1) - \frac{(s_1+s_2-2)}{(p-r-2)}\left[\frac{r}{m} - r - 2v_{r+1}\right] + 2v_{r+1} \\ &= \frac{r}{m} - (s_1 + s_2 - 1) - \frac{(s_1+s_2-2)}{r - 1}\left[\frac{r}{m} - r - \frac{2(r+1)r}{(2r+1)m} + 1 + \frac{r}{(2r+1)m}\right] \\ &\qquad - \frac{2r(r+1)}{(2r+1)m} + 1 + \frac{r}{(2r+1)m} = 0 \quad .\end{aligned}$$

If we let $p = 2r + 2$, then $q = s_1 + s_2 - 2$ and we obtain similarly

$$\begin{aligned}\Delta &= r/_m - (s_1 + s_2 - 2) - \frac{(s_1+s_2-2)}{r}\left[\frac{r}{m} - r - \frac{2(r+1)r}{2(r+1)m} + \frac{2r}{2r+1}\right. \\ &\qquad \left. + \frac{2r^2}{(2r+1)2(r+1)m}\right] - \frac{2(r+1)r}{2(r+1)m} + \frac{2r}{2r+1} \\ &\qquad + \frac{2r^2}{(2r+1)2(r+1)m} \\ &= \frac{r}{m} - (s_1 + s_2 - 2)\left\{1 + \frac{1}{m} - 1 - \frac{1}{m} + \frac{2}{2r+1} + \frac{r}{(2r+1)(r+1)m}\right\} \\ &\qquad - \frac{r}{m} + \frac{2r}{2r+1} + \frac{r^2}{(2r+1)(r+1)m} \\ &= -(s_1 + s_2 - 2)\left\{\frac{1}{2r+1}\left[2 + \frac{r}{(r+1)m}\right]\right\} + \frac{r}{2r+1}\left\{2 + \frac{r}{(r+1)m}\right\} \\ &= \frac{1}{2r+1}\left[-\left(2 + \frac{r}{(r+1)m}\right)(s_1 + s_2 - 2) + r\left\{2 + \frac{r}{(r+1)m}\right\}\right] \\ &= \frac{1}{2r+1}\left[\left(2 + \frac{r}{(r+1)m}\right)(r - (s_1 + s_2 - 2))\right] \geq 0,\end{aligned}$$

since $s_1 + s_2 = q + 2 \leq r + 2$.

CASE (b): $r = 1$. If $r = 1, p = 3$ then we define v_t as follows:

$$v_t = \begin{cases} -1 & , t = 1 \\ \left[2 - \frac{3m+1}{2}\right]\frac{1}{3m} = -\frac{1}{2} + \frac{1}{2m} \geq -\frac{1}{2} > -2, & t = 2 \\ \frac{1}{m} & , t = 3 \\ \frac{k}{m} + v_e & , t = 3k + e,\ 0 \leq k \leq m, \\ & \quad 0 \leq e \leq 2 \quad . \end{cases}$$

As before we let $t_1 = 3k_1 + e_1$, $t_2 = 3k_2 + e_2$.

Ad (i), (ii), (iii) and (iv): Proofs of superadditivity for v_t given in the general case are still valid. (v) cannot obtain.

Ad (vi): We must have $e_1 = 1$, $e_2 = 2$.

$$v_{t_1+t_2} = \frac{(k_1+k_2+1)}{m} ,$$

$$v_{t_1} = \frac{k_1}{m} - 1 ,$$

$$v_{t_2} = \frac{k_2}{m} + \frac{1 - m}{2m} ,$$

$$v_{t_1} + v_{t_2} = \frac{1}{m} (k_1 + k_2) + \frac{1 - m - 2m}{2m} = \frac{(k_1+k_2)}{m} + \frac{1}{2m} - \frac{3}{2} .$$

Recall that $\Delta = v_{t_1+t_2} - v_{t_1} - v_{t_2}$ and we have

$$\Delta = \frac{1}{m} - \frac{1}{2m} + 3/2 = \frac{1}{2m} + 3/2 > 0 \quad \text{for all } m .$$

Ad (vii): In this case $e_1 = 2$, $e_2 = 2$.

$$v_{t_1+t_2} = \frac{(k_1+k_2+1)}{m} - 1 ,$$

$$v_{t_1} = \frac{k_1}{m} + \frac{1 - m}{2m} ,$$

$$v_{t_2} = \frac{k_2}{m} + \frac{1 - m}{2m} .$$

Thus

$$\Delta = \frac{1}{m} - 1 - \frac{1 - m}{m} = \frac{1 - m - 1 + m}{m} = 0 .$$

Hence v_t is a true characteristic function and Γ lies in (D). This completes the proof of Theorem 4.3.

The above theorems show:

(a) For all <u>odd</u> n, there is a symmetric, n-person, zero-sum game Γ which has a finite, symmetric solution, (Theorem 4.1, $p = 2$, $r = 1$, m an arbitrary positive integer).

(b) For all n of the form $3m + 2$, there is a symmetric, n-person, zero-sum game Γ which has a symmetric solution, (Theorems 3.2, 4.1, $p = 3$, $r = 2$; here v_2 may be arbitrary so long as 3-person coalitions are minimax).

(c) For all n of the form $3m + 1$, there is a symmetric, n-person, zero-sum game Γ which has a finite, symmetric solution, (Theorem 4.3, $p = 3$, $r = 1$, $M(r) = 1$).

(d) The result of Theorem 3.3 applied to $n = 4 = 3.1 + 1$ yields one of the known solutions to the symmetric, 4-person, zero-sum game. (See bibliography.)

BIBLIOGRAPHY

[1] von NEUMANN, J., and MORGENSTERN, O., Theory of Games and Economic Behavior, Princeton, 2nd ed., 1947.

B. R. Gelbaum

University of Minnesota

SYMMETRIC SOLUTIONS FOR GENERAL-SUM
SYMMETRIC 4-PERSON GAMES

Evar D. Nering

INTRODUCTION

We shall provide symmetric solutions for all normalized general-sum symmetric 4-person games. We shall consider the games to be so normalized that the value of a 1-person coalition is -1 and the value of a 4-person coalition is 0. A symmetric 4-person game in normalized form is determined by two parameters, the value v_2 of a 2-person coalition and the value v_3 of a 3-person coalition. Clearly, v_2 and v_3 are subject to the conditions $-2 \leq v_2 \leq 0$ and $v_2 - 1 \leq v_3 \leq 1$.

For most games we shall give a family of solutions depending on a parameter λ. The cases where a parameter λ is not explicitly given may be regarded as limiting cases where the dependence on λ is not continuous. The solutions will be made up of six types of components, some of which may be empty for certain values of the parameters while some may overlap slightly.

Since we shall be dealing with symmetric solutions of symmetric games, we may give the components of an imputation in any convenient order. In particular we shall give the solutions by giving the components of the imputations in a solution in one representative order. The imputations will usually be given with the components in decreasing order.

§1. THE SOLUTIONS

Here we list the various types of solutions and the ranges of the parameters for which they are solutions. In each case the solution is the symmetrized union of the components given.

$$\text{I:}\quad V_1 = \left\{(q,\ q,\ q,\ -3q)\ \text{ where }\ \lambda \leq q \leq \tfrac{1}{3}\right\}$$

$$V_2 = \left\{(\lambda,\ \lambda,\ \tfrac{v_2}{2},\ -\tfrac{v_2}{2} - 2\lambda)\right\}$$

$$V_3 = \left\{ (s,\ s,\ -s,\ -s) \text{ where } 0 \le s \le -\frac{v_2}{2} \right\}.$$

Range: $-\frac{5v_2}{2} \le v_3;\ \ \frac{v_3}{3} < \lambda,\ \ -\frac{3v_2}{2} \le \lambda,\ \ \lambda \le \frac{2v_3 - v_2}{4}$.

NOTE. For $v_2 = 0$ and $v_3 = 1$, i.e., for the zero-sum symmetric 4-person game, we get $\frac{1}{3} < \lambda \le \frac{1}{2}$ and $V_1 = \emptyset$, $V_2 = \{(\lambda, \lambda, 0, -2\lambda)\}$, $V_3 = \{(0, 0, 0, 0)\}$, which are solutions given by von Neumann and Morgenstern in [1], page 321.

$$\text{II:}\ \ V_1 = \left\{ (q,\ q,\ q,\ -3q) \text{ where } \lambda \le q \le \tfrac{1}{3} \right\}$$
$$V_2 = \left\{ (\lambda,\ \lambda,\ r,\ -r - 2\lambda) \text{ where } -\lambda \le r \le -v_2 - \lambda \right\}$$
$$V_3 = \left\{ (s,\ s,\ -s,\ -s) \text{ where } 0 \le s \le v_2 + \lambda \right\}.$$

Range: $-\frac{3v_2}{2} < v_3 < -\frac{9v_2}{2};\ \ \frac{v_3}{3} < \lambda \le v_2 + v_3,\ \ \lambda < -\frac{3v_2}{2}$.

$$\text{III:}\ \ V_1 = \left\{ (q,\ q,\ q,\ -3q) \text{ where } \frac{v_3}{3} \le q \le \tfrac{1}{3} \right\}$$
$$V_2 = \left\{ \left(\frac{v_3}{3},\ \frac{v_3}{3},\ r,\ -\frac{2v_3}{3} - r \right) \text{ where } -\frac{v_3}{3} \le r \le -v_2 - \frac{v_3}{3} \right\}$$

Range: $-\frac{3v_2}{2} \le v_3$.

NOTE. For $v_2 = 0$ and $v_3 = 1$ we have

$$V_1 = \left\{ \left(\tfrac{1}{3},\ \tfrac{1}{3},\ \tfrac{1}{3},\ -1 \right) \right\} ,$$
$$V_2 = \left\{ \left(\tfrac{1}{3},\ \tfrac{1}{3},\ -\tfrac{1}{3},\ -\tfrac{1}{3} \right) \right\},$$

which is a solution given in [1], pp. 317-318.

$$\text{IV:}\ \ V_1 = \left\{ (q,\ q,\ q,\ -3q) \text{ where } \lambda \le q \le \tfrac{1}{3} \right\}$$
$$V_2 = \left\{ (\lambda,\ \lambda,\ r,\ -r - 2\lambda) \text{ where } -\lambda \le r \le -v_2 - \lambda \right\}$$
$$V_3 = \left\{ (s,\ s,\ -s,\ -s) \text{ where } 0 \le s \le 2\lambda - v_3 \right\}$$

Range: $-\frac{v_2}{2} < v_3 < -\frac{5v_2}{2}$;

$$\max\left\{ -\frac{v_2}{2},\ v_2 + v_3 \right\} \le \lambda < \frac{2v_3 - v_2}{4} .$$

$$\text{V:}\ \ V_1 = \left\{ (q,\ q,\ q,\ -3q) \text{ where } \frac{2v_3 - v_2}{4} \le q \le \tfrac{1}{3} \right\}$$

$$V_2 = \left\{ \left(\frac{2v_3 - v_2}{4}, \frac{2v_3 - v_2}{4}, r, - r - \frac{2v_3 - v_2}{2} \right) \right.$$

where

$$\left. \frac{v_2}{2} \leq r \leq \frac{- 3v_2 - 2v_3}{4} \right\}$$

$$V_3 = \left\{ (s, s, - s, - s) \text{ where } 0 \leq s \leq - \frac{v_2}{2} \right\}.$$

Range: $- \frac{v_2}{2} \leq v_3 \leq - \frac{5v_2}{2}$.

VI: $V_1 = \left\{ (q, q, q, - 3q) \text{ where } \lambda \leq q \leq \frac{1}{3} \right\}$

$V_2 = \left\{ (\lambda, \lambda, r, - r, - 2\lambda) \text{ where } - v_3 \leq r \leq - v_2 - \lambda \right\}$

$V_3 = \left\{ (s, s, - s, - s) \text{ where } 0 \leq s \leq v_3 \right\}$

$V_4 = \left\{ (t, t, x, y) \text{ where } v_3 \leq t \leq - \frac{v_2}{2} \text{ and } x, y \leq - v_3 \right\}$

$V_5 = \left\{ (u, u, - v_3, v_3 - 2u) \text{ where } - \frac{v_2}{2} < u < \lambda \right\}.$

Range: $0 \leq v_3 \leq - \frac{v_2}{2}; \quad - \frac{v_2}{2} \leq \lambda \leq v_3 - v_2$.

VII: $V_1 = \left\{ (q, q, q, - 3q) \text{ where } \lambda \leq q \leq \frac{1}{3} \right\}$

$V_2 = \left\{ (\lambda, \lambda, r, - r - 2\lambda) \text{ where } - v_3 \leq r \leq - v_2 - \lambda \right\}$

$V_4 = \left\{ (t, t, x, y) \text{ where } - v_3 \leq t \leq - \frac{v_2}{2} \text{ and } x, y \leq - v_3 \right\}$

$V_5 = \left\{ (u, u, - v_3, v_3 - 2u) \text{ where } - \frac{v_2}{2} < u < \lambda \right\}$

$V_6 = \left\{ (\alpha_1, \alpha_2, \alpha_3, \alpha_4) \text{ where } \alpha_i + \alpha_j \geq v_2 \right.$

and

$$\left. \alpha_i + \alpha_j + \alpha_k \geq v_3 \right\} .$$

Range:

$$v_3 \leq 0; \quad - \frac{v_2}{2} \leq \lambda \leq \max \left\{ - \frac{v_2}{2}, v_3 - v_2 \right\} .$$

NOTE. Actually, for $v_3 \leq \frac{v_2}{2}$ the solution consists of $V_1 \cup V_6$ alone.

§2. DISCUSSION

The ranges of validity for the various types of solutions may be effectively visualized by plotting for a fixed value of $v_2 < 0$ the range for λ as a function of v_3. Figure 1 shows the areas in which solutions I, II, IV, VI, and VII are valid. For values of

$$v_3 < \frac{v_2}{2}$$

region VII is reduced to the line

$$\lambda = -\frac{v_2}{2}$$

extending to the left.

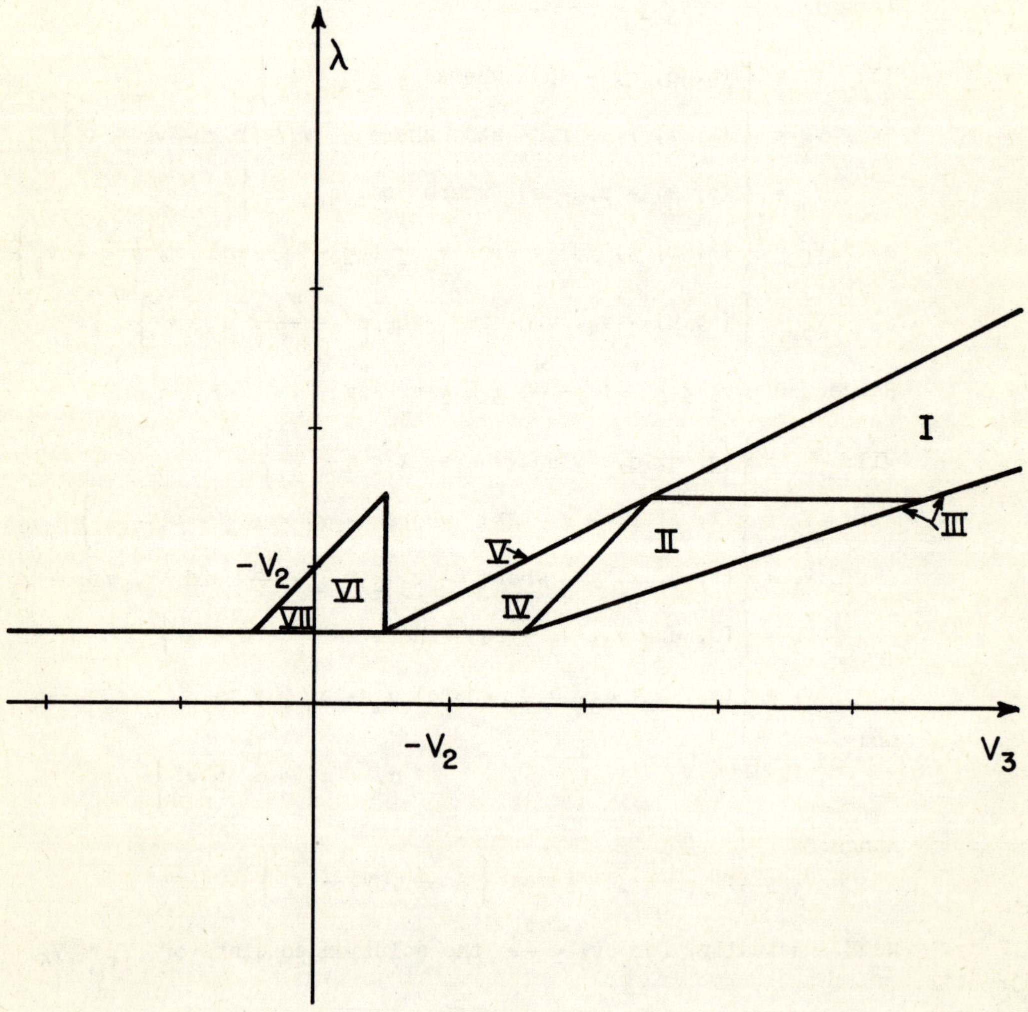

FIGURE 1

The various regions include their boundaries except the boundaries of regions I and II along the line

$$\lambda = \frac{v_3}{3} ,$$

the boundary of region IV along the line

$$\lambda = \frac{2v_3 - v_2}{4} ,$$

and the boundary between regions I and II which belongs to region I. Although the parameter λ does not appear explicitly in solutions III and V, they may be thought of as filling the missing boundaries, as is indicated in Figure 1.

For $v_2 = 0$, regions II and IV vanish, regions V and VI degenerate to the same point, and region VII degenerates into a straight line.

The factors which determine the ranges of validity for the various solutions can be seen geometrically. The imputations can be thought of as barycentric coordinates representing points in a regular tetrahedron, called the fundamental tetrahedron. Taking $v_2 < 0$ as the more interesting situation, the set of imputations $\vec{\alpha} = (\alpha_1, \alpha_2, \alpha_3, \alpha_4)$ with $\alpha_i + \alpha_j \geq v_2$ form a cube C situated symmetrically so that four of the vertices, which we shall call the main vertices of C, point to the four faces of the fundamental tetrahedron. The cube may be so large that it projects through the faces of the fundamental tetrahedron, but its orientation is always the same. If $v_3 < 0$, there is a tetrahedron, oriented dually with respect to the fundamental tetrahedron, in which $\alpha_i + \alpha_j + \alpha_k \geq v_3$. In this case there exist detached imputations and this is the case in which solution VII is valid. If $v_3 > 0$, there is a tetrahedron T in which $\alpha_i + \alpha_j + \alpha_k \leq v_3$, and this tetrahedron is oriented in the same manner as the fundamental tetrahedron.

The simplest configuration occurs when

$$v_3 = -\frac{3v_2}{2} .$$

In this case the four main vertices of the cube C lie in the four faces of the tetrahedron T. Solution III is valid in this case. Component V_2 forms the diagonals joining the four main vertices of C. Component V_1 consists of straight lines running from each of the main vertices of C perpendicularly to the nearest face of the fundamental tetrahedron. The imputations corresponding to the four main vertices of the cube dominate the interior of the cube C and all the faces, except the diagonals corresponding to the component V_2, on 3-element sets. The component V_2 dominates on 2-element sets all the exterior of the cube C, except for four small tetrahedrons, each with a base in a face of the fundamental tetrahedron and a vertex at a main vertex of C. Component V_1 forms the altitudes of these tetrahedrons and dominates the complement of V_1 in them on 2-element sets.

Now it happens that this configuration remains a solution if the parameter v_3 is changed slightly. As v_3 decreases this configuration remains consistent and domination on 2-element sets is unchanged, but domination on 3-element sets becomes more difficult. This is the configuration described as solution IV with

$$\lambda = -\frac{v_2}{2} .$$

As long as

$$-v_2 < v_3 \leq -\frac{3v_2}{2}$$

this configuration remains a solution, and it is seen that for this range of v_3 the component V_3 of solution IV is empty. For

$$-\frac{v_2}{2} < v_3 \leq -v_2$$

there is an undominated region, a tetrahedron τ inside C oriented the same way the fundamental tetrahedron is oriented. Domination in this region is effected by adding to the proposed solution the component V_3 which is made up of lines joining the midpoints of opposite edges of τ.

For

$$v_3 = -\frac{v_2}{2} ,$$

T and τ coincide and their edges join the minor vertices of C. For

$$0 \leq v_3 \leq -\frac{v_2}{2} ,$$

T is inside C and component V_4 must be added for solutions of type VI. (For

$$\lambda = -\frac{v_2}{2} ,$$

V_5 is empty.)

All else is obtained by varying the parameters to determine the range in which the configuration still yields a solution, and mutilating the basic configurations in one manner or another as necessary. To see this, consider as described above the solutions of type IV with

$$\lambda = -\frac{v_2}{2} \left(\text{and therefore } -\frac{v_2}{2} < v_3 \leq -\frac{3v_2}{2} \right) .$$

Keep this configuration fixed and take $v_2' = v_2 + \epsilon$ $(\epsilon \geq 0)$ to be the value of a 2-person coalition. Then $(\lambda, \lambda - \epsilon, \lambda, -3\lambda + \epsilon) \in V_2$ dominates all $(\lambda, \lambda, r, -r - 2\lambda)$ with $\lambda - \epsilon < r < \lambda$. This section, and those symmetric to it, must be removed from V_2. Since $\lambda - \epsilon = - v_2' - \lambda$, the component which is left is

$$V_2' = \{(\lambda, \lambda, r, - r - 2\lambda) \text{ where } - \lambda \leq r \leq - v_2' - \lambda\} \quad .$$

Figure 1 represents a plane section for a constant value of $v_2 < 0$ of a 3-dimensional diagram of the range of validity for the solutions. Since the relations given depend on the parameters in a linear homogeneous manner, to increase v_2 and leave v_3 and λ fixed, it would be equivalent to think of increasing v_3 and λ proportionally and leaving v_2 fixed. Thus we may think of a point in Figure 1 as moving out along a ray through the origin as ϵ increases.

The mutilated configuration will remain consistent as long as ϵ is not increased to the point where domination is possible from V_3 on 2-element sets. For

$$- v_2 < v_3 \leq - \frac{3v_2}{2}$$

component V_3 is empty and this consideration is no restriction. In fact, all the solutions of type III may be obtained in this way. For

$$- \frac{v_2}{2} < v_3 \leq - v_2 \quad ,$$

component V_3 is not empty and the restriction $\epsilon < 2v_3 + v_2$ is necessary. This forces the omission of region V from region IV. For $\epsilon = 2v_3 + v_2$, or

$$\lambda = \frac{2v_3 - v_2'}{4} \quad ,$$

the endpoints of the lines in V_3 can dominate on 2-element sets. This necessitates the removal of the middle portions of the lines of component V_2' and results in configurations of the type which appear in solutions of type V.

The mutilated configuration (if consistent) will remain a solution as long as the parts removed are not essential for domination. For sufficiently small ϵ this is the case and yields the condition

$$\epsilon \leq - \frac{3v_2}{2} - v_3 \quad ,$$

or $\lambda \geq v_2' + v_3$, which determines the dividing line between region IV and region II. For

$$\epsilon > - \frac{3v_2}{2} - v_3$$

component V_3 may have to be expanded, and this consideration yields the component V_3 in solutions of type II. (We note that for $\lambda < v_2' + v_3$ we have $2\lambda - v_3 < v_2' + \lambda$ and there is an actual expansion only if $\lambda \geq - v_2'$.) In order that this expanded configuration remain consistent we must impose the condition $- 2v_2' - 2\lambda > v_2'$, which determines the dividing line

$$\lambda = - \frac{3v_2'}{2}$$

between region II and region I. For

$$\lambda = - \frac{3v_2'}{2}$$

the interiors of the line segments in component V_2' must be removed and the result is a configuration of a solution of type I.

All the solutions described were discovered by these geometric methods. It is also on the basis of these geometric methods that the analytic proofs, of which a sample will be given, can be constructed quite easily.

The geometric viewpoint has also been quite effective in uncovering solutions for a large range of unsymmetric general-sum 4-person games. In fact, for every such game considered solutions were easily found. Although it seems possible that by these geometric methods solutions for all 4-person games can be found, the complexity of the effort has discouraged an attempt to provide solutions for them. Except for some pathology in the form of some of the solutions, the progress that was made along these lines produced little of essential interest.

§3. THE PROOFS

The analytic proofs necessary to establish the proposed solutions are tedious, but fortunately the various types of solutions have similar constituent parts and the proofs both of consistency and domination are quite similar and easily supplied. Type VI is the most complicated, so we shall give the proof for that type with an indication of some of the modifications needed to give the corresponding proofs for the other types.

CONSISTENCY OF SOLUTION VI. We first determine the effective sets for each V_1 and discuss consistency within each V_1.

For $(q, q, q, - 3q) \in V_1$, no 2-element or 3-element set excluding player 4 is effective unless $\lambda = v_2 = v_3 = 0$ because

$$3q \geq 3\lambda \geq - \frac{3v_2}{2} \geq 3v_3 \geq v_3$$

and $2q \geq 2\lambda \geq - v_2 \geq v_2$. (In case $\lambda = v_2 = v_3 = 0$, $V_1 \cap V_2 \cap \ldots \cap V_5 = V_1$, which may easily be shown to be a solution.) Thus V_1 is seen to be consistent within itself.

The effective sets for $(\lambda, \lambda, r, - r - 2\lambda) \in V_2$ are (3, 4), (1, 3, 4), (2, 3, 4), and (1, 2, 4). (1, 4) and (2, 4) are effective only if $r = - v_2 - \lambda$ and (1, 2, 3) is effective only if

$$- r = \lambda = - \frac{v_2}{2} = v_3 \quad .$$

Since a 2-element set is effective only if the components are minimal, we see that V_2 is consistent within itself.

The effective sets for $(s, s, - s, - s) \in V_3$ are all 3-element sets and (3, 4) only for

$$s = v_3 = - \frac{v_2}{2} \quad .$$

Other 2-element sets are effective only if $v_2 = v_3 = 0$, a case discussed previously. Thus V_3 is consistent within itself.

The effective sets for $(t, t, x, y) \in V_4$ are (1, 3, 4) and (2, 3, 4) with (1, 2, 4) effective only if $x = - v_3$ and (3, 4) effective only if

$$t = - \frac{v_2}{2} \quad .$$

Thus it is easily seen that V_4 is consistent.

The effective sets for $(u, u, - v_3, v_3 - 2\lambda) \in V_5$ are (1, 3, 4), (2, 3, 4), (1, 2, 4), and (3, 4), and we see that V_5 is consistent.

Finally we have

$$\begin{aligned} - 3q \leq - 3\lambda \leq - r - 2\lambda \leq v_3 - 2u \leq v_2 + v_3 \leq y,\ x \leq - v_3\ (\leq r) \\ \leq - s \leq s \leq v_3 \leq t \leq - \frac{v_2}{2} \leq u \leq \lambda \leq q \quad , \end{aligned}$$

which means that proper inequalities for domination between the various V_i cannot exist except on ineffective sets.

DOMINATION BY SOLUTION VI. Let $\overrightarrow{\alpha} = (\alpha_1, \alpha_2, \alpha_3, \alpha_4)$ be any imputation not in solution VI. Because of the symmetry we may assume without loss of generality that $\alpha_i \geq \alpha_{i+1}$. We shall break the argument into several cases.

A. $\alpha_3 \geq \lambda$: Then $\alpha_1 + \alpha_2 + \alpha_3 > 3\alpha_3$ since equality is possible only if $\vec{\alpha} \in V_1$. Let $-\alpha_4 = \alpha_1 + \alpha_2 + \alpha_3 = 3\alpha_3 + 4\epsilon$, where $\epsilon > 0$. Then $(\alpha_3 + \epsilon, \alpha_3 + \epsilon, \alpha_3 + \epsilon, \alpha_4 + \epsilon) \in V_1$ dominates $\vec{\alpha}$ on $(3, 4)$ since $\alpha_3 + \alpha_4 + 2\epsilon = -2(\alpha_3 + \epsilon) \leq -2\lambda - 2\epsilon \leq v_2 - 2\epsilon$.

B. $\alpha_3 < \lambda$, $\alpha_4 < v_2 - \lambda$: Then $(\lambda, -v_2 - \lambda, \lambda, v_2 - \lambda) \in V_2$ dominates $\vec{\alpha}$ on $(3, 4)$.

C. $\alpha_3 < \lambda$, $\alpha_4 \geq v_2 - \lambda$, $\alpha_3 < -v_3$, $\alpha_2 \geq -\frac{v_2}{2}$: Then $\alpha_3 + \alpha_4 = -(\alpha_1 + \alpha_2) \leq -2\alpha_2 \leq v_2$, with equality possible only if $\vec{\alpha} \in V_4$. Let

$$\epsilon = \min\left\{-v_3 - \alpha_3, \frac{v_2 - \alpha_3 - \alpha_4}{2}\right\} > 0 \quad .$$

Then $\alpha_3 < \alpha_3 + \epsilon \leq -v_3$ and $\alpha_4 < \alpha_4 + \epsilon \leq v_2 - \alpha_3 - \epsilon$. Now $v_2 - \alpha_3 - \epsilon = \alpha_4 + \epsilon$ unless $\epsilon = -v_3 - \alpha_3$ so that

$$v_2 - \alpha_3 - \epsilon \leq \min\left\{\alpha_4 + \epsilon, v_2 - \alpha_3 + v_3 + \alpha_3\right\} \leq -v_3.$$

Thus

$$\left(-\frac{v_2}{2}, -\frac{v_2}{2}, \alpha_3 + \epsilon, v_2 - \alpha_3 - \epsilon\right) \in V_4$$

dominates $\vec{\alpha}$ on $(3, 4)$.

D. $\alpha_3 < \lambda$, $\alpha_4 \geq v_2 - \lambda$, $\alpha_3 < -v_3$, $\alpha_2 < -\frac{v_2}{2}$: Then

$$\alpha_2 < \frac{\alpha_1 + \alpha_2}{2}$$

since equality would mean $\vec{\alpha} \in V_4$. Thus, excluding the simple special case in which

$$v_3 = -\frac{v_2}{2} \quad ,$$

there is a t such that

$$\max\left\{\alpha_2, v_3\right\} < t < \min\left\{\frac{\alpha_1 + \alpha_2}{2}, -\frac{v_2}{2}\right\}.$$

This implies that $\alpha_3 + \alpha_4 < -2t$, so that there is an x such that

$$\max\left\{\alpha_3, v_3 - 2t\right\} < x < \min\left\{-v_3, -\alpha_4 - 2t\right\} .$$

Thus $\vec{\alpha}$ is dominated by $(t, t, x, y) \in V_4$ on $(2, 3, 4)$.

E. $\alpha_3 < \lambda$, $\alpha_4 \geq v_2 - \lambda$, $\alpha_3 \geq -v_3$, $\alpha_2 \geq \lambda$: Then $\alpha_3 + \alpha_4 = -(\alpha_1 + \alpha_2) \leq -2\alpha_2 \leq -2\lambda$ with equality possible only if $\vec{\alpha} \in V_2$.

Let $\alpha_3 + \alpha_4 + 2\epsilon = -2\lambda$ where $\epsilon > 0$. Then $(\lambda, \lambda, \alpha_3 + \epsilon, \alpha_4 + \epsilon) \in V_2$ dominates $\vec{\alpha}$ on $(3, 4)$.

F. $\alpha_3 < \lambda$, $\alpha_4 \geq v_2 - \lambda$, $\alpha_3 \geq -v_3$, $\alpha_2 < \lambda$, $\alpha_4 < -v_3$: Then $2\alpha_2 + \alpha_4 \leq \alpha_1 + \alpha_2 + \alpha_4 = -\alpha_3 \leq v_3$ and equality holds only if $\alpha_1 = \alpha_2$ and $\alpha_3 = -v_3$. In this case $\vec{\alpha} = (\alpha_2, \alpha_2, -v_3, \alpha_4) \in V_4 \cup V_5$ since $\alpha_4 \leq \alpha_3 = -v_3$ and

$$v_3 < \frac{v_3 - \alpha_4}{2} = \alpha_2 < \lambda .$$

Thus

$$\alpha_2 < \frac{v_3 - \alpha_4}{2} .$$

Excluding the simple special case in which $v_3 = -\frac{v_2}{2} = \lambda$, there is a t such that

$$\max\left\{\alpha_2, v_3\right\} < t < \min\left\{\frac{v_3 - \alpha_4}{2}, \lambda\right\} .$$

Then $(-v_3, t, t, v_3 - 2t) \in V_4 \cup V_5$ dominates $\vec{\alpha}$ on $(2, 3, 4)$.

G. $\alpha_3 < \lambda$, $\alpha_4 \geq v_2 - \lambda$, $\alpha_3 \geq -v_3$, $\alpha_2 < \lambda$, $\alpha_4 \geq -v_3$: Since $\alpha_2 + \alpha_4 = -(\alpha_1 + \alpha_3) \leq -(\alpha_2 + \alpha_4)$ we see that $\alpha_2 + \alpha_4 \leq 0$ with equality possible only if $\vec{\alpha} \in V_3$. Let $\epsilon > 0$ be determined so that $0 < \epsilon < \min\{-(\alpha_2 + \alpha_4), -\alpha_4\}$. Then $(\alpha_4 + \epsilon, -\alpha_4 - \epsilon, -\alpha_4 - \epsilon, \alpha_4 + \epsilon) \in V_3$ dominates $\vec{\alpha}$ on $(2, 3, 4)$.

Consistency and domination for the remaining solutions can be proved in a similar fashion. For them we shall give only the breakdown of the set of all imputations with $\alpha_i \geq \alpha_{i+1}$ into regions dominated by various parts of the proposed solutions. Where a region is denoted by a letter A through G, domination is similar to that proved under the case with the same letter for solution VI.

I. A. $\alpha_3 \geq \lambda$.

B. $\alpha_3 < \lambda$, $\alpha_4 < -\frac{v_2}{2} - 2\lambda$.

H. $\alpha_3 < \lambda$, $\alpha_4 \geq -\frac{v_2}{2} - 2\lambda$, $\alpha_2 \geq \lambda$.

G. $\alpha_3 < \lambda$, $\alpha_4 \geq -\frac{v_2}{2} - 2\lambda$, $\alpha_2 < \lambda$, $\alpha_4 \geq \frac{v_2}{2}$.

J. $\alpha_3 < \lambda$, $\alpha_4 \geq -\frac{v_2}{2} - 2\lambda$, $\alpha_2 < \lambda$, $\alpha_4 < \frac{v_2}{2}$.

In region H domination is accomplished by

$$\left(-\frac{v_2}{2}, -\frac{v_2}{2}, \frac{v_2}{2}, \frac{v_2}{2}\right) \in V_3$$

on (3, 4). Region J is dominated by

$$\left(- \frac{v_2}{2} - 2\lambda,\ \lambda,\ \lambda,\ \frac{v_2}{2}\right) \in V_2$$

on (2, 3, 4).

II. A. $\alpha_3 \geq \lambda$.

B. $\alpha_3 < \lambda,\ \alpha_4 < v_2 - \lambda$.

E. $\alpha_3 < \lambda,\ \alpha_4 \geq v_2 - \lambda,\ \alpha_2 \geq \lambda$.

G. $\alpha_3 < \lambda,\ \alpha_4 \geq v_2 - \lambda,\ \alpha_2 < \lambda,\ \alpha_4 \geq - v_2 - \lambda$.

J. $\alpha_3 < \lambda,\ \alpha_4 \geq v_2 - \lambda,\ \alpha_2 < \lambda,\ \alpha_4 < - v_2 - \lambda$.

Although region E does not seem to resemble region E for solution VI, domination is accomplished in the same way.

III. Let $\lambda = \frac{v_3}{3}$.

A. $\alpha_3 \geq \lambda$.

B. $\alpha_3 < \lambda,\ \alpha_4 < v_2 - \lambda$.

E. $\alpha_3 < \lambda,\ \alpha_4 \geq v_2 - \lambda,\ \alpha_2 \geq \lambda$.

K. $\alpha_3 < \lambda,\ \alpha_4 \geq v_2 - \lambda,\ \alpha_2 < \lambda$.

Region K is dominated by $(- v_3,\ \lambda,\ \lambda,\ \lambda) \in V_1$.

IV. A. $\alpha_3 \geq \lambda$.

B. $\alpha_3 < \lambda,\ \alpha_4 < v_2 - \lambda$.

E. $\alpha_3 < \lambda,\ \alpha_4 \geq v_2 - \lambda,\ \alpha_2 \geq \lambda$.

G. $\alpha_3 < \lambda,\ \alpha_4 \geq v_2 - \lambda,\ \alpha_2 < \lambda,\ \alpha_4 \geq v_3 - 2\lambda$.

J. $\alpha_3 < \lambda,\ \alpha_4 \geq v_2 - \lambda,\ \alpha_2 < \lambda,\ \alpha_4 < v_3 - 2\lambda$.

V. Let $\lambda = \frac{2v_3 - v_2}{4}$.

A. $\alpha_3 \geq \lambda$.

B. $\alpha_3 < \lambda,\ \alpha_4 < v_2 - \lambda$.

H. $\alpha_3 < \lambda,\ \alpha_4 \geq v_2 - \lambda,\ \alpha_2 \geq \lambda,\ \alpha_3 < \frac{v_2}{2}$.

E. $\alpha_3 < \lambda,\ \alpha_4 \geq v_2 - \lambda,\ \alpha_2 \geq \lambda,\ \alpha_3 \geq \frac{v_2}{2}$.

G. $\alpha_3 < \lambda,\ \alpha_4 \geq v_2 - \lambda,\ \alpha_2 < \lambda,\ \alpha_4 \geq \frac{v_2}{2}$.

J. $\alpha_3 < \lambda,\ \alpha_4 \geq v_2 - \lambda,\ \alpha_2 < \lambda,\ \alpha_4 < \frac{v_2}{2}$.

VII. A. $\alpha_3 \geq \lambda$.

B. $\alpha_3 < \lambda,\ \alpha_4 < v_2 - \lambda$.

C. $\alpha_3 < \lambda,\ \alpha_4 \geq v_2 - \lambda,\ \alpha_3 < -v_3,\ \alpha_2 \geq -\frac{v_2}{2}$.

D. $\alpha_3 < \lambda,\ \alpha_4 \geq v_2 - \lambda,\ \alpha_3 < -v_3,\ \alpha_2 < -\frac{v_2}{2},\ \alpha_2 \geq -v_3$.

L. $\alpha_3 < \lambda,\ \alpha_4 \geq v_2 - \lambda,\ \alpha_3 < -v_3,\ \alpha_2 < -v_3$.

E. $\alpha_3 < \lambda,\ \alpha_4 \geq v_2 - \lambda,\ \alpha_3 \geq -v_3,\ \alpha_2 \geq \lambda$ ..

F. $\alpha_3 < \lambda,\ \alpha_4 \geq v_2 - \lambda,\ \alpha_3 \geq -v_3,\ \alpha_2 < \lambda$.

Region L is dominated by imputations in V_6.

BIBLIOGRAPHY

[1] von NEUMANN, J. and MORGENSTERN, O., Theory of Games and Economic Behavior, Princeton, 2nd ed. 1947.

Evar D. Nering

University of Minnesota

THE FOUR PERSON GAME - FINITE SOLUTIONS ON THE FACE OF THE CUBE[1]

W. H. Mills

Ever since von Neumann and Morgenstern introduced their theory of the n-person game, there has been a tendency to place special emphasis on finite solutions. In part this is because they do not exhibit the bewildering variety that has been observed for arbitrary solutions, in part because they are frequently easier to interpret, and in part because of the inspiration of the three person game. Unfortunately, even for the four-person game, our knowledge of finite solutions is limited. It is known, however, that those four-person zero-sum games that correspond to the vertices of the von Neumann cube Q have unique finite solutions of a simple nature [1], while those games that correspond to the interior points of the edges of Q have no finite solutions whatsoever [2].

In the present paper all finite solutions are determined for those zero-sum four-person games that correspond to the faces of the cube Q. Each face of the cube Q has two opposite vertices that correspond to the three person game with a dummy [1]. Let $\mathfrak{D}$ be the diagonal joining these two vertices. Then each interior point of $\mathfrak{D}$ corresponds to a game with a unique finite solution, and this finite solution consists of seven points. Except for the points of $\mathfrak{D}$ and the vertices there are no games whatsoever on the face of the cube that have finite solutions. Thus a new seven point solution is discovered, and the class of games for which it is known that there are no finite solutions is enlarged from a one dimensional set to a two dimensional one.

§1. GENERAL FINITENESS CONDITION

Let Γ be an essential zero-sum n-person game with characteristic function $v(\mathfrak{A})$, let the players be denoted by $1, 2, \ldots, n$, and suppose that the von Neumann-Morgenstern normalization has been applied

[1] This work was supported (in part) by the Office of Naval Research.

so that $v(\{i\}) = -1$ for any one element set $\{i\}$. We shall use capital letters to denote imputations and the corresponding small letters to denote components. Thus if P is an imputation, then p_i denotes the payoff to Player i, so that $P = (p_1, p_2, \ldots, p_n)$. Letters in the first half of the alphabet will represent the same imputations throughout the paper, while the imputations represented by letters in the second half of the alphabet will vary from proof to proof. We will say that a set $\mathfrak{A}$ is properly effective for an imputation P if the complement of $\mathfrak{A}$ is not effective for P. If a set $\mathcal{S}$ contains an imputation that dominates P, then we will say that P is dominated by $\mathcal{S}$, and we will write, as usual, $\mathcal{S} \succ P$.

Let $\mathcal{V}$ be a solution of Γ. We begin by establishing a necessary and sufficient condition that $\mathcal{V}$ be finite.

THEOREM 1. The solution $\mathcal{V}$ of Γ is finite if and only if for every $P \in \mathcal{V}$ and every i for which $p_i > -1$, there exist an imputation $R \in \mathcal{V}$ and a subset $\mathfrak{A}$ of the set of players such that

(i) $\mathfrak{A}$ is effective for R and properly effective for P,

(ii) $i \in \mathfrak{A}$ and $r_i = p_i$,

(iii) $r_j > p_j$ if $j \in \mathfrak{A}$, $j \neq i$.

PROOF. Suppose that $\mathcal{V}$ is finite, $P \in \mathcal{V}$, $p_i > -1$. Put $p_i' = p_i - \epsilon$, $p_j' = p_j + \epsilon/(n-1)$ for $j \neq i$. Then, for sufficiently small positive ϵ, $P' = (p_1', p_2', \ldots, p_n')$ is an imputation. Since $\mathcal{V}$ is finite we can choose ϵ small enough so that $p_i' > s_i$ for every imputation $S \in \mathcal{V}$ for which $s_i < p_i$. Then $P' \notin \mathcal{V}$, and so P' is dominated by an imputation $R \in \mathcal{V}$ with respect to a suitable set $\mathfrak{A}$. If $j \in \mathfrak{A}$ and $j \neq i$, then we have $r_j > p_j' > p_j$. Since R does not dominate P it follows that $i \in \mathfrak{A}$ and $p_i \geq r_i > p_i - \epsilon$. Hence $r_i = p_i$ by the choice of ϵ. Finally

$$\sum_{j \in \mathfrak{A}} p_j < \sum_{j \in \mathfrak{A}} r_j \leq v(\mathfrak{A}) ,$$

and so $\mathfrak{A}$ is properly effective for P, as well as effective for R.

To prove the converse let P be an imputation of the solution $\mathcal{V}$ such that, for every i for which $p_i > -1$, there exist an imputation $R_i \in \mathcal{V}$ and a subset $\mathfrak{A}_i$ of the set of players satisfying (i), (ii), and (iii). Let $\mathcal{N}_i$ be the set of all imputations that are dominated by R_i with respect to $\mathfrak{A}_i$, and let $\mathcal{N}$ be the union of the sets $\mathcal{N}_i$. Since the imputations R_i are elements of $\mathcal{V}$ it follows that $\mathcal{N}$ and $\mathcal{V}$ have no points in common. If N is an imputation sufficiently close to P and $N \neq P$, then $N \in \mathcal{N}$, because $p_i > n_i \geq -1$ for some i and $R_i \succ N$ with respect to $\mathfrak{A}_i$. Hence P is not a point of accumulation of $\mathcal{V}$. Therefore if $\mathcal{V}$ is a solution for which the condition of Theorem 1 holds, then $\mathcal{V}$ has no points of accumulation. Since $\mathcal{V}$ is closed and bounded it follows that $\mathcal{V}$ is finite. This completes the proof of Theorem 1.

If two imputations, P and R, satisfy (i), (ii), and (iii) of Theorem 1, we will say that P_{i-} is dominated by R with respect to $\mathfrak{A}$, or that P_{i-} is $\mathfrak{A}$ dominated by R. We will write, of course, $R \succ P_{i-}$. If a set $\mathcal{S}$ contains an imputation that dominates P_{i-}, then we will say that P_{i-} is dominated by $\mathcal{S}$, and we will write $\mathcal{S} \succ P_{i-}$.

If r is a real number and i one of the n players, we let $\mathcal{C}_i(r)$ denote the set of all imputations $P \in \mathcal{V}$ such that $p_i = r$. The following result is an immediate consequence of Theorem 1:

> COROLLARY. If $\mathcal{V}$ is finite and if $\mathcal{C}_i(r)$ consists of a single imputation P, then $r = -1$.

§2. PRELIMINARY LEMMAS

We will restrict ourselves, throughout the remainder of the paper, to the case where $n = 4$, Γ corresponds to a point on the face of the von Neumann cube Q, and $\mathcal{V}$ is a finite solution of Γ. Since any face of Q can be mapped onto any other face by a suitable permutation of players, we need only study one particular face. We suppose henceforth that

$$v(\{1, 2\}) = 2, \quad v(\{1, 3\}) = t, \quad v(\{1, 4\}) = u \ ,$$

where $-2 \leq t \leq 2$, $-2 \leq u \leq 2$.

Let $\mathcal{A}$ be the set of all imputations for which $\{1, 3\}$ and $\{1, 4\}$ are both properly effective, and $\mathcal{B}$ be the set of all imputations for which $\{2, 3\}$ and $\{2, 4\}$ are both properly effective, i.e.,

$$\mathcal{A} = \{P \mid p_1 + p_3 < t, \quad p_1 + p_4 < u\} \ ,$$

$$\mathcal{B} = \{P \mid p_2 + p_3 < -u, \quad p_2 + p_4 < -t\} \ .$$

Let $\bar{\mathcal{A}}$ be the closure of $\mathcal{A}$ and $\bar{\mathcal{B}}$ the closure of $\mathcal{B}$. Thus $\bar{\mathcal{A}}$ is the set of all imputations for which both $\{1, 3\}$ and $\{1, 4\}$ are effective, and $\bar{\mathcal{B}}$ is the set of all imputations for which both $\{2, 3\}$ and $\{2, 4\}$ are effective. We put

$$\alpha = \operatorname*{Min}_{P \in \mathcal{V}} p_1, \qquad \alpha^* = \operatorname*{Max}_{P \in \mathcal{V}} p_1 ,$$

$$\beta = \operatorname*{Min}_{P \in \mathcal{V}} p_2, \qquad \beta^* = \operatorname*{Max}_{P \in \mathcal{V}} p_2 .$$

Suppose that $P \in \mathcal{V}$ and that P dominates an imputation R with respect to $\{2, 3, 4\}$. If $P \notin \mathcal{A}$, then either $\{2, 3\}$ or $\{2, 4\}$ is effective for P, and $P \vdash R$ with respect to either $\{2, 3\}$ or $\{2, 4\}$. If $P \in \mathcal{A}$ and $p_1 > \alpha$, let $S \in \mathcal{C}_1(\alpha)$. Since P does not dominate S we have $s_2 \geq p_2$, $s_3 \geq p_3$, $s_4 \geq p_4$ and so $S \vdash R$ with respect to $\{2, 3, 4\}$. Either $S \in \mathcal{A}$, or $S \vdash R$ with respect to $\{2, 3\}$ or $\{2, 4\}$. Thus we see that if $\mathcal{V}$ dominates an imputation R with respect to $\{2, 3, 4\}$, then either $(\mathcal{C}_1(\alpha) \cap \mathcal{A}) \vdash R$ with respect to $\{2, 3, 4\}$, or $\mathcal{V} \vdash R$ with respect to $\{2, 3\}$ or $\{2, 4\}$. Similarly if $\mathcal{V} \vdash R$ with respect to $\{1, 3, 4\}$, then either $(\mathcal{C}_2(\beta) \cap \mathcal{B}) \vdash R$ with respect to $\{1, 3, 4\}$, or $\mathcal{V} \vdash R$ with respect to $\{1, 3\}$ or $\{1, 4\}$.

The set $\{1, 2,\}$ is always effective so that $\{3, 4\}$ domination never occurs and $\{1, 2, 3\}$ and $\{1, 2, 4\}$ domination need not be considered. Thus we have the following result:

LEMMA 1. If R is an imputation not in $\mathcal{V}$, then either $\mathcal{V} \vdash R$ with respect to one of the five sets

(1) $\{1, 2\}, \ \{1, 3\}, \ \{1, 4\}, \ \{2, 3\}, \ \{2, 4\}$,

or $(\mathcal{C}_1(\alpha) \cap \mathcal{A}) \vdash R$ with respect to $\{2, 3, 4\}$, or $(\mathcal{C}_2(\beta) \cap \mathcal{B}) \vdash R$ with respect to $\{1, 3, 4\}$. Similarly if $R \in \mathcal{V}$ and $r_i > -1$, then either $\mathcal{V} \vdash R_{i-}$ with respect to one of the five sets (1), or $(\mathcal{C}_1(\alpha) \cap \mathcal{A}) \vdash R_{i-}$ with respect to $\{2, 3, 4\}$, or $(\mathcal{C}_2(\beta) \cap \mathcal{B}) \vdash R_{i-}$ with respect to $\{1, 3, 4\}$.

LEMMA 2. $\mathcal{C}_1(\alpha)$ is a proper subset of $\mathcal{C}_2(\beta^*)$, and $\mathcal{C}_2(\beta)$ is a proper subset of $\mathcal{C}_1(\alpha^*)$.

PROOF. Suppose $\mathcal{C}_2(\beta^*) \subseteq \mathcal{C}_1(\alpha)$. Let $R \in \mathcal{C}_2(\beta^*)$ with maximal r_3. If $\beta^* = -1$, then $\mathcal{V} = \mathcal{C}_1(\alpha) \cap \mathcal{C}_2(-1)$, which implies that $\mathcal{C}_3(r_3)$ and $\mathcal{C}_4(r_4)$ each consists of R alone. By the Corollary of Theorem 1 this implies $r_3 = r_4 = -1$, and so $\mathcal{V}$ consists of $R = (3, -1, -1, -1)$ alone, which is impossible. Therefore $\beta^* > -1$. By Theorem 1, R_{2-} is dominated by an imputation $S \in \mathcal{V}$ with respect to a suitable set $\mathfrak{A}$. We have $2 \in \mathfrak{A}$, $s_2 = r_2 = \beta^*$, and our assumption

$\mathcal{C}_2(\beta^*) \subseteq \mathcal{C}_1(\alpha)$ implies $s_1 = r_1$. By the maximality of r_3 we have $s_3 \leq r_3$. Therefore $\mathfrak{A} = \{2, 4\}$. We suppose that S is chosen with s_4 maximal subject to $S \in \mathcal{C}_2(\beta^*)$ and $s_2 + s_4 \leq -t$. Then S_{2-} is dominated by an imputation $T \in \mathcal{V}$ with respect to a set $\mathfrak{B}$, $2 \in \mathfrak{B}$. Since $t_2 = \beta^*$, we have $t_1 = \alpha = s_1$. This implies that $1 \notin \mathfrak{B}$ and that $\mathfrak{B} \neq \{2, 3, 4\}$. Now $\mathfrak{B} \neq \{2, 4\}$ by the maximality of s_4. Hence $\mathfrak{B} = \{2, 3\}$. We suppose T chosen with t_3 maximal subject to $T \in \mathcal{C}_2(\beta^*)$ and $t_2 + t_3 \leq -u$. We note that $s_2 + s_3 < t_2 + t_3$ and $s_4 > t_4 \geq -1$. S_{4-} is not $\{1, 4\}$ or $\{2, 4\}$ dominated by $\mathcal{V}$ since T is not, and S_{4-} is not $\{2, 3, 4\}$ dominated by $\mathcal{C}_1(\alpha)$ because $s_1 = \alpha$. Therefore S_{4-} must be $\{1, 3, 4\}$ dominated by an imputation $U \in \mathcal{C}_2(\beta) \cap \mathcal{R}$, $u_1 > s_1 = t_1 = \alpha$, $u_2 < s_2 = t_2 = \beta^*$, $u_3 > s_3$, and $u_4 = s_4 > t_4$. Now U does not dominate T and T does not dominate U. Therefore $u_3 = t_3$. Put $w_2 = -\alpha - t_3 - s_4$. Then, since $t_2 = -t_1 - t_3 - t_4$ and $u_2 = -u_1 - u_3 - u_4$, we have $t_2 > w_2 > u_2 \geq -1$, and $W = (\alpha, w_2, t_3, s_4)$ is an imputation. If $P \in \mathcal{V}$ such that $w_2 \leq p_2 < \beta^*$, then, since neither S nor T dominates P, we have $p_1 \geq \alpha$, $p_3 \geq t_3$, $p_4 \geq s_4$, and hence $P = W$. Combining this with $\mathcal{C}_2(\beta^*) \subseteq \mathcal{C}_1(\alpha)$ and the maximality of t_3 and s_4 we see that W is not $\{1, 2\}$, $\{2, 3\}$, or $\{2, 4\}$ dominated by $\mathcal{V}$. Clearly W is not $\{2, 3, 4\}$ dominated by $\mathcal{V}$. It is not $\{1, 4\}$ dominated by $\mathcal{V}$ since S is not, and it is not $\{1, 3\}$ or $\{1, 3, 4\}$ dominated by $\mathcal{V}$ since T is not. Therefore $W \in \mathcal{V}$ and hence $\mathcal{C}_2(w_2)$ consists of W alone, contradicting the Corollary of Theorem 1. It follows that $\mathcal{C}_2(\beta^*) \not\subseteq \mathcal{C}_1(\alpha)$.

We now see that there exists an imputation $X \in \mathcal{V}$ such that $x_1 > \alpha$ and $x_2 = \beta^*$. If $Y \in \mathcal{C}_1(\alpha)$, then since X does not dominate Y it follows that $y_2 \geq x_2 = \beta^*$ and hence $y_2 = \beta^*$. Therefore $\mathcal{C}_1(\alpha) \subseteq \mathcal{C}_2(\beta^*)$. It follows at once that $\mathcal{C}_1(\alpha)$ is a proper subset of $\mathcal{C}_2(\beta^*)$. Interchanging Players 1 and 2 we see that $\mathcal{C}_2(\beta)$ is a proper subset of $\mathcal{C}_1(\alpha^*)$. This completes the proof of Lemma 2.

As an immediate consequence of Lemma 2 we obtain

$$\alpha^* > \alpha, \qquad \beta^* > \beta \quad . \tag{2}$$

If $\mathcal{C}_1(\alpha)$ consists of a single imputation we shall denote this imputation by A. If $\mathcal{C}_1(\alpha)$ contains more than one imputation we let $B \in \mathcal{C}_1(\alpha)$ with maximal b_3 and $C \in \mathcal{C}_1(\alpha)$ with maximal c_4. It follows from Lemma 2 that in the former case $a_2 = \beta^*$. In the latter case $b_2 = c_2 = \beta^*$, which implies that B and C are uniquely determined, $b_3 > c_3$, and $b_4 < c_4$.

We are now in a position to prove a key lemma:

LEMMA 3. $\mathcal{C}_1(\alpha)$ contains at most two imputations. If

$\mathcal{C}_1(\alpha)$ contains two imputations, then $\mathcal{C}_2(\beta)$ consists of the single imputation $(\alpha^*, -1, b_3, c_4)$.

PROOF. Suppose that $\mathcal{C}_1(\alpha)$ contains more than one imputation. It follows from the choice of B and C and from Lemma 2 that all points of $\mathcal{C}_1(\alpha)$ lie on the closed straight line segment BC. By Theorem 1, $\mathcal{V} \vdash B_{3-}$ with respect to a suitable set $\mathfrak{A}$, $3 \in \mathfrak{A}$. Since $\mathcal{V}$ does not dominate C it follows that $4 \in \mathfrak{A}$, and since $b_2 = \beta^*$ we have $2 \notin \mathfrak{A}$. Therefore B_{3-} must be $\{1, 3, 4\}$ dominated by an imputation $R \in \mathcal{C}_2(\beta)$, $r_3 = b_3$, $r_4 > b_4$. It follows that $r_1 + r_2 < b_1 + b_2 = p_1 + p_2$ for any $P \in \mathcal{C}_1(\alpha)$. Interchanging the roles of Players 1 and 2 it follows that if $\mathcal{C}_2(\beta)$ contains more than one imputation, then $\mathcal{C}_1(\alpha)$ contains an imputation S such that $s_1 + s_2 < r_1 + r_2$. Therefore $\mathcal{C}_2(\beta)$ is the one element set $\{R\}$. Applying the Corollary of Theorem 1 we see that $r_2 = \beta = -1$. We already have $r_3 = b_3$. Interchanging Players 3 and 4 this gives us $r_4 = c_4$. From Lemma 2 and (2) we obtain $r_1 = \alpha^* > \alpha$. It follows that $R = (\alpha^*, -1, b_3, c_4)$ and that R dominates the interior points of the straight line segment BC. Thus Lemma 3 is established.

If $\mathcal{C}_2(\beta)$ consists of a single imputation we denote that imputation by D. If $\mathcal{C}_2(\beta)$ contains more than one imputation we let E and F be imputations of $\mathcal{C}_2(\beta)$ with e_3 and f_4 maximal. In this case E and F are uniquely determined, $e_1 = f_1 = \alpha^*$, $e_3 > f_3$, and $e_4 < f_4$. Interchanging Players 1 and 2 in Lemma 3 we obtain the following:

LEMMA 4. $\mathcal{C}_2(\beta)$ contains at most two imputations. If $\mathcal{C}_2(\beta)$ contains two imputations, then $\mathcal{C}_1(\alpha) = \{A\}$ and $A = (-1, \beta^*, e_3, f_4)$.

By Lemmas 3 and 4 there are only three possibilities for $\mathcal{C}_1(\alpha)$ and $\mathcal{C}_2(\beta)$:

1. $\mathcal{C}_1(\alpha) = \{A\}$, $\mathcal{C}_2(\beta) = \{D\}$,
2. $\mathcal{C}_1(\alpha) = \{B, C\}$, $\mathcal{C}_2(\beta) = \{D\}$, $D = (\alpha^*, -1, b_3, c_4)$,
3. $\mathcal{C}_1(\alpha) = \{A\}$, $\mathcal{C}_2(\beta) = \{E, F\}$, $A = (-1, \beta^*, e_3, f_4)$.

§3. REDUCTION TO THE CASE $\gamma = \gamma'$, $\delta = \delta'$

Put

$$\gamma = \begin{cases} a_3 & \text{if } \mathcal{C}_1(\alpha) = \{A\} \\ b_3 & \text{if } \mathcal{C}_1(\alpha) = \{B, C\}, \end{cases}$$

$$\delta = \begin{cases} a_4 & \text{if } \mathcal{C}_1(\alpha) = \{A\} \\ c_4 & \text{if } \mathcal{C}_1(\alpha) = \{B, C\} \end{cases} ,$$

$$\gamma' = \begin{cases} d_3 & \text{if } \mathcal{C}_2(\beta) = \{D\} \\ e_3 & \text{if } \mathcal{C}_2(\beta) = \{E, F\} \end{cases} ,$$

$$\delta' = \begin{cases} d_4 & \text{if } \mathcal{C}_2(\beta) = \{D\} \\ f_4 & \text{if } \mathcal{C}_2(\beta) = \{E, F\} \end{cases} .$$

Then we have

$$\gamma = \underset{P \in \mathcal{C}_1(\alpha)}{\text{Max}} \, p_3, \qquad \delta = \underset{P \in \mathcal{C}_1(\alpha)}{\text{Max}} \, p_4 ,$$

$$\gamma' = \underset{P \in \mathcal{C}_2(\beta)}{\text{Max}} \, p_3, \qquad \delta' = \underset{P \in \mathcal{C}_2(\beta)}{\text{Max}} \, p_4 .$$

The main purpose of this section is to eliminate the cases $\gamma \neq \gamma'$ and $\delta \neq \delta'$. In so doing we will obtain some information about the other possibility $\gamma = \gamma'$ and $\delta = \delta'$.

Suppose that there is an imputation $P \in \mathcal{V}$ such that $p_3 > \gamma$ and $p_3 > \gamma'$. Then $p_1 > \alpha$ and $p_2 > \beta$. Since P does not dominate the imputations of $\mathcal{C}_1(\alpha)$ and $\mathcal{C}_2(\beta)$ it follows that neither $\{1, 3\}$ nor $\{2, 3\}$ is effective for P. Therefore, by Lemma 1, either $\mathcal{C}_1(\alpha) \leftarrow P_{3-}$ with respect to $\{2, 3, 4\}$ or $\mathcal{C}_2(\beta) \leftarrow P_{3-}$ with respect to $\{1, 3, 4\}$. Since these are both impossible we have

$$\underset{P \in \mathcal{V}}{\text{Max}} \, p_3 = \text{Max} \{\gamma, \gamma'\} \quad . \tag{3}$$

Similarly

$$\underset{P \in \mathcal{V}}{\text{Max}} \, p_4 = \text{Max} \{\delta, \delta'\} \quad . \tag{4}$$

It follows from (2) that there exists at least one imputation $P \in \mathcal{V}$ such that $p_1 > \alpha$, and at least one imputation $R \in \mathcal{V}$ such that $r_2 > \beta$. We put

$$\bar{\alpha} = \underset{P \in \mathcal{V}, p_1 > \alpha}{\text{Min}} \, p_1 \qquad \text{and} \qquad \bar{\beta} = \underset{P \in \mathcal{V}, p_2 > \beta}{\text{Min}} \, p_2 \quad .$$

Our next step is to study the sets $\mathcal{C}_1(\bar{\alpha})$ and $\mathcal{C}_2(\bar{\beta})$. We shall prove a series of lemmas about these two sets.

LEMMA 5. $\mathcal{C}_1(\bar{\alpha}) \cap \mathcal{C}_2(\beta^*)$ is not vacuous.

PROOF. By Lemma 2, $\mathcal{C}_2(\beta^*)$ contains an imputation R such that $r_1 > \alpha$. We have $r_1 \geq \bar{\alpha}$. If $r_1 = \bar{\alpha}$, then $R \in \mathcal{C}_1(\bar{\alpha}) \cap \mathcal{C}_2(\beta^*)$. If $r_1 > \bar{\alpha}$ let $S \in \mathcal{C}_1\bar{\alpha})$. Then, since R does not dominate S, we have $s_2 \geq r_2 = \beta^*$, and hence $s_2 = \beta^*$. This implies $S \in \mathcal{C}_1(\bar{\alpha}) \cap \mathcal{C}_2(\beta^*)$.

LEMMA 6. If $\mathcal{C}_1(\alpha) = \{A\}$, then $(\bar{\alpha}, -\bar{\alpha} - a_3 - a_4, a_3, a_4) \notin \mathcal{V} \cap \bar{\mathcal{I}}$.

PROOF. Put $R = (\bar{\alpha}, -\bar{\alpha} - a_3 - a_4, a_3, a_4)$ and suppose $R \in \mathcal{V} \cap \bar{\mathcal{I}}$. Since $r_1 > a_1$, $r_3 = a_3$, and $r_4 = a_4$, we have $r_2 < a_2$. By Lemma 5 there is an imputation $S \in \mathcal{C}_1(\bar{\alpha}) \cap \mathcal{C}_2(\beta^*)$. Since $s_1 > a_1$ and $s_2 = a_2$ it follows that either $s_3 < a_3$ or $s_4 < a_4$. Without loss of generality suppose $s_3 < a_3$.

Now let P be any imputation of $\mathcal{C}_1(\bar{\alpha}) \cap \mathcal{C}_2(\beta^*)$ with $p_3 < a_3$. We observe that R_{3-} is not $\{1, 3\}$ dominated by $\mathcal{V}$ since P is not, and that R_{3-} is not $\{2, 3\}$ dominated since $R \in \bar{\mathcal{I}}$. Clearly R_{3-} is not $\{2, 3, 4\}$ dominated by A, and hence R_{3-} must be $\{1, 3, 4\}$ dominated by $\mathcal{C}_2(\beta)$. Therefore $\delta' > r_4 = a_4$, and Lemma 4 yields $\mathcal{C}_2(\beta) = \{D\}$. We have $d_1 > r_1 = \bar{\alpha} = p_1$, $d_3 = r_3 > p_3$, and hence $p_4 \geq d_4 = \delta' > a_4 = \delta$. Hence $p_4 = d_4$ by (4) and so $P = (\bar{\alpha}, \beta^*, -\bar{\alpha} - \beta^* - d_4, d_4)$. It follows that $P = S$ and there is only one imputation $P \in \mathcal{C}_1(\bar{\alpha}) \cap \mathcal{C}_2(\beta^*)$ with $p_3 < a_3$. Furthermore since D does not dominate A we have $a_3 \geq d_3$.

Interchanging the roles of Players 3 and 4 we see that if there is an imputation $P' \in \mathcal{C}_1(\bar{\alpha}) \cap \mathcal{C}_2(\beta^*)$ with $p_4' < a_4$, then $d_3 > a_3$, a contradiction. Therefore $\mathcal{C}_1(\bar{\alpha}) \cap \mathcal{C}_2(\beta^*)$ is the one element set $\{S\}$. Now S_{2-} is not $\{1, 2\}$ dominated by $\mathcal{V}$ since R is not. S_{2-} is not $\{2, 3, 4\}$ or $\{2, 4\}$ dominated by $\mathcal{V}$ since $s_4 = \delta' > \delta$. Therefore S_{2-} is $\{2, 3\}$ dominated by an imputation $T \in \mathcal{C}_2(\beta^*)$. Since $t_2 > r_2$ we have $t_1 \leq r_1 = \bar{\alpha}$, and hence $t_1 = \bar{\alpha}$ or α. Now $T \neq S$ and hence $t_1 \neq \bar{\alpha}$. Therefore $t_1 = \alpha$ which implies $T = A$. Now

$$t_1 + t_4 \geq u \geq r_1 + r_4 > a_1 + a_4 ,$$

a contradiction which establishes Lemma 6.

If $\mathcal{C}_1(\alpha) = \{A\}$, then, under certain conditions, $\mathcal{C}_1(\bar{\alpha})$ contains two imputations that play a role somewhat similar to the role of B and C in the other case. It will be convenient to use the same letters, B and C, to denote them.

LEMMA 7. If $\mathcal{C}_1(\alpha) = \{A\}$, and if either
(i) $\gamma = \gamma' > -1$ and $\delta = \delta' > -1$ or
(ii) $\gamma > \gamma'$, then $A \in \mathcal{A}$, $\mathcal{C}_1(\bar{\alpha}) \cap \mathcal{C}_3(\gamma)$ consists of a single imputation B, $B \in \bar{\mathcal{A}}$, $\mathcal{C}_1(\bar{\alpha}) \cap \mathcal{C}_4(\delta)$ consists of a single imputation C, $C \in \bar{\mathcal{A}}$, $b_3 > c_3$, $b_4 < c_4$, and

$$b_2 = c_2 = \operatorname*{Min}_{P \in \mathcal{C}_1(\bar{\alpha})} p_2 \quad . \tag{5}$$

PROOF. By Lemma 5 there is an imputation $R \in \mathcal{C}_1(\bar{\alpha}) \cap \mathcal{C}_2(\beta^*)$. Now suppose that $\mathcal{C}_1(\alpha) = \{A\}$ and that (i) or (ii) holds. Then $a_3 = \gamma \geq \gamma'$, $a_3 > -1$, $a_2 = \beta^*$, and A_{3-} must be $\{1, 3\}$ dominated by an imputation $S \in \mathcal{V}$. If $s_1 = \bar{\alpha}$ we put $B = S$. If $s_1 > \bar{\alpha}$, then $s_1 > r_1$, $a_3 = s_3 \leq r_3$, (3) yields $r_3 = a_3$, and we put $B = R$. In either case $\{1, 3\}$ is effective for B, $\bar{\alpha} + \gamma \leq t$, and $B \in \mathcal{C}_1(\bar{\alpha}) \cap \mathcal{C}_3(\gamma)$.

If (i) holds, then from (4) we obtain $b_4 \leq \delta$, and the hypothesis is symmetric in Players 3 and 4 so that $\bar{\alpha} + \delta \leq u$, and $\{1, 4\}$ is effective for B. If (ii) holds, then $b_3 > \gamma'$ and $b_2 > \beta$, so that $\{1, 4\}$ is properly effective for B. In either case we have $B \in \bar{\mathcal{A}}$.

Since B does not dominate A we have $b_4 \leq a_4$ and Lemma 6 yields $b_4 < a_4$. In particular $a_4 > -1$. By (3) and $a_2 = \beta^*$ we see that A_{4-} must be $\{1, 4\}$ dominated by an imputation $C \in \mathcal{V}$. Since $c_4 = a_4 > b_4$, we have $c_1 \leq b_1 = \bar{\alpha}$, and therefore $c_1 = \bar{\alpha}$. Thus $C \in \mathcal{C}_1(\bar{\alpha}) \cap \mathcal{C}_4(\delta)$. Using (3) we see that $c_1 + c_3 \leq \bar{\alpha} + \gamma \leq t$. Furthermore $a_1 + a_4 < c_1 + c_4 \leq u$ and $a_1 + a_3 < b_1 + b_3 \leq t$. Thus we have $C \in \bar{\mathcal{A}}$ and $A \in \mathcal{A}$.

Let $T \in \mathcal{C}_1(\bar{\alpha}) \cap \mathcal{C}_3(\gamma)$ with maximal t_2. Then $t_1 + t_3 = b_1 + b_3 \leq t$ and $t_2 + t_3 \geq b_2 + b_3 \geq -u$. Therefore T_{2-} cannot be $\{2, 4\}$ or $\{2, 3\}$ dominated. Clearly T_{2-} is not $\{2, 3, 4\}$ dominated by $\mathcal{C}_1(\alpha)$. It follows that either $t_2 = -1$ or T_{2-} is $\{1, 2\}$ dominated by $\mathcal{V}$. Therefore

$$t_2 = \operatorname*{Min}_{P \in \mathcal{C}_1(\bar{\alpha})} p_2 \quad .$$

It follows that $T = B$ and $\mathcal{C}_1(\bar{\alpha}) \cap \mathcal{C}_3(\gamma) = \{B\}$. Similarly $\mathcal{C}_1(\bar{\alpha}) \cap \mathcal{C}_4(\delta) = \{C\}$, and

$$c_2 = \operatorname*{Min}_{P \in \mathcal{C}_1(\bar{\alpha})} p_2 = b_2 \quad .$$

We have already $b_1 = c_1$ and $b_4 < a_4 = c_4$. Therefore $b_3 > c_3$ and the proof of Lemma 7 is complete.

LEMMA 8. If $\gamma > \gamma'$, then $\mathcal{C}_1(\alpha) = \{A\}$, $\mathcal{C}_2(\beta) = \{D\}$, $b_2 = c_2 < a_2$, and $d_3 \geq c_3$, where B and C are the imputations defined by Lemma 7.

PROOF. Suppose $\gamma > \gamma'$. Then by Lemmas 3 and 4 we have $\mathcal{C}_1(\alpha) = \{A\}$ and $\mathcal{C}_2(\beta) = \{D\}$. By Lemma 7, $b_1 = c_1$, $b_2 = c_2$, $b_3 > c_3$, and $b_3 > d_3$. Now B_{3-} is not $\{1, 3\}$ or $\{2, 3\}$ dominated by $\mathcal{V}$ since C is not, and B_{3-} is not $\{1, 3, 4\}$ dominated by D. Therefore B_{3-} is $\{2, 3, 4\}$ dominated by A and hence $a_2 > b_2$.

Suppose further that $c_3 > d_3$. Then $c_3 > -1$ and C_{3-} is neither $\{1, 3, 4\}$ dominated by D nor $\{2, 3, 4\}$ dominated by A. Since $C \in \bar{\mathcal{A}}$ it follows that C_{3-} is $\{1, 3\}$ dominated by an imputation $R \in \mathcal{V}$, $a_3 > r_3 = c_3 > d_3$, $R \neq D$, $r_2 > d_2$, and $\{1, 4\}$ is properly effective for R. Now $r_1 > c_1 = b_1$. Hence $r_2 \leq b_2 < a_2$ and $r_4 \leq b_4 < a_4$. Therefore $A \succ R$ with respect to $\{2, 3, 4\}$, a contradiction. This completes the proof of Lemma 8.

Interchanging Players 1 and 2 and Players 3 and 4 in Lemmas 7 and 8 we obtain the following:

LEMMA 9. If $\mathcal{C}_2(\beta) = \{D\}$, and if either (i) $\gamma = \gamma' > -1$ and $\delta = \delta' > -1$ or (ii') $\delta' > \delta$, then $D \in \bar{\mathcal{B}}$, $\mathcal{C}_2(\bar{\beta}) \cap \mathcal{C}_3(\gamma')$ consists of a single imputation that we denote by E, $E \in \bar{\mathcal{B}}$, $\mathcal{C}_2(\bar{\beta}) \cap \mathcal{C}_4(\delta')$ consists of a single imputation that we denote by F, $F \in \bar{\mathcal{B}}$, $e_3 > f_3$, $e_4 < f_4$, and

$$e_1 = f_1 = \min_{P \in \mathcal{C}_2(\bar{\beta})} p_1 \quad . \tag{6}$$

Furthermore $\delta' > \delta$ implies that $\mathcal{C}_2(\beta) = \{D\}$, $\mathcal{C}_1(\alpha) = \{A\}$, $e_1 = f_1 < d_1$, and $a_4 \geq e_4$.

We are now in a position to prove the main result of this section:

THEOREM 2. $\gamma = \gamma'$ and $\delta = \delta'$.

PROOF. Because of the symmetries between Players 1 and 2 and between Players 3 and 4 it is sufficient to prove the impossibility of $\delta' > \delta$. We suppose $\delta' > \delta$. Then, by Lemma 9, $\mathcal{C}_1(\alpha) = \{A\}$, $\mathcal{C}_2(\beta) = \{D\}$, $e_1 < d_1$, $a_4 \geq e_4$, $E \in \bar{\mathcal{B}}$, and $F \in \bar{\mathcal{B}}$. Furthermore $d_4 > a_4$, $d_1 = \alpha^* > a_1$, and since D does not dominate A we have $a_3 \geq d_3$. We distinguish two cases:

CASE 1. $a_3 = d_3$ and $a_4 < d_4$. Here $a_3 = e_3$, $a_4 \geq e_4$, and hence $a_1 + a_2 \leq e_1 + e_2 = f_1 + f_2$. Put $r_3 = -a_1 - a_2 - d_4$. Then $r_3 \geq -f_1 - f_2 - f_4 = f_3 \geq -1$, and therefore $R = (a_1, a_2, r_3, d_4)$ is an imputation. Since $d_4 > a_4$ it follows that $R \neq A$ and hence $R \notin \mathcal{V}$. Since $a_2 = \beta^*$ and $d_4 = \delta' > \delta$ it follows that R is $\{1, 3\}$ dominated by an imputation $S \in \mathcal{V}$. Now $s_3 > r_3 \geq f_3$ and $s_1 + s_3 \leq t \leq f_1 + f_3$. Therefore $s_1 < f_1$ and (6) yields $s_2 \neq \bar{\beta}$. Clearly $S \neq D$. Hence $s_2 > \bar{\beta} = f_2$. It follows that $\{1, 4\}$ is properly effective for S and thus $S \in \mathcal{A}$. Therefore $s_4 \leq a_4 < d_4$. Since $s_1 < f_1 < d_1$, we have $s_3 \geq d_3 = \gamma' = \gamma$. Now (3) yields $s_3 = d_3 = a_3$.

Suppose $s_1 > \bar{\alpha}$ and let T be any imputation of $\mathcal{C}_1(\bar{\alpha})$. We have $t_3 \geq s_3 = \gamma' = \gamma$ so that $t_3 = s_3$, $t_1 + t_3 < s_1 + s_3 \leq t$, and $T \vdash R$ with respect to $\{1, 3\}$. Thus without loss of generality we suppose $s_1 = \bar{\alpha}$. Lemma 6 now yields $s_4 \neq a_4$, and hence $s_4 < a_4$.

We see that $a_4 > -1$ and hence A_{4-} must be $\{1, 4\}$ dominated by an imputation $U \in \mathcal{V}$. Now $u_4 = a_4 > s_4$, and hence $u_1 \leq s_1 = \bar{\alpha}$, which implies $u_1 = \bar{\alpha}$. Thus $u_1 = \bar{\alpha} = s_1 < d_1$. Since $u_4 < d_4$ and D does not dominate U we have $u_3 \geq d_3 = a_3$. Applying (3) again we obtain $u_3 = a_3 = s_3$. Now $u_1 + u_3 = s_1 + s_3 \leq t$. Therefore $\{1, 3\}$, as well as $\{1, 4\}$, is effective for U, and $U \in \mathcal{A}$. This contradicts Lemma 6 and Case 1 is eliminated.

CASE 2. $a_3 > d_3$ and $a_4 < d_4$. Here (ii) of Lemma 7 and (ii') of Lemma 9 both hold. Furthermore Lemma 8 applies. Now $E \neq A$ and therefore $e_1 \geq \bar{\alpha} = b_1$. Similarly $B \neq D$ and $b_2 \geq \bar{\beta} = f_2$. We now have $c_1 = b_1 \leq e_1 < d_1$ and $c_4 = a_4 < d_4$. Hence, since D does not dominate C, we have $c_3 \geq d_3$. Comparison with Lemma 8 gives us $c_3 = d_3 = e_3$. Similarly $e_4 = a_4 = c_4$. Therefore $c_1 + c_2 = e_1 + e_2 = f_1 + f_2$ and $W = (c_1, c_2, f_3, f_4)$ is an imputation. Since $w_4 = \delta' > \delta$, it follows from (4) that W is not $\{1, 3, 4\}$, $\{2, 3, 4\}$, $\{1, 4\}$, or $\{2, 4\}$ dominated by $\mathcal{V}$. W is not $\{1, 2\}$ dominated by $\mathcal{V}$ since C is not. Now $c_2 = b_2 \geq f_2$ and F is not $\{2, 3\}$ dominated by $\mathcal{V}$. Hence W is not $\{2, 3\}$ dominated by $\mathcal{V}$. Thus either $W \in \mathcal{V}$ or W is $\{1, 3\}$ dominated by $\mathcal{V}$.

Suppose now that W is $\{1, 3\}$ dominated by an imputation $X \in \mathcal{V}$. Then $x_1 > c_1 = b_1$, so that $x_3 \leq c_3 < a_3$ and $x_2 \leq c_2 < a_2$. Therefore $x_4 \geq a_4 > b_4$ and $\{2, 3\}$ is properly effective for X. Since $D \in \mathcal{B}$ we have $X \neq D$ and hence $x_2 \geq \bar{\beta}$. Now $x_3 > f_3$ so that $x_2 \leq f_2 = \bar{\beta}$, and therefore $x_2 = \bar{\beta}$. By (6) this implies $x_1 \geq e_1 = f_1$. Now $x_1 + x_3 \leq t \leq f_1 + f_3$ which contradicts $x_1 \geq f_1$ and $x_3 > f_3$. Therefore W is not $\{1, 3\}$ dominated by $\mathcal{V}$. Thus $W \in \mathcal{V}$.

Now $w_4 = f_4 = d_4 > a_4$, $w_1 = \bar{\alpha} > a_1$, and W does not dominate

A. Therefore $\bar{\alpha} + d_4 > u$. Interchanging the roles of Players 1 and 2 and of Players 3 and 4 this becomes $\bar{\beta} + a_3 > -u$. From these two inequalities we obtain

$$(7) \qquad \bar{\alpha} + a_3 + \bar{\beta} + d_4 > 0 \quad .$$

Now $B \in \bar{\mathcal{A}}$ and $F \in \bar{\mathcal{B}}$. Therefore $\bar{\alpha} + a_3 = b_1 + b_3 \leq t$ and $\bar{\beta} + d_4 = f_2 + f_4 \leq -t$, which contradict (7). Thus Case 2 is also impossible. This completes the proof of Theorem 2.

§4. THE DISCRIMINATORY CASE

Suppose $\gamma = -1$. Then $\gamma' = -1$ by Theorem 2, and therefore $p_3 = -1$ for all $P \in \mathcal{V}$ by (3). Thus $\mathcal{V}$ lies entirely in one face of the tetrahedron of imputations. It follows, from the theory of the general-sum three-person game, that $t = -u = -2$, and that $\mathcal{V}$ is the well known three point solution that consists of $(-1, 1, -1, 1)$, $(1, -1, -1, 1)$, and $(1, 1, -1, -1)$.

Similarly if $\delta = -1$, then $t = -u = 2$ and $\mathcal{V}$ is the three point solution consisting of $(-1, 1, 1, -1)$, $(1, -1, 1, -1)$, and $(1, 1, -1, -1)$.

§5. B, C, E, F

In §3 we proved that $\gamma = \gamma'$ and $\delta = \delta'$. In §4 we treated the cases $\gamma = -1$ and $\delta = -1$. Thus in all remaining cases we have

$$(8) \qquad \gamma = \gamma' > -1, \qquad \delta = \delta' > -1 \quad .$$

We will suppose, throughout this section and §6, that (8) holds. Then (3) and (4) simplify to

$$(9) \qquad \gamma = \operatorname*{Max}_{P \in \mathcal{V}} p_3, \qquad \delta = \operatorname*{Max}_{P \in \mathcal{V}} p_4 \quad .$$

We note that (8) is the condition (i) of Lemmas 7 and 9. By Lemma 7, the paragraph following (2), Lemma 9, and the paragraph preceding Lemma 4 we see that the imputations B, C, E, and F are defined and belong to $\mathcal{V}$, and

$$b_1 = c_1 = \begin{cases} \alpha & \text{if } \mathcal{C}_1(\alpha) = \{B, C\} \\ \bar{\alpha} & \text{if } \mathcal{C}_1(\alpha) = \{A\} \end{cases} ,$$

$$e_2 = f_2 = \begin{cases} \beta & \text{if} \quad \mathcal{C}_2(\beta) = \{E, F\} \\ \bar{\beta} & \text{if} \quad \mathcal{C}_2(\beta) = \{D\} \end{cases},$$

$b_3 = e_3 = \gamma$, $c_4 = f_4 = \delta$, $b_3 > c_3$, and $e_3 > f_3$. Furthermore from Lemma 2 and (5) and (6) we obtain

$$(10) \qquad b_2 = c_2 = \operatorname*{Min}_{P \in \mathcal{C}_1(b_1)} p_2, \qquad e_1 = f_1 = \operatorname*{Min}_{P \in \mathcal{C}_2(e_2)} p_1 .$$

LEMMA 10. If $B = E$, then $C = F$. If $B \neq E$, then $b_1 < e_1$ and $b_2 > e_2$.

PROOF. We observe that b_1 is always the smallest number such that $\mathcal{C}_1(b_1)$ contains more than one imputation. Hence $e_1 \geq b_1$. Similarly $b_2 \geq e_2$. It follows from (10) that $e_1 = b_1$ if and only if $e_2 = b_2$. Thus, since $b_3 = e_3$, we see that either $e_1 = b_1$ or $e_2 = b_2$ implies $B = E$. Therefore $B \neq E$ implies $e_1 > b_1$ and $b_2 > e_2$. Finally if $B = E$, then $c_1 = f_1$, $c_2 = f_2$, and $c_4 = f_4$, which imply $C = F$.

LEMMA 11. $b_4 = e_4$ and $c_3 = f_3$.

PROOF. Because of the symmetries between Players 1 and 2 and between Players 3 and 4 it is sufficient to prove the impossibility of $b_4 > e_4$. Suppose $b_4 > e_4$.

Let R be any imputation of $\mathcal{C}_4(b_4)$. Since $e_4 < r_4 = b_4 < \delta'$, it follows that $r_2 > \beta$, and that R_{4-} is not $\{1, 3, 4\}$ dominated by $\mathcal{C}_2(\beta)$. Furthermore $r_2 \geq e_2$. R_{4-} is not $\{2, 4\}$ dominated by $\mathcal{V}$ since E is not. Therefore R_{4-} is either $\{1, 4\}$ dominated by $\mathcal{V}$ or $\{2, 3, 4\}$ dominated by $\mathcal{C}_1(\alpha)$.

Now $B \in \mathcal{C}_4(b_4)$ and $b_3 = \gamma$. Therefore B_{4-} is $\{1, 4\}$ dominated by an imputation $S \in \mathcal{C}_4(b_4)$. We suppose S chosen with s_1 maximal subject to the conditions $S \in \mathcal{V}$, $S \succ B_{4-}$. By the maximality of s_1 it follows that S_{4-} is not $\{1, 4\}$ dominated by $\mathcal{V}$. Hence S_{4-} is $\{2, 3, 4\}$ dominated by $\mathcal{C}_1(\alpha) \cap \mathcal{C}_4(b_4)$. Since $b_4 < \delta$, this implies that $\mathcal{C}_1(\alpha) = \{B, C\}$ and that S_{4-} is $\{2, 3, 4\}$ dominated by B. Lemmas 4 and 9 give us $\mathcal{C}_2(\beta) = \{D\}$ and $E \in \mathcal{B}$. Now $s_4 = b_4 < d_4$ and $s_3 < b_3 = d_3$. Therefore $s_1 \geq d_1 \geq e_1$. Since $\{1, 4\}$ is effective for S and $E \in \mathcal{B}$ we obtain

$$e_1 + e_4 \geq u \geq s_1 + s_4 \geq e_1 + b_4 ,$$

which contradicts the assumption $b_4 > e_4$.

LEMMA 12. No interior points of the straight line segment EF belong to $\mathcal{V}$.

PROOF. Since F is not dominated by $\mathcal{V}$ it follows that E_{3-} is neither {1, 3} nor {2, 3} dominated by $\mathcal{V}$. Hence either E_{3-} is {2, 3, 4} dominated by $\mathcal{C}_1(\alpha)$ or {1, 3, 4} dominated by $\mathcal{C}_2(\beta)$. By Lemma 11 we have $b_4 = e_4$ and $c_3 = f_3 < e_3$. Hence E_{3-} is not {2, 3, 4} dominated by B or C. Therefore if E_{3-} is {2, 3, 4} dominated by $\mathcal{C}_1(\alpha)$ then $\mathcal{C}_1(\alpha) = \{A\}$, $a_3 = \gamma = e_3$, $a_4 = \delta = f_4$, and A dominates all interior points of the straight line segment EF. If E_{3-} is {1, 3, 4} dominated by $\mathcal{C}_2(\beta)$, then $\mathcal{C}_2(\beta) = \{D\}$ and D dominates all interior points of the segment EF.

LEMMA 13. $\alpha^* + \beta^* \geq 2$.

PROOF. Suppose $\alpha^* + \beta^* < 2$. Then, by Lemma 1, $R = (\alpha^*, 2 - \alpha^*, -1, -1)$ is an imputation undominated by $\mathcal{V}$. Therefore $R \in \mathcal{V}$. Since $r_1 > \alpha$ and $r_2 > \beta^*$, R dominates the imputations of $\mathcal{C}_1(\alpha)$, which is impossible.

§6. THE CASE B = E

In this section we will treat the case B = E. Because of the symmetry between Players 1 and 2, it is sufficient to discuss this case in detail only when $\mathcal{C}_1(\alpha) = \{A\}$.

Suppose B = E and $\mathcal{C}_1(\alpha) = \{A\}$. Then C = F by Lemma 10, and B and C are elements of $\bar{\mathcal{A}}$ by Lemma 7. We have

$$u \geq c_1 + c_4 > b_1 + b_4 = e_1 + e_4 .$$

It follows that $E \notin \bar{\mathcal{B}}$. Therefore $\mathcal{C}_2(\beta) = \{E, F\} = \{B, C\}$ by Lemma 9. If $b_4 > -1$, then, since $b_3 = \gamma$ and $b_1 = \alpha^*$, B_{4-} must be {2, 4} dominated by $\mathcal{V}$, which is impossible because $B \in \bar{\mathcal{A}}$. Therefore $b_4 = -1$. If $b_2 > -1$, then B_{2-} is {2, 4} dominated by $\mathcal{V}$ which is also impossible. Therefore $b_2 = -1$. Similarly $c_3 = -1$. Now $-t \leq b_2 + b_4 = -2$ and $-u \leq c_2 + c_3 = -2$. Therefore $t = u = 2$. By Lemma 5 there exists an imputation $R \in \mathcal{C}_1(b_1) \cap \mathcal{C}_2(a_2)$. Hence $\alpha^* + \beta^* = b_1 + a_2 \leq 2$. Therefore $b_1 + a_2 = 2$ by Lemma 13. It follows that $R = (b_1, a_2, -1, -1) \in \mathcal{V}$. Since $A = (-1, a_2, a_3, a_4)$, $B = (b_1, -1, a_3, -1)$, and $C = (b_1, -1, -1, a_4)$ we have $a_2 = a_3 = a_4 = 1/3$ and $b_1 = 5/3$.

It can be shown directly [1] that the four imputations $(-1, 1/3, 1/3, 1/3)$, $(5/3, 1/3, -1, -1)$, $(5/3, -1, 1/3, -1)$, and

(5/3, - 1, - 1, 1/3) are a solution if $t = u = 2$.

Interchanging Players 1 and 2 we obtain the corresponding four point solution on the vertex $t = u = -2$. Thus we see that if $B = E$, then $t = u = \pm 2$ and $\mathcal{V}$ is the well known four point solution described by von Neumann and Morgenstern.

§7. THE CASE $B \neq E$

In order to treat the remaining case, $B \neq E$, we need the equality $\alpha^* + \beta^* = 2$. To establish this we need two additional lemmas.

LEMMA 14. If $\alpha^* + \beta^* > 2$, then there does not exist an imputation $P \in \mathcal{V}$ such that $p_1 > b_1$, $p_2 > e_2$, $p_3 < \gamma$, $p_4 < \delta$.

PROOF. Suppose $\mathcal{V}$ contains such an imputation P. Without loss of generality we suppose further that $\mathcal{C}_1(\alpha) = \{A\}$. Then $p_3 < \gamma = a_3$, $p_4 < \delta = a_4$, and therefore $p_2 = a_2 = \beta^*$. Hence $p_1 = -p_2 - p_3 - p_4 \leq 2 - \beta^* < \alpha^*$. Now $\mathcal{C}_2(\beta) = \{D\}$ implies $D \succ P$, and therefore $\mathcal{C}_2(\beta) = \{E, F\}$. It follows that B_{3-} must be {2, 3, 4} dominated by A and $b_2 < a_2 = p_2$. Hence $P \succ B$ with respect to {1, 2}, a contradiction.

LEMMA 15. If $\alpha^* + \beta^* > 2$ and $B \neq E$, then $\mathcal{C}_2(b_2) \cap \mathcal{C}_1(e_1)$ is empty.

PROOF. Let $R \in \mathcal{C}_2(b_2) \cap \mathcal{C}_1(e_1)$. From Lemma 10 we obtain $r_1 = e_1 > b_1$ and $r_2 = b_2 > e_2$. By Lemma 14 we have either $r_3 = \gamma$ or $r_4 = \delta$. Since $\alpha^* + \beta^* > 2$ we have either $r_1 < \alpha^*$ or $r_2 < \beta^*$. Without loss of generality suppose $r_3 = \gamma$ and $r_2 < \beta^*$. Then $b_2 < \beta^*$ which implies $\mathcal{C}_1(\alpha) = \{A\}$. By Lemma 5 there exists an imputation $S \in \mathcal{C}_1(\bar{\alpha}) \cap \mathcal{C}_2(\beta^*)$. Now $S \neq B$ and $S \neq C$. Therefore, by Lemma 7, $s_3 < \gamma$ and $s_4 < \delta$. Since $s_1 = b_1 < e_1 \leq \alpha^*$ and $\mathcal{C}_2(\beta)$ does not dominate S we must have $\mathcal{C}_2(\beta) = \{E, F\}$. Now $s_1 < e_1$, $s_3 < e_3$, and hence $b_4 = e_4 \leq s_4$. We have $r_3 = \gamma$ and $r_1 = e_1 = \alpha^*$. Therefore

$$0 = r_1 + r_2 + r_3 + r_4 \geq \alpha^* + r_2 + \gamma - 1$$

or

$$\alpha^* \leq 1 - r_2 - \gamma \quad .$$

Furthermore, since $s_1 = b_1$ and $s_4 \geq b_4$, we have

$$r_2 + \gamma = b_2 + b_3 \geq s_2 + s_3 \geq \beta^* - 1$$

or

$$\beta^* \leq 1 + r_2 + \gamma \quad .$$

It follows at once that $\alpha^* + \beta^* \leq 2$, a contradiction.

THEOREM 3. $\alpha^* + \beta^* = 2$.

PROOF. If $\gamma = -1$ or $\delta = -1$, then the discussion of §4 gives us $\alpha^* = \beta^* = 1$. If $B = E$, then by §6, we have $\alpha^* + \beta^* = 5/3 + 1/3 = 2$.

Suppose $\alpha^* + \beta^* \neq 2$. Then $\gamma = \gamma' > -1$, $\delta = \delta' > -1$, and $B \neq E$. From Lemma 13 we obtain $\alpha^* + \beta^* > 2$. Thus Lemmas 14 and 15 apply. E_{1-} is $\{1, 2\}$ dominated by an imputation $R \in \mathcal{V}$, $r_1 = e_1 > b_1$, $b_2 \geq r_2 > e_2$. Lemma 15 yields $r_2 < b_2$, and by Lemma 14 we have either $r_3 = \gamma$ or $r_4 = \delta$.

Suppose, for the moment, that $r_3 = \gamma$. Then $r_3 = e_3$, and hence $r_4 < e_4$. E_{4-} is not $\{1, 4\}$ dominated by $\mathcal{V}$ since R is not. Hence, since $e_3 = \gamma$, E_{4-} is $\{2, 4\}$ dominated by an imputation $S \in \mathcal{V}$, $s_4 = e_4 > r_4$, $b_2 > r_2 \geq s_2 > e_2$, $s_1 \geq b_1$. From (10) we obtain $s_1 > b_1$. Now $s_4 = e_4 < \delta$ and Lemma 14 gives us $s_3 = \gamma = e_3$. Therefore $s_1 < e_1 = r_1$. Since R does not dominate S we have $s_2 = r_2$. Now R_{1-} is not $\{1, 2\}$ dominated by $\mathcal{V}$ since S is not. Therefore, since $r_3 = \gamma$, it follows that R_{1-} is $\{1, 4\}$ dominated by an imputation $T \in \mathcal{V}$, $t_1 = r_1 = e_1 > s_1 > b_1$, $t_2 \leq s_2 < b_2$, $r_4 < t_4 \leq s_4 < \delta$. Since $t_4 < \delta$ we have $t_2 \geq e_2$. If $t_3 < \gamma$, then Lemma 14 yields $t_2 = e_2 = f_2$ and T is an interior point of the straight line segment EF, contradicting Lemma 12. It follows that $t_3 = \gamma = e_3 = r_3$ and therefore $t_2 < r_2 = s_2$. Since $\{2, 4\}$ is effective for S we have $t_4 \geq s_4$. Hence $t_4 = s_4 = e_4$. It follows that $T = E$, and therefore $e_1 + e_4 \leq u$. Furthermore, since $b_2 > r_2$ and $b_4 = e_4 > r_4$, we have $b_2 + b_4 > -t$.

We have now established that $r_3 = \gamma$ implies

$$e_1 + e_4 \leq u \quad \text{and} \quad b_2 + b_4 > -t \quad . \tag{11}$$

Interchanging Players 3 and 4 we see that $r_4 = \delta$ implies

$$f_1 + f_3 \leq t \quad \text{and} \quad c_2 + c_3 > -u \quad . \tag{12}$$

Since either $r_3 = \gamma$ or $r_4 = \delta$, we see that either (11) or (12) holds. Now interchanging the roles of Players 1 and 2 we must have either

$$b_2 + b_4 \leq -t \quad \text{and} \quad e_1 + e_4 > u \tag{13}$$

or

(14) $$c_2 + c_3 \leq - u \quad \text{and} \quad f_1 + f_3 > t \quad .$$

If (11) and (14) each hold, then

$$e_1 + c_2 + c_3 + e_4 < f_1 + b_2 + f_3 + b_4 \quad ,$$

which is impossible since $e_1 = f_1$, $c_2 = b_2$, $c_3 = f_3$, and $e_4 = b_4$. Thus (11) contradicts both (13) and (14). Similarly (12) contradicts both (13) and (14). These contradictions establish Theorem 3.

It follows from Theorem 3 that $G = (\alpha^*, \beta^*, -1, -1)$ is an imputation. G is clearly undominated by $\mathcal{V}$. Therefore $G \in \mathcal{V}$ and $\mathcal{V} = \mathcal{C}_1(\alpha^*) \cup \mathcal{C}_2(\beta^*)$ in all cases.

As a result of the last few sections there remains only the case $\gamma > -1$, $\delta > -1$, $B \neq E$ to be treated. We assume, throughout the remainder of this section, that $\gamma > -1$, $\delta > -1$, and $B \neq E$. Then by Lemma 10 we have $b_1 < e_1 \leq \alpha^*$ and hence $b_2 = \beta^*$. It follows that B_{3-} is $\{1, 3, 4\}$ dominated by $\mathcal{C}_2(\beta)$ which implies $\mathcal{C}_2(\beta) = \{D\}$. Similarly $e_1 = \alpha^*$ and $\mathcal{C}_1(\alpha) = \{A\}$. The Corollary of Theorem 1 gives us

(a) $$a_1 = d_2 = -1 \quad .$$

Hence $A = (-1, \beta^*, \gamma, \delta)$ and $D = (\alpha^*, -1, \gamma, \delta)$. Thus $\alpha^* = \beta^* = 1$ and we have

(b) $$d_1 = e_1 = f_1 = g_1 = a_2 = b_2 = c_2 = g_2 = -g_3 = -g_4 = 1 \quad .$$

Now if $P \in \mathcal{V}$, $P \neq G$, then we have either $p_1 < 1 = d_1$ or $p_2 < 1 = a_2$. Therefore, since neither A nor D can dominate P we have either $p_3 = \gamma$ or $p_4 = \delta$. Hence

(15) $$\mathcal{V} = \{G\} \cup \mathcal{C}_3(\gamma) \cup \mathcal{C}_4(\delta) \quad .$$

Now suppose $c_3 > -1$. Then C_{3-} is $\{1, 3\}$ dominated by an imputation $R \in \mathcal{V}$, $r_1 > c_1$, $r_3 = c_3 = f_3 < \gamma$. By (15) we have $r_4 = \delta = c_4 = f_4$. Hence $r_2 < c_2 = \beta^*$ and thus $\underline{r}_1 = \alpha^* = f_1$. Therefore $R = F$, and $f_1 + f_3 \leq t$. By Lemma 9, $F \in \mathcal{B}$ which implies $f_1 + f_3 \geq t$. Therefore $f_1 + f_3 = t$. Now $f_3 = c_3 > -1$, and interchanging the roles of Players 1 and 2 we obtain $c_2 + c_3 = -u$. Hence $t = f_1 + f_3 = 1 + c_3 = c_2 + c_3 = -u$. Thus we have established that $c_3 > -1$ implies

(16) $$t = -u = 1 + c_3 \quad .$$

Interchanging Players 3 and 4 we see that $b_4 > -1$ implies

(17) $$u = -t = 1 + b_4 .$$

If $c_3 > -1$ and $b_4 > -1$ both hold, then adding (16) and (17) we obtain $0 = 2 + c_3 + b_4$, a contradiction. Therefore either $c_3 = -1$ or $b_4 = -1$.

If $c_3 = b_4 = -1$, then, since $E, F \in \bar{\mathcal{B}}$ and $B, C \in \bar{\mathcal{J}}$, we have

$$t \leq f_1 + f_3 = 1 + c_3 = 0 = -b_2 - b_4 = b_1 + b_3 \leq t ,$$

and

$$-u \leq c_2 + c_3 = 0 = -e_1 - e_4 = e_2 + e_3 \leq -u .$$

It follows that $c_3 = b_4 = -1$ implies both (16) and (17).

We now see that $c_3 = -1$ implies (17), and that $b_4 = -1$ implies (16). Hence, in both cases, $c_3 - b_4 = t = -u$. Now $b_1 = c_1$ and $b_2 = c_2$. Therefore $\gamma + b_4 = b_3 + b_4 = c_3 + c_4 = c_3 + \delta$. Thus $\gamma - \delta = c_3 - b_4 = t$. Since $A = (-1, 1, \gamma, \delta)$ we have $\gamma + \delta = 0$. Therefore $\gamma = -\delta = t/2$, and

(c) $$a_3 = b_3 = d_3 = e_3 = t/2 ,$$

(d) $$a_4 = c_4 = d_4 = f_4 = -t/2 .$$

If $c_3 = -1$, then $b_4 = -1 - t \geq -1$, $t \leq 0$, and $\bar{\alpha} = c_1 = -c_2 - c_3 - c_4 = t/2$. If $b_4 = -1$, then $c_3 = -1 + t \geq -1$, $t \geq 0$, and $\bar{\alpha} = b_1 = -b_2 - b_3 - b_4 = -t/2$. Hence, in both cases, $\bar{\alpha} = -|t|/2$. Now $\bar{\alpha} > a_1 = -1$ and thus $|t| < 2$. This gives us $-2 < t = -u < 2$.

Now $B = (\bar{\alpha}, 1, b_3, b_4)$ and $E = (1, \bar{\beta}, b_3, b_4)$. Hence $\bar{\beta} = \bar{\alpha} = -|t|/2$. Thus

(e) $$b_1 = c_1 = e_2 = f_2 = -|t|/2 .$$

Finally we have

(f) $$b_4 = e_4 = \text{Max}\{-1, -1 - t\}$$

and

(g) $$c_3 = f_3 = \text{Max}\{-1, -1 + t\} .$$

We have now shown that $-2 < t = -u < 2$ and that $\mathcal{V}$ contains the seven points A, B, C, D, E, F, G which are given by (a), (b), (c), (d), (e), (f), and (g). A straightforward argument shows that if $-2 < t = -u < 2$, then these seven points do indeed form a solution of $\mathcal{V}$.

§8. SUMMARY

Summarizing our results we see that the game Γ has at most one finite solution. If Γ has a finite solution $\mathcal{V}$, then either $t = u = \pm 2$ or $t = -u$.

If $t = u = \pm 2$, then $\mathcal{V}$ is the four point von Neumann-Morgenstern solution described in §6. If $t = -u = \pm 2$, then $\mathcal{V}$ is the three point von Neumann-Morgenstern discriminatory solution discussed briefly in §4. If $-2 < t = -u < 2$, then $\mathcal{V}$ is the seven point solution obtained in §7. It is interesting to note that as t approaches 2 or -2, the seven point solution approaches the appropriate three point discriminatory solution.

It is also worthwhile observing that, in all three cases, each point of $\mathcal{V}$ lies either on one of the faces of the imputation tetrahedron or on one of the planes $p_1 + p_3 = t$, $p_1 + p_4 = u$. Thus each imputation of $\mathcal{V}$ corresponds to the formation of two complementary coalitions. In each of these two coalitions the profit, or loss, is divided in a prescribed manner.

BIBLIOGRAPHY

[1] VON NEUMANN, J., and MORGENSTERN, O., Theory of Games and Economic Behavior, Princeton 1944, 2nd ed. 1947.

[2] MILLS, W. H., "The four person game - edge of the cube," Annals of Mathematics 59 (1954), pp. 367-378.

W. H. Mills

Yale University

THE SOLUTIONS OF A SYMMETRIC MARKET GAME

L. S. Shapley

INTRODUCTION

This paper is directed to the problem of determining the full sets of solutions to certain multi-person games that display a rudimentary competitive pattern typical of many economic models. The players are divided into two groups M and N — we may think of them as buyers and sellers of some commodity — and the payoff functions are so constructed that players of opposite types are complementary (i.e., can enter into mutually profitable arrangements), while players of the same type are not; in fact, being perfectly interchangeable in coalitions, they find themselves in relentless competition for the chance to "do business" with their opposites. The number of players of each type is unrestricted.

Our highly symmetrical characteristic function:

$$v(S) = \min(|S \cap M|, \; |S \cap N|) \tag{1.1}$$

(the number of elements in a set X is denoted by $|X|$), emphasizes the basic complementarity/substitutability pattern to the exclusion of other features of the market process that might have been included, such as asymmetrical resources, elastic demand functions, indivisible goods, etc.,[1] and the rather exceptional regularity that we shall observe in the solution sets reflects the structural simplicity of (1.1). Nevertheless, our close analysis of a special class of market games can be expected to point the way for more general types. Also, quite apart from the remote practical significance of the results, the conquest of this class of large-sized games should prove of some theoretical interest, especially since a technical device of considerable generality is developed along the way.[2]

[1] Characteristic functions embodying a number of these factors are formulated and discussed briefly in [9] (see the bibliography at the end of the paper); we intend to treat them more fully in a future publication.

[2] We refer to the "skew sets", defined in §§5 and 7 below.

As Gillies remarked on a similar occasion: "The intensive study of a particular class of games provides empirical data on the nature of solutions, methods which may be applied to other games, and may suggest or disprove conjectures on solutions in general."[3]

Surprisingly, large market games have been generally neglected by game theorists since the initial work of von Neumann and Morgenstern on the subject.[4] Markets meet the underlying assumptions of complete information, transferable utility, etc., better than most economic phenomena, and ought to provie the material for some good tests of the von Neumann-Morgenstern solution theory. A really decisive confrontation of the theory is not easy to arrange, but in such well-suited applications, as opposed to more artificially-derived examples, one feels that the critics are fully justified in insisting that the solutions make economic sense: that they satisfy or perhaps extend — but do not contradict — the expectations and intuitions based on observed experience.

The one-parameter sets of imputations that make up our solutions are closely related to the "bargaining curves" that have been observed in many other game solutions.[5] The parameter in the present case can be interpreted as the average net market price of the commodity, or (transformed) as the total profit of the sellers as a group. It varies continuously in each solution from zero profit to the sellers ("cut throat" pricing) to zero gain for the buyers ("all the traffic will bear"). At any particular parameter value, the solution tells exactly how the individual gains are to be imputed among the players. In other words, if the solution is known and the average price is known, then the outcome, financially speaking, is completely determined.

There is just one symmetrical solution; it corresponds to the "free trade" or "same-price-to-all-comers" standard of behavior. More generally, the solution can be regarded as a description of the institutionalized modes of collusion — premiums, rebates, class discrimination,

---

3 [3], p. 325.

4 [13], §64; see also [9], [10], [12]. Our (1.1) can be obtained from (64:6) of [13] by specializing the parameters of the latter.

5 See [3], [5], [7], and [13], §§47.8.2 and 62.6.2; also [4], p. 209. Bargaining curves are commonly found combined with other point-sets in the solution; the purity of the present occurrences can be ascribed to the very direct "complementary/substitutable" structure of (1.1). There is a close connection between the latter and the (m+n)-person simple game $B_m^* \times B_n^*$, with (non-superadditive) characteristic function:

$$v(S) = \min(|S \cap M|, |S \cap N|, 1) \quad ,$$

whose solutions are <u>all</u> of the bargaining curves connecting the zero faces of the two groups of players (see [8], [11]).

boycotts, etc. — in terms of their net effect on the outcome. A nonsymmetric solution expresses a stable, self-consistent departure from the "free trade" norm.

Meanwhile, the basic function of the actual market institution — that of establishing and maintaining the general price equilibrium — must presumably be carried out by means of "noncollusive" bargaining tactics: bids, prices, concessions, counter-proposals, etc., insofar as they are available among the formally permitted moves of the game. This sets the stage for a rather remarkable division of labor between "cooperative" game theory and its "noncooperative" cousin. We make no attempt in the present paper to solve the noncooperative bargaining problems posed by our cooperative solutions; to do so would require preconstructing the strategy spaces and payoff functions that underlie (1.1), and would lead us far afield. But the possibility of being able to "solve the solution" is not an unnatural one, when we reflect on how much of the problem is left out of the cooperative approach: most of the formal bargaining moves of the extensive-form game are rendered superfluous by the added, free coalition-forming process implicit in the characteristic function; likewise, all of the details of price and money transfer are swallowed up by the hypothesis of unrestricted side-payments. Indeed, the fact that something identifiable as "average price" appears as a parameter in the solutions, after so much has been apparently suppressed, speaks well for the validity of the cooperative solution concept.

Our formal results fall short of a complete list of solutions, such as obtained by von Neumann and W. H. Mills on comparable occasions;[6] however, we bring that ultimate goal within reach. We prove that all solutions are monotonic arcs spanning the simplex of imputations, as discussed above; we also obtain a bound on their location in the simplex and determine explicitly a major, centrally-located subclass of solutions, which in certain cases turns out to be complete (see below, Theorems 3, 2, 1, respectively). As already noted, there is a unique solution possessing the full symmetry of the game.[7]

The main body of this paper is concerned exclusively with the mathematical problem.[8] We have tried to keep the presentation self-contained, and have adopted a mildly expository tone at first, with the idea of easing the way for readers not versed in the intricacies of solution theory.

---

[6] See [13], §55, and [4].

[7] This corresponds to Bott's result for the (n,k)-games [1], [3] — another comparable, but incompletely solved, class of large-sized games.

[8] The heuristic account of the game and its solutions, given in [10], still applies for the most part, although the mathematical results therein are now superseded.

§1. PRELIMINARIES

Let M and N be the two groups of players, having respectively $m = |M|$ and $n = |N|$ elements. The $(m+n)$-person game to be considered is given by the characteristic function:

$$v(S) = \min(|S \cap M|, |S \cap N|), \qquad \text{all } S \text{ in } M \cup N. \tag{1.1}$$

Define $g = v(M \cup N) = \min(m, n)$. Vectors on $M \cup N$ will be written as

$$x \text{ or } \langle x'; x'' \rangle \text{ or } (x_1', \ldots, x_m'; x_1'', \ldots, x_n'') .$$

Sums over their components will generally be abbreviated

$$x(S) \quad \text{for} \quad \sum_{S \cap M} x_\mu' + \sum_{S \cap N} x_\nu'' .$$

An <u>imputation</u> is a nonnegative vector on $M \cup N$ such that $x(M \cup N) = g$; the space of all imputations is a simplex of $m + n - 1$ dimensions, denoted by A. The "face" A_S is the $(|S| - 1)$-dimensional set of imputations x such that $x(S) = g$; these are the vectors that impute a total of g to the members of S, and nothing to the other players. The opposing complementary faces A_M and A_N will figure prominently in our analysis.

In general, x is said to <u>dominate</u> y <u>via</u> S provided that

$$x - y \text{ is strictly positive on } S, \tag{1.2}$$

and

$$x(S) \leq v(S) . \tag{1.3}$$

In the present instance it will suffice to consider only domination via sets of the form $\{\mu, \nu\}$, $\mu \in M$, $\nu \in N$, since other dominations always imply a domination of this kind. Then (1.2) and (1.3) become simply:

$$x_\mu' > y_\mu' \quad \text{and} \quad x_\nu'' > y_\nu'' , \tag{1.4}$$

$$x_\mu' + x_\nu'' \leq 1 . \tag{1.5}$$

The <u>dominion</u> of a set X of imputations, written "dom X", is the (open) set of imputations dominated by elements of X. The set A - dom A of undominated imputations is called the <u>core</u> of the game; it comprises just those imputations x in which all coalitions are "satisfied": $x(S) \geq v(S)$.

A solution of the game is defined to be any set V of imputations that dominates its complement in A, and nothing else: $V = A - \text{dom } V$. Every solution, being the complement of a dominion, is a closed set and contains the core; no solution contains another. The two properties:

$$V \cap \text{dom } V = \emptyset \qquad \text{and} \qquad V \cup \text{dom } V = A \quad ,$$

that combine to characterize a solution, will be referred to sometimes as internal and external stability, respectively.

LEMMA 1. No solution of (1.1) contains an open set, unless $m = n = 1$.

PROOF. For every $x \in A$, there is a pair μ^*, ν^* for which (1.5) holds; otherwise we could sum g inequalities of the form $x'_\mu + x''_\nu > 1$, involving $2g$ distinct indices, and obtain an absurdity $x(S) > g$. Furthermore, if x is interior to A, every neighborhood of x will contain a y for which (1.4) holds, with respect to μ^*, ν^*, unless $m = n = 1$. Hence every open set in A is internally unstable. But no solution can contain an internally unstable subset. QED

§2. THE CASE $m = n$

The case where M and N are of equal size can be disposed of quickly. Let V be the set of imputations of the form:

$$(2.1) \qquad z_p = (p, \ldots, p; 1 - p, \ldots, 1 - p), \qquad 0 \leq p \leq 1.$$

For any $S \subset M \cup N$ we have:

$$\begin{aligned} z_p(S) &= p|S \cap M| + (1 - p)|S \cap N| \\ &\geq \min(|S \cap M|, |S \cap N|) = v(S) \quad . \end{aligned}$$

Thus z_p is undominated, and V is at least contained in the core. On the other hand, any $x \in A$ not in V must have $x'_{\mu^*} + x''_{\nu^*} < 1$ for at least one pair μ^*, ν^*, and hence is dominated via $\{\mu^*, \nu^*\}$ by some $y \in A$. Thus V is precisely the core. What is more, we can take y in the preceding argument to be one of the elements of V, namely z_{p^*} where p^* satisfies

$$x'_{\mu^*} < p^* < 1 - x''_{\nu^*} \quad .$$

We conclude that V dominates all of $A - V$, making V both the core and a solution of the game. Under these circumstances V is necessarily the unique solution. Geometrically, V is the straight line joining the centers of gravity of the opposing faces A_M and A_N.

§3. THE CASE min(m, n) = 1

The case of monopoly or monopsony,[9] $\min(m, n) = g = 1$, will now be considered. We may assume $m = 1, n > 1$. Since v takes on only the values 0 and 1 we have what is called a simple game: the "winning" coalitions are those consisting of the single member of M and one or more players from N. This game happens to be factorable into one-person simple games, and a complete description of its solutions is therefore available (see [11]; also footnote 5 above). They turn out to be monotonic curves running from the face A_N to the opposite vertex of the simplex. Stated precisely, a solution is any set of points of the form:

$$z_p = (p; f_1(p), \ldots, f_n(p)), \qquad 0 \leq p \leq 1 , \tag{3.1}$$

where the functions $\{f_\nu\}$ satisfy $f_\nu \geq 0$, $\Sigma f_\nu(p) = 1 - p$, and are continuous and nonincreasing. In contrast with the preceding case, the core, which consists of the single imputation $(1; 0, \ldots, 0)$, does not dominate even a part of its complement.

The proof that the curves (3.1) are solutions, and that they are the only solutions, is omitted in deference to the more general results proved later on (see Theorem 1 and the second Corollary to Theorem 3). Figure 1 illustrates two solutions for the 3-person case[10] $m = 1, n = 2$;

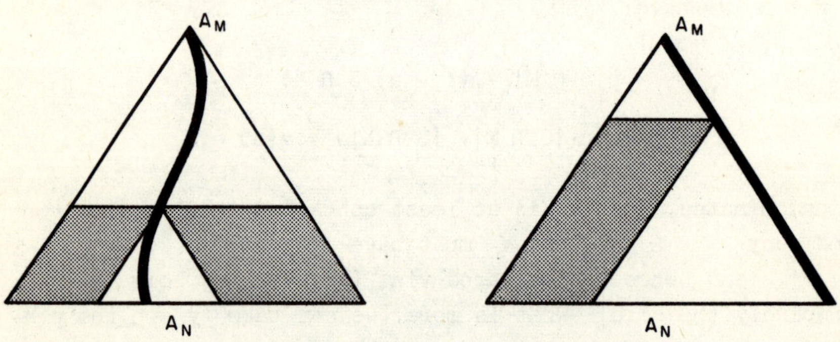

FIGURE 1

[9] Compare [13], §64.2.2.

[10] These solutions are included, of course, in the complete analysis of general-sum 3-person games in [13], §§60.3 and 47.5.

the shaded areas represent the dominions of typical points on the curves, and make it more or less apparent why the functions f_ν must be monotonic to avoid internal instability — i.e., self-domination.

§4. THE GENERAL CASE

We now drop the restrictions on m and n. By a <u>monotonic arc</u> we shall mean a one-parameter family of imputations of the form:

$$(4.1) \qquad z_p = (f'_1(p), \ldots, f'_m(p); f''_1(p), \ldots, f''_n(p)) \qquad \text{all } p \in R,$$

where R is some real interval, and the functions f'_μ, f''_ν are continuous, nonnegative, and respectively nondecreasing and nonincreasing. Without loss of generality, we can choose the parameter p so that

$$(4.2) \qquad \sum_M f'_\mu(p) = p, \qquad \sum_N f''_\nu(p) = g - p,$$

thereby making R a subinterval of $[0, g]$.[11]

Let E denote the subset of A delimited by the mn inequalities:

$$(4.3) \qquad x'_\mu + x''_\nu \leq 1, \qquad \text{all } \mu \in M, \nu \in N.$$

Clearly E is nonempty, closed, and convex, and has a nonempty intersection with both A_M and A_N. It is the subset of A in which every pair $\{\mu, \nu\}$ is an "effective" coalition, in the sense of (1.3).

THEOREM 1. Every monotonic arc in E connecting A_M and A_N is a solution of the game.

REMARK 1. The theorem is not vacuous, since the straight line joining the centers of gravity of A_M and A_N is a monotonic arc and is contained in E.

REMARK 2. If $m = n$ then E is exactly the core (see (2.1)), and if $g = 1$ then $E = A$. Thus, all of the solutions described in §§2 and 3 are included in the theorem.

---

[11] We can now interpret p as the total profit realized by the sellers' group, M, under the imputation z_p. Adding their costs (assumed constant) and dividing by g gives us the average net market price (see the discussion in the Introduction).

REMARK 3. When $m \neq n$, one of the "contacts" $E \cap A_M$ or $E \cap A_N$ is a single point, namely $(1, \ldots, 1; 0, \ldots, 0)$ or $(0, \ldots, 0; 1, \ldots, 1)$, which pins down one end of the solutions of the theorem. The explanation for this lies in the fact that point in question is the core, and is necessarily contained in all solutions.

PROOF OF THEOREM. Let V be a monotonic arc in E running from A_M to A_N and parametrized according to (4.1), (4.2). To show that V is externally stable, take any $x \in A$, and let p_1 be the greatest $p \in R$ such that

$$(4.4) \qquad f'_\mu(p) \leq x'_\mu \ , \qquad\qquad \text{all} \quad \mu \in M \ ,$$

and p_2 the least $p \in R$ such that

$$(4.5) \qquad f''_\nu(p) \leq x''_\nu \ , \qquad\qquad \text{all} \quad \nu \in N \ .$$

The existence of these extrema is assured by the fact that V touches A_M and A_N, i.e., that $R = [0, g]$. We distinguish two cases.

CASE A: $p_1 < p_2$. Let $p_1 < p^* < p_2$. Then for some μ^*, ν^* we have both

$$f'_{\mu^*}(p^*) > x'_{\mu^*} \qquad \text{and} \qquad f''_{\nu^*}(p^*) > x''_{\nu^*} \ .$$

Using the fact that $V \subset E$, we see that z_{p^*} dominates x via $\{\mu^*, \nu^*\}$.

CASE B: $p_1 \geq p_2$. Inserting p_1 in (4.4) and p_2 in (4.5) and summing, we obtain

$$p_1 + g - p_2 \leq x(M \cup N) = g \ ,$$

with the aid of (4.2). This means that $p_1 = p_2$ and that equality holds in all of the $m + n$ inequalities (4.4), (4.5), with p_1 and p_2 inserted. Hence we have:

$$x = \langle f'(p_1); f''(p_2) \rangle = z_{p_1} \in V \ .$$

This completes the proof of external stability, our two cases having shown that every imputation x is either in V or in $\text{dom}\, V$. As for internal stability, it is obvious that the contrary direction of the two sets of monotonic functions $\{f'_\mu\}$, $\{f''_\nu\}$ rules out the possibility of domination within a monotonic arc. This completes the proof of the theorem.

It is easily verified that the solutions given by Theorem 1 fill up all of E. An example, given below in §6, shows that solutions exist that are not contained in E. We now define a larger subset of A, denoted by F, that contains all solutions, giving us the chain:

$$A \supset F \supset \text{union of all solutions} \supset E \supset \text{core}.$$

In fact, let F be the set of all imputations x such that

$$\begin{cases} \max_{M} x'_{\mu} + \min_{N} x''_{\nu} \leq 1, \quad \text{and} \\ \\ \min_{M} x'_{\mu} + \max_{N} x''_{\nu} \leq 1 \quad . \end{cases}$$

A comparison with (4.3) shows that F contains E. Note that F may not be convex.

THEOREM 2. Every solution of the game is contained in F.

PROOF. Let V be any solution, let x be any element of V, and let $\mu*$ be any element of M such that

$$(4.6) \qquad x'_{\mu*} + x''_{\nu} > 1 \qquad \text{all} \quad \nu \in N \quad ,$$

supposing that such a $\mu*$ exists. The case $x'_{\mu*} = 0$ leads at once to the absurdity: $x(N) > n \geq g$. If $x'_{\mu*} > 0$ we can find $z \in A$ that majorizes x in all components but $x'_{\mu*}$, and that is so near to x that the strict inequalities of (4.6) remain valid for z. Then any imputation that dominates z dominates x as well. Therefore z is undominated by V, and hence must actually be in V. By the same reasoning, a small neighborhood of z is in V. But this is impossible, by Lemma 1. Hence the existence of $\mu*$ fulfilling (4.6) is refuted, and

$$\max_{M} x'_{\mu} + \min_{N} x''_{\nu} \leq 1$$

is established. A symmetrical argument completes the proof.

COROLLARY. No component of any imputation of any solution exceeds 1.

This is an instance of the general location theorem of Gillies and Milnor,[12] which states that the components of all imputations in a solution must satisfy the inequality:

$$x_i \leq \max_{S \ni i} [v(S) - v(S - \{i\})] \quad .$$

A stronger general theorem ([2], p. 21) is available; it would improve on the corollary but not on Theorem 2 itself.

§5. SKEWNESS OF SOLUTIONS

In this section we shall prove that every solution of the game (1.1) is a monotonic arc connecting A_M and A_N, though not necessarily contained in E. Our main tool will be a property called "skewness", which is a special kind of internal stability. Once we have proved that every solution is skew (Lemma 5) the main result (Theorem 3) follows quickly.

For vectors a and b the notation $a \geq b$ [$a \leq b$] will denote that every component of $a - b$ is nonnegative [nonpositive]. By sup(a, b) [inf(a, b)] we shall mean the least upper [greatest lower] bound of a and b. By med(a, b, c) we shall mean the vector each of whose components is the median of the corresponding components of a, b, c. A vector c will be said to lie between the two vectors a and b if c = med(a, b, c). We observe that the median of any three vectors is between every two of them.

We call a pair x, y of imputations skew if either

$$\left\{ \begin{array}{l} x' \geq y' \text{ and} \\ x'' \leq y'' \end{array} \right\} , \quad \text{or} \quad \left\{ \begin{array}{l} x' \leq y' \text{ and} \\ x'' \geq y'' \end{array} \right\} .$$

A skew set is one in which every pair of elements is skew. A skew set is obviously internally stable, in view of (1.4). Given any three elements of a skew set, one of them will be found to lie between the other two. A monotonic arc is an obvious example of a skew set.

While there is no *a priori* reason that a solution must be skew, the hypothesis of non-skewness has far-reaching implications that eventually prove contradictory. Our entering wedge is the next lemma; note that it is trivial for skew solutions.

LEMMA 2. If x and y are elements of a solution V, then the two vectors

---

[12] [2] and [6]; or see [4], p. 239.

$$u = \langle \sup(x', y'); \quad \inf(x'', y'') \rangle$$

and

$$v = \langle \inf(x', y'); \quad \sup(x'', y'') \rangle$$

are also elements of V.

PROOF. If u and v are in A, that is, if

$$u(M \cup N) = v(M \cup N) = g \quad , \tag{5.1}$$

then it is easily seen that they are in V, since any $z \in V$ that might dominate u (say) would also have to dominate either x or y. Therefore our task is to prove (5.1). Suppose that

$$u(M \cup N) < g \quad .$$

Take a vector w that is strictly greater than u in every component, with $w(M \cup N) = g$. Then w must be in V, since anything dominating it must also dominate either x or y. By the same reasoning, a small neighborhood (in A) of w must likewise be in V, which is impossible, by Lemma 1. Hence we have

$$u(M \cup N) \geq g \quad . \tag{5.2}$$

Similarly, we have

$$v(M \cup N) \geq g \quad . \tag{5.3}$$

Adding, and using the identity $u + v = x + y$, we obtain the expression $2g \geq 2g$. This means that we actually have equality in (5.2) and (5.3), as was to be shown.

COROLLARY 1. The median of any three elements of a solution is itself in the solution.

PROOF. Apply the lemma repeatedly to the identity:

$$\mathrm{med}(x, y, z) \equiv$$

$$\langle \inf[\sup(x',y'),\sup(y',z'),\sup(z',x')]; \ \sup[\inf(x'',y''),\inf(y'',z''),\inf(z'',x'')] \rangle,$$

exploiting the fact that the "med" function can be expanded in terms of "inf" and "sup" in two different, anti-symmetric ways.

COROLLARY 2. If x and y are elements of the same solution V, then $x' \geq y'$ implies $x'' \leq y''$. Equivalently, if $x, y \in V$ are not skew, then four indices $\mu_1, \mu_2 \in M$, $\nu_1, \nu_2 \in N$ can be found such that:

$$x'_{\mu_1} > y'_{\mu_1}, \qquad x'_{\mu_2} < y'_{\mu_2}, \qquad x''_{\nu_1} > y''_{\nu_1}, \qquad x''_{\nu_2} < y''_{\nu_2} .$$

PROOF. Suppose $x' \geq y'$ but not $x'' \leq y''$. Then $u' = x'$ and $u'' \leq x''$, with $u \neq x$ (defining u as in the lemma). Then $u(M \cup N) < x(M \cup N) = g$, contradicting (5.1).

LEMMA 3. If two distinct elements of a solution V are skew, then there is a third, distinct element of V between them.

PROOF. Let x, y be a distinct, skew pair of imputations belonging to V, with $x' \geq y'$, $x'' \leq y''$, and let z be any other imputation between them. If z is in V then we are finished; if not, we can find a $w \in V$ that dominates z via some $\{\mu^*, \nu^*\}$, thus:

$$w'_{\mu*} > z'_{\mu*} \geq y'_{\mu*} ,$$

$$w''_{\nu*} > z''_{\nu*} \geq x''_{\nu*} ,$$

and

$$w'_{\mu*} + w''_{\nu*} \leq 1 .$$

Since w is not permitted to dominate either x or y, we must also have

$$x'_{\mu*} \geq w'_{\mu*} \qquad \text{and} \qquad y''_{\nu*} \geq w''_{\nu*} .$$

Let $t = \text{med}(x, y, w)$. By Corollary 1 above, t is in V. Clearly t is between x and y. Inspecting the above inequalities, we see that

$$t'_{\mu*} = w'_{\mu*} \neq y'_{\mu*} \qquad \text{and} \qquad t''_{\nu*} = w''_{\nu*} \neq x''_{\nu*} .$$

Hence t is distinct from both x and y, as required.

LEMMA 4. Every pair of skew points in a solution V can be connected by a monotonic arc lying entirely within V.

PROOF. Define a partial ordering of imputations by the relation:

(5.4) $s \succ t$ if and only if $s' \geq t'$, $s'' \leq t''$, $s \neq t$.

Let $x, y \in V$ be skew, with $x \succ y$. By Zorn's Lemma, the set V_{xy} of elements of V that lie between x and y (including x and y) contains at least one maximal chain (maximal linearly-ordered subset). Denote this chain by C. Clearly C contains x and y. Since V_{xy} is a closed set, and the closure of a chain is still a chain, C is a closed set. The continuous function $\phi(s) = s(M)$ maps C onto a closed subset $\phi(C)$ of the real interval $I = [y(M), x(M)]$. The mapping is $1:1$ as far as it goes, since ϕ is strictly order-preserving, and it covers the endpoints of I. If it failed to cover the interior of I, then $I - \phi(C)$ would contain an open subinterval with endpoints belonging to $\phi(C)$. The inverse images of these endpoints in C would be skew to each other. By Lemma 3 there would be a distinct element $z \in V - C$ between them. But such a z could be added to the chain C, contradicting the latter's assumed maximality. Therefore we conclude that $\phi(C)$ covers I — i.e., that ϕ is "$1:1$ onto". The inverse of ϕ, mapping I continuously back onto C, then provides a parametrization of C as a monotonic arc joining x and y, as required.

LEMMA 5. Every solution is a skew set.

PROOF. Suppose the contrary, and let x, y be a nonskew pair in a solution V. The associated points $u = \langle \sup(x', y'); \inf(x'', y'') \rangle$ and $v = \langle \inf(x', y'); \sup(x'', y'') \rangle$ (see Lemma 2) are then distinct from x and y, but skew to them both. (In the notation of (5.4), we have $u \succ x \succ v$ and $u \succ y \succ v$.) Lemma 4 then reveals the existence of four monotonic arcs $\widehat{ux}$, $\widehat{uy}$, $\widehat{xv}$, $\widehat{yv}$, all contained in V. From this configuration we shall derive a contradiction.

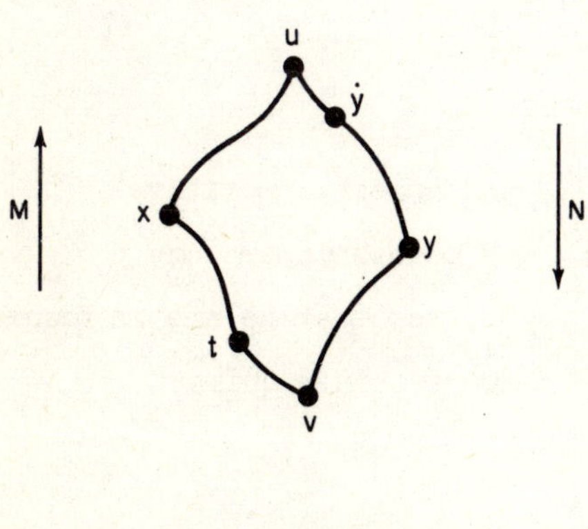

FIGURE 2

Define four disjoint sets of indices as follows:

$$M_1 = \{\mu \in M \mid x'_\mu > y'_\mu\}$$

$$M_2 = \{\mu \in M \mid x'_\mu < y'_\mu\}$$

$$N_1 = \{\nu \in N \mid x''_\nu > y''_\nu\}$$

$$N_2 = \{\nu \in N \mid x''_\nu < y''_\nu\}$$

By the second corollary to Lemma 2, none of these is empty. Our first task will be to prove:

$$\min_{\mu \epsilon M_2} x'_\mu = \min_{\mu \epsilon M_1} y'_\mu \quad . \tag{5.5}$$

To do this, we start with a fixed $\nu_1 \in N_1$, and select a positive $\epsilon < x''_{\nu_1} - y''_{\nu_1}$. Define $z \in A$ by

$$\begin{cases} z'_\mu = x'_\mu + \epsilon/(m + n - 1) & \text{all } \mu \in M \\ z''_\nu - x''_\nu + \epsilon/(m + n - 1) & \text{all } \nu \in N - \{\nu_1\} \\ z''_{\nu_1} = x''_{\nu_1} - \epsilon \quad . \end{cases}$$

Suppose first that $z \in V$. Take any $\nu_o \in N - \{\nu_1\}$ and let μ_o be such that $z'_{\mu_o} = \min_M z'_\mu$. By Theorem 2 we have $z'_{\mu_o} + z''_{\nu_o} \leq 1$. Hence z dominates x via $\{\mu_o, \nu_o\}$, contradicting the internal stability of V. Therefore $z \notin V$, and there is a $w \in V$ that dominates z via some $\{\mu^*, \nu^*\}$, with

$$w'_{\mu^*} + w''_{\nu^*} \leq 1 \quad .$$

But z "almost" majorizes x; to keep w from dominating x as well as z we must have $\nu^* = \nu_1$. Therefore:

$$w''_{\nu^*} = w''_{\nu_1} > z''_{\nu_1} = x''_{\nu_1} - \epsilon > y''_{\nu_1} \quad .$$

But now w threatens to dominate y; to prevent this we must have:

$$y'_{\mu^*} \geq w'_{\mu^*} > z'_{\mu^*} > x'_{\mu^*} \quad ,$$

from which we conclude that $\mu^* \in M_2$. Hence:

$$\begin{aligned} \min_{M_2} x'_\mu \leq x'_{\mu^*} < w'_{\mu^*} \leq 1 - w''_{\nu^*} \\ < 1 - x''_{\nu_1} + \epsilon \quad . \end{aligned} \tag{5.6}$$

Now, take a $\mu_1 \in M_1$ so that $y'_{\mu_1} = \min_{M_1} y'_\mu$, and select a positive $\eta < x'_{\mu_1} - y'_{\mu_1}$. There will be a point t on the monotonic arc $\widehat{xv} \subset V$ with $t'_{\mu_1} = y'_{\mu_1} + \eta$. Note that $t''_{\nu_1} = x''_{\nu_1} > y''_{\nu_1}$, so that we are in danger of having t dominate y via $\{\mu_1, \nu_1\}$. To avert this requires $t'_{\mu_1} + t''_{\nu_1} > 1$. Thus we have:

$$\begin{aligned}(5.7)\qquad \min_{M_1} y'_\mu = y'_{\mu_1} &= t'_{\mu_1} - \eta \\ &> 1 - t''_{\nu_1} - \eta = 1 - x''_{\nu_1} - \eta \ .\end{aligned}$$

Since (5.6) and (5.7) are valid for arbitrarily small ϵ and η, we conclude that $\min_{M_1} y'_\mu \geq \min_{M_2} x'_\mu$. A symmetrical argument establishes the reverse inequality, and (5.5) follows.

To complete the proof of the lemma, choose a point $\dot{y} \in \widehat{uy} \subset V$ far enough from y so that $\dot{y}'_\mu > y'_\mu$ for all $\mu \in M_1$, but not so far that $\dot{y} = u$. (see Figure 2.) Then x and $\dot{y}$ are nonskew. Defining $\dot{M}_1$ and $\dot{M}_2$ as above, with respect to the pair $x, \dot{y}$, we see at once that $\dot{M}_1 \subset M_1$ and $\dot{M}_2 \supset M_2$, since $\dot{y}' \geq y'$. Hence

$$\min_{\dot{M}_2} x'_\mu \leq \min_{M_2} x'_\mu \quad ,$$

and

$$\min_{\dot{M}_1} \dot{y}'_\mu \geq \min_{M_1} \dot{y}'_\mu > \min_{M_1} y'_\mu \quad ,$$

and (5.5) must fail for one of the pairs x, y or $x, \dot{y}$. This is the desired contradiction.

THEOREM 3. Every solution of the game is a monotonic arc connecting A_M and A_N.

PROOF. Let V be a solution, and let $x \in V$ be its "nearest approach" to the face A_M in the sense that $x(M)$ is maximized. Since V is a skew set (Lemma 5), we have

$$w' \leq x', \qquad w'' \geq x'' \qquad \text{all} \quad w \in V \ .$$

The imputation $z \in A_m$, given by

$$\begin{cases} z'_\mu = x'_\mu + x(N)/m & \text{all} \quad \mu \in M \\ z''_\nu = 0 & \text{all} \quad \nu \in N \ , \end{cases}$$

is skew to every element of V, and hence is undominated by V. Therefore it belongs to V, and is in fact the element x. In other words, V actually touches A_M. By a similar argument, V touches A_N. Therefore

(Lemma 4) V includes a monotonic arc C connecting the two faces. But C is obviously a maximal skew set; hence V = C. This completes the proof.

COROLLARY 1. The only solution possessing the full symmetry of the game is the set of imputations of the form

$$(p, \ldots, p; q, \ldots, q), \qquad mp + nq = g \quad ,$$

— that is, the line joining the midpoints of A_M and A_N.

Thus, the unique symmetric solution is precisely the set of symmetric imputations.[13]

COROLLARY 2. If min(m; n) = g = 1, then all the solutions of the game are given by Theorem 1. (See §3.)

§6. TWO EXAMPLES

Theorems 2 and 3, while they narrow the class of candidates to a fairly concise family of sets — the maximal monotonic arcs in F — nevertheless fall short of a complete characterization of the solutions of the game. Maximal monotonic arcs lying entirely in E are certainly solutions, by Theorem 1, but the status of those that enter the region F - E remains in doubt. We give two examples to show that the problem has no simple, all-or-none resolution.

Let M = {1, 2}, N = {3, 4, 5}. (Any smaller game would come under the special cases m = n or g = 1.) Let V_1 and V_2 be polygonal arcs, with vertices connected in the order listed:

Vertices of V_1	Vertices of V_2
(1, 1; 0, 0, 0)	(1, 1; 0, 0, 0)
(3/4, 3/4; 1/6, 1/6, 1/6)	(2/3, 2/3; 1/3, 1/6, 1/6)
(3/4, 1/4; 1/2, 1/4, 1/4)	(2/3, 1/3; 2/3, 1/6, 1/6)
(1/4, 1/4; 1/2, 1/2, 1/2)	(1/3, 1/3; 2/3, 1/3, 1/3)
(0, 0; 2/3, 2/3, 2/3)	(0, 0; 2/3, 2/3, 2/3)

[13] Contrast this with the situation for the familiar three-person simple majority game: the unique symmetric solution consists of three asymmetric imputations; the unique symmetric imputation belongs to three asymmetric solutions (see [13], §§32, 33).

In each case, the third vertex is in $F - E$, by virtue of $x_1' + x_3'' > 1$. We assert that V_1 is a solution, and that V_2 is not.

To verify the former, we refer to the proof of Theorem 1, and observe that the only statement therein that does not apply to V_1 is the claim of domination via $\{\mu^*, \nu^*\} = \{1, 3\}$ for certain values of p^* — a claim that is invalidated by the failure of condition (1.5) in the vicinity of the third vertex. However, the only imputations whose domination depends on this are those with $1/2 \leq x_1' < 3/4$, $1/4 \leq x_3'' < 1/2$, and a straightforward argument shows that all such imputations are dominated by V_1 via other pairs $\{\mu, \nu\}$, for which (1.5) always holds. We omit the details.

To verify that V_2 is not a solution, we observe simply that the imputation $(1/3, 2/3; 1/3, 1/3, 1/3)$, among others, is undominated by V_2.

§7. GENERAL FORM OF THE SKEWNESS CONCEPT

It was remarked in §5 that skewness is a special form of internal stability. It may be of interest to have this relationship made precise in the context of general n-person games, and to suggest why it works so well in the present case.

Following Gillies [2], let us call a coalition S vital if there does not exist a nontrivial partition $(S_1, \dots, S_r)$ of S such that $v(S) = \Sigma v(S_1)$. (In the game (1.1) the only vital coalitions are the one-element sets and those of the form $\{\mu, \nu\}$ with $\mu \in M$, $\nu \in N$.) Let us call two imputations x and y skew if neither $x - y$ nor $y - x$ is strictly positive on any vital coalition having more than one member.

Our previous definition (§5) is obviously included in the above. It is easy to verify in general that skew sets are internally stable, and that for simple games the solutions are always maximal skew sets. That the converse of the latter is not true may be seen from the example of three-point configurations in the essential zero-sum 3-person game which are maximal skew, but not solutions. An indication of a posible role which skewness might play in the general theory may be gleaned from [13], §30.3 (esp. §30.3.6). However, there are grave difficulties to be overcome in the general approach outlined there, and the best immediate prospect lies in applications to restricted classes of games, especially those with relatively few vital coalitions.

In the present application, the key property is the fact that skew sets are linearly ordered chains in the partial ordering (6.4) defined in the proof of Lemma 4 (but used elsewhere implicitly); that is, skewness of a three-point set implies that one of the points is between the other two.[14] For other classes of games we may hope to be able to discover and exploit other, equally decisive, special properties.

---

[14] This is a converse to the general theorem: if z is between x and y, and if $\{x, y\}$ is skew, then $\{x, y, z\}$ is skew.

BIBLIOGRAPHY

[1] BOTT, R., "Symmetric solutions to majority games," Annals of Mathematics Study No. 28 (Princeton, 1953), pp. 319-323.

[2] GILLIES, D. B., "Some theorems on n-person games," PhD Thesis, Department of Mathematics, Princeton University, Princeton, 1953; see also his paper in this Study.

[3] GILLIES, D. B., "Discriminatory and bargaining solutions to a class of symmetric n-person games," Annals of Mathematics Study No. 28 (Princeton, 1953), pp. 325-342.

[4] LUCE, R. D. and RAIFFA, H., Games and Decisions: Introduction and Critical Survey, New York, John Wiley and Sons, 1957.

[5] MILLS, W. H., "The four person game — edge of the cube," Annals of Mathematics 59 (1954), pp. 367-378.

[6] MILNOR, J., "Reasonable outcomes for n-person games," Research Memorandum RM-916, The RAND Corporation, Santa Monica, 1952.

[7] SHAPLEY, L. S., "Quota solutions of n-person games," Annals of Mathematics Study No. 28 (Princeton, 1953), pp. 343-359.

[8] SHAPLEY, L. S., "Simple games: an outline of the descriptive theory," Research Memorandum RM-1384, The RAND Corporation, Santa Monica, 1954.

[9] SHAPLEY, L. S., "Markets as cooperative games," Paper P-629, The RAND Corporation, Santa Monica, 1955; see also Report of an Informal Conference on Recent Developments in the Theory of Games, Department of Mathematics, Princeton University, Princeton, 1955, pp. 18-19.

[10] SHAPLEY, L. S., "A symmetric market game," Research Memorandum RM-1533, The RAND Corporation, Santa Monica, 1955.

[11] SHAPLEY, L. S., "Compound simple games," Paper P-1027-A (abstract), The RAND Corporation, Santa Monica, 1957; see also Report of the Third Conference on Games, Department of Mathematics, Princeton University, Princeton, 1957, pp. 19-20.

[12] SHUBIK, M., "Edgeworth market games," this Study.

[13] von NEUMANN, J. and MORGENSTERN, O., Theory of Games and Economic Behavior, Princeton, 1944, 2nd ed. 1947, 3rd ed. 1953.

L. S. Shapley

The RAND Corporation

GENERALIZED QUOTA SOLUTIONS OF n-PERSON GAMES[1]

G. K. Kalisch

§1. INTRODUCTION

The kinds of solutions we shall be concerned with are generalizations of Shapley's quota solutions (see [2][2]). We are considering solutions V with the important property that for given $m \geq 2$, V dominates every imputation $\alpha \notin V$ on a set of m elements, while any two imputations α and β in V have the property that at most m components of one are greater than the corresponding components of the other. The treatment in [2] is concerned only with the case $m = 2$. We are thus led first to construct "m-solutions", i.e., sets V_m which dominate every $\alpha \notin V_m$ on an m-element set ("m-domination"), and whose internal consistency is likewise confined to the failure of internal m-domination to obtain. In order to take care of internal inconsistencies arising from m'-domination for $m' < m$, it is possible to employ "local m-solutions" in various parts of V_m; this case will be discussed in a future paper. The case $m' \geq m$ is taken care of more simply (see Section 3). We wish to remark that in addition to the piecewise linear solutions which we find on the basis of the above principles, it is possible to find curvilinear ones as Shapley does in [2]. We shall, however, not enter into these generalizations here.

In Section 2 we carry out various generalizations of [2] valid for the case $m = 2$.

We refer the reader to [1] for an exposition of the relevant concepts and notations. We shall be dealing with (not necessarily zero-sum) essential n-person games Γ normalized as follows (where v is the characteristic function of Γ): $v((i)) = -1$ for all i, $v(I) = 0$.

---

[1] The preparation of this paper was sponsored (in part) by the Office of Naval Research.

[2] Numbers in square brackets refer to the bibliography at the end of this paper.

§2. THE CASE $m = 2$

We shall say that a game Γ has a 2-quota solution V_2 if there exists an n-dimensional real vector $\omega = \{\omega_i\}$ which is not necessarily an imputation such that $V_2 = L$ where L is the union of certain line segments ℓ^i where i runs over a certain nonempty subset of I; these ℓ^i are defined as follows:

(i) If $\omega_i - \Sigma_{k\epsilon I}\omega_k > -1$, there exists an integer $b_i \epsilon I$, $b_i \neq i$ and

$$\ell^i = \left[\left\{\ldots\ \omega_i - (\Omega + t)\ \ldots\ \omega_{b_i} + t\ \ldots\right\}\right] = [\lambda^i] = [\lambda^i(t)] \quad , \tag{1}$$

where the λ^i are imputations, $t \geq 0$,

$$\Omega = \sum_{k\epsilon I}\omega_k \quad ,$$

and the dots ... mean that the components not specifically noted are equal to the corresponding ω_i's, a convention which we shall use throughout the paper.

(ii) If $\omega_i - \Omega = -1$, ℓ^i as just defined contains a single point:

$$\ell^i = [\{\ldots\ \omega_i - \Omega\ \ldots.\}] = [\lambda^i] \quad . \tag{2}$$

In this case it is, of course, unnecessary to define b_i.

While we could formulate our results in terms of the distinction $\omega_i < -1$ vs. $\omega_i \geq -1$, it seems more appropriate in view of the form of the imputations λ^i, to assume that

$$\omega_i \geq -1 \text{ for all } i \quad .$$

Otherwise, if, say, $\omega_{i_o} < -1$, we could replace ω_{i_o} by -1 and set $b_i = i_o$ for those i for which b_i is defined. It is inherent in our theory that at most one of the ω_i's can be so distinguished.

We shall now investigate various necessary conditions which must obtain in order that L be a solution V_2.

THEOREM 1. If L is a solution of some game Γ, then $\Omega \geq 0$.

PROOF. $\Omega < 0$ would imply that the imputation $\alpha = \{\omega_i + \epsilon_i\}$ for $\epsilon_i > 0$, which is not in L, is undominatable by L, a contradiction.

It will be important in the sequel to distinguish two fundamentally different cases:

CASE 1. $\Omega = 0$,

CASE 2. $\Omega > 0$.

If we examine the imputations λ^i, we observe that in order that they be indeed imputations, it is necessary that

$$\omega_i - \Omega \geq - 1 \quad .$$

These inequalities give rise to the following partition of I:

$$(3) \qquad \begin{cases} I_1 = (i;\ \Omega - 1 < \omega_i) \ , \\ I_2 = (i;\ \Omega - 1 = \omega_i) \ , \\ I_3 = (i;\ \Omega - 1 > \omega_i) \ . \end{cases}$$

The $i \in I_1$ give rise to line segments ℓ^i which contain more than one point (see (1)); for these and these alone is it necessary to define b_i. The $i \in I_2$ give rise to ℓ^i containing a single point (see (2)). Finally, the $i \in I_3$ give rise to no ℓ^i. I_3 is empty if $\Omega = 0$ (Case 1). These remarks imply the following:

THEOREM 2. If L is a solution of some game Γ, then $I_1 \cup I_2 \neq \emptyset$.

PROOF. The preceding remarks, together with the fact that no game has an empty solution, constitute a proof of our theorem.

In connection with (3) we make the following two observations, where $|S|$ is the number of elements in the subset $S \subset I$ and L is to be a solution of some game Γ:

(i) If $|I_1| = \emptyset$ then $|I_2| > 2$ and $\Omega > 0$.

PROOF. If $\Omega = 0$, L would consist of the single imputation ω which would fail to dominate, for example, such imputations as $\{\dots\ \omega_i - \epsilon\ \dots .\ \omega_j + \epsilon\ \dots .\}$ which are admissible since not all the ω_i's can equal -1. Our hypothesis implies that L consists of a finite set of λ^i's: $\lambda^i = \{\dots -1 \dots\}$, "$-1$" in the ith place for all $i \in I_2$.

To complete the proof, suppose that $|I_2| < 3$, say, $I_2 = (1)$ or $(1, 2)$. In that case, L would fail to dominate such imputations as $\{-1, \omega_2 - \epsilon, \omega_3 + \epsilon, \dots\}$ for $\epsilon > 0$, which, again, are admissible since not all ω_i's can equal -1.

The example of "Corner VIII" for 4-person games, and the solution $[\{-1, 1, 1, -1\}$ and all permutations of the first three components$]$ (see [1], p. 307) show that $|I_1| = 0$ and $|I_2| = 3$ is possible with $\omega = \{1, 1, 1, -1\}$.

(ii) $|I_1| = 1$, $|I_2| = 0$ is possible.

PROOF. Let $n = 3$, $\omega_1 = 3$, $\omega_2 = \omega_3 = -1$; $\Omega = 1$, $\Omega - 1 = 0 < \omega_1 = 3$; $\omega_2 = \omega_3 = -1 < \Omega - 1 = 0$. L consists of ℓ^1 which has the following form if we take $b_1 = 2$:

$$\ell^1 = [\{\omega_1 - t_1, \omega_2 + t_2, \omega_3\}], \; t_1 \geq 0,$$

which is nothing but a discriminatory solution of a zero-sum essential three-person game.

We resume the discussion of necessary conditions for L to be a solution of a game Γ.

THEOREM 3. L is consistent if and only if

$$\omega_i + \omega_{b_j} \geq v((i, b_j)) \tag{4}$$

for all i and j such that: $j \in I_1$; $|(i, j, b_j)| = 3$; and, if $\Omega = 0$, $i \in I_1$; if $\Omega > 0$, $i \in I_1 \cup I_2$.

PROOF. Consider the two imputations $\lambda^j(t)$ and $\lambda^i(s)$ where $s = 0$ if $\omega_i - \Omega = -1$ (i.e., if $i \in I_2$).

Since now $\Omega \geq 0$, $i \in I_1$ implies that $\lambda^j(t) \succ \lambda^i(s)$ on (i, b_j) if it is effective and if $s > 0$, $t > 0$, providing $b_j \neq b_i$ or $t > s$ (this latter always being possible). Thus the consistency of L is equivalent with the ineffectivity of (i, b_j) for $\lambda^j(t)$ (all sufficiently small t), i.e., with (4). The case $\Omega > 0$, $i \in I_2$ is handled in much the same way.

THEOREM 4. If L is a solution of some game Γ, then it is necessary that

$$\omega_i + \omega_j \leq v((i, j)) \tag{5}$$

for all i and j such that $|(i, j)| = 2$ and:

if $\Omega = 0$ then $(i, j) \subset I_1$; if $\Omega > 0$ then $(i, j) \subset I_1 \cup I_2$.

PROOF. (1) Let $\Omega = 0$ and try to dominate the imputation

$$\alpha = \left\{\ldots \omega_i - \epsilon_1 \ldots \omega_j - \epsilon_2 \ldots \omega_{b_i} + \epsilon_1 \ldots \omega_{b_j} + \epsilon_2 \ldots\right\}, \epsilon_i > 0 .$$

Theorem 3 implies that the domination $\lambda^k \succ \alpha$ on (i, b_k) for $|(i, k, b_k)| = 3$ is impossible because (i, b_k) is not effective for λ^k; a similar remark applies to (j, b_k); domination on (i, b_i) or (j, b_j) by λ^i or λ^j is patently impossible even if $b_i = b_j$. Thus the only possibility that remains is domination in (i, j); since the ϵ_i can be arbitrarily small, (5) follows.

(2) Let $\Omega > 0$. We first note that if L is to be a solution of some game Γ, $|I_1 \cup I_2| > 2$, for otherwise if $(i, j) = I_1 \cup I_2$,

$$\alpha = \{\ldots \omega_i - \Omega \ldots \omega_j - \Omega \ldots \omega_k + \Omega \ldots\}$$

would be undominated by L. We now try to dominate α. Domination by λ^i is impossible on (j, b_i) by Theorem 3; it is also impossible on any set involving i since the maximum of the ith component of λ^i is $\omega_i - \Omega$. A similar statement is true for j. Thus we can dominate only (i, j); again, as above, domination by λ^i or λ^j is impossible; we must dominate on (i, j) by means of λ^k where $k \in I_1 \cup I_2 - (i, j)$; we showed at the beginning of this paragraph that such a k must exist. It makes no difference now whether $k \in I_1$ or not; or whether $b_k \in (i, j)$; effectivity of (i, j) implies (5).

THEOREM 5. If L is a solution of some game Γ, then it is necessary that

$$\omega_i + \omega_{b_i} - \Omega \leq v((i, b_i)) \tag{6}$$

for all $i \in I_1$.

PROOF. We now try to dominate the imputation

$$\alpha = \left\{\ldots \omega_i - (\Omega + \epsilon_1 + \epsilon_2) \ldots \omega_k + \epsilon_1 \ldots \omega_{b_i} + \epsilon_2 \ldots\right\}, \epsilon_i > 0 .$$

Theorem 3 implies that this is possible only on (i, b_i) by λ^i. Thus (i, b_i) must be effective for it and this implies (6).

We next turn to the question of whether given $\omega = \{\omega_i\}$ there exist games satisfying (4), (5), (6). In order to construct a characteristic function with the requisite properties, we find it convenient to introduce the following terminology.

A subset $S \subset I$ will be called regular if it can be partitioned into a set of admissible pairs (i, j) which are of two types and are defined as follows:

Type (i): if $\Omega = 0$, $(i, j) \subset I_1$; if $\Omega > 0$, $(i, j) \subset I_1 \cup I_2$;
Type (ii): (i, b_i) not of type (i).

Given a subset $S \subset I$ and a regular subset $S_r \subset S$, let

$$(7) \qquad f_{S_r}(S) = \sum_{i \in S_r} \omega_i - |S| + |S_r| - k_{S_r}\Omega$$

where k_{S_r} is the number of pairs of type (ii) in S_r. We finally define

$$(8) \qquad v(S) = \max_{S_r \subset S} f_{S_r}(S),\ S \neq I;\ v(I) = 0 \ .$$

These definitions enable us to state and prove the following Theorem.

THEOREM 6. Given ω such that $\omega_i \geq -1$, $\Omega = \Sigma\omega_i \geq 0$. Then a necessary and sufficient condition that there exist a game Γ satisfying (4), (5), (6) is

$$(9) \qquad \sum_{i \in S_r} \omega_i \leq n - |S_r| + k_r\Omega$$

for all regular sets S_r.

PROOF. Necessity. Suppose we have a game Γ as described in the statement of the theorem. Let S_r be a regular subset of I with k_{S_r} pairs of type (ii). Let $S_r = S_1 \cup S_2$ where S_1 is the union of the pairs in S_r which are of type (i), S_2 is the union of the pairs of type (ii) in S_r. Note that $|S_2| = 2k_{S_r}$. Then (5) and (6) imply that

$$\begin{aligned}\sum_{i \in S_r} \omega_i &\leq \sum_{(i,j) \subset S_1} v((i, j)) + \sum_{(i,j) \subset S_2} v((i, j)) + k_{S_r}\Omega \\ &\leq v(S_r) + k_{S_r}\Omega \\ &\leq n - |S_r| + k_{S_r}\Omega \ .\end{aligned}$$

Sufficiency. (1) v satisfies (4), (5), (6). To prove (4), note that if (i, b_j) is of type (i), $v((i, b_j)) = \omega_i + \omega_{b_j}$, if (i, b_j) is of type (ii), $v((i, b_j)) = \omega_i + \omega_{b_j} - \Omega$, while in the remaining case $v((i, b_j)) = -2$; thus (4) is true in all these cases. The truth of (5) and (6) is implied by the fact that the relevant pairs are admissible; (7) and (8) then imply that $v((i, j)) = \omega_i + \omega_j$, or $v((i, b_i)) = \omega_i + \omega_{b_i} - \Omega$, respectively, i.e., (5) and (6) are verified.

(2) We now show that v as defined by (8) is the characteristic function of some game. Since (7) and (8) imply that $v((i)) = -1$ for all $i \in I$, and since $v(I) = 0$ by (8), all we need show is that if $S \cap T = \emptyset$, then $v(S \cup T) \geq v(S) + v(T)$. Note first that if $S \cap T = \emptyset$ and S and T are regular, so is $S \cup T$ and $k_{S \cup T} = k_S + k_T$.

Consider now first the case where $S \cup T \neq I$. We have

$$v(S) + v(T) = \sum_{i \in S_r \cup T_r} \omega_i - |S \cup T| + |S_r \cup T_r| - k_{S_r \cup T_r}\Omega$$

for suitable S_r and T_r. The right hand side of this equation, which we shall denote by R, is one of the terms occurring in the competition for the "max" in the definition of $v(S \cup T)$ and hence $R \leq v(S \cup T)$, $v(S) + v(T) \leq v(S \cup T)$ is true in this case.

We finally consider the case where $S \cup T = I$. We now must prove that $R \leq 0$. But (9) implies that

$$\sum_{i \in S_r \cup T_r} \omega_i \leq n - |S_r \cup T_r| + k_{S_r \cup T_r}\Omega$$

and hence $R \leq n - |S_r \cup T_r| + k_{S_r \cup T_r}\Omega - |S \cup T| + |S_r \cup T_r| - k_{S_r \cup T_r}\Omega = 0$. This completes the proof of the theorem.

Let us note that if $\Omega = 0$, $v(S) = \Sigma_{i \in S}\omega_i$ for all S such that $|S| \geq 2$, $v((i)) = -1$ for all $i \in I$ will give us a characteristic function satisfying (4), (5), and (6).

Our last concern in this section is to verify that L is indeed a solution for our games Γ. A complete catalogue of all possibilities would be too lengthy. The following theorem indicates some of the possibilities.

THEOREM 7. Let $\omega = \{\omega_i\}$ and the characteristic function v of a game Γ satisfy the following: $\omega_i \geq -1$; $\Omega = \Sigma\omega_i \geq 0$; inequalities (4), (5), (6). Then:

(1) If $\Omega = 0$, and $|I_1 \cup I_2| \geq 1$, then L is

a solution V_2 of Γ;

(2) if $\Omega > 0$ and $|I_1 \cup I_2| \geq 3$ and if we strengthen (5) to

(10) $$\omega_i + \omega_j \leq v((i, j))$$

for all i and j such that $|(i, j)| = 2$, $\omega_i > -1$, $\omega_j > -1$, then L is a solution V_2 of Γ;

(3) if $\Omega > 0$, $I_1 \neq \emptyset$, $|I_1 \cup I_2| \geq 3$, if $\max_{i \in I_1 \cup I_2} \omega_i = \omega_{k_o}$ and $\omega_{k_o} > |I_3| + \max_{i \in I_1 \cup I_2 - (k_o)} \omega_i$ and if we then select $b_{k_o} \in I_1 \cup I_2$ and further demand that

(11) $$\omega_{k_o} + \omega_{b_{k_o}} \leq v((j, b_{k_o}))$$
$$\omega_j + \omega_{k_o} \leq v((j, k_o)).$$

for all $j \in I_3$ such that $\omega_j > -1$, then L is a solution V_2 of Γ.

PROOF. (1) Inequality (4) implies that L is consistent. To show that L dominates all imputations in its complement, let (i) α be an imputation two or more of whose components are less than the corresponding ω_i's. Let i and j be two distinct elements of I such that $\alpha_i < \omega_i$ and $\alpha_j < \omega_j$. This implies that $(i, j) \subset I_1$ and ω which is in L will dominate α on (i, j); effectivity of the latter for ω is assured by (5).

Let now (ii) α be such that just one of its components is less than the corresponding component of ω. Thus suppose that $\alpha_{i_1} = \omega_{i_1} - \epsilon_{i_1} < \omega_{i_1}$; $\alpha_k = \omega_k + \epsilon_k$ (all $\epsilon_i \geq 0$); $\epsilon_{i_1} = \Sigma_{k \neq i_1} \epsilon_k$. Since $\alpha \notin V$, $\epsilon_{b_{i_1}} < \epsilon_{i_1}$ and hence there exists t such that $\epsilon_{b_{i_1}} < t < \epsilon_{i_1}$ and $\lambda^{i_1}(t) \succ \alpha$ on (i_1, b_{i_1}) which is effective for $\lambda^{i_1}(t)$ by (6).

(2) Consistency is proved as in (1). (i) Suppose that two or more components of an imputation α are less than the corresponding components of ω. Our purpose in strengthening (5) to (10) is to enable us to dominate α on (i, j) by means of some $\lambda^k (k \notin (i, j))$.

(ii) Suppose now that just one of the components of an imputation $\alpha \notin L$ is less than the corresponding component of ω. Since

$\alpha_k = \omega_k + \epsilon_k (\epsilon_k \geq 0)$ for all but one k, say, for $k \neq i_1$, we must have $\alpha_{i_1} = \omega_{i_1} - \Omega - \epsilon_{i_1}$ since $\omega_{i_1} - \Omega$ is the maximum of the i_1 th component of an imputation all of whose other components are greater than, or equal to, the corresponding ω_i's. We have $\epsilon_{i_1} = \Sigma_{k \neq i_1} \epsilon_k$, and, since $\alpha \notin L$, $\epsilon_{i_1} > 0$, and $i_1 \in I_1$. The reason for the last statement is that our hypotheses on α imply that $i_1 \in I_1 \cup I_2$; if now i_1 were in I_2, α would be in L, a contradiction. Another consequence of $\alpha \notin V$ is that $\epsilon_{b_{i_1}} < \epsilon_{i_1}$ and the proof now proceeds as in Case (1).

(3) Consistency as above. (i) Let $\alpha \notin V$ be such that two of its components are less than the corresponding components of ω; say, $\alpha_i < \omega_i$, $\alpha_j < \omega_j$ $(i \neq j)$. Suppose first that $(i, j) \subset I_1 \cup I_2$. Then as in Case (2) we can dominate α by λ^k $(k \notin (i, j))$ on (i, j).

We next suppose that the inequality $\alpha_i < \omega_i$ is true for at most one $i \in I_1 \cup I_2$. We desire to dominate α by λ^{k_o} on (b_{k_o}, j) for suitable $j \in I_3$. To that end we determine $\max \alpha_{b_{k_o}}$ under the assumption that $i \neq k_o$. It is convenient to write $\alpha_{b_{k_o}} = \omega_{b_{k_o}} + \beta$ and to determine $\max \beta$ instead. The max occurs when $b_{k_o} \neq i$, $\alpha_j = -1$ for all $j \in I_3$ and if there exists $i \in I_1 \cup I_2$ such that $\alpha_i < \omega_i$. We have

$$\max \beta = - \sum_{j \in I_1 \cup I_2} \omega_j + |I_3| + \omega_i + 1$$

$$< - \sum_{j \in I_1 \cup I_2} \omega_j + \omega_{k_o} + 1$$

$$\leq \omega_{k_o} + 1 - \Omega \quad .$$

The reason for the first inequality is that $\omega_{k_o} > |I_3| + \omega_i$ by our hypothesis. The reason for the second inequality is as follows. We have to show that $-\Sigma_{i \in I_1 \cup I_2} \omega_i \leq -\Omega$ or $\Sigma_{i \in I_3} \omega_i \leq 0$. If $\Omega \leq 1$, $\omega_i < 0$ for $i \in I_3$ and $\Sigma_{i \in I_3} \omega_i \leq 0$ as desired. In order to show the truth of the inequality when $\Omega > 1$, we observe first that $\omega_{k_o} > |I_3| + \Omega - 1$. For our hypothesis implies that $\omega_{k_o} > |I_3| + \omega_i$ for all $i \in I_1 \cup I_2 - (k_o)$ whence $\omega_{k_o} > |I_3| + \Omega - 1$ and hence $\Sigma_{i \in I_3} \omega_i = \Omega - \omega_{k_o} - \Sigma_{i \in I_1 \cup I_2 - (k_o)} \omega_i < \Omega - |I_3| - \Omega + 1 - \Sigma_{i \in I_1 \cup I_2 - (k_o)} \omega_i$. But when $\Omega > 1$, $\omega_i > \Omega - 1 > 0$ for all $i \in I_1 \cup I_2$ and again $\Sigma_{i \in I_3} \omega_i \leq 0$ as desired. This establishes the inequality $\max \beta < \omega_{k_o} + 1 - \Omega$ and permits us to dominate α by $\lambda^{k_o}(\omega_{k_o} + 1 - \Omega)$ on (b_{k_o}, j) for suitable $j \in I_3$. The effectivity of (b_{k_o}, j) for this λ^{k_o} is assured by (11): $\omega_{b_{k_o}} + \omega_{k_o} - \Omega + 1 + \omega_j < \omega_{b_{k_o}} + \omega_{k_o} \leq v((b_{k_o}, j))$ since $j \in I_3$.

If, finally, $\alpha_{k_o} < \omega_{k_o}$ we take $k \neq k_o$ and see that $\lambda^k \succ \alpha$ on (k_o, j) for suitable $j \in I_3$; effectivity is again assured by (11).

(ii) The case where just one component of α is less than the corresponding component of ω is handled as in (2).

COROLLARY. If Γ has the property that for all $i \neq j$, $v((i, j)) = v_2 \geq 0$, then Γ has a solution V_2.

PROOF. (1) Let n be even. Then $\frac{1}{2} n v_2 \leq v(I) = 0$, $v_2 = 0$ and if we set $\omega_i = 0$ for all $i \in I$, we see that the hypotheses of Theorem 7, Part 1, are satisfied and Γ has a solution V_2.

(2) Let n be odd. Then $\frac{1}{2}(n - 1)v_2 - 1 \leq v(I) = 0$, $v_2 \leq 2/(n - 1)$. If we now set $\omega_i = \frac{1}{2} v_2$, we see that the hypotheses of Theorem 7, Part 2, are satisfied and Γ has a solution of type V_2.

As may be seen from (4), (5), (6), and Theorem 7, a judicious choice of the function b_i can give us considerable freedom in the application of our quota games and quota solutions. Suppose, for example, that $\Omega = 0$ and $\omega_i > -1$ for all $i \in I$ (i.e., $I = I_1$), then the choice $b_1 = 2$, $b_i = 1$ for all $i > 1$ gives us the following conditions on ω and v:

$$\omega_1 + \omega_i = v((1, i)) \text{ for } i > 1 \quad ,$$

$$\omega_2 + \omega_i = v((2, i)) \text{ for } i \neq 2 \quad ,$$

$$\omega_i + \omega_j \leq v((i, j)) \text{ for } i \neq j \quad .$$

§3. NO RESTRICTION ON m

The generalization of 2-quota games Q_2 to "m-quota games" Q_m is based on the desire to find solutions V_m of suitable games Γ such that every imputation not in V_m is dominated by an imputation in V_m on an m-element set and such that no imputation in V_m dominates any other imputation in V_m on a p-element set for $p \geq m$. From the standpoint of game theory, we could alter the concept of solution to that of an "m-solution";

we prefer to consider our solutions as in [1], to generalize (1) and to assume that for $r < m$, v is so small as never to be effective for any imputation in V_m (for example, by assuming that all r-element sets are flat) so that consistency of V_m need not be considered for r-element sets. We will indicate in a subsequent paper how this restriction might be removed.

In what follows we shall once more deal with a "quota point" (see [2]) $\omega = \{\omega_i\}$ which we shall here assume to be an imputation. Generalizations to the case where ω is no longer an imputation will be considered in a subsequent paper. We still find it convenient to use the partition (3) of I; since ω is an imputation, $I_1 \neq \emptyset$ and $I_3 = \emptyset$.

We shall say that a game Γ has an m-quota solution V_m, if, first of all,

$$\sum_{i \in S} \omega_i = v(S) \tag{12}$$

for all subsets $S \subset I$ such that $|S| = m$ and where v is the characteristic function of the game Γ; V_m is to be the union L of certain sets ℓ^i (for each $i \in I_1$) which are defined as follows: for each $i \in I_1$ there exist integers $b_i^j (j = 2, \ldots, m)$ such that $S_i = (i, b_i^2, \ldots, b_i^m) \subset I$ and $|S_i| = m$, and finally

$$\ell^i = [\lambda^i] = \left[\left\{\ldots \omega_i - \sum_2^m x_k \ldots \omega_{b_i^2} + x_2 \ldots \right.\right.$$
$$\left.\left. \ldots \omega_{b_i^m} + x_m \ldots \right\}\right] , \tag{13}$$

where the λ^i are imputations, $x_k \geq 0$ for $k = 2, \ldots, m$, and the dots ... mean that components not specifically noted are equal to the corresponding ω_i's. We now have the following analogue of the first part of Theorem 7.

THEOREM 8. Let $\omega = \{\omega_i\}$ be an imputation; let the characteristic function v of a game Γ and the ω_i's be related by (12). Let the values of v on r-element sets for $r < m$ be so small that we can disregard consistency of L on r-element sets (see the introductory paragraph of this section). Then L is a solution V_m of Γ for $m > 2$ if and only if

$$|S_i \cap S_j| \leq 1 \tag{14}$$

for all i and j such that $|(i, j)| = 2$, $(i, j) \subset I_1$.

PROOF. 1. <u>Sufficiency</u>. (i) By hypothesis the consistency of L need be considered only for m-element sets: for sets with fewer elements this is specifically mentioned in the statement of the theorem; for sets with more elements, we observe that at most m components of any imputation of L can be greater than the corresponding components of any other imputation of L, so that the case of <u>more</u> than m elements cannot arise at all. The expression (13) for ℓ^i shows that $\lambda^i \rightarrow \lambda^j$ is certainly impossible for $i = j$; for $i \neq j$, the set $(b_i^2, \ldots, b_i^m, j)$ would have to be effective for λ^j (with all $x_{b_i^j} > 0$), contradicting (12). Note that we have not yet used condition (14).

(ii). Consider now an imputation $\alpha \notin L$. If α has m or more components less than the corresponding ω_i's, then there exists an imputation in L which dominates α on an m-element set, viz., ω itself; (12) assures the effectivity of every m-element set for ω. If, at the other extreme, α has just one component less than the corresponding component of ω, then it is not difficult to find an element in L which will dominate α: Suppose that (to simplify our notations) $\alpha_1 < \omega_1$, say, $\alpha_1 = \omega_1 - \epsilon_1 < \omega_1$, whence $1 \in I_1$; $\alpha_k = \omega_k + \epsilon_k$ for $k \neq 1$, $\epsilon_i \geq 0$, $\Sigma_{k\neq 1}\epsilon_k = \epsilon_1$. Since α is not in L,

$$\sum_{k\in S_1-(1)} \epsilon_k < \sum_{k\neq 1} \epsilon_k = \epsilon_1$$

and hence there exist positive real numbers x_k for $k \in S_1 - (1)$ such that $\epsilon_k < x_k$ for all $k \in S_1 - (1)$ and

$$\sum_{k\in S_1-(1)} x_k < \sum_{k\neq 1} \epsilon_k = \epsilon_1 \quad .$$

The imputation λ^1 so defined will dominate α on S_1 which is effective for it by (12). Note again that (14) has not yet been used.

(iii) Suppose finally that the imputation $\alpha \notin L$ is such that the number z_α of its components which are less than the corresponding components of ω lies between 1 and m. We shall complete our proof by induction. Suppose that if $z_\alpha < k < m$, we know that α can be dominated by imputations in L. Consider now an imputation $\alpha \notin L$ such that $z_\alpha = k$. To simplify our notation, suppose that the subset $K \subset I$, on which the components of α are less than the corresponding components of ω, consists of $(1, \ldots, k)$. We can now express the components of α thus: $\alpha_i = \omega_i - \epsilon_i$ $(i \in K)$; $\alpha_i = \omega_i + \epsilon_i$ $(i \in -K)$ where $\epsilon_i > 0$ for $i \in K$ and $\epsilon_i \geq 0$ otherwise. We note, incidentally, that $K \subset I_1$.

Let us now define sets U_i $(i \in K)$ as follows:

$$U_1 = S_1 \cap (-K),$$
$$U_2 = S_2 \cap (-K) \cap (-U_1), \ \ldots \ ,$$
$$U_k = S_k \cap (-K) \cap (-\bigcup_{h<k} U_h) \ .$$

A simple induction argument shows that

$$U_i = S_i \cap (-K) \cap (-\bigcup_{h<i} S_h) \quad \text{for all} \quad i \in K \ .$$

We wish to show that

$$(15) \qquad |U_i| \geq m - k \quad \text{for all} \quad i \in K \ .$$

This is seen as follows. Suppose S_i has exactly $m - j$ of its elements outside of K. Using now condition (14) we observe that at most $k - j$ of the sets $S_{i-1}, \ldots, S_1$ can have their "overlap" with S_i (if there is any such overlap) outside of K. Since this overlap is by (14) at most 1, $|U_i| \geq m - j - (k - j) = m - k$, proving (15).

In addition to (15) we need the following: there exists $j_o \in K$ such that

$$(16) \qquad |U_{j_o}| > m - k \ .$$

For suppose that $|U_1| = m - k$. This implies that $K \subset S_1$. Consider now $U_2 = S_2 \cap (-K) \cap (-U_1)$. If $S_1 \cap S_2 = \emptyset$, $U_2 = S_2$, and (16) is true for $j_o = 2$. Otherwise (14) implies that $|S_1 \cap S_2| = 1$; let $x \in S_1 \cap S_2$. Then $U_2 = S_2 \cap (-K) \cap (-U_1) = S_2 - (x)$ and $|U_2| = m - 1 > m - k$ (the case $k = 1$ was taken care of in (ii)). Thus (16) is proved.

Note now that if $U_{j'} \subset U_j$ is such that $|U_{j'}| = m - k$, and if our imputation α is such that

$$\sum_{i \in U_{j'}} \epsilon_i < \epsilon_j \ ,$$

then there exists $\lambda^j \in \ell^j$ such that $\lambda^j \vdash \alpha$ on $K \cup U_{j'}$: for $i \in K$, $i \neq j$, $\omega_i > \omega_i - \epsilon_i$; for $i = j \in K$ we need $\omega_j - \Sigma_{s \in U_{j'}} x_s > \omega_j - \epsilon_j$; on $U_{j'}$ we need $\omega_i + x_i > \omega_i + \epsilon_i$ whence finally $\epsilon_i < x_i$ $(i \in U_{j'})$, $\Sigma_{i \in U_{j'}} x_i < \epsilon_j$ and if $\Sigma_{i \in U_{j'}} \epsilon_i < \epsilon_j$, we can indeed find suitable x's; if $\Sigma_{i \in U_{j'}} \epsilon_i \geq \epsilon_j$ we cannot do so. Thus if α were undominated by L, then

$$(17) \qquad \epsilon_j \leq \sum_{i \in U_{j'}} \epsilon_i$$

for all $j \in K$ (here we use (15) to ensure the existence of a $U_{j'} \subset U_j$ with $m - k$ elements in order to come to our conclusion) and for all $U_{j'} \subset U_j$ such that $|U_{j'}| = m - k$. We also note that since α is an imputation,

$$(18) \qquad \sum_{j \in K} \epsilon_j = \sum_{i \in -K} \epsilon_i \ .$$

Taking advantage of the fact that the U_j are disjoint, we see on combining (17) and (18) that

$$(19) \qquad \sum_{j \in K} \epsilon_j \leq \sum_{i \in \bigcup_{j \in K} U_{j'}} \epsilon_i \ ,$$

obtained by summing the expressions occurring in (17), implies that equality obtains in (19) and

$$(20) \qquad \epsilon_i = 0$$

for all $i \notin \bigcup_{j \in K} U_{j'} \cup K$. Referring now to (16), we have $j_0 \in K$ such that $|U_{j_0}| > m - k$. Since we continue to assume that α is undominated by L, we may use (17):

$$\epsilon_{j_0} \leq \sum_{i \in U_{j_0'}} \epsilon_i \ .$$

Equation (20) is now applicable and we obtain

$$(21) \qquad \epsilon_i = 0$$

for all $i \in U_{j_0} \cap (-U_{j_0'})$. Since $|U_{j_0}| > m - k$ and $|U_{j_0'}| = m - k$, each $i \in U_{j_0}$ can be made to play the role of the i in (21) which implies that $\epsilon_i = 0$ for all $i \in U_{j_0}$, hence for all $i \in U_{j_0'}$; therefore $\epsilon_{j_0} = 0$ and α can be dominated by L by virtue of our induction hypothesis. This completes the proof of the sufficiency of (14).

2. Necessity. The proof that (14) is necessary for $m \geq 3$ is based on a straightforward but tedious verification of the following. Suppose that the "overlap" $|S_i \cap S_j| \geq 2$ $(i \neq j)$. If, then, we try to dominate imputations two of whose components are less than the corresponding components of ω, we discover, after a discussion of a fairly large number of special cases, that there exist undominatable such imputations. It is sufficient to carry through the discussion for the case where the "overlap"

is 2. We know from the theory of 2-quota games (Section 2 and [2]) that (14) is not necessary when $m = 2$. The reason is that in the latter case 2-domination can (and, indeed, must) be used, while we had to exclude it for $m > 2$. This completes the discussion of the necessity of (14).

A final word about the existence of m-quota games and their solutions. The situation is much more complicated than in the 2-quota case, and I hope to discuss some of the problems in a subsequent paper. One of the difficulties is the necessity of satisfying (14); if e.g., $m = 3$, a possible assignment for the b_i^j can be read from the following diagram where $-$ and $+$ in the ith row stand for i and b_i^j respectively:

$$\begin{array}{ccccccc}
- & + & + & \cdot & \cdot & \cdot & \cdot \\
\cdot & - & \cdot & + & + & \cdot & \cdot \\
\cdot & \cdot & - & + & \cdot & + & \cdot \\
+ & \cdot & \cdot & - & \cdot & \cdot & + \\
+ & \cdot & \cdot & \cdot & - & + & \cdot \\
\cdot & + & \cdot & \cdot & \cdot & - & + \\
\cdot & \cdot & + & \cdot & + & \cdot & -
\end{array} ,$$

which, by the way, is a minimal arrangement, i.e., $m = 3$ and (14) imply that $n \geq 7$. A closer investigation of the consequences of (14) lead to interesting number theoretic problems which we shall not pursue at this time. We conclude by mentioning the fact that there exist quota games for each $m \geq 2$.

BIBLIOGRAPHY

[1] von NEUMANN, J. and MORGENSTERN, O., Theory of Games and Economic Behavior, Princeton, 2nd ed., 1947.

[2] SHAPLEY, L. S., "Quota solutions of n-person games," Annals of Mathematics Study No. 28 (Princeton 1953), pp. 343-359.

G. K. Kalisch

University of Minnesota

FIVE-PERSON, CONSTANT-SUM, EXTREME GAMES[1]

Herbert M. Gurk

§1. INTRODUCTION

von Neumann and Morgenstern have pointed out [1] that the set of all four-person, constant-sum (normalized) games can be represented as a cube, the vertices of which are the four-person, strong (i.e., constant-sum) simple games. They also have shown that the dimension of the set of n-person, constant-sum games is $2^{n-1} - n - 1 (n \geq 3)$. However, questions of the nature and number of vertices or extreme points for these general sets are not answered in their book. Only the fact that the n-person strong simple games are extreme is given.

This note will answer the question for the next simplest case, the five-person games. We shall indicate a necessary and sufficient condition that a five-person, constant-sum game be extreme, and from this we will exhibit all five-person, constant-sum, extreme games. We shall show that every such game has a solution which is a generalization of the completely discriminatory solution [1] of simple games. We shall also show that these results, in their present form, cannot be generalized for all n-person games.

Throughout the paper we use the $(0, 1)$ normalization, i.e., the value of one-player coalitions is 0, the value of the set of all players, 1. Also, all games considered are constant-sum. For definitions of n-person games, see [1].

§2. STRUCTURE OF THE GAMES

We are interested in the following type of games:

DEFINITION 1. An n-person game G is extreme provided

[1] The material in this paper is included in a thesis presented in partial fulfilment of the requirements for a Ph.D. at the University of Pennsylvania. The paper was presented to the American Mathematical Society, at the Summer Meeting in Ann Arbor, August 30 - September 2, 1955.

there do not exist two different games G_1 and G_2 such that

$$(1) \qquad v(S) = av_1(S) + (1 - a)v_2(S) \text{ for all } S \subset (1, \ldots, n)$$

where $0 < a < 1$, and v, v_1, and v_2 are the characteristic functions of the games.

[The notion of an extreme game does not depend on the normalization. The general definition is:

An n-person game G is <u>extreme</u> provided that if there exist games G_1 and G_2 with characteristic functions v_1 and v_2 such that

$$(1') \qquad v(S) = v_1(S) + v_2(S) \text{ for all } S \subset (1, \ldots, n) \quad ,$$

then G, G_1, and G_2 are strategically equivalent [1].

However, since we use the normalization throughout the paper, we shall use the first definition.]

Obviously, all simple games (i.e., games such that $v(S) = 0$ or 1 for $S \subset (1, \ldots, n)$) are extreme. For, otherwise, G_1 and G_2 would have sets with values greater than one, an impossibility. The converse is true for $n \leq 4$, but fails for $n > 4$. We are interested in these failures and we shall exhibit them all for $n = 5$. Let us first note a rather useful, though simple, lemma related to equation (1).

LEMMA 1. Suppose G, G_1 and G_2 are games satisfying equation (1), and $v(S_1) + v(S_2) = v(S_1 \cup S_2)$, where $S_1 \cap S_2 = \emptyset$. Then $v_i(S_1) + v_i(S_2) = v_i(S_1 \cup S_2)$ for $i = 1, 2$.

PROOF. Suppose G, G_1 and G_2 are games satisfying equation (1), and $v(S_1) + v(S_2) = v(S_1 \cup S_2)$. Then

$$av_1(S_1) + (1 - a)v_2(S_1) + av_1(S_2) + (1 - a)v_2(S_2)$$
$$= av_1(S_1 \cup S_2) + (1 - a)v_2(S_1 \cup S_2).$$

If $v_1(S_1) + v_1(S_2) < v_1(S_1 \cup S_2)$, then $v_2(S_1) + v_2(S_2) > v_2(S_1 \cup S_2)$, which is impossible since the characteristic functions of all n-person games are superadditive [1], i.e., $v(S) + v(T) \leq v(S \cup T)$ for $S \cap T = \emptyset$.

Intuitively, we see that except for such obvious cases as simple games, some sort of chain of coalitions is needed to show that (1) cannot hold for a given game, and thus, that it is extreme. The following chain is the one we need for five-person extreme games:

DEFINITION 2. A <u>K-chain</u> of coalitions for a game G is a chain $T_1, \ldots, T_{2K}$ of subsets of $(1, \ldots, n)$ such that

i) $T_1 = T_{2K}$
ii) $T_j \cap T_{j+1} = \emptyset \qquad (j = 1, \ldots, 2K - 1)$
iii) $v(T_j) + v(T_{j+1}) = 1 \qquad (j = 1, \ldots, 2K - 1)$.

It follows immediately from the definition that $v(T_j) = 1/2$ for $j = 1, \ldots, 2K$.

Now we can give a necessary and sufficient condition for five-person extreme games.

THEOREM 1. G is an extreme, five-person (constant-sum) game $\Longleftrightarrow v(S) \neq 0, 1$ for $S \subset (1, \ldots, 5) \Longrightarrow$ there exists a K-chain such that $S = T_1 = T_{2K}$.

PROOF. <u>Sufficiency</u>. Suppose G were not extreme. Then there exist G_1, G_2 and $\underline{a}$ satisfying (1). Thus, there exists an S_o such that $v(S_o) = 1/2$ and $v_1(S_o) < 1/2$ or $v_2(S_o) < 1/2$ (since by the lemma, $v(S_o) = 0$ or 1 implies $v_i(S_o) = 0$ or 1). By the hypothesis, there exists a K-chain starting and ending with S_o.

Without loss of generality, assume $v_1(S_o) < 1/2$. Therefore, $v_2(S_o) > 1/2$, which implies $v_2(T_2) < 1/2 \Longrightarrow v_1(T_2) > 1/2 \Longrightarrow v_1(T_3) < 1/2 \Longrightarrow \ldots \Longrightarrow v_2(T_{2K-1}) > 1/2 \Longrightarrow v_2(T_{2K}) < 1/2 \Longrightarrow v_1(T_{2K}) > 1/2$, i.e., $v_1(S_o) > 1/2$, a contradiction.

<u>Necessity</u>. Suppose the theorem is false. Then there exists an extreme five-person game G and a set $S_o \subset (1, \ldots, 5)$ such that $v(S_o) \neq 0, 1$, and there does not exist a K-chain starting and ending with S_o.

Form all chains $C = \{P_1, P_2, \ldots\}$ such that

1) $P_1 = S_o$,
2) $P_i \cap P_{i+1} = \emptyset$, for $i = 1, 2, \ldots,$ and
3) $v(P_i) + v(P_{i+1}) = 1$ for $i = 1, 2, \ldots$.

Since the game is normalized, all P_i's are two or three person coalitions.

Let

$$\alpha = \min_{\substack{S,T \\ T \subset S \\ v(S) > v(T) + v(S-T)}} \{v(S) - v(T) - v(S - T)\} .$$

Thus $\alpha > 0$ since there are only a finite number of such pairs S, T. Let

$$v_1(P) = \begin{cases} v(P) + \alpha/4 & \text{if } P \text{ is an odd link in a chain,} \\ v(P) - \alpha/4 & \text{if } P \text{ is an even link in a chain,} \\ v(P) & \text{otherwise} . \end{cases}$$

$$v_2(P) = \begin{cases} v(P) - \alpha/4 & \text{if } P \text{ is an odd link in a chain,} \\ v(P) + \alpha/4 & \text{if } P \text{ is an even link in a chain,} \\ v(P) & \text{otherwise} . \end{cases}$$

First let us show $v_i(P)$ $(i = 1, 2)$ is uniquely defined for i, P. Obviously, we need only look at v_1 since v_2 is similarly defined. The only possible trouble would occur if there exists $P \subset (1, \ldots, 5)$ which is even in one chain, odd in another. But if this were so and if we adjoined at P the portions of the two chains from S_o to P, then we have a K-chain. Thus, we have uniqueness.

Now $v(P) = 1/2\{v_1(P) + v_2(P)\}$ for $P \subset (1, \ldots, 5)$. Thus, if v_1 and v_2 are characteristic functions of five-person games, then G is not extreme. But this can be shown as follows:

(a) By definition of α, v_1, and v_2, $0 \leq v_1(P)$, $v_2(P) \leq 1$.

(b) $v_1(\emptyset) = v_2(\emptyset) = v(\emptyset) = 0$ since $\emptyset$ is in no chain.

(c) $v_i(- S) = 1 - v_i(S)$.

For, if S is not in a chain, neither is $-S$, and $v_i(S) = v(S)$. If S is in a chain, there exists another chain with S and $-S$ as adjacent links (and so of different parity); thus we have $v_i(S) + v_i(-S) = 1 + \alpha/4 - \alpha/4 = 1$

(d) $v_i(S \cup T) \geq v_i(S) + v_i(T)$ if $S \cap T = \emptyset$

For, if $v(S \cup T) > v(S) + v(T)$, then $v_i(S \cup T) > v_i(S) + v_i(T)$ by definition of v_i and α. On the other hand, if $v(S \cup T) = v(S) + v(T)$, then we consider three cases:

(1) If $|S \cup T| = 2$ where "$|R|$" means "the number of players in R", then S or T is void, in which case (d) is trivially satisfied, or $|S| = |T| = 1$, $v(S) = v(T) = v(S \cup T)$, and neither S nor T is in a chain.

(2) If $|S \cup T| = 3$, then either $|S|$ or $|T| = 1$, say $|S|$

(the void case has been taken care of in (1)). Thus $v(S) = 0$, and $v(S \cup T) = v(T)$. But then, $v(T) + v(-(S \cup T)) = 1$, and if either T or $(S \cup T)$ is in a chain, then so is the other, and their link numbers are of the same parity. Thus $v_i(T) = v_i(T \cup S)$.

(3) If $|S \cup T| = 4$ or 5, then $v(S \cup T) = 1$. Either S and T are both in a chain and of opposite parity, and so $v_i(S) + v_i(T) = 1 + \alpha/4 - \alpha/4$, or neither is in a chain, in which case $v = v_i$ for S, T.

Thus, v_1 and v_2 are characteristic functions of five-person games [1], and G is not extreme, a contradiction.

Note that in the proof of the sufficiency part of the theorem, we did not use the fact that the game had five players. Thus, we have

COROLLARY 1. If G is an n-person game and $v(S) \neq 0, 1$ for $S \subset (1, \ldots, n) \Longrightarrow$ there exists a K-chain starting with S, then G is extreme.

Unfortunately, the converse of this is not true as can be seen by the six-person game given by:

$$\begin{aligned}
v(S) &= 0 && \text{for } |S| = 0 \text{ or } 1\\
v(S) &= 1/3 && \text{for } |S| = 2\\
v(1jk) &= 1/3 && \\
v(rmn) &= 2/3 && \text{for } r, m, n \neq 1\\
v(S) &= 2/3 && \text{for } |S| = 4\\
v(S) &= 1 && \text{for } |S| = 5 \text{ or } 6 .
\end{aligned}$$

That this game is extreme is easily seen. For if not, there exist G_1 and G_2 and an $S \subset (1, \ldots, 6)$ such that $v_1(S) < v(S)$. Without loss of generality, $|S| = 2$. For if $|S| = 4$, we can use $-S$ (and G_2 not G_1), if $S = (1jk)$, then by the lemma $v_1(jk) = v_1(1jk) < v(1jk) = v(jk)$, and if $S = (rmn)$, then we use $-S$ as above. Now by repeated use of the lemma, $v_1(jk) < 1/3$ for all $j \neq k$. But this contradicts the lemma since $1 > v_1(12) + v_1(34) + v_1(56) = v_1(12) + v_1(3456) = v_1(1, \ldots, 6) = 1$.

Actually Corollary 1 is only one of many, even more general, sufficiency conditions. However, no non-trivial one that we know now is also necessary, and therefore, we shall not record them all here.

Theorem 1 also gives the following corollary which enables us to write down immediately all five-person extreme games:

COROLLARY 2. G is an extreme, non-simple five-person game $\Longleftrightarrow v(S) = 0, 1/2,$ or 1 for $S \subset (1, \ldots, 5)$, and there exists a K-chain having six links (thus five different ones).

PROOF. Sufficiency. Every K-chain contains a link disjoint from every two-person set. For if not and (12) is the set, then all odd links in the chain contain 1, and all even ones contain 2, or vice versa. Thus the first and last link (the 2Kth) cannot be the same. Thus, if $T_1, \ldots, T_6$ is our K-chain and S is a two-person set such that $v(S) = 1/2$, then if $S \notin (T_1, \ldots, T_6)$, but $S \cap T_j = \emptyset$, then a K-chain starting with S is $S, T_j, T_{j-1}, \ldots, T_1, T_5, T_4, \ldots, T_j, S$ since $T_1 = T_6$. If $S = T_j$, then we have $T_j, T_{j-1}, \ldots, T_1, T_5, T_4, \ldots, T_j$ as our K-chain. If $|S| = 3$, then we use $-S$ in place of S above, and then add S to the beginning and end of the resulting K-chain. Thus by Theorem 1, G is extreme.

Necessity. Suppose G is extreme, and there is no six-link K-chain. By Theorem 1 we know that if the game is non-simple, there exists a K-chain. We can see that any three-person set in the chain can be deleted along with one of its neighbors since there is a unique set (having at least two players) disjoint from a three-person set. Thus, we need only consider K-chains with links of two players. Likewise, if any interior link is repeated, we can delete all the links between the repetitions and still have a K-chain. Any K-chain of this simplified form will start: (ab), (cd), (ae), (bc), ... where a, b, c, d, e are the numbers 1, ..., 5 in some order. Thus, we see that at least two more links are needed to finish the chain. Since (de) for the fifth link would permit a six-link chain ending (ab), the fifth link must be (ad). Likewise, the next link must be (ce) since the only other possibility, (be), would allow a six-link chain from (cd). Thus, we have so far (ab), (cd), (ae), (bc), (ad), (ce). Since we don't allow repeats, the seventh link is (bd). But now we can make the six-link chain, (bd), (ae), (cd), (ab), (ce), (bd), which contradicts our assumption.

Since, in the above proof, we showed that every two-element set is disjoint from at least one link in a K-chain, and since every extreme, non-simple game has a six link K-chain with 1/2-valued links, we see that no two-person coalition in such a game has value 1. Thus, in a non-simple extreme game, the only coalitions with fewer than four players that may have the value 1 are the three-person coalitions.

Suppose that G is a five-person extreme game, and (ab), (cd), (ae), (bd), (ce), (ab) is a K-chain of this game (where a, b, c, d, e are 1, 2, 3, 4, 5 in some order). The only possible all winning (i.e., value 1) three-person sets are (acb), (cbe), (bed), (eda), and (dac). And by Corollary 2, any or all of these sets may have value 1, and G will still be extreme. But note that if the players are arranged in a ring in the order a, c, b, e, d, a, ..., then these coalitions are just the sets of three consecutive players in the ring. Thus, we have:

> COROLLARY 3. Any five-person, non-simple, extreme game can be described as follows:
>
> The players can be arranged in a particular order around a ring so that some sets (possibly none) of three consecutive players have value 1, and all other undetermined two- and three-person sets have value 1/2.

Thus, we see that there are eight different types of non-simple, extreme, five-person games (not counting permutations). Adding these to the six different five-person simple games [1], we get 14 different types of five-person extreme games.

We can also count all extreme, five-person games (including permutations) by means of Corollary 3. We find that there are <u>218</u> non-simple games and <u>76</u> simple games. Thus, in all, there are <u>294 extreme points</u> in the ten-dimensional polyhedron of five-person games. For comparison, recall that the four-person games form a three-dimensional cube (8 extreme points).

§3. A GENERAL SOLUTION

The completely discriminatory solutions of simple games are the sets of imputations $V_S = \{\alpha \mid \Sigma_{i \in S}\alpha_i = 1\}$ where S is a minimal winning coalition [1]. We will show a generalization of this solution which holds for all five-person extreme games.

> DEFINITIONS. If v is the characteristic function of a game G, then the set of <u>minimal winning coalitions of</u> G is

$$W^m = \{S \mid v(S) = 1 \quad \text{and} \quad v(T) < 1 \quad \text{for} \quad T \subset S \quad \text{and} \quad T \neq S\} .$$

> Let the <u>indecomposable minimal winning sets</u> be the elements of

$$I^m = \{S \mid S \in W^m \quad \text{and} \quad T \subset S, \quad T \neq S, \quad \text{and} \quad T \neq \emptyset \Longrightarrow v(S - T) < 1 - v(T)\} .$$

From these definitions we can now get the following:

THEOREM 2. Let G be an extreme five-person game. Let

$$V_S = \{\alpha \mid \Sigma_S \alpha_i = 1 \quad \text{and} \quad \Sigma_T \alpha_i \geq v(T) \quad \text{for} \quad T \subset S\} \quad \text{for } S \in W^m .$$

Then

(i) if $S \in I^m$, V_S is a solution of G;

(ii) if $I^m = \emptyset$, then $V = \bigcup_{S \in W^m} V_S$ is a solution of G.

PROOF. For simple games, (i) holds since it is the completely discriminatory solution [1]. For non-simple games, if $S \in I^m$, then $|S| = 3$. Consider such an S in such a game. Without loss of generality, $S = (123)$. For V_S to be a solution, we must have:

a) $V_S \cap \text{dom } V_S = \emptyset$. This follows from the fact that $\alpha \in V_S$, $\alpha \underset{T}{\succ} \beta \Longrightarrow T \subset (123)$, and if $\beta \in V$, then $\Sigma_T \alpha_i > \Sigma_T \beta_i \geq v(T)$, and thus, T is not effective.

b) $V_S \cup \text{dom } V_S =$ the set of all imputations. For this we need only consider $\gamma \notin V_S$ such that $\Sigma_{i=1}^3 \gamma_i = 1$ since if $\Sigma_{i=1}^3 \gamma_i < 1$, γ is dominated via (123) by an imputation δ with the desired property. Any domination of δ via a subset of (123) is also a domination of γ. But for $\gamma \notin V_S$, there exists $T \subset (123)$ such that $\Sigma_T \gamma_i < v(T)$. Thus, T must be a two-person set, say (12). But then, $\gamma_3 > 1/2$ (since $v(T) = 1/2$), and there exists $\alpha \in V_S$ such that $\alpha_1 > \gamma_1$, $\alpha_2 > \gamma_2$ and $\alpha_3 = 1/2$. Thus $\alpha \underset{T}{\succ} \gamma$, and V_S is a solution.

For (ii), there is only one five-person extreme game with a void I^m. This is the game with all two- and three-player sets having value $1/2$. The minimal winning sets are the four-player sets, and each V_S consists only of the imputation having $\alpha_i = 1/4$ for $i \in S$. Thus $V = \{\alpha \mid \alpha = (1/4, 1/4, 1/4, 1/4, 0)$ and all permutations of components$\}$. It is easy to see that there is no inner domination, and since all imputations not in V have at least two components less than $1/4$, V dominates them. Thus V is a solution of this game.

It can also be seen that Theorem 2 does not hold for extreme games with more players. Indeed, neither of (i) and (ii) need hold. For we can show that (ii) is not satisfied by means of the extreme six-person game that was also the counterexample given for the generalization of Theorem 1. We immediately see that this game has no indecomposable minimal winning coalitions, and the minimal winning coalitions are the five-player sets. Let S_i be the five-player coalition not containing i, and V_i, the corresponding set of imputations. We see that V_1 is void, and V_i, for $i \neq 1$, is the one imputation given by $\alpha_i = 0$, $\alpha_1 = 1/9$, $\alpha_j = 2/9$,

$j \neq 1, i$. The union of these imputations is not a solution although it is contained in the following solution:

$$V = \begin{cases} (1/9,\ 2/9,\ 2/9,\ 2/9,\ 2/9,\ 0) \\ \text{and all permutations of the} \\ \text{last five components,} \\ \\ (1/9,\ 1/9,\ 1/9,\ 2/9,\ 2/9,\ 2/9) \\ \text{and all permutations of the} \\ \text{last five components.} \end{cases}$$

We can also show that (1) does not hold for a certain eight person game. Define the game as follows:

$$\begin{aligned}
&v(S) = 0 \quad \text{for} \quad |S| = 0, 1 \\
&v(ij) = 2/5 \quad \text{for} \quad (ij) \subset (12..5) \\
&v(im) = 1/5 \quad \text{for} \quad i \in (12..5),\ m \in (678) \\
&v(mn) = 0 \quad \text{for} \quad (mn) \subset (678) \\
&v(678) = 0 \\
&v(ijk) = 2/5 \quad \text{for} \quad (ij) \subset (12..5), \quad \text{any} \quad k \\
&v(imn) = 1/5 \quad \text{for} \quad i \in (12..5),\ (mn) \subset (678) \\
&v(ijk\ell) = 4/5 \quad \text{for} \quad (ijk\ell) \subset (12..5) \\
&v(ijkm) = 3/5 \quad \text{for} \quad (ijk) \subset (12..5),\ m \in (678) \\
&v(S) = 1 - v(-S) \quad \text{for all other} \quad s \subset (12..8).
\end{aligned}$$

That this game is extreme also follows from repeated use of the lemma. For, if equation (1) is satisfied, then by superadditivity, for some $(ij) \subset (12..5)$, $v_1(ij)$ or $v_2(ij) = 2/5 + \epsilon$, say $v_1(ij)$. But since $v(k\ell m) = 2/5$ for all $(k\ell m) \subset (12..5)$, then by the lemma $v_1(k\ell) = 2/5 + \epsilon$ for all $(k\ell) \subset (12..5)$. Hence, $v_1(ijk\ell) = 4/5 + 2\epsilon$ for all $(ijk\ell) \subset (12..5)$, and therefore $v_1(im) = 1/5 - 2\epsilon$ for all $i \in (12..5)$, $m \in (678)$. But then $v_1(12) + v_1(36) + v_1(47) + v_1(58) = 2/5 + \epsilon + 3/5 - 6\epsilon = 1 - 5\epsilon < 1$, which is impossible (by the lemma). Hence the game is extreme. That (1) is not satisfied is seen by considering V_S for $S = (12345)$, an indecomposable minimal winning coalition. $V_S = \{(1/5, 1/5, 1/5, 1/5, 1/5, 0, 0, 0)\}$, a single imputation which obviously is not a solution.

The solution of the non-simple game with no three-person winning sets, given by Theorem 2, is unique among the solutions given by that theorem in that it is finite (i.e., it contains only a finite number of imputations). It is rather interesting to note that finite solutions can be found for all the five-person extreme games but two, and even for these two we have not proved that such solutions do not exist. If we consider the players as being arranged in numerical order 1, 2, 3, 4, 5 (for

Corollary 3), then the game having only the two three-person winning sets (123) and (512) and the game having only the three three-person winning sets (123), (234), and (345) have no known finite solutions. The simple five-person games all have main simple solutions which, of course, are finite [1]. The other extreme five-person games all have finite solutions based on values for the components of the imputations of 0, 1/4, and 1/2 or 0, 1/9, 1/6, and 1/3. Indeed, the extreme game with four three-person winning sets and the other extreme game with just two such sets (e.g., (123) and (345) if we use the numerical order) each have a finite solution based on each set of values. However, we do not attempt here to solve the problem of finite solutions for these games.

Acknowledgement. The author wishes to thank Professor Harold Kuhn of Bryn Mawr College for his suggestions and helpful criticism.

BIBLIOGRAPHY

[1] von NEUMANN, J. and MORGENSTERN, O., Theory of Games and Economic Behavior, Ch. X, Princeton 1944, Second edition, 1947.

Herbert M. Gurk

University of Pennsylvania

EXTREME GAMES WITH THREE VALUES[1]

James H. Griesmer

§1. INTRODUCTION

In any fixed normalization, the set of n-person games can be represented by a convex polyhedron with the characteristic function of a game providing the coordinates of the corresponding point. A game is <u>extreme</u> if it is represented by an extreme point of this polyhedron. Since we shall use the (0, 1)-normalization exclusively, we shall use the following

> DEFINITION 1.1. A game v in (0, 1)-normalized form is extreme if and only if v cannot be written as a sum
>
> $$v = 1/2v_1 + 1/2v_2$$
>
> where $v \neq v_1, v_2$, and v_1, v_2 are games in (0, 1)-normalized form.

In general, the following definition obtains if no normalization is assumed:

> DEFINITION 1.2. A game v is extreme if and only if v cannot be written as a sum
>
> $$v = v_1 + v_2$$
>
> where v_1, v_2 are games and v_1, v_2 are not S-equivalent to v.

We use Definition 1.2 in the next section when we introduce the

---

[1] This is part of the author's doctoral dissertation, Department of Mathematics, Princeton University, which was supported (in part) by the Office of Naval Research Logistics Project, Princeton University.

flow assignment method for testing extremeness. Thereafter, we shall deal with normalized games, either partially normalized so that one-element coalitions have value 0 or (0, 1)-normalized. In particular, we will study constant-sum extreme games which have three values after normalization.

Extreme games with two values, or simple games, have been studied extensively. (See [6], [7], [8].) For $n = 3$ and 4, simple games comprise the entire set of extreme games. However, for $n \geq 5$, the class of extreme games is a much larger class of games. Gurk [5] characterized the five-person case completely, and showed that the set of extreme games consists of simple games, plus some games with values $\{0, 1/2, 1\}$. It is this class of extreme games, the set of extreme games with three values, which we study in this paper.

In Section 3, we characterize constant-sum extreme games with three values, and give an interpretation of the condition which guarantees extremeness. A geometric model is introduced in Section 4 whose properties can be used to decide the extremeness question for these games. In Section 5, properties of the system of inequalities, which define the set of all (0, 1)-normalized, n-person, constant-sum games, are given which characterize the combined set of extreme games with two values and extreme games with three values. Solutions for extreme games with three values are discussed in Section 6. Discriminatory solutions of two types are described: in some cases, a minimal winning coalition can completely discriminate against the remaining players; in other cases, two disjoint minimal intermediate coalitions discriminate against the complement of their union. Additional solutions for each game in the five-person case, and at least one solution for each six-person game with three values are obtained. A finite solution for the case when an extreme game with three values is a symmetric game is also given. Again, the vital sets [2], or coalitions on which the solutions are based, are the minimal winning coalitions and the minimal intermediate coalitions.

§2. THE FLOW ASSIGNMENT METHOD FOR TESTING EXTREMENESS

This method of testing whether or not a game is extreme was suggested by Professor H. W. Kuhn. Although we will require it for constant-sum games only, it can be applied to general-sum games and we shall prove it for this case. The games considered need not be normalized. Before introducing the method, we need the following lemma which is a more general form of Gurk's key lemma [5], and which is a direct consequence of Lemma 4 of [3].

LEMMA 2.1. If v, v_1, v_2 are n-person games such that

$$v = v_1 + v_2 \ ,$$

then

$$v(S) + v(T) = v(S \cup T),\ S \cap T = \emptyset \Longrightarrow v_i(S) + v_i(T) = v_i(S \cup T) \ ,$$

for $i = 1, 2$.

DEFINITION 2.1. By an <u>assignment of flows</u> to the non-empty coalitions of an n-person game v, we mean a real-valued function I which satisfies the following conditions:

(i) $I(I_n) = 0$, where I_n is the whole player set,

(ii) $I(\{i\}) = 0 \qquad i \in I_n$,

(iii) $I(S) + I(T) = I(S \cup T)$ if $v(S) + v(T) = v(S \cup T)$ and $S \cap T = \emptyset$.

THEOREM 2.1. A n-person game v is extreme $\Longleftrightarrow$ the only possible assignment of flows is the zero flow, i.e., $I(S) \equiv 0$.

PROOF. Suppose v is not extreme, i.e., $v = v_1 + v_2$, where v is not S-equivalent to v_1, v_2. We shall show v has a non-zero flow assignment. Consider v and v_1 in (0, 1)-normalized form, say m and m_1, respectively. Since v is not S-equivalent to v_1, $m \neq m_1$, and hence there exists S_o such that $m_1(S_o) \neq m(S_o)$. Let

$$I(S) = m(S) - m_1(S) \qquad S \subseteq I_n \ .$$

Clearly, $I(S) \not\equiv 0$, and

(i) $I(I_n) = 0$,

(ii) $I(\{i\}) = 0 \qquad i \in I_n$.

We now show

(iii) $v(S) + v(T) = v(S \cup T),\ S \cap T = \emptyset \Longrightarrow I(S) + I(T) = I(S \cup T)$.

By Lemma 2.1, $v(S) + v(T) = v(S \cup T) \Longrightarrow v_1(S) + v_1(T) = v_1(S \cup T)$. Also, $v(S) + v(T) = v(S \cup T) \Longrightarrow m(S) + m(T) = m(S \cup T)$, and $v_1(S) + v_1(T) = v_1(S \cup T) \Longrightarrow m_1(S) + m_1(T) = m_1(S \cup T)$. Hence if

$v(S) + v(T) = v(S \cup T)$, then

$$\begin{aligned} I(S \cup T) &= m(S \cup T) - m_1(S \cup T) \\ &= m(S) + m(T) - m_1(S) - m_1(T) \\ &= I(S) + I(T) \quad . \end{aligned}$$

Therefore, $I(S)$ is a non-zero flow assignment.

Suppose v has a non-zero flow assignment I. We show v is not extreme. Define

$$M = \max_{S \subseteq I_n} |I(S)|$$

$$\alpha = \min_{S \supset Q} \{v(S) - v(Q) - v(S - Q)\}$$

$$v(S) > v(Q) + v(S - Q) \quad .$$

Let

$$v_1(S) = 1/2v(S) + \frac{I(S)}{8M}\alpha, \; v_1(\emptyset) = 0$$

$$v_2(S) = 1/2v(S) - \frac{I(S)}{8M}\alpha, \; v_2(\emptyset) = 0 \quad .$$

Clearly $v = v_1 + v_2$. We show v_1 is a game not S-equivalent to v. The proof for v_2 is similar.

(i) $v_1(\emptyset) = 0$.

(ii) $v_1(S) + v_1(T) \leq v_1(S \cup T)$, $\qquad S \cap T = \emptyset$.

If $v(S) + v(T) = v(S \cup T)$, then since $I(S) + I(T) = I(S \cup T)$, we have

$$\begin{aligned} v_1(S) + v_1(T) &= \tfrac{1}{2} v(S) + \frac{I(S)}{8M}\alpha + \tfrac{1}{2} v(T) + \frac{I(T)}{8M}\alpha \\ &= \tfrac{1}{2} v(S \cup T) + \frac{I(S \cup T)}{8M}\alpha = v_1(S \cup T) \quad . \end{aligned}$$

If $v(S) + v(T) < v(S \cup T)$, then, since $v(S) + v(T) \leq v(S \cup T)$ we have

$$v_1(S) + v_1(T) = \frac{1}{2} v(S) + \frac{I(S)}{8M}\alpha + \frac{1}{2} v(T) + \frac{I(T)}{8M}\alpha$$

$$\leq \frac{1}{2} v(S \cup T) + \frac{I(S)}{8M}\alpha + \frac{I(T)}{8M}\alpha - \frac{\alpha}{2}$$

$$= v_1(S \cup T) - \frac{I(S \cup T)}{8M}\alpha + \frac{I(S)}{8M}\alpha + \frac{I(T)}{8M}\alpha - \frac{\alpha}{2}$$

$$\leq v_1(S \cup T) + \frac{\alpha}{8} + \frac{\alpha}{8} + \frac{\alpha}{8} - \frac{\alpha}{2}$$

$$< v_1(S \cup T) .$$

(iii) If v is constant-sum, so is v_1, using (ii).

Hence, v_1 is a game. We now show v_1 is not S-equivalent to v by showing their (0,1)-normalized forms, say m and m_1, respectively, are not equal. Since $I(S) \neq 0$, there exists $S_0 \subset I_n$ such that $v_1(S_0) \neq 1/2\ v(S_0)$. Since $I(\{i\}) = 0$, for $i \in I_n$, and $I(I_n) = 0$, $v_1(\{i\}) = 1/2\ v(\{i\})$, for $i \in I_n$, and $v_1(I_n) = 1/2\ v(I_n)$. Hence, $m(S_0) \neq m_1(S_0)$ which implies that v_1 is not S-equivalent to v.

Similarly, v_2 is a game not S-equivalent to v, so v is not extreme.

We think of an assignment of flows in the following way: each triple $S, T, S \cup T$ such that

$$v(S) + v(T) = v(S \cup T) \qquad S \cap T = \emptyset$$

corresponds to a node with three directed edges incident to it; two of the edges are directed toward the node and are assigned to S and T; the remaining edge is directed away from the node and is assigned to $S \cup T$.

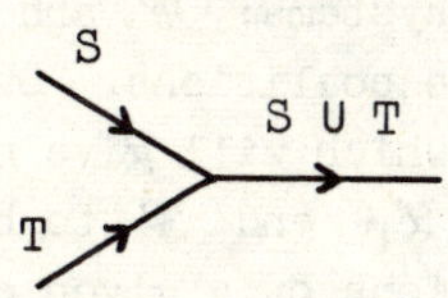

Since the equation

$$v(S) + v(T) = v(S \cup T)$$

holds, we know a like equation must hold for the function I:

$$I(S) + I(T) = I(S \cup T)$$

and this we think of as a condition regarding flows in edges incident to the given node, $I(S)$, $I(T)$, and $I(S \cup T)$ being the respective flows. We make such a correspondence for each such triple, S, T, $S \cup T$, and attempt to find a non-zero flow assignment.

In the following example, the flow assignment method is used to decide the extremeness question for a five-person game.

EXAMPLE 2.1.

$$v(S) = \begin{cases} 0 & (13)\ (14)\ (15)\ (24)\ (45)\ (145); \quad |S| = 0,\ 1 \\ 1/5 & (12)\ (23)\ (34)\ (123)\ (124)\ (234) \\ 2/5 & (35)\ (345)\ (135)\ (134)\ (1234)\ (1345) \\ 3/5 & (25)\ (235)\ (125)\ (1235) \\ 1 & (245)\ (2345)\ (1245)\ (12345) \quad . \end{cases}$$

We have immediately $v(S) = v(T)$ implies $I(S) = I(T)$. Let I_0, $I_{1/5}$, $I_{2/5}$, $I_{3/5}$, I_1 denote the respective flow assignments. I_0 and I_1 are both 0. Also $v(\{12\}) + v(\{34\}) = v(\{1234\}) \Longrightarrow 2\, I_{1/5} = I_{2/5}$; $v(\{12\}) + v(\{35\}) = v(\{1235\}) \Longrightarrow I_{1/5} + I_{2/5} = I_{3/5}$; $v(\{25\}) + v(\{134\}) = v(\{12345\}) \Longrightarrow I_{2/5} + I_{3/5} = I_1 = 0$. Hence $I(S) \equiv 0$ is the only possible flow assignment and v is extreme.

§3. CHARACTERIZATION THEOREMS

Just as the study of constant-sum games with two values, or strong simple games, began with a characterization of the two systems: $\mathcal{W}$ and $\mathcal{L}$, the systems of winning coalitions and losing coalitions, respectively [7], so in the case of constant-sum games with three values, we begin with a characterization of the three systems: $\mathcal{W}$ and $\mathcal{L}$, and the third system, $\mathcal{B}$, the system of intermediate coalitions. Our characterization will be given as part of Theorem 3.1, which will give necessary and sufficient conditions for three systems $\mathcal{W}$, $\mathcal{L}$, and $\mathcal{B}$ to be the systems of winning, losing and intermediate coalitions of a given constant-sum, essential game v. We define these systems which, for a given v, we denote by $\mathcal{W}_v$, $\mathcal{L}_v$, and $\mathcal{B}_v$, respectively. As in [7, p. 423], the losing coalitions are the flat coalitions:

$$S \in \mathcal{L}_v \Longleftrightarrow \sum_{i \in S} v(\{i\}) = v(S) \quad .$$

A winning coalition is one whose complement is flat:

$$S \in \mathcal{W}_v \Longleftrightarrow \sum_{i \in I_n - S} v(\{i\}) = v(I_n - S) \quad .$$

An intermediate coalition is one whose "cohesive power" or "strength" is equal to that of its complement. If a coalition S forms, then the cumulative benefit to the players of S is the value of S less the amounts each player could have obtained separately, i.e.,

$$v(S) - \sum_{i \in S} v(\{i\}) \quad .$$

If the benefit corresponding to the formation of S is the same as that corresponding to the formation of $I_n - S$, then we say S is an intermediate coalition. Thus,

$$S \in \mathcal{B}_v \Longleftrightarrow v(S) - \sum_{i \in S} v(\{i\}) = v(I_n - S) - \sum_{i \in I_n - S} v(\{i\}) \quad .$$

We now give Theorem 3.1, but omit the proof, which is a straightforward generalization of the proof given in [7] characterizing the systems of winning and losing coalitions.

THEOREM 3.1. In order that three given systems $\mathcal{W}$, $\mathcal{L}$, and $\mathcal{B}(\subseteq \mathcal{P}(I_n)$, the power set of the whole player coalition) be the $\mathcal{W}_v$, $\mathcal{L}_v$, and $\mathcal{B}_v$ of a constant-sum, essential game v, the following conditions are necessary and sufficient:

(A) (i) Complementation carries a winning coalition into a losing coalition, and vice versa.

(ii) Complementation carries an intermediate coalition into an intermediate coalition.

(iii) Every subset of a losing coalition is losing, or equivalently, every superset of a winning coalition is winning.

(iv) If S_1 and S_2 are intermediate, so is every T such that $S_1 \subset T \subset S_2$.

(v) The union of two disjoint intermediate coalitions is winning.

(B) $\mathcal{W} \cap \mathcal{L} = \emptyset, \quad \mathcal{W} \cap \mathcal{B} = \emptyset, \quad \mathcal{L} \cap \mathcal{B} = \emptyset$.

(C) The empty coalition and the one-element coalitions are losing.

Theorem 3.1 permits us to give an exact definition of an essential, constant-sum game with three values, viz., it is one for which

$$\text{(D)} \qquad \mathscr{W}_v \cup \mathscr{L}_v \cup \mathscr{B}_v = \mathscr{P}(I_n) \quad .$$

Since we are restricting ourselves to this class of games, our study could also be carried out as a study of those triples $\mathscr{W}$, $\mathscr{L}$, and $\mathscr{B}$ which satisfy (A), (B), (C), (D). Such a triple determines a game v up to S-equivalence, so to make the correspondence unique, we require v to be (0, 1)-normalized. The unique v will assign value 0 to the coalitions of $\mathscr{L}$, value 1/2 to the coalitions of $\mathscr{B}$, and value 1 to the coalitions of $\mathscr{W}$. We shall assume hereafter that this normalization has been carried out.

Although all constant-sum games with two values are extreme ($S \in \mathscr{L} \cup \mathscr{W} \Longrightarrow I(S) = 0$), not all constant-sum games with three values are extreme. We already have a four-person game (the "center of the cube") which has three values, but is not extreme. Hence, a condition must be added in order to characterize the extreme games with three values. This condition is the chain condition used by Gurk to characterize all five-person, constant-sum, extreme games [5]. Gurk also showed that it is a sufficient condition for extremeness for n-person games with $n \geq 6$. We complete the picture by showing it is a necessary condition when the extreme games are assumed to have three values.

DEFINITION 3.1. [5] A K-chain of coalitions for a game v is a chain or sequence of subsets of I_n,

$$T_1, \ldots, T_{2R}$$

such that

$$\text{(i)} \quad T_1 = T_{2R}$$

$$\text{(ii)} \quad T_i \cap T_{i+1} = \emptyset \qquad i = 1, \ldots, 2R-1$$

$$\text{(iii)} \quad v(T_i) + v(T_{i+1}) = 1 \qquad i = 1, \ldots, 2R-1 \quad .$$

Clearly, if T_i is in a K-chain, then $v(T_i) = 1/2$ and $I(T_i) = 0$, for any flow assignment I.

THEOREM 3.2. A necessary and sufficient condition that an n-person constant-sum game in (0, 1)-normalized form with values {0, 1/2, 1} be extreme is that each set S with value 1/2 has a K-chain which begins and ends with S.

PROOF. Sufficiency. Clearly, if each set with value $1/2$, i.e., each set in $\mathcal{B}$, has a K-chain beginning and ending with it, then any flow assignment I must assign zero flow to any intermediate coalition. Since $S \in \mathcal{L} \cup \mathcal{W}$, i.e., $v(S) = 0$ or 1, implies $I(S) = 0$ for any flow assignment, we have $I(S) \equiv 0$ as the only possible flow assignment and hence v is extreme.

Necessity. If there exists an intermediate coalition S_o such that no K-chain begins and ends with S_o, then, by forming all possible sequences

$$S_o, S_1, S_2, \dots ,$$

where

$$\text{(i)} \quad S_i \cap S_{i+1} = \emptyset$$

$$\text{(ii)} \quad v(S_i) + v(S_{i+1}) = 1 \quad ,$$

and defining

$$I(S) = \begin{cases} 1 & \text{if } S \text{ is an even-indexed coalition in some sequence,} \\ -1 & \text{if } S \text{ is an odd-indexed coalition in some sequence,} \\ 0 & \text{otherwise,} \end{cases}$$

it is easy to show that $I(S)$ is a well-defined flow assignment, and hence v is not extreme.

What interpretation can be given to this chain condition? It is a condition which restricts (or partially discourages) coalition formation for the system of intermediate coalitions. An extreme game v is one in which a minimum number of inequalities: $v(S) + v(T) \leq v(S \cup T)$ are satisfied as strict inequalities $(<)$; if any more equalities $(=)$ were demanded, v would be inessential. If v is an extreme game with three values, considering only intermediate coalitions, we have some coalition changes advantageous to these coalitions, but not every coalition change. In general, if an intermediate coalition S forms, we would expect it to try to unite with a losing coalition to form a winning coalition. (There would be no advantage to S to unite with another intermediate coalition T to form the winning coalition $S \cup T$, since $v(S) + v(T) = v(S \cup T)$.) If we consider only additions and deletions of losing coalitions, certain coalition changes — which we can represent as a sequence of intermediate coalitions, $S = S_1, S_2, \dots, S_p$, each differing from the preceding by a losing coalition, i.e.,

$$S_{i+1} = S_i \cup T \qquad S_i, S_{i+1} \in \mathcal{B}, T \in \mathcal{L}$$

or

$$S_{i+1} = S_i - T \qquad S_i, S_{i+1} \in \mathcal{B}, T \in \mathcal{L}$$

— can be seen by each member of the intermediate coalition S to be disadvantageous to himself. We can guarantee that such an outlook will be held regarding a certain sequence of coalition changes if we show this sequence leads from the given intermediate coalition S to its complement $S^{\sim}$, so that each player regards himself as being in danger of "being discarded" and, hence, will veto such a change. The chain condition of Gurk insures that every intermediate coalition S has at least one sequence of changes which takes it into its complement, viz., if

$$S = S_1, S_2, S_3, \ldots, S_{2R-1}, S_{2R} = S$$

is a K-chain beginning and ending with S, then

$$S = S_1, S_2^{\sim}, S_3, \ldots, S_{2R-1}, S_{2R}^{\sim} = S^{\sim}$$

is such a sequence.

If we consider only minimal intermediate coalitions, i.e., intermediate coalitions all of whose proper subsets are losing, since these are the intermediate coalitions most likely to form, then the extent to which these coalitions will try to add players will vary among extreme games with three values. In the extreme game

$$v(S) = \begin{cases} 0 & |S| = 0, 1 \\ 1/2 & |S| = 2, 3 \\ 1 & |S| = 4, 5 \end{cases}$$

as noted in [7, p. 333], the object of the players is to form two-element or minimal intermediate coalitions, and to attempt no further additions. On the other hand, in the extreme game

$$v(S) = \begin{cases} 0 & |S| = 0, 1;\ S = (12)(15)(24)(34)(35) \\ 1/2 & S = (13)(14)(45)(23)(25)(245)(235) \\ & \qquad (123)(145)(134) \\ 1 & |S| = 4, 5;\ S = (345)(234)(135)(125)(124) \end{cases}$$

while there are certain coalition changes which no two-element coalition will consider, there is also a coalition change which is clearly advantageous, viz., for each intermediate coalition $\{i, j\}$, there is a one-element (losing) coalition $\{k\}$, such that $\{i, j, k\}$ is winning.

§4. A GEOMETRIC MODEL

A linear graph can be associated with each constant-sum game with three values, whose properties can be used to decide the extremeness question for the game. The model was suggested by Professor H. W. Kuhn to explain the fact that, for the case $n = 5$, the extreme games can only assume values $\{0, 1\}$ or values $\{0, 1/2, 1\}$. However, just as the Gurk chain condition could be extended for $n \geq 6$, so can this model.

We work only with $\mathcal{B}$, the set of intermediate coalitions. Construct a linear graph $\mathcal{G}$, whose nodes are equal in number to the cardinality of $\mathcal{B}$. To each $S \in \mathcal{B}$, assign a node of $\mathcal{G}$. We connect the nodes assigned to S_1 and S_2 by an edge if and only if $S_1 \cap S_2 = \emptyset$.

EXAMPLE 4.1.
To the five-person game

$$v(S) = \begin{cases} 0 & |S| = 0, 1; \quad S = (15)(24)(34)(35) \\ 1/2 & S = (13)(14)(25)(23)(45)(12)(245) \\ & \quad (235)(134)(145)(123)(345) \\ 1 & |S| = 4, 5; \quad S = (234)(135)(125)(124) \end{cases}$$

we associate the following graph:

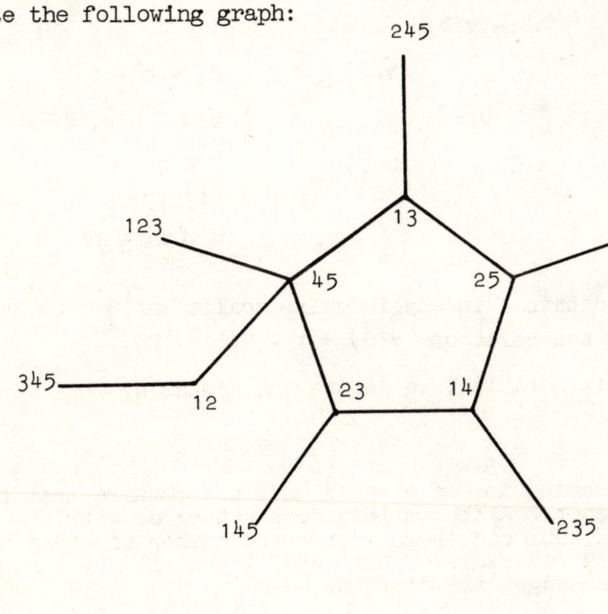

We have immediately:

THEOREM 4.1. Given v, a constant-sum game with three values, and associating a linear graph $\mathcal{G}$ with the set of intermediate coalitions $\mathcal{B}$ (nodes of $\mathcal{G} \longleftrightarrow$ sets of $\mathcal{B}$; edges of $\mathcal{G} \longleftrightarrow$ disjoint pairs of coalitions in $\mathcal{B}$), then in order that v be extreme it is necessary and sufficient that each node of $\mathcal{G}$ be part of a cycle[1] containing an odd number of edges.

PROOF. The proof follows immediately from the fact that $S_0 \in \mathcal{B}$ has a K-chain which begins and ends with it if and only if the node assigned to S_0 is part of a cycle with an odd number of edges.

§5. PROPERTIES OF THE SYSTEM OF INEQUALITIES

We now examine the system of inequalities which defines the (convex) set of all n-person constant-sum games in (0, 1)-normalized form. In particular, we establish properties of the system for the combined class of extreme games with two values and extreme games with three values.

The following system of linear inequalities defines the set of all n-person constant-sum games in (0, 1)-normalized form, for $n \geq 4$:

$$\text{(I)}\quad \begin{array}{lrl} \text{(i)} & -v(S) \leq 0 & |S| = 2 \\ \text{(ii)} & v(S) - v(S \cup \{i\}) \leq 0 & i \notin S,\ 2 \leq |S| \leq n-3 \\ \text{(iii)} & v(S) \leq 1 & |S| = n-2 \\ \text{(iv)} & v(S) + v(T) \leq 1 & S \cap T = \emptyset,\ |S \cup T| = n-1 \\ \text{(v)} & v(S) + v(T) - v(S \cup T) \leq 0 & S \cap T = \emptyset \\ & & |S|,\ |T| \geq 2 \\ & & |S \cup T| \leq n-2 \end{array}$$

where $\{n\}$ is not contained in any indexing coalition S. We determine $v(S)$ for $n \in S$ by the relation $v(S) = 1 - v(I_n - S)$.

Using matrix notation, we denote the system by

[1] We note that our definition of a cycle does not require that the edges of the cycle be distinct. Although this does not agree with the usual definition of a cycle, our requirement that the number of edges be odd insures that, within any odd cycle of our type, there will be a non-trivial cycle, i.e., one whose edges are distinct.

(I) $$Av \leq c \quad .$$

The system is a set of N linear inequalities in $2^{n-1} - n - 1$ unknowns, where

$$N = \binom{n-1}{2}\left[\frac{1}{2}\binom{2}{1}\right] + \binom{n-1}{3}\left[\binom{3}{1}\right] + \binom{n-1}{4}\left[\binom{4}{1} + \frac{1}{2}\binom{4}{2}\right] + \cdots$$

$$+ \begin{cases} \binom{n-1}{n-1}\left[\binom{n-1}{1} + \binom{n-1}{2} + \cdots + \frac{1}{2}\binom{n-1}{(n-1)/2}\right] & n \text{ odd} \\ \binom{n-1}{n-1}\left[\binom{n-1}{1} + \binom{n-1}{2} + \cdots + \binom{n-1}{(n-2)/2}\right] & n \text{ even} \quad . \end{cases}$$

The matrix of coefficients A is composed of entries which are either $+1$, 0, or -1. Any row of A has at most three non-zero entries.

If, for a given constant-sum game v, we restrict our attention to the subsystem of those inequalities which are satisfied as equations ($=$), we denote the "truncated" system by

$$A'v = c' \quad .$$

We now consider the class of constant sum games which have either two values or three values. The first theorem does not require the games to be extreme.

THEOREM 5.1. Let v be a constant-sum game in (0,1)-normalized form with values $\{0, 1\}$ or $\{0, 1/2, 1\}$. In the coefficient matrix A' of the subsystem of those inequalities which are satisfied as equations ($=$), any row which has three non-zero entries is linearly dependent on a set of rows, each having at most two non-zero entries.

PROOF. The existence of a row in A' with three non-zero entries means there exists an equation

$$v(S) + v(T) - v(S \cup T) = 0 \qquad S \cap T = \emptyset \quad .$$

One of the following four possibilities must obtain for the triple S, T, $S \cup T$:

Case 1: $S, T, S \cup T \in \mathcal{L}$.

Case 2: $S \in \mathcal{L}$; $T, S \cup T \in \mathcal{B}$.

Case 3: $S \in \mathcal{L}$; $T, S \cup T \in \mathcal{W}$.

Case 4: $S, T \in \mathcal{B}$; $S \cup T \in \mathcal{W}$.

We prove Case 1; the remaining cases can be demonstrated in a similar manner. Pick a two-element set $S_o \subseteq S$. Form a sequence

$$S_o,\ S_1 = S_o \cup \{i\},\quad S_2 = S_o \cup \{i, j\},\ \ldots,\ S_p = S_o \cup \{i, j, \ldots, m\} = S$$

where each succeeding S_q contains a new player from S not included in S_{q-1}. $S \in \mathcal{L} \implies S_q \in \mathcal{L}$, $0 \leq q \leq p - 1$. We thus have the following equations in the system $A'v = c'$:

$$\begin{aligned} - v(S_o) &= 0 \\ v(S_o) - v(S_1) &= 0 \\ &\vdots \\ v(S_{p-1}) - v(S) &= 0 \end{aligned}$$

and, clearly, adding the rows of the matrix A' corresponding to these equations, we obtain a vector with exactly one non-zero entry, a -1 in the "Sth" component. Call this vector E_S. Similarly, we can obtain a vector with -1 in the "Tth" component, say E_T, and a vector with -1 in the "S U Tth" component, say E_{SUT}. $- E_S - E_T + E_{SUT}$ is the original row in the matrix A', and it is clearly dependent on a set of rows, each having at most two non-zero entries.

REMARK. This theorem relates directly to the definition of constant-sum games with three values and with two values, and to the characterization of the systems $\mathcal{W}$, $\mathcal{L}$, and $\mathcal{B}$. Theorem 3.1 gives a characterization for the three systems, and reduces to a characterization for the systems $\mathcal{W}$ and $\mathcal{L}$ if $\mathcal{B} = \emptyset$. The characterization involves only properties of single sets and relations between pairs of sets, and Theorem 5.1 states that these are the independent relations of these games. (Although (A) (iv) of Theorem 3.1 is a condition relating three sets, it is dependent on (A) (iii) when (D) is assumed, i.e., when every coalition is either winning, losing, or intermediate.)

COROLLARY 5.1A. Suppose v is an extreme game in $(0, 1)$-normalized form with values either $\{0, 1\}$ or $\{0, 1/2, 1\}$, i.e., v is an extreme point of the convex polyhedron determined by the inequality system (I), and each component of v has value either 0, $1/2$, or 1. Then, since v is the unique solution to some fundamental subsystem of $2^{n-1} - n - 1$ linearly independent equations chosen from the system $A'v = c'$, we can choose this subsystem so that each row of the coefficient matrix has at most two non-zero entries.

As a converse, we have

THEOREM 5.2. Let v be an extreme game and, hence, the unique solution of some fundamental subsystem of linearly independent equations chosen from the subsystem $A'v = c'$. Suppose that this subsystem can be so chosen that each row of its coefficient matrix has at most two non-zero entries. Then v has values either $\{0, 1\}$ or $\{0, 1/2, 1\}$.

PROOF. Let this fundamental subsystem of $2^{n-1} - n - 1$ equations be denoted by

$$A''v = c'' \quad .$$

We first transform the matrix A'' by interchange operations. Bring any row which has only one non-zero entry to the upper part of A'' by a row interchange, and then bring the non-zero entry to the main diagonal by a column interchange. If there is any row which has a non-zero entry in a column which has been previously interchanged, bring its other non-zero entry to the main diagonal, after it has been moved to the upper part of A''. Continue these interchanges until we have remaining a $h \times h$ submatrix B in the lower right corner of A'' with two non-zero entries in each row, and all these entries appearing in the last h columns of A''. If any column in B has only one non-zero entry, bring this column to the left side of B, and the non-zero entry to the main diagonal. Continue this process (reducing by one the order of the matrix considered) until we have remaining a $k \times k$ submatrix C, each of whose rows and columns contains exactly two non-zero entries. Transform C by interchange operations into block diagonal form, each block C_i having the appearance

$$\begin{bmatrix} x & x & & & & \\ & x & x & & & \\ & & \cdot & & & \\ & & & \cdot & & \\ & & & & \cdot & \\ & & & & x & x \\ x & & & & & x \end{bmatrix}$$

where x marks a non-zero entry, either $+1$ or -1.

We now show that the only possible values a variable $v(S)$ can assume are 0, 1/2, or 1. To do this we could use Cramer's Rule, since v is the unique solution of the system $A''v = c''$. (It can be shown that the determinant of A'' is $\pm 2^m$, where m is the number of C_i blocks, and that the determinants of the matrices which result from successive substitution of the constant column c'' for each column of A'' are either 0, $\pm 2^{m-1}$, or $\pm 2^m$.) However, it is easier to evaluate the variables $v(S)$ directly, assuming that A'' is in the equivalent form we obtained. Each column of A'' is associated with a given variable $v(S)$, and we also associate the corresponding row with the variable.

(a) If a row has only one non-zero entry, clearly the corresponding variable $v(S)$ has value 0 or 1, according as the non-zero entry is -1 or $+1$.

(b) If a row has two non-zero entries, one on the main diagonal and one to the left of the main diagonal, but the row does not pass through a C_i submatrix, then the corresponding variable $v(S)$ has value either 0 or 1, since $v(S)$ is connected to a variable $v(T)$, corresponding to a preceding row, by one of the following relations:

$$v(S) = v(T)$$

or

$$v(S) = 1 - v(T) ,$$

and we already have $v(T) = 0$ or 1.

(c) If a row passes through a C_i block then we show the corresponding variable $v(S)$ has value 1/2. With no loss of generality, we may assume this row is the first row of C_i. The submatrix C_i must contain an odd number of rows having $(+1, +1)$ as non-zero entries; if C_i had an even number of such rows, then determinant $(C_i) = 0$, which would imply that determinant $(A'') = 0$ and would contradict the assumption that the subsystem $A''v = c''$ is linearly independent. If C_i is of order p, and the sequence of coalitions

$$S = S_1, S_2, \ldots, S_p$$

indexes the successive rows of C_i, we have two possibilities for the relations which link pairs of successive variables:

(i) If the row corresponding to $v(S_i)$ has non-zero entries $(+1, +1)$, then

$$v(S_i) + v(S_{i+1}) = 1 \quad .$$

(ii) If the row corresponding to $v(S_i)$ has non-zero entries $(+1, -1)$, then

$$v(S_i) = v(S_{i+1}) \quad ,$$

where $i = 1, \ldots, p$, and $S_{p+1} = S_1 = S$. Using the fact that the number of relations of type (i) is odd, we obtain, by successive elimination, a relation $v(S) = 1 - v(S)$, and hence $v(S) = 1/2$.

(d) Finally, if a row has two non-zero entries, one on the main diagonal and one to the right of the main diagonal, but the row does not pass through a C_i block, then the row occurs above the first block C_1; the corresponding variable $v(S)$ has value $1/2$, since $v(S)$ is connected to a variable $v(T)$, corresponding to a succeeding row, by one of the following relations:

$$v(S) = v(T)$$

or

$$v(S) = 1 - v(T) \quad ,$$

and we already have $v(T) = 1/2$.

Thus we have shown that $v(S)$ can only assume the values 0, 1/2, or 1, i.e., v is either an extreme game with values $\{0, 1\}$ or an extreme game with values $\{0, 1/2, 1\}$.

For the five-person case, we have the following system of inequalities defining the set of constant-sum games in (0,1)-normalized form:

6 inequalities of the form:

$$-v(S) \leq 0 \qquad |S| = 2,\ \{5\} \not\subset S.$$

12 inequalities of the form:

$$v(S) - v(S \cup \{i\}) \leq 0 \qquad |S| = 2,\ \{5\} \not\subset S \cup \{i\} \quad .$$

4 inequalities of the form:

$$v(S) \leq 1 \qquad |S| = 3,\ \{5\} \not\subset S \quad .$$

3 inequalities of the form:

$$v(S) + v(T) \leq 1 \qquad |S| = |T| = 2,\ \{5\} \not\subset S \cup T \quad .$$

Since no row of the coefficient matrix has more than two non-zero entries, we have

COROLLARY 5.2A. (Gurk) The set of five-person, constant-sum, extreme games in (0, 1)-normalized form consists of games whose values are either $\{0, 1\}$ or $\{0, 1/2, 1\}$.

§6. SOLUTIONS

In this section we give a family of discriminatory solutions for a class of extreme games with three values. The class contains all five-person and six-person extreme games with three values, and higher games where the number of intermediate coalitions and winning coalitions contained in a given winning coalition is "small". We also obtain a finite solution for every symmetric extreme game with three values. We assume all games are in (0, 1)-normalized form.

We first make the following definitions:

DEFINITION 6.1. [7] Let Δ be the <u>set of imputations</u>, i.e.,

$$\Delta = \left\{ \alpha = (\alpha_1, \ldots, \alpha_n) \;\middle|\; \sum_{i=1}^{n} \alpha_i = 1,\ \alpha_i \geq 0 \right\} .$$

We say α <u>dominates</u> β, or $\alpha \in\!\!\rightarrow \beta$, via a coalition S if

(i) $\alpha_i > \beta_i \qquad i \in S$

(ii) $\sum_{i \in S} \alpha_i \leq v(S)$.

Then $\mathcal{V}$ is a <u>solution</u> for a given game v if

$$\mathcal{V} = \Delta - \operatorname{dom} \mathcal{V}$$

where $\operatorname{dom} \mathcal{V} = \{\beta \mid \alpha \in\!\!\rightarrow \beta$ via $S \subset I_n$ for some $\alpha \in \mathcal{V}\}$.

DEFINITION 6.2. The set of minimal winning coalitions $\mathcal{W}^{m}$, is the set of all $S \in \mathcal{W}$ satisfying

$$T \subset S \text{ and } T \neq S \Longrightarrow v(T) < 1 .$$

DEFINITION 6.3. The set of minimal intermediate

coalitions, $\mathscr{B}^m$, is the set of all $S \in \mathscr{B}$ satisfying

$$T \subset S \text{ and } T \neq S \Longrightarrow v(T) < 1/2 \quad .$$

In extreme games with three values, we have found two types of discriminatory solutions: first, when a coalition in $\mathscr{W}^m$ discriminates against its complement; second, when a pair of disjoint coalitions in $\mathscr{B}^m$ discriminate against the complement of their union. In the first case, such discrimination occurs when a coalition in $\mathscr{W}^m$ is indecomposable [5], i.e., when there exists $S \in \mathscr{W}^m$ such that $T \subset S$, $T \neq \emptyset$ and S implies $v(T) + v(S - T) < 1$; the discriminatory solution has the form

$$\mathscr{V}_S = \left\{ \alpha \;\middle|\; \sum_{i \in S} \alpha_i = 1; \quad \sum_{i \in T} \alpha_i \geq v(T), \; T \subset S \right\} .$$

However, the condition that S be indecomposable is not sufficient to guarantee that $\mathscr{V}_S$ is a solution. Only if S is a three-element or a four-element coalition, can it be shown that $\mathscr{V}_S$ is always a solution. For example, in the following extreme game with three values:

$$v(S) = \begin{cases} 0 & |S| = 0, 1, 2 \\ 1/2 & |S| = 3, 4 \\ 1 & |S| = 5, 6, 7 \end{cases}$$

each five-element coalition S is indecomposable, but $\mathscr{V}_S$ is not a solution. Gurk gave a solution $\mathscr{V}_S$ for each five-person extreme game (with two, as well as with three, values) with the single exception of the symmetric game in which $\mathscr{B}$ consists of all two-element and three-element coalitions [5].

The second kind of discrimination occurs when two disjoint coalitions of $\mathscr{B}^m$ form. We give a family of solutions (see [4] for proofs) as Theorem 6.1, which serves to provide at least one solution for each five-person and six-person extreme game with three values, and can be used for higher games if the number of ways in which the members of the two disjoint coalitions can form intermediate or winning coalitions is "small". Let S and T be the two disjoint coalitions in $\mathscr{B}^m$; then $S \cup T \in \mathscr{W}$. We shall assume that $R \in \mathscr{W}$, $R \subset S \cup T$ implies either that $R \supset S$ or $R \supset T$. Let U and V be disjoint sets such that $U \cup V = S \cup T$. We define the following sets of imputations:

$$\mathscr{V}_1 = \left\{ \alpha \;\middle|\; \sum_{i\in S} \alpha_i = \sum_{i\in T} \alpha_i = 1/2,\ \sum_{i\in U} \alpha_i \geq v(U),\ U \subset S \cup T,\ U \in \mathscr{B} \right\}$$

$$\mathscr{V}_2 = \left\{ \alpha \;\middle|\; \sum_{i\in S} \alpha_i = 1/2,\ \sum_{i\in T\cap U} \alpha_i = 1/4,\ \sum_{i\in T\cap V} \alpha_i = 1/4 \right\}$$

$$\mathscr{V}_3 = \left\{ \alpha \;\middle|\; \sum_{i\in S\cap U} \alpha_i = 1/4,\ \sum_{i\in S\cap V} \alpha_i = 1/4,\ \sum_{i\in T} \alpha_i = 1/2 \right\}$$

$$\mathscr{V}_4 = \left\{ \alpha \;\middle|\; \sum_{i\in S} \alpha_i = 1/2,\ \sum_{i\in S\cap V} \alpha_i \geq 1/4,\ \sum_{i\in T\cap U} \alpha_i = 1/4,\ \sum_{i\in T\cap V} \alpha_i = 1/4 \right\}$$

$$\mathscr{V}_5 = \left\{ \alpha \;\middle|\; \sum_{i\in T} \alpha_i = 1/2,\ \sum_{i\in T\cap U} \alpha_i \geq 1/4,\ \sum_{i\in S\cap U} \alpha_i = 1/4,\ \sum_{i\in S\cap V} \alpha_i = 1/4 \right\}.$$

THEOREM 6.1. Let v be an extreme game with three values. Suppose S and T are disjoint members of $\mathscr{B}^m$ such that $R \in \mathscr{W}$, $R \subset S \cup T$ implies either $R \supset S$ or $R \supset T$. If the minimal intermediate

coalitions contained in $S \cup T$ are	then a solution for v is
S, T	$\mathscr{V}_1$
S, T, U	$\mathscr{V}_1$
S, T, U, V	$\mathscr{V}_1$
$S, T, U, \{S \cap U\} \cup \{T \cap V\}$	$\mathscr{V}_1 \cup \mathscr{V}_2$
$S, T, U, V, \{S \cap U\} \cup \{T \cap V\}$	$\mathscr{V}_1 \cup \mathscr{V}_4 \cup \mathscr{V}_5$
$S, T, U, V, \{S \cap U\} \cup \{T \cap V\}$, and $\{S \cap V\} \cup \{T \cap U\}$	$\mathscr{V}_2 \cup \mathscr{V}_3$

It is easy to verify that Theorem 6.1 provides a solution for each five-person and six-person extreme game with three values using the following lemmata. Lemma 6.1 implies that all winning coalitions have at least three players, and so, using the disjoint two-element coalitions of $\mathscr{B}^m$ as S and T, which exist by Lemma 6.2, we have $R \in \mathscr{W}$, $R \subset S \cup T$ implies either $R \supset S$ or $R \supset T$. Hence, we can apply Theorem 6.1 which covers all possible cases when $n = 5$ and 6.

LEMMA 6.1. Let v be an extreme game with three values. Then no two-element coalition is in $\mathscr{W}$.

PROOF. Suppose this is not the case, i.e., suppose that v is extreme, and there exists $\{i, j\} \in \mathscr{W}$. Then $I_n - \{i, j\} \in \mathscr{L}$, as well as all subsets of $I_n - \{i, j\}$. Hence, every $S \in \mathscr{B}$ contains either player i or player j. The following non-zero flow assignment contradicts the assumption that v is extreme:

$$I(S) = \begin{cases} +1 & S \in \mathscr{B}, \quad \{i\} \subset S \\ -1 & S \in \mathscr{B}, \quad \{j\} \subset S \\ 0 & \text{otherwise} \end{cases} .$$

Hence, no winning two-element coalition exists.

LEMMA 6.2. Let v be a five-person or a six-person extreme game with three values. Then there exists a pair of disjoint two-element coalitions in $\mathscr{B}$.

PROOF. If v is a five-person game, then there exists at least one two-element coalition, say $\{i, j\}$, in $\mathscr{B}$. If no two disjoint two-element coalitions occur in $\mathscr{B}$, then the following non-zero flow assignment contradicts the assumption that v is extreme:

$$I(S) = \begin{cases} +1 & S = \{i, j\} \\ -1 & S = I_5 - \{i, j\} \\ 0 & \text{otherwise} \end{cases} .$$

If v is a six-person extreme game with three values, then the only coalitions which can belong to $\mathscr{B}$ are those with two, three, or four elements. If no two-element coalition is in $\mathscr{B}$, then we can use the following non-zero flow assignment to contradict the assumption that v is extreme:

$$I(S) = \begin{cases} +1 & S = \{i, j, k\} \\ -1 & S = I_6 - \{i, j, k\} \\ 0 & \text{otherwise} \end{cases}$$

where $\{i, j, k\} \in \mathscr{B}$. If there is at least one two-element coalition in $\mathscr{B}$, say $\{i, j\}$, but no pairs of disjoint two-element coalitions are in $\mathscr{B}$, then one of the following two possibilities obtains:

(a) All other two-element coalitions have the same player in common with $\{i, j\}$, say player i. Using the non-zero flow assignment:

$$I(S) = \begin{cases} +1 & S \in \mathscr{B}, \ i \in S \\ -1 & S \in \mathscr{B}, \ i \notin S \\ 0 & \text{otherwise} \end{cases}$$

we have a contradiction to the assumption that v is extreme.

(b) $\{i, k\}$, $\{j, k\}$ are also in $\mathscr{B}$. Using the non-zero flow assignment:

$$I(S) = \begin{cases} +1 & S \in \mathscr{B}; \ \{i, j\} \subseteq S, \ \{i, k\} \subseteq S, \ \text{or} \ \{j, k\} \subseteq S \\ -1 & S \in \mathscr{B}; \ \{i, j\}, \{i, k\}, \{j, k\} \not\subseteq S \\ 0 & \text{otherwise} \end{cases}$$

we have a contradiction to the assumption that v is extreme.

Only in the case where an extreme game with three values is symmetric, were we able to find a general result on finite solutions.

DEFINITION 6.4. [7] An n-person game v is symmetric if there exist n numbers $v_1, \ldots, v_n$ such that

$$v(S) = v_s \quad S \subseteq I_n$$

where $|S| = s$.

We are concerned only with those symmetric games where v_i is either 0, $1/2$, or 1, $1 \leq i \leq n$. For such games we define

$$s = \min \{q \mid v_q = 1\}$$
$$r = \min \{q \mid v_q = 1/2\}$$

so that the s-element coalitions are the minimal winning coalitions and the r-element coalitions are the minimal intermediate ones. Clearly, $s + r = n + 1$. The following lemma characterizes constant-sum, symmetric, extreme games with three values, in terms of the values r can assume. Since we are concerned with extreme games with three values, we require $n \geq 5$.

LEMMA 6.3. Let v be an n-person, constant-sum, symmetric game with three values. In order that v be extreme, it is necessary and sufficient that

$$\frac{n+1}{3} \leq r < \frac{n}{2} \quad .$$

PROOF. Necessity. If $r < (n + 1)/3$, then $3r < n + 1$ or $3r \leq n$, which would violate superadditivity; if $r = n/2$, then the following non-zero flow assignment shows that v is not extreme:

$$I(S) = \begin{cases} +1 & S = \{1, \ldots, r\} \\ -1 & S = \{r+1, \ldots, n\} \\ 0 & \text{otherwise} \quad . \end{cases}$$

Sufficiency. $r \geq (n + 1)/3$ implies that all superadditivity conditions are satisfied. $r < n/2$ or $2r + 1 \leq n$ implies that a K-chain can be formed beginning and ending with each r-element coalition, and, hence, with each (r+i)-element coalition, $1 \leq i \leq n - 2r$, so that v is extreme by Theorem 3.2.

In order to discuss finite solutions for these games, we consider two cases.

CASE I. $r = (n + 1)/3$, or equivalently, $s = 2r$.

In this case, the r-element coalitions are the only vital sets for domination [2]. This case includes the five-person, symmetric, extreme game with three values and gives the solution obtained by Gurk [5]. The solution given is also an atomic solution as defined by Gurk and Isbell [6], viz., the r-element coalitions are the atoms. Finally, the solution in this case is the same as that given by Gelbaum [1]; in his terminology, the r-element coalitions are the minimax coalitions. We give the following theorem without proof.

THEOREM 6.2. Let v be an n-person, constant-sum, symmetric, extreme game with values $\{0, 1/2, 1\}$. Let $r = \min\{q \mid v_q = 1/2\}$ and suppose $r = (n + 1)/3$. Then a finite solution for v is

$$\mathcal{V} = \{\alpha \mid \alpha_{i_1} = \cdots = \alpha_{i_{2r}} = 1/2r;\ \alpha_{i_{2r+1}} = \cdots = \alpha_{i_n} = 0;$$

$$\text{and all permutations of the } n \text{ indices}\} \quad .$$

CASE II. $r > (n + 1)/3$, or equivalently, $s < 2r$.

In this case, both the s-element coalitions and the r-element coalitions are vital. This case first obtains for $n = 7$, when $r = 3$ and $s = 5$. We merely state the following theorem; for a proof, see [4].

THEOREM 6.3. Let v be an n-person, constant-sum, symmetric, extreme game with values $\{0, 1/2, 1\}$. Let $r = \min\{q \mid v_q = 1/2\}$ and $s = \min\{q \mid v_q = 1\}$. Suppose $2r > s$. Then a finite solution for v is

$$\mathcal{V} = \mathcal{V}_1 \cup \mathcal{V}_2$$

$$= \{\alpha \mid \alpha_{i_1} = \dots = \alpha_{i_s} = \tfrac{1}{s};\ \alpha_{i_{s+1}} = \dots = \alpha_{i_n} = 0\ ;$$

and all permutations of the n indices$\}$

$$\cup\ \{\alpha \mid \alpha_{i_1} = \dots = \alpha_{i_r} = \tfrac{1}{s};\ \alpha_{i_{r+1}} = \dots = \alpha_{i_{2s-r}} = \tfrac{1}{2s}\ ;$$

$$\alpha_{i_{2s-r+1}} = \dots = \alpha_{i_n} = 0;$$ and all permutations of the n indices$\}$.

BIBLIOGRAPHY

[1] GELBAUM, B. R., "Symmetric zero-sum n-person games," this Study.

[2] GILLIES, D. B., "Solutions to general non-zero-sum games," this Study.

[3] GOLDMAN, A. J., and TUCKER, A. W., "Polyhedral convex cones," Annals of Mathematics Study No. 38 (Princeton, 1956), pp. 19-40.

[4] GRIESMER, J. H., "On extreme games," Ph. D. Thesis, Department of Mathematics, Princeton University, 1958.

[5] GURK, H. M., "Five-person, constant-sum, extreme games," this Study.

[6] GURK, H. M., and ISBELL, J. R., "Simple solutions," this Study.

[7] von NEUMANN, J., and MORGENSTERN, O., Theory of Games and Economic Behavior, Princeton 1944, 2nd ed. 1947.

[8] SHAPLEY, L. S., Lecture Notes on Simple Games, Department of Mathematics, Princeton, 1954.

James H. Griesmer

Princeton University

and

International Business Machines Corporation

SELF-POLICING PROPERTIES OF CERTAIN IMPUTATION SETS

William Vickrey

MOTIVATION

In presenting their notion of a "solution", von Neumann and Morgenstern suggest that a solution can be regarded as in some sense a standard of behavior to which the players might in some way be constrained, without, however, any specification of the nature or strength of the constraint.[1] Now in general it is conceivable that a society or a community might by custom, development of the mores, common assent, or in some other way, develop some criterion by which a certain set of imputations[2] is selected as forming a pattern of approved ways of distributing the proceeds of the game among the players while the remainder of the imputations not conforming to this pattern are disapproved; moreover if the social approval and disapproval are strong enough, and particularly if they become entrenched in formal rules of law, then adherence to any given set of imputations can be enforced and all others proscribed. Indeed, the proscription of certain imputations may have as much force as the rules of the game from which the characteristic function[3] is determined; but if this is the case then it would seem hardly worth the trouble to work through the characteristic function when these social norms could be appealed to directly.

A theory of games may still have something to contribute, however, if the problem is approached from the standpoint of inquiring what sets of imputations, if adopted as approved standards or patterns of distribution, stand the best chance of being adhered to, or alternatively, what sets of imputations require sanctions of minimum strength in order to compel adherence. Of course, if the set of all possible imputations

---

1 John von Neumann and Oskar Morgenstern, Theory of Games and Economic Behavior, second edition, Princeton, 1947 (hereafter cited TGEB), pp. 40-45, 4.6-4.8.

2 TGEB, p. 263, 30.1; p. 519, 56.9.5.

3 TGEB, pp. 238-243, 25.

is admitted as an approved pattern, there is no possibility of non-compliance; potential standards or patterns will be the more significant as they are more and more restrictive. A form of sanction that arises in some cases from the game structure itself is the tendency of a disapproved imputation to be displaced by an approved imputation in a way disadvantageous to one or more of those responsible for the disapproved imputation.

The problem of determining which patterns or types of patterns are easiest to enforce, relative to their restrictiveness, is considerably simplified if we assume that the standard is to be a simple black-and-white matter, so that there is no social preference expressed as between any two approved imputations, nor as between any two of the non-approved imputations. Just what form this preference takes between an approved and a non-approved imputation and how it makes its influence felt we will leave unspecified for the moment. Rather, we shall approach the problem from the opposite direction: we will suppose a standard to have been adopted and an imputation conforming to the standard to have been reached, and we will attempt to consider under what conditions and for what types of standards the strength of the external sanctions required to preserve this conforming imputation from infringement will be minimized. It turns out that in some cases a barely perceptible tendency for moves from non-conforming to conforming imputations to take place more readily than other shifts will be sufficient to result in the persistence of any conforming imputation that is once reached.

PRELIMINARY DEFINITIONS: CONFORMING, HERETICAL AND POLICING IMPUTATIONS

Let $\underline{c}$ be an imputation conforming to some specified approved pattern. There will in general be one or more sets of players H such that the set H collectively has the power, assuming concerted action by its members, and each of the members of H individually have a (short-run) motive, to upset the arrangements leading to $\underline{c}$ in some particular way, with an imputation $\underline{h}$ as outcome which the coalition H has the power to control to the extent of keeping $\underline{h}$ within some (non-empty) range $\underline{R}$ ($\underline{R}$ may, of course, consist of a single imputation $\underline{h}$ only, in which case we will say that H is "strictly effective" for $\underline{h}$).[4] In general, the short run motive can be said to exist for each player whose share in $\underline{h}$ exceeds his share in $\underline{c}$, and the power can be said to exist when the value of the characteristic function for the set H (or, simply, the power of H) equals or exceeds the sum of the shares of its members

[4] Compare TGEB, p. 264.

in $\underline{h}$.[5] We will assume that the coalition H can control completely the distribution among its members of any sum not in excess of its power, independently of what happens to non-members of H, and that a decision as to the manner of this sharing is part of the process of the formation of the coalition H and is to be the same throughout $\underline{R}$, so that the share of a member of H in the imputations $\underline{h}$ in $\underline{R}$ has a unique meaning. The imputations $\underline{h}$ in a given $\underline{R}$ will then differ at most in the way the balance not going to members of H is shared among non-members. When H has $n - 1$ players, or when the members of H get a total of 1 in imputations in $\underline{R}$ (in the (0, 1) normalization), then $\underline{R}$ will contain but a single imputation.

If the range $\underline{R}$ determined by the action of H contains only non-conforming imputations (as will always be the case if the approved pattern is such as to admit no internal domination, and in particular if it is also a solution), we can term the set H a "heretical coalition", the forming of H and the setting up of the conditions driving the distribution of the value of the game from the conforming imputation $\underline{c}$ to within a particular range $\underline{R}$ a "heretical move", and any particular $\underline{h}$ that could result from this move a "heretical imputation", or, briefly, "heresy". A coalition is "heretical" by virtue of the action it takes, of course, and not by virtue of its mere composition.

The particular $\underline{h}$ that does result may in turn be subject to attack by a new coalition P capable of enforcing a shift away from $\underline{h}$ to some other imputation within a range $\underline{S}$ determined by the action of P. $\underline{S}$ may in some cases contain only a single imputation; in any case it does not contain $\underline{h}$; if some imputation $\underline{p}$ in $\underline{S}$ is a conforming imputation, and at the same time $\underline{S}$ does not intersect $\underline{R}$, then P can be called a "policing coalition" and any conforming imputation $\underline{p}$ in $\underline{S}$ a "policing imputation". Again, a coalition is "policing" by virtue of its behavior, not of its composition. If in a given policing imputation $\underline{p}$ some member j of H is worse off than he was in the original imputation $\underline{c}$ (i.e., $\exists\, j \in H,\ p_j < c_j$), then we can term the imputation $\underline{p}$ a penalizing imputation for this member j of H, or say briefly that $\underline{p}$ penalizes j.

If for a given heretical move $(\underline{c}, H, \underline{R})$ from the conforming imputation $\underline{c}$ as a result of action by H to a range of imputations $\underline{R}$ determined by the action of H, there is a member j of H such that all policing imputations $\underline{p}$ that police any $\underline{h}$ in $\underline{R}$ penalize j, and if at least one such penalizing imputation exists, then $(\underline{c}, H, \underline{R})$ is termed a suicidal heresy for j. If every heresy from a conforming

[5] J. C. C. McKinsey, Introduction to the Theory of Games (McGraw Hill, 1952); p. 327.

imputation $\underline{c}$ is suicidal for some heretic, then $\underline{c}$ is termed a strong imputation in relation to the pattern of which it is a part. If every imputation in a pattern of approved imputations is strong in relation to that pattern, the pattern is called a self-policing pattern. A solution that is also a self-policing pattern will be called a strong solution.

THE SELF-POLICING PRINCIPLE

A self-policing pattern, so defined, will have the property that if it is adopted as an approved pattern, and if as a result of this approval there develops even a barely perceptible tendency for changes from non-conforming to conforming imputations to be easier to make than changes from conforming to non-conforming imputations, then a conforming imputation, once arrived at, will tend to persist indefinitely, and heresies will not occur. For any proposed heresy there will always be at least one member of the proposed heretical coalition whose cooperation is required in order to bring about the proposed change, but who can anticipate that when the heresy is eventually broken up by a policing action, he will be penalized. The policing action is to be expected, for by hypothesis such a change is assumed to be easier to organize than the heresy. Moreover, if the heretics, either from lack of foresight, ignorance, or overweening optimism, nevertheless go ahead, the eventual penalizing of one of their members would tend to serve as an object lesson in preventing further heresies from $\underline{p}$, thus perpetuating the penalty. The structure of the approved pattern thus turns even the slightest obstacle to movement away from the pattern into a prohibitive barrier.

STRONG AND WEAK SOLUTIONS OF THE THREE-PERSON CONSTANT-SUM GAME

It is fairly easy to see that the symmetrical solution of the three-person constant-sum game, consisting of the three midpoints of the sides of the fundamental triangle[6] constitutes a self-policing pattern. To avoid awkward fractions, let us consider the game in a (0, 24) normalization (or alternatively, the elements of the imputations as shown can be considered to be expressed in units of 1/24). The 3 imputations of the symmetrical solution are then $\underline{c}^1 = (0, 12, 12)$, $\underline{c}^2 = (12, 0, 12)$ and $\underline{c}^3 = (12, 12, 0)$. Consider a heresy by $H = (1, 2)$ from $\underline{c}^1$ to $\underline{h}$ where we must have $0 < h_1 < 12$, $12 < h_2 < 24$, and $0 < h_3 < 12$; the policing set then can only be $P = (1, 3)$ with $\underline{p} = \underline{c}^2$ and player 2 is thus likely to be left with $c_2^2 = 0$ as his final share, instead of

---

[6] TGEB, pp. 285-287, 32.2.1.

$c_1^2 = 12$ if he allows himself to be lured into the heresy.

The asymmetrical solutions are not self-policing however, and in fact lie at the other extreme of the scale in this respect.[7] For any imputation in one of the asymmetric solutions, there exist "safe" heresies in which there is no immediate danger of any of the heretics being penalized.

Let player 3 be the one getting the constant amount k in the solution, with $k < 12$. Consider an initial conforming imputation $\underline{c} = (a, b, k)$ with $a \geq b$, permuting the players if necessary. The imputation $\underline{h} = (0, 12, 12)$ is then a safe heresy for $H = (2, 3)$ (except where the initial imputation $\underline{c} = (12, 12, 0)$ in which case we can substitute $\underline{h} = (0, 18, 6)$), for since no policing imputation can give player 3 anything but k, the policing set must be $(1, 2)$ which in turn implies that the policing set must give player 2 in the policing imputation more than he had in the heresy, which in turn must be more than he had originally. Thus even if there should be a return to a conforming imputation after only an extremely brief period of persistence of the heresy, players 2 and 3 responsible for the heresy would both stand to gain from the excursion, 3 temporarily and 2 more permanently.

Even if a return from the heresy to a conforming imputation is made more indirectly, say via an intervening non-conforming imputation such as $(4, 20, 0)$ or $(10, 0, 14)$ so that it is possible for player 2 to be worse off in an eventual return to, say, $\underline{p} = (a + d, b - d, k)$, it is by no means clear from the characteristics of the game that player 2 would not be able to avoid such an eventual worsening of his position. And even if after one particular heretical excursion player 2 finds his position in the resulting conforming imputation worse than in the original one, there is now nothing to prevent him from trying again, since player 3 whose cooperation he needs for the perpetration of a safe heresy has nothing to lose by the heresy in any event and stands to gain temporarily for as long as the heresy can be made to persist. In effect, if either of players 1 or 2 is prepared to engage in heretical excursions, he will have an advantage, in jockeying for position among the approved imputations, over one who is wedded to orthodoxy and eschews such tactics. For this pattern, if adherence is to be enforced it appears that the sanctions required would have to be very nearly as powerful as the rules of the game itself.

---

[7] TGEB, pp. 287-288, 32.2.2.

STRENGTH OF SOLUTIONS IN THREE-PERSON VARIABLE-SUM GAMES

It is also possible, if we are content to specify merely that a conforming imputation, once reached, shall maintain itself, to exhibit self-policing patterns that are not solutions. Some interesting and fairly simple cases of this occur with three-person variable-sum games. Let x, y, and z respectively, be the values of the characteristic function for the three two-person sets in the $(0, 24)$ normalization and let $\underline{E}$ be the set of all efficient imputations, an efficient imputation being one in which the sum of the shares of the players has the maximum value, in this case 24. Then in the case where $x + y + z \leq 48$, there is a triangle $\underline{C} \subset \underline{E}$ of imputations such that in each imputation in $\underline{C}$ no set of players gets less than its power; such an imputation set is termed the "core" of the game.[8] If $\underline{c}$ is a conforming imputation in $\underline{C}$, no heretical move away from $\underline{c}$ will then be possible, so that any subset of $\underline{C}$, as well as $\underline{C}$ itself, will be a self-policing pattern. To have a solution, however, we must take the entire triangle $\underline{C}$ and append to it an arbitrary monotonic curve from each of its three vertices to the corresponding side of $\underline{E}$ where one of the players gets a share of zero, the curve from a given vertex being arbitrary within the restriction that it never make an angle of less than 60 degrees with the side of $\underline{E}$. All such solutions are strong with the possible exception of those in which the 60 degree limit is actually reached by one of these arbitrary curves over a finite length, so that some player gets the same share in two or more distinct imputations along this curve. Self-policing patterns can also be obtained by taking a strong solution and curtailing the lines outside $\underline{C}$ either completely or to an arbitrary partial extent at the outer ends. The only property lost by this curtailment is the ability of the approved imputations to dominate all non-conforming imputations directly. But since the imputations not dominated by the curtailed pattern could never in any case be reached by any series of heresies from an imputation within the curtailed pattern, the existence of these non-dominated imputations constitutes no threat to the stability of the pattern, though it may be considered to impair the speed or probability of reaching an imputation within the pattern *ab initio*. If $x + y + z < 48$, it is also possible to develop further self-policing patterns by the deletion of parts of $\underline{C}$.

The case where $x + y + z > 48$ differs slightly. Here the core is empty, but there is a triangle $\underline{T}$ in which no two-person set gets more than its power. Among imputations in $\underline{T}$, the domination relations are as for the three-person constant-sum game, and strong solutions in this case contain the three midpoints of the sides of $\underline{T}$, (corresponding

[8] The shaded triangle of Figure 86, p. 415, TGEB.

to the strong solution of the constant-sum game) plus three curves from these points to the corresponding sides of E, each of these curves being arbitrary within the same monotonic restriction as before.[9] These solutions are strong provided the arbitrary curves are again strictly monotonic, i.e., the 60 degree limit is not reached over any finite length. Again, self-policing patterns result from the arbitrary curtailment of any of these curves at their outer ends. Unlike the case with a core, however, no part of the pattern belonging to the triangle T can be suppressed without destroying the self-policing property.

Solutions also exist corresponding to the weak solutions of the constant-sum game, consisting of a straight line across triangle T parallel to one of its sides, supplemented by arbitrary lines to the sides of E,[10] but as these solutions are also weak, they are of little interest here.

MINIMAL SELF-POLICING PATTERNS

Faced with such an abundance of patterns, one is tempted to try to select from among them by some further criteria. If we consider that, as a trivial case, the set of all imputations is a self-policing pattern, in that no imputation is heretical, it becomes apparent that patterns are more significant as they are more restrictive. In the absence of a meaningful measure over sets of imputations, the only principle of selection this yields is one which rejects any self-policing pattern which is a superset of another self-policing pattern in favor of the latter. This procedure leaves us with "minimal self-policing patterns" which are those self-policing patterns that contain no proper subsets that are also self-policing patterns. In the case with a core, the minimal self-policing patterns are the single imputations in the core; in the coreless case, the unique minimal self-policing pattern is the set of three imputations at the midpoints of the sides of T. For $x + y + z = 48$, the two cases converge to leave as the unique minimal self-policing pattern a single point.

In the coreless case, the minimal self-policing pattern is also the intersection of all known self-policing patterns and this may be adduced as an additional reason for considering outcomes in this minimal pattern as being more likely, or easier to enforce, than outcomes in the "tails" of the more inclusive patterns. For even if the individual players may be assumed, for the purposes of the theory, to be complete game

---

[9] Cf., TGEB, Figure 82, p. 412.

[10] Cf., TGEB, Figure 83, p. 412.

theorists themselves, knowledge of game theory is not the same thing as knowledge of which precise pattern, among the many self-policing patterns available, has been selected by the powers that be as the socially approved one. Disagreement or uncertainty as to which precise form of the tails has been selected as the approved pattern is likely to result in a certain amount of jockeying, the effect of which must always be to move the imputation toward triangle T, never away from it as long as shifts from one imputation to another require the corresponding domination. An ultimate result in triangle T, therefore, seems much more likely than in one of the uncertainly placed tails.

For the case with a core, the situation is less clear cut. Jockeying from imputations near one of the tails of the solution will tend to shift the result toward the corresponding vertex of C, but no further once this vertex is reached. The sides of C can be reached by direct domination from the areas near the corners of E, but the interior of C can be reached from other points in E only via imputations outside E (i.e., via "inefficient" imputations in which the shares of the players sum to less than the maximum), if at all. Nevertheless, it would seem improper to attribute outstanding stability to an approved pattern that admits of non-conforming imputations that cannot be dominated by any other imputation. Accordingly it seems reasonable to designate the core itself, in this case, as the self-policing pattern on which attention might well be focused. However, no point in C can be reached from any point in the tails of a self-policing pattern by any finite sequence of shifts each of which corresponds to a domination of the preceding by the succeeding imputation.

SELF-POLICING PATTERNS WITH INTERNAL DOMINATION

If thus we can have self-policing patterns that leave some imputations undominated, one may also ask if it is possible to have self-policing patterns that admit of domination of one imputation in the pattern by another in the pattern. If one does not insist on the pattern being a minimal one, then as a trivial case the set of all imputations is a self-policing pattern, there being no heresies possible. If, however, one conforming imputation c dominates another conforming imputation d, then the pattern, if it is to be self-policing, must contain all imputations on the line cd, otherwise these imputations would be non-suicidal heresies from d with c as a non-penalizing policing imputation. As an example slightly less trivial than the set of all possible imputations, one can offer for the 3-person constant-sum game the convex set spanned by the imputations of the symmetrical solution, or in the variable sum case the set formed by the union of all solutions, as self-policing patterns with internal domination. None of these are minimal, however, and are thus still of relatively little interest. It does not appear very easy to find minimal self-policing patterns that have internal domination. At the very least, such patterns would seem more difficult to analyze and their imputations would seem to be individually not as stable, even though the pattern as a whole might be as well adhered to, as patterns that admit of no such internal domination.

In what follows, we will concentrate our attention on self-policing patterns that are also solutions, it being noted, however, that such patterns may be non-minimal.

SIMPLE CONSTANT-SUM GAMES

Simple games[11] are of special interest for the study of self-policing patterns because in such games, by definition, each given set can secure for itself either nothing or everything, and therefore every set that is effective at all is strictly effective for any domination in which all members of the set gain. This fact sharpens the distinction between strong and weak imputations and solutions, since we do not have to deal with cases such as arise in some non-simple games in which a heretic coalition cannot be sure of the exact immediate outcome of its heretical action, nor with cases in which remnants of a heresy could partially frustrate a policing action by preventing such action from resulting in a conforming imputation.

We shall show, in what follows, that for any "majority game" that is also "homogeneous" there is a unique strong solution. A "majority game" is a simple constant-sum game in which the players can be assigned weights in such a way that the winning coalitions are precisely those with a majority of the total weight, no coalition having exactly half of the total weight. A "homogeneous game" is a simple constant-sum game in which there exists for each player i a quota $q_i > 0$ such that for each minimal winning set M, $\Sigma_M q_i = 1$. It should be noted that the restriction to positive quota elements is essential, for there exist "quasi-homogeneous" games having a quota with zero elements for which important theorems applicable to homogeneous games fail: an example is the seven person game with minimal winning sets

$$\begin{array}{l}
(1,2,3,7), \\
(4,5,6,7), \\
(1,2,4), \\
(1,3,5), \\
(2,3,6), \\
(1,2,5), \\
(1,3,6), \\
(2,3,4), \\
(1,4,6), \\
(2,5,6), \quad \text{and} \\
(3,4,5);
\end{array}$$

[11] TGEB, pp. 420-428, 49.1-49.4.

if we put $q_7 = 0$, $q_i = 1/3$, $i \neq 7$, the quotas sum to 1 over each minimal winning set.

The above definition of a homogeneous game is broader than the corresponding definition of a "homogeneous majority game" in TGEB (50.2, 50.8) in that it applies to non-majority as well as to majority games. It is obvious that a "homogeneous majority game" as defined in TGEB is a homogeneous game as defined here, as well as a majority game, for it is only necessary to put $q_i = kw_i$, where the w_i are the homogeneous weights normalized to a majority of 1 as described in TGEB (50.2.2) and

$$k = \frac{2}{1 + \Sigma_I w_i}$$

to obtain the required quota. Conversely, if a homogeneous game is also a majority game, we shall show that there must exist a quota with $\Sigma_I q_i < 2$, and these q_i will serve as weights for the game which will be homogeneous weights in the sense of TGEB, so that the two definitions of homogeneous majority game are equivalent.

LEMMA 1. If a homogeneous game is also a majority game, there exists a quota q with $\Sigma_I q_i < 2$.

Let $\underline{w}$ be a vector of weights, and let j be a player with minimum weight w_j among all players that are members of one or more minimal winning sets, and let M be a minimal winning set containing j. Then $M - \{j\}$ is losing, and $P = I - M + \{j\}$ is winning. Let N be a minimal winning set contained in P, and let $D = P - N$. Since M and N both have a majority of the total weight, $\Sigma_M w_i + \Sigma_N w_i > \Sigma_I w_i$, and $\Sigma_D w_i = \Sigma_I w_i - \Sigma_M w_i - \Sigma_N w_i + w_j < w_j$, so that none of the members of D can belong to any minimal winning set, and the quotas assigned to the members of D are arbitrary and inconsequential. We can then choose these quotas so that $\Sigma_D q_i < q_j$; we have from the definition of homogeneous games that $\Sigma_M q_i = \Sigma_N q_i = 1$, so that since $I = M + N + D - (j)$, $\Sigma_I q_i = 2 + \Sigma_D q_i - q_j < 2$.

We proceed now to establish the existence of a unique strong solution to homogeneous majority games by first proving some preliminary lemmas.

LEMMA 2. For any quota q of a homogeneous game, a "main" solution constructed by taking for each minimal winning set M the corresponding imputation in which each member of M gets his quota, exists and is strong.

Consider a heresy perpetrated by the heretic set H against the conforming imputation $\underline{c}^M$ corresponding to a minimal winning set M. H must be winning if it is to have the power to effect a heresy, and so must intersect M in some non-empty set R,[12] with $h_r > c_r^M = q_r$ for all players r in R, since each heretic must gain. Consider the set S of all players for which $h_s \geq q_s$. Then $R \subseteq S$. S cannot be winning, since $\Sigma_S q_s < \Sigma_S h_s < 1$ and for no subset T of S could we have $\Sigma_T q_t = 1$, $T \subseteq S$, and accordingly I - S must be winning. If there is to exist a policing set P effective for the domination of $\underline{h}$ by some conforming $\underline{p}$ we must have $p_p = q_p > h_p$ for all players p in P, $p_i = 0$ for all i not in P, so P must be disjoint from S. Such a P must exist, since S is losing, I - S therefore winning, and I - S therefore contains at least one minimal winning set P, disjoint from S and hence from R, so $p_r = 0$ and $p_r < c_r^M = q_r > 0$ for all r in R, so that the players R are necessarily penalized and the heresy is suicidal for them.

Note that this theorem does not hold for quasi-homogeneous games with zero elements in the quota, since here we might have $q_r = 0$ and the corresponding players would not be penalized. But even here a main solution exists, and the conforming imputations $\underline{c}^N$ corresponding to minimal winning sets N containing one or more zero-quota players do not dominate any other imputations, since one of the members of the required winning set can gain nothing. The imputation set $\{\underline{c}^M\} - \{\underline{c}^N\}$ can accordingly be thought of as a self-policing pattern, the $\underline{c}^N$ being merely imputations which if reached would not be dominated by the conforming imputations, but which are themselves virtually unattainable. Thus though there is no strong solution to this game in the strict sense, there is a pattern of imputations with very strong self-policing properties.

LEMMA 3. Apart from variations in the arbitrary and inconsequential quotas assigned to "dummy" players, the quota of a homogeneous game is unique.

Suppose p and q are distinct quotas; let $\underline{U}$ and $\underline{V}$ be the corresponding main solutions; let R be the set of players for whom $p_r > q_r$, S the set of players for whom $p_s = q_s$, T the set of players for whom $p_t < q_t$, and X, Y be an arbitrary partition of S.

Then if $T \cup X$ is winning, there exists a minimal winning subset M of $T \cup X$, with a corresponding imputation $\underline{q}^M$ belonging to the solution $\underline{V}$ which considered in relation to $\underline{U}$ is undominated by any

[12] TGEB, p. 447, (51; D:a); p. 446, note 2.

imputation in $\underline{U}$, since to be effective for this domination, a set N would have to consist only of players for which $q_i^M < p_i$; but $q_i^M = q_i \geq p_i$ for all $\underline{i}$ in M, so N would have to be contained in I - M, but since M is winning, I - M is losing and no such N can be effective. Therefore if U is to be a solution, q^M must belong to it, which is possible only if M is contained in X and X is therefore winning. Thus if $T \cup X$ is winning, X must be winning.

Similarly, if $R \cup Y$ is winning, Y must be winning. But as $R \cup Y \cup T \cup X = I$, either $R \cup Y$ or $T \cup X$ must be winning, hence either Y or X must be winning. Then if any set W is winning, $L = I - W$ is losing, and $X = L \cap S$ is losing since it is a subset of L, hence $Y = S - X = W \cap S$ is winning; no set containing any members of T or R can be a minimal winning set, and T and R contain only dummies, so that $\underline{p}$ and $\underline{q}$ differ only in the quotas assigned to these dummies which quotas are obviously inconsequential since they appear in none of the conforming imputations. The uniqueness of this lemma would appear to apply also to quasi-homogeneous games.

LEMMA 4. In any game, no imputation in a strong solution can have zeros over any set with a positive power.

If it did, then any heresy by this "non-flat" set would be safe, none of its members having anything to lose. Note that this proposition depends on the lack of internal domination in a solution as compared with a self-policing pattern.

LEMMA 5. Let $\underline{x}$ and $\underline{y}$ be two imputations in a strong solution of a simple constant-sum game; let the sets U and V be any arbitrary partition of the set D consisting of those players who get the same positive share in x as in y, including (U, V) = (D, 0) and (U, V) = (0, D); then

(1) the set of players getting zero in $\underline{x}$,
(2) the set of players getting zero in $\underline{y}$,
(3) the set of players getting more in $\underline{x}$, than in $\underline{y}$, and
(4) the set of players getting more in $\underline{y}$ than in $\underline{x}$, either
(a) each form winning coalitions with $\underline{U}$, or
(b) each form winning coalitions with $\underline{V}$.

In relation to any two given imputations $\underline{x}$ and $\underline{y}$, a partition of the players and a third imputation $\underline{h}$ can be specified as follows:

A: $0 = y_a = x_a$; $0 < h_a$
B: $0 = y_b < x_b$; $0 < h_b < x_b$
C: $0 < y_c < x_c$; $y_c < h_c < x_c$
D: $0 < y_d = x_d$;
U: $U \subseteq D$; $x_u < h_u$
V: $V = D - U$; $h_v < x_v$
E: $0 < x_e < y_e$; $h_e < x_e$
F: $0 = x_f < y_f$; h_f unconstrained.

Now if $D \cup A \cup F$ is not empty, imputations $\underline{h}$ exist for every U such that $U \cup A \cup F$ is non-empty, for since $E \cup F$ cannot be empty, (otherwise $\underline{x}$ and $\underline{y}$ would coincide), a positive amount can be taken from the shares of players $E \cup F$ in $\underline{y}$ to be added to the shares of the players in $A \cup B \cup C \cup U$ in $\underline{y}$ so as to form $\underline{h}$; this is sufficient, since for no player is h_i required to exceed y_i by any specified amount, and on the other hand the amount required to be taken from the players E cannot be excessive, since there is no upper bound (short of 1) on the amount to be given any of the players in $U \cup A \cup F$.

If $\underline{x}$ and $\underline{y}$ are conforming imputations, $D \cup A \cup F$ is not empty, and $A \cup B \cup C \cup U$ is winning, then $\underline{h}$ is a heresy from $\underline{y}$ with $A \cup B \cup C \cup U$ as the heretic set. If in addition $B \cup C \cup V \cup E$ is winning then $\underline{x}$ dominates $\underline{h}$ and is thus a policing imputation. But $\underline{x}$ is non-penalizing for any of the heretics, which are the members of $A \cup B \cup C \cup U$, so that $\underline{y}$ cannot be strong. Thus if $\underline{y}$ is strong and $U \cup A \cup F$ is non-empty, either $A \cup B \cup C \cup U$ is losing or $B \cup C \cup V \cup E$ is losing; that is, either $V \cup E \cup F$ or $U \cup A \cup F$ is winning. On the other hand if $U \cup A \cup F$ is empty, $V \cup E$ must be winning, otherwise $B \cup C$ would be winning and $\underline{x}$ would dominate $\underline{y}$, so that in either case, if $\underline{y}$ is strong, then either $V \cup E \cup F$ or $U \cup A \cup F$ is winning.

By interchanging U and V, and also interchanging, independently, $\underline{x}$ and $\underline{y}$, (which implies putting F, E, C, B for B, C, E, F, respectively), we get the following four propositions:

Either $V \cup E \cup F$ or $U \cup A \cup F$ is winning,
either $U \cup E \cup F$ or $V \cup A \cup F$ is winning,
either $U \cup B \cup C$ or $V \cup A \cup B$ is winning, and
either $V \cup B \cup C$ or $U \cup A \cup B$ is winning.

Suppose $V \cup E \cup F$ is winning; then $U \cup A \cup B$ cannot be winning since it does not intersect $V \cup E \cup F$, so $V \cup B \cup C$ must be winning. Similarly $U \cup E \cup F$ and $U \cup B \cup C$ cannot be winning, so $V \cup A \cup F$ and $V \cup A \cup B$ must be. Similarly, if $U \cup A \cup F$ is winning, $U \cup E \cup F$, $U \cup B \cup C$, and $U \cup A \cup B$ must be winning also. Thus we have that either $U \cup A \cup B$, $U \cup A \cup F$, $U \cup B \cup C$, and $U \cup E \cup F$ are all winning, and $V \cup A \cup B$, $V \cup A \cup F$, $V \cup B \cup C$, and $V \cup E \cup F$ are all losing, or vice versa, Q.E.D.

COROLLARY. $D \cup A \cup B$, $D \cup A \cup F$, $D \cup B \cup C$, and $D \cup E \cup F$ are all winning, and $A \cup B$, $A \cup F$, $B \cup C$, and $E \cup F$ are all losing, since if $U = D$, then $V = 0$, and $U \cup A \cup B \supset V \cup A \cup B$; since just one of these two sets is winning, it must be $U \cup A \cup B = D \cup A \cup B$.

COROLLARY. D is non-empty, i.e., every two strong imputations must have at least one common element.

LEMMA 6. In any constant-sum simple game, in any strong imputation the players who get positive shares form a minimal winning set.

Let S be the set of players getting positive amounts in a strong imputation $\underline{c}$. Suppose S to be a non-minimal winning set, and let u be one of the dispensable players in S, so that $S - (u) = W$ is a winning set. Form a heresy by $I - (u)$ from $\underline{c}$ to $\underline{h}$, with $h_u = 0$ and $h_i > c_i$ for $i \neq u$ and let $\underline{p}$ be an imputation policing $\underline{h}$. In Lemma 5, put $\underline{y} = \underline{c}$, and $\underline{x} = \underline{p}$. Let P be the set of all players who get more in $\underline{p}$ than in $\underline{h}$, and V the set of all players in W who get the same share in $\underline{p}$ as in $\underline{c}$. Then the set $V \cup A \cup F$ consisting of V plus the players who get zero in $\underline{p}$ has no intersection with P, which must be winning, and hence must be losing; by Lemma 5, the set of players other than V who get the same positive amount in $\underline{p}$ as in $\underline{c}$ (which can comprise only the single player u) must combine with the players getting zero in $\underline{c}$ to form a winning set (i.e., $U \cup A \cup B$); but this set is $I - W$, which contradicts the assumption that W was winning. By Lemma 4, S cannot be losing, hence must be a minimal winning set.

LEMMA 7. In any given strong solution of a simple constant-sum game, any given player gets the same amount in all conforming imputations in which he gets more than zero.

Applying Lemma 6 to the partition of Lemma 5, $C \cup D \cup E \cup F$ must be a minimal winning set; but since by Lemma 5 itself $D \cup E \cup F$ must be winning, C must be empty, and similarly for E. The constant amount for each player will be termed the "ration" corresponding to the given solution. The lemma can then be stated: "Every strong solution to a simple constant-sum game has a ration."

THEOREM I. The main solution is the only strong solution for homogeneous majority games.

Suppose that for a homogeneous majority game there exists a solution $\underline{V}$ satisfying the conditions of Lemma 7, and let r_i be the ration of player i in solution $\underline{V}$, let q_i be the quota of the player i and $\underline{U}$ the corresponding main solution which by Lemmas 2 and 3 exists and is unique and strong. Let R be the set of players for which $r_i > q_i$, S the set for which $r_i = q_i$, and T the set for which $r_i < q_i$, and let X, Y be any arbitrary partition of S. If $T \cup X$ is winning, there exists a minimal winning subset M of $T \cup X$ and the corresponding imputation $\underline{m}$ in the main solution $\underline{U}$ is undominated by any imputation in $\underline{V}$, since no member of M can gain by a return to a conforming imputation and no policing set can be winning without containing a member of M. Therefore $\underline{m}$ must belong to $\underline{V}$, which implies that M is contained in X, and hence that X is winning. If on the other hand $T \cup X$ is losing, then $\Sigma_{T \cup X} q_i < 1$ and we can construct a non-conforming imputation in which $h_i > q_i$ for all i in $T \cup X$; the policing set P for this $\underline{h}$ must then be contained in $Y \cup R$, but since any winning P in $Y \cup R$ containing a player in R would have $\Sigma_P r_i > 1$, and thus be ineffective to any imputation in $\underline{V}$, P must be in Y and Y must be winning. Thus for any partition of S into X and Y, either X or Y must be winning. Now let G be a set not contained in S, and put $X = S \cap G$, $Y = S - X$. Then if X is winning it is a subset of G, so that G cannot be a minimal winning set, while if X is losing Y must be winning and G must be losing since $G \cap V = 0$; thus there are no minimal winning sets not contained in S, and $\Sigma_M r_i = \Sigma_M q_i = 1$ for every minimal winning set M. Every imputation $\underline{h}$ with $h_j > 0$ for any j in $R \cup T$ is dominated by some conforming imputation so that all r_j with $r_j \neq q_j$ become irrelevant and $\underline{V}$ must coincide with the main solution $\underline{U}$.

THEOREM II. Non-homogeneous majority games have no strong solution.

Any strong solution must still satisfy the conditions of Lemma 7, but by definition it is not possible to choose the r_i so as to make $\Sigma_M r_i = 1$ for every minimal winning set M, since if this were so the r_i would constitute a set of homogeneous weights. If for some winning coalition W, $\Sigma_W r_i < 1$, then W could obtain for each of its members more than they could obtain within the solution, so that the heresy thus formed would be undominated, as no member of W could then be tempted into any policing set P. Thus we must have $\Sigma_M r_i \geq 1$ for all minimal winning sets M, and $\Sigma_S r_i > 1$ for some minimal winning set S. If in addition $\Sigma_I r_i < 2$, then for $T = I - S$, $\Sigma_T r_i < 1$, and we can form a heresy in which $h_i > r_i$ for all i in T; this heresy is undominated by any conforming imputation since no member of T can gain in a policing action, no proper subset of S is winning, and S itself is ineffective to any conforming imputation.

On the other hand if $\Sigma_I r_i > 2$, consider the set A of all players who belong to one or more minimum winning sets M with $\Sigma_M r_i = 1$. If in any imputation $\underline{h}$ there is some player k not in A with $h_k > 0$, and if $\underline{h}$ is undominated by any conforming imputation, then the imputation $\underline{g}$ with $0 < g_k < h_k$; $g_i > h_i$; $i \neq k$ is likewise undominated; by Lemma 6, $\underline{h}$ and $\underline{g}$ cannot both be strong imputations, hence $\underline{h}$ cannot be a conforming imputation of a strong solution. Thus no player k not in A can ever get more than 0 in any strong imputation, and we can put $r_k = 0$ for such players, so that $\Sigma_A r_i > 2$.

Let j be a player with minimum weight among those in A, let B be a minimal winning set containing j with $\Sigma_B r_i = 1$, and let $\underline{b}$ be the corresponding imputation. Consider a non-conforming imputation $\underline{h}$ with $h_j < b_j = r_j$, and $h_i > b_i$ for all $i \neq j$. Any policing set must be contained in $C = A - B + \{j\}$, since $h_i > r_i$ for all i not in C. C itself cannot be effective to any conforming imputation, for $\Sigma_C r_i = \Sigma_A r_i - \Sigma_B r_i + r_j > 2 - 1 + r_j > 1$, while any proper subset D of C has $\Sigma_D w_i = \Sigma_{A-B} w_i + w_j - \Sigma_{C-D} w_i \leq \Sigma_{A-B} w_i$ where the w_i are the majority weights so that since B has a majority, $A - B$ cannot have, nor can D, so that D must be losing. Thus $\underline{h}$ is undominated by any conforming imputation and no strong solution is possible. Note however that $\underline{h}$ was merely a non-conforming imputation and not an actual heresy dominating a conforming imputation, so that it is still possible to have patterns with fairly strong self-policing properties as illustrated above under Lemma 2.

Theorems I and II do not hold for non-majority non-homogeneous games. Such games may have several strong solutions. Consider a thirteen person game which includes among its minimal winning sets the following:

$$M_i = (i, i + 1, i + 3, i + 9), \; i = 1, 2, 3, \ldots, 13,$$
the addition being modulo 13.

It is readily verified that all other sets of up to four persons are disjunct from at least two of the M_i and somewhat more tediously that all 5-person sets are either disjunct from or a superset of one of the M_i. There are a large number of 6 and 7 person sets for which this does not hold, however, and it is fairly obvious from this fact that no quota exists for such games, since the quota conditions for the M_i alone would require $q_i = 1/4$ for all i. A strong solution exists for $r_i = 1/4$, however, with one imputation for each of the M_i, the 6 or 7 person minimal winning sets being ineffective to any conforming imputation. Any heresy can provide $h_i > r_i$ for at most four players, who must not constitute an M_i otherwise $\underline{h}$ would be a conforming imputation. This four player set is then disjunct from at least two of the M_i, one of which will be effective for the domination of $\underline{h}$ by $\underline{c}^i$, and since it is readily apparent that the $\underline{c}^i$ cannot dominate each other, the $\underline{c}^i$ form a solution. The solution is strong, as can be seen directly or by applying the following more general lemma, which is the converse of Lemmas 6 and 7.

> LEMMA 8. A solution to a simple game in which each player i gets a constant share r_i in all imputations in which he gets more than zero, and in which the players getting more than zero in any given imputation form a minimal winning set, is strong.

Let $\underline{c}$ be a conforming imputation, and M the corresponding minimal winning set, $\underline{h}$ a heresy by H. Let j be a member of $H \cap M$; j must exist, since H and M must both be winning and hence cannot be disjoint, so that $h_j > c_j = r_j > 0$. If P is the policing set effective for the domination of $\underline{h}$ by a conforming imputation p, for every k in P we must have $p_k > h_k$; but p_i can have only the values r_i or 0, so $p_j < h_j$, j cannot be in P, p_j is therefore 0 and j is penalized.

There are other strong solutions possible for the above 13-person game, so that in this game the uniqueness of the strong solution fails. For we can put $r_i = 1/4 + 3d$ for the members of $M_1 = (1, 2, 4, 10)$, $r_j = 1/4 - d$ for the other players, with the result that M_1 becomes ineffective to any conforming imputation, while the other M_i remain effective, with $\Sigma_{M_i} r_j = 1$ for $i \neq 1$.

Since there were in the former case always at least two possible policing sets for each heresy, the elimination of M_1 leaves at least one of them to effect the policing and the set of 12 imputations remaining constitutes a strong solution provided that $0 < d < 1/20$ so that $r_j > 1/5$ and it remains impossible for more than four players to secure $h_j > r_j$ simultaneously.

It is conjectured that multiple strong solutions of this type may also occur for homogeneous non-majority games of 19 or more players though the construction of such a game has thus far proven too tedious and elusive. No such game appears to exist for less than 15 players.

FOUR-PERSON CONSTANT-SUM GAMES NEAR THE CENTER

For non-simple games the nature and existence of strong solutions is a much more difficult and complex problem. The simplest of the non-simple games is perhaps the symmetrical four-person constant-sum game in which all two-person sets have the same power. It can be shown that no strictly strong solutions exist for this game or others in its neighborhood, but on the other hand some of the solutions that are not strictly strong still have characteristics that lead to expectations of stability that are considerably greater than other solutions, and may even approach the stability of strong solutions.

Part of the difficulty in applying the self-policing concept to non-simple games lies in the concept of domination. As long as we are dealing with coalitions that are completely winning, as is always the case in simple games, it is reasonable to suppose that if a coalition is effective at all for the domination of $\underline{a}$ by $\underline{b}$ as defined by von Neumann and Morgenstern, this implies the power not only to upset $\underline{a}$ but also to enforce $\underline{b}$, so that the range of the heresy, $\underline{R}$, in each case consists of but a single imputation. Where an effective coalition has a power $v(S) < 1$, however, it may indeed have the power to upset $\underline{a}$ in a direction desired by each of its members, and even to insure that each of its members secures the amounts specified in imputation $\underline{b}$, but in general it will be unable to control the distribution of the residue $1 - v(S)$ among the non-members of S. It will be convenient to say that where the effective set for a domination has the power 1, it is "strictly effective", otherwise it can be termed partially effective, and the domination can similarly be termed strict or loose.

DEGREES OF VULNERABILITY OF HERESIES

Compounded with this difficulty is the further difficulty that even for a specific heretical imputation there may be more than one policing imputation, and different policing imputations may penalize different heretics, or some policing imputations may be penalizing and others not. Where there are both non-penalizing and penalizing policing imputations for a given heretical imputation, that imputation may be termed "risky" in the heresy in question; when all policing imputations are penalizing, but no one heretic is penalized in all of them, the heretical imputation may be termed "fratricidal", as distinguished from the case where the penalized heretic is always the same player, which we have called "suicidal". At the other extreme, when none of the policing imputations penalize any heretic, the heresy or the heretical imputation is "safe".

In the case of fratricidal or risky heresies, further distinctions can be drawn between cases where each heretic, ex ante, can anticipate the possibility of gain in the policing imputation as well as the possibility of loss, and cases where if he escapes being penalized he can expect nothing better than a return to his former state after a (relatively brief?) period of gain while the heresy persists.

Another question arises where the policing coalition is only loosely effective: the range to within which the policing coalition can enforce a shift may include both conforming and non-conforming imputations. One could, perhaps, postulate that the social forces toward conformity would be sufficient to induce a conforming rather than a non-conforming outcome, and thus permit the non-conforming outcomes to be neglected at this stage in the analysis.

For the heretical moves, however, all heretical imputations are to be considered equally violative of the social norms, so that we get no direction of the outcome from this source. Nevertheless, some types of loose domination seem particularly unlikely to represent actual potentialities for change. For example, if a given heresy $\underline{h}$ saturates the heretic set H, that is, gives the members of H in the aggregate an amount equal to the power of H, then it may appear as unlikely that any of the non-heretics would get more in $\underline{h}$ than in the conforming imputation $\underline{c}$ from which the heresy departs. It may appear as particularly unlikely that any non-heretic i should obtain more in $\underline{h}$ than the maximum amount that his adherence adds to the power of any subset of the non-heretics, and we might therefore require $h_i \leq \max_{S \subset (I-H)} [v(S + (i)) - v(S)]$, $i \in (I - H)$. Yet even this is perhaps a somewhat rash postulate to make without going behind the characteristic function into the detailed procedure of the underlying game; certainly where H is completely effective, there seems to be no obvious reason why a set H with $v(H) = 1$ should not be able to control the distribution among the non-members of whatever portion of the value of the game it does not appropriate to its own members. On the other hand it is not at all clear what rule would be most reasonable to apply where the heretical imputation does not saturate an incompletely effective heretical coalition $(\Sigma_H h_i < v(H) < 1)$.

THE ABSENCE OF STRICTLY STRONG SOLUTIONS

If to avoid all these difficulties we stick to requiring imputations in self-policing solutions to be strong in the strictest sense, then many games lack such solutions altogether. In particular, four-person constant-sum games in the neighborhood of the center, including the symmetrical game, have none, as can be shown, somewhat tediously, as follows:

> THEOREM III. No player can ever get more in any imputation of a solution than the maximum of the amounts that his addition to a set adds to the power of the set: $c_i \leq \underset{S}{\text{Max}}[v(S + (i)) - v(S)] = d_i$.

Suppose that in some conforming imputation, c, $c_i - d_i = x > 0$, for some player i. Form an imputation $\underline{h}$ with $h_i = d_i$,

$$h_j = c_j + \frac{x}{n - 1}, \quad j \neq i.$$

Then $\underline{h}$ dominates $\underline{c}$ with effective set $I - (i)$, and if there is a $\underline{p}$ which dominates $\underline{h}$ with the effective set P, then $\underline{p}$ must dominate $\underline{c}$ with the effective set $P - (i)$ if $i \in P$, and with the effective set P itself otherwise. Thus $\underline{p}$ cannot belong to the same solution as $\underline{c}$, so $\underline{h}$ is undominated by any imputation in a solution including $\underline{c}$, and no solution including $\underline{c}$ exists.

> LEMMA 9. In any interior four-person game (i.e., a constant-sum game in which there are no flat 2-person sets), for any two imputations of a strong solution there must be a player who gets the same share in both imputations.

If $\underline{a}$ and $\underline{b}$ belong to a solution and three players each get more in $\underline{a}$ than in $\underline{b}$, then $\underline{a}$ dominates $\underline{b}$, which is inadmissible, and similarly with $\underline{a}$ and $\underline{b}$ interchanged. Therefore if no player gets the same share in $\underline{a}$ and $\underline{b}$, we can by appropriately designating the players write: $a_1 < b_1$; $a_2 < b_2$; $b_3 < a_3$; $b_4 < a_4$; also, by Lemma 4, b_3 and b_4 cannot both be zero, and by appropriate permutation we can write $b_3 > 0$. Now $a_3 + a_4 > v((3, 4))$, otherwise $\underline{a}$ would dominate $\underline{b}$ with effective set $(3, 4)$, hence $a_1 + a_2 < v((1, 2))$ and we can consider a heresy from $\underline{a}$ by $H = (1, 2)$ with an outcome $\underline{h} = (a_1 + \epsilon, a_2 + \epsilon, 0, x)$. (Note that this heresy involves loose domination, and that while the particular outcome considered, $\underline{h}$, does indeed dominate $\underline{a}$ in the von Neumann and Morgenstern sense, it does not necessarily constitute a particularly likely outcome of any particular action that H might take to upset $\underline{a}$.) But if ϵ is sufficiently small this $\underline{h}$ is dominated by a non-penalizing policing imputation within the solution at $\underline{b}$, with $P = (1, 2, 3)$.

Such a non-penalizing policing imputation dominating a given heretical imputation $\underline{h}$ reached from a given $\underline{c}$ by action of a heretical set H will be termed a "refuge" for this heresy. Existence of such a heresy obviously makes $\underline{h}$ safe and shows that $\underline{c}$ cannot be strong. Accordingly, for a strictly strong solution, some player must get the same amount in both imputations.

> LEMMA 10. In any interior four-person game, if for any two imputations of a strong solution there is exactly one player getting the same share in both imputations,
>
> (a) this share must be positive, and
>
> (b) the imputation that gives two of the three other players each more than the other imputation must give them together more than the power of the set which they constitute, and must give the remaining player zero.

By appropriately designating the players we can write: $a_1 = b_1$; $a_2 < b_2$; $a_3 < b_3$; and $b_4 < a_4$. If $b_4 > 0$, then a heresy from $\underline{a}$ by $H = (1, 2, 3)$ to $\underline{h} = (a_1 + x, a_2 + \epsilon, a_3 + \epsilon, 0)$ has, for sufficiently small ϵ, a refuge in $\underline{b}$ by $P = (2, 3, 4)$. Hence $b_4 = 0$. If $b_2 + b_3 < v((2, 3))$, then $\underline{b}$ dominates $\underline{a}$, therefore $b_2 + b_3 > v((2, 3))$. Since $(1, 4)$ is not flat, b_1 and b_4 cannot both be zero, by lemma, hence $b_1 = a_1 > 0$.

> COROLLARY. In a strong solution any two strong imputations having no zero elements must have two elements in common.

> LEMMA 11. In any interior four-person game, if any two imputations in a strong solution have two elements in common, $a_i = b_i$ and $a_j = b_j$, then either $a_i + a_j = b_i + b_j < v((i, j))$ or the imputations are of the form $(x, y, 0, z)$ and $(x, y, z, 0)$, or permutations thereof.

Let the indifferent players be 1 and 2, and let 4 be the player getting more in $\underline{a}$. Suppose $b_1 + b_2 \geq v((1, 2))$ and $b_4 > 0$. Then a heresy by $H = (1, 2, 3)$ from $\underline{a}$ to $\underline{h} = (a_1 + z, a_2 + z, a_3 + \epsilon, 0)$ has, for sufficiently small ϵ, a refuge in $\underline{b}$ by $P = (3, 4)$, since $a_3 < b_3$ and $b_3 + b_4 = 1 - b_1 - b_2 \leq 1 - v((1, 2))$ and by permuting $\underline{a}$ with $\underline{b}$, and 3 with 4, in the above, we obtain $a_3 = 0$.

LEMMA 12. If any two imputations in a strong solution of an interior four-person game have a minimum element in common (a minimum element being one equal to the smallest amount that the corresponding player can get in any conforming imputation) they must have another element in common.

Suppose there is no other common element. Then we can write, by Lemma 10: $a_1 = b_1 = m_1$; $a_2 < b_2$; $a_3 < b_3$; $b_4 < a_4$; $b_4 = 0$; $b_1 < v((1, 4))$. Then $H = (1, 4)$ is unsaturated in $\underline{b}$, and can produce heresies, all of which are safe, since neither of the players in H can get less in any conforming imputation than in $\underline{b}$.

LEMMA 13. For any four-person game in the neighborhood $\underline{C}$ of the center defined by $1/3 \leq v((i, j)) \leq 2/3$ for all two-person sets, any imputation in a strong solution with a zero element can have no other minimum element.

Suppose $\underline{a} = (0, m_2, a_3, a_4)$ is a conforming imputation; if $m_2 < v((1, 2))$, there will be safe heresies by $H = (1, 2)$; if $m_2 \geq v((1, 2))$, then the imputation $\underline{h} = (1/3, 0, 1/3, 1/3)$ is undominated, since $m_2 \geq 1/3$, which requires $p_i > 1/3$ for all i in P.

LEMMA 14. If any imputation in a strong solution of an interior four-person game has no zero element, then for each player there must be some other imputation in which that player gets the same share as in the given imputation.

Consider an imputation $\underline{a} = (a_1, a_2, a_3, a_4)$ with positive elements and the heresy by $H = (2, 3, 4)$ from $\underline{a}$ to $h = (0, a_2 + \epsilon, a_3 + \epsilon, a_4 + \epsilon)$. If a policing imputation is not to have $p_1 = a_1$, then if $\underline{p}$ is to have two elements in common with $\underline{a}$ it must be of the form $\underline{p} = (x, a_2, a_3, y)$ with $a_2 + a_3 < v((2, 3))$ or symmetrically by permutation of players 2, 3 and 4, by Lemma 11. But this implies $x + y > v((1, 4))$ so policing imputations of this type fail to dominate $\underline{h}$. If $\underline{p}$ is to have only one element in common with $\underline{a}$ it must, by Lemma 10, be of the form $\underline{p} = (x, 0, a_3, y)$ with $x + y > v((1, 4))$ or a permutation thereof for which $p_1 \neq a_1$; domination of $\underline{h}$ is again impossible. Thus for $\underline{h}$ to be dominated we must have $p_1 = a_1$, and similarly there must exist conforming imputations other than $\underline{a}$ matching the shares of players 2, 3, and 4, respectively.

COROLLARY. If for any four-person game in the neighborhood $\underline{C}$ of the center a strong solution gives some player a minimum share greater than zero, there must be at least two distinct imputations giving this player this minimum, for by Lemma 13, Lemma 14 must apply.

LEMMA 15. In any strong solution of a game in C, every player gets a minimum share of zero.

Suppose some player has a minimum share of more than zero. Call him player 1 and the minimum m_1. By the Corollary to Lemma 14, there must be at least two imputations with first elements m_1; by Lemma 13 they can have no zero elements, and by Lemma 10 they must have two elements in common. Let player 2 be the other player getting the same share in the two imputations; they can then be written $\underline{a} = (m, n, c, d)$, $\underline{b} = (m, n, f, g)$. Let players 3, 4 be permuted, if necessary so that $c < f$; then $g < d$. By Lemma 11, $m + n < v((1, 2))$, so $f + g = c + d > v((3, 4))$. By Lemmas 9, 10, and 11, any further imputations belonging to the same strong solution as $\underline{a}$ and $\underline{b}$ must have one of the following forms:

$$(y,0,f,d) \text{ with } y + d > v((1, 4)), \quad y + f > ((1, 3));$$
$$(z,n,0,w) \text{ with } z + w > v((1, 4));$$
$$(t,n,u,0) \text{ with } t + u > v((1, 3));$$
$$(m,n,r,s) \text{ with } r + s > v((3, 4)); \text{ or}$$
$$(p,n,c,g) \text{ with } p + g > v((1, 4)).$$

Since $m + n < v((1, 2))$, there exist heresies by $H = (1, 2)$ from $\underline{a}$ or $\underline{b}$ to $\underline{h} = (m + \epsilon, n + \epsilon, f, g - 2\epsilon)$; but $\underline{h}$ is undominated by any of these possible conforming imputations. Thus no positive minimum share is possible.

COROLLARY. A strong solution to a game in $\underline{C}$ must include four imputations in each of which a different player gets a share of zero, and some player must get the same share in three of these four imputations.

For there are six pairs of imputations, for each pair of which some players must get the same share, by Lemma 9. If no player got the same share in more than two of the imputations, each player could take care of only one of these six relations, and two would be left unsatisfied.

In general, any set of n imputations in a solution to an n-person game which includes for each of the n players an imputation in which that player is the only one to get zero, will be termed a "basis" of the solution, denoted by $\underline{B}$, the imputation in $\underline{B}$ in which player i gets zero being denoted by B^i.

THEOREM IV. No strong solution exists for any game in $\underline{C}$.

The following proof of this theorem is more tedious than elegant, unfortunately, and proceeds by the separate elimination of special cases.

CASE 1. There exists a basis in which there are three players each of whom gets the same share in three of the four imputations, the constant share being equal for the three players. The basis must then be (in (0, 24) normalization):

$$\underline{B} = \begin{matrix} (0, & 8, & 8, & 8) \\ (8, & 0, & 8, & 8) \\ (8, & 8, & 0, & 8) \\ (8, & 8, & 8, & 0) \end{matrix}$$

If any further imputation $\underline{c}$ in the solution is to have an element in common with each of the B^i as required by Lemma 9, it must have two elements equal to 8; moreover since either (4, 8, 4, 8) or (8, 4, 8, 4) is undominated by any B^i (as we cannot have both $v((1, 3)) \geq 16$ and $v((2, 4)) \geq 16$), some such imputation $\underline{c}$ must be included in the solution. But since any such imputation agrees with some B^i in two elements equal to 8, and since $\underline{c}$ can have no zero elements (otherwise it would coincide with a B^i), by Lemma 11 we must have $v((i, j)) = 16$, which is impossible in $\underline{C}$, so no strong solution in this class exists.

CASE 2. There exists a basis in which each of three players gets a constant positive share in three of the four imputations, this share not being the same for all three players. It can then be written:

$$\underline{B} = \begin{matrix} (0, & b, & c, & a) \\ (a, & 0, & c, & b) \\ (a, & b, & 0, & c) \\ (a, & b, & c, & 0) \end{matrix}$$

where the players are permuted so as to make $a \leq b \leq c$, with $a < c$. Then $a < 8 \leq v((1, 3))$, and an unpenalized heresy by $H = (1, 3)$ from B^3 to $\underline{h} = (a + \epsilon, b, \epsilon, c - 2\epsilon)$ exists. For if $\underline{p}$ is penalizing, $p_1 < a$, and if $\underline{p}$ is to have a common element with each B^i as called for by Lemma 9, either (i) $p_2 = b$ and $p_3 = c$, (ii) $p_3 = p_4 = c$, or (iii) $p_2 = p_4 = b$; (i) requires $p_4 \leq 1 - b - c = a < c - 2e = h_4$, so that $\underline{p}$ cannot dominate $\underline{h}$; (ii) implies $p_2 \leq 1 - 2c < b = h_2$, hence $P = (3, 4)$, but since $p_3 + p_4 = 2c > 16 \geq v((3, 4))$, P is ineffective to $\underline{p}$; finally with (iii), if $b < c$, then for small enough ϵ, $b < c - 2\epsilon$. and $\underline{p}$ again fails to dominate $\underline{h}$, while if $b = c$, then $\underline{p}$ has two elements in common with B^3, and since $2b > 16 \geq v((2, 4))$, then by Lemma 11, $\underline{p} = (0, b, a, b)$, since $(a, b, 0, b)$ is non-penalizing; applying Lemma 11 again to $\underline{p}$ and B^1 gives $a + b > v((3, 4))$, so that $P = (3, 4)$ is ineffective to $\underline{p}$ and $\underline{p}$ cannot dominate $\underline{h}$.

CASE 3. Exactly two of the players get a constant amount in three of the four imputations of the basis. We can write the basis as:

$$\underline{B} = \begin{matrix} (0, & b, & d, & g) \\ (a, & 0, & d, & f) \\ (a, & b, & 0, & c) \\ (a, & b, & c, & 0) \end{matrix}$$

with $b \leq a$, and hence $f \leq g$. The two d's are needed as a common element for the first two imputations. Also $c \neq d$, since if $c = d$, the case becomes one of the preceding ones.

Suppose $d < c$; then by applying Lemma 10 to B^1 and B^4 we get $b < v((2, 4))$, and there exists an unpenalized heresy by $H = (2, 4)$ from B^4 to $\underline{h} = (a, b + \epsilon, c - 2\epsilon, \epsilon)$. For if $\underline{p}$ is penalizing, $p_2 < b$; to comply with Lemma 9, $p_1 = a$ and either $p_3 = d$ or $p_4 = g$; $p_4 = g$ implies $p_3 \leq b + d - a \leq d$; either way, $p_3 \leq h_3$ and $\underline{p}$ does not dominate $\underline{h}$.

Alternatively, if $c < d$, then comparing B^1 and B^4 we obtain $f \leq g < a$; applying Lemma 10 to B^1 and B^4 we have $b < v((1, 2))$, and there exist heresies by $H = (1, 2)$ to $\underline{h} = (\epsilon, b + \epsilon, d, g - 2\epsilon)$, and also to $\underline{h}' = (\epsilon, b + \epsilon, d - 2\epsilon, g)$. If $\underline{p}$ is penalizing for either of these, $p_2 < b$; Lemma 9 then requires $p_1 = a$ and either $p_3 = d$ or $p_4 = g$. If $p_3 = d$, then $p_4 \leq g$; if $p_3 = d$ and $p_4 < g$, then for ϵ sufficiently small, $\underline{p}$ would fail to dominate $\underline{h}$, so we must have $p_4 = g$, implying $p_3 \leq d$, $P = (1, 4)$, and $2f < a + g = p_1 + p_4 \leq v((1, 4)) \leq 16$, whence $f < 8$, and $a + d > 16$. Similarly if $\underline{p}'$ is to dominate $\underline{h}'$, we must have $p_3' = d$, since $p_4' = g$ implies $p_3' \leq d$, while $p_4' = g$ and $p_3' < d$ would leave $\underline{h}$ undominated for small enough ϵ; $p_3' = d$ implies $p_4' \leq g$ and $P = (1, 3)$; but then $p_1' + p_3' = a + d > 16 \geq v((1, 3))$, so if $\underline{h}$ is penalized by $\underline{p}$, then $\underline{h}'$ is unpenalized.

CASE 4. There is a basis in which only one player gets the same amount three times. Let this player be 1; imputation B^1 must have an element in common with each of the other three, and since players 2, 3, and 4 cannot get the same share more than twice each, they cannot bear more than one of these relationships each and must therefore each bear exactly one of them. Let player 2 be a player having a lowest positive share in imputation 1; let B^3 be the imputation in which player 2 gets the same share as in B^1, player 3 being the one getting zero in B^3. The basis can then be written:

$$\underline{B} = \begin{matrix}(0,b,c,d)\\(a,0,f,d)\\(a,b,0,g)\\(a,j,c,0)\end{matrix}$$

with $b \neq j,\ c \neq f,\ d \neq g,\ b \leq c \leq d$.

Suppose (a) that b, c, and d are not all equal; then $b < 8$, $c + d > 16 \geq v((3, 4))$. By Lemma 10 applied to imputations B^3 and B^4, if $b > j$, then $a < v((1, 3))$, while if $b < j$, $a < v((1, 4))$. In the first case a heresy by $H = (1, 3)$ from $(a, b, 0, g)$ to $(a + \epsilon, b, \epsilon, g - 2\epsilon)$ is unpenalized for small ϵ, since the only imputation compatible with $\underline{B}$ and having a first element less than a is $B^1 = (0, b, c, d)$ and $P = (3, 4)$ is ineffective to B^1 because $c + d > v((3, 4))$. In the second case similarly a heresy by $H = (1, 4)$ from $(a, j, c, 0)$ to $(a + \epsilon, j, c - 2\epsilon, \epsilon)$ is unpenalized.

If on the other hand (b) we have $b = c = d$, then $b = 8$ and $f = g = j$, so that the basis becomes

$$\underline{B} = \begin{matrix}(0,8,8,8)\\(a,0,f,8)\\(a,8,0,f)\\(a,f,8,0)\end{matrix}\ .$$

If $f < 8$, then applying Lemma 6 to B^2 and B^3 gives $a < v((1, 2))$, while if $f > 8$, B^2 and B^4 give the same result. If $f = 8$ we have Case 1. Thus we can have a heresy by $H = (1, 2)$ from $(a, 0, f, 8)$ to $\underline{h} = (a + \epsilon, \epsilon, f - 2\epsilon, 8)$; again the only possible penalizing imputation is B^1. If $f > 8$, B^1 fails to dominate, while if $f < 8$, the domination requires $v((2, 3)) \geq 16$, in which case B^1 would dominate B^2.

Since these cases cover all possibilities, there is no strong solution to any game in $\underline{C}$.

The number of times the values 8 and 16 turn up in the proof makes one suspect that the situation may be different outside of $\underline{C}$, but the examination of the possibilities for region $\underline{C}$ was already sufficiently difficult, even piloted as it was by prior consideration of the symmetrical game by itself, to make examination of the outer regions appear a fairly formidable task. On the other hand the absence of strictly strong solutions for such a relatively simple and significant class of games prompts further examination to see if some solutions may not have interesting stability properties that might distinguish them from others, even though they do not meet the more stringent requirements of the strictly strong solution. Note that the proof shows the existence of non-penalizing policing imputations, so that even relaxing the requirements to admit solutions wherein fratricidal heresies would occur would still not produce strong solutions in this region.

SOLUTIONS THAT SEEM ALMOST STRONG

Of the solutions of the four-person symmetrical game thus far examined, the one that appears strongest is the following symmetrical one (expressed, still, in (0, 24) normalization):

$\underline{V}_S$: the four permutations of $\underline{a}^4 = (8, 8, 8, 0)$, plus
the six permutations of $\underline{b}^{34} = (8, 8, 4, 4)$.

Consider heresies by $H = (1, 2, 4)$ against $(8, 8, 8, 0)$. We have $h_1 > 8$ and $h_2 > 8$, so since 8 is the maximum conforming share, $P = (3, 4)$. The policing imputations are either $(8, 4, 4, 8)$ and $(4, 8, 4, 8)$ if $h_3 < 4$, or $(8, 4, 8, 4)$ and $(4, 8, 8, 4)$ if $h_4 < 4$; one of the two conditions must obtain, and both may. Whatever be the actual $\underline{h}$ selected, the way is open to the penalizing of 1 or 2 but not both. Players 1 and 2 can thus each hope to get out of the heresy with a whole skin, but assuming that a conforming policing imputation eventually results, both cannot succeed. And since both 1 and 2 are getting as much in $\underline{a}^4$ as they can ever hope to get in any conforming imputation, the prospects of improving their position over the long run by engaging in such fratricidal heresies do not seem inviting.

However, while the members of P have the power collectively and the motive individually to secure for themselves the conforming shares (.. 8, 4) or (.. 4, 8), depending on $\underline{h}$, there is no reason to presume that they also have any direct power to insure that players 1 and 2 will divide the remainder of the cake among themselves in a conforming manner, particularly as the shares of 3 and 4 in $\underline{p}$ exhaust the power of P; if conformity is to be enforced upon 1 and 2 it must be by direct or indirect operation of the social pressures. One could suppose, for example, that the players 1 and 2 might have agreed, in the process of setting up the heresy, that upon its policing by P they would share their portion equally, thus leading to a quasi-policing imputation $\underline{q} = (6, 6, 8, 4)$ or $(6, 6, 4, 8)$. Even this prospect would at first glance seem sufficiently unattractive to 1 and 2 to inhibit the formation of the heresy. But the imputations $\underline{q}$ are in turn non-conforming imputations dominated by the conforming imputations (8, 8, 0, 8) or (8, 8, 8, 0) with effective sets R = (1, 2, 4) or (1, 2, 3) respectively, in which players 1 and 2 return essentially to the status quo ante. Whether the heresy seems attractive to 1 and 2 will then depend on whether the gains in $\underline{h}$ relative to the initial imputation $\underline{a}^4$ outweigh the losses expected in $\underline{q}$ (in magnitude and in "duration" combined); if for example players 1 and 2 succeed in persuading 4 to accept (11, 11, 0, 2) as $\underline{h}$, $\underline{h}$ may not need to "last" as long as $\underline{q}$ for there to be a net gain from the excursion as a whole. Also a crucial matter is whether there is any method of enforcing an understanding between 1 and 2 at the time H is formed to share the loss when the policing action takes place. This may be especially difficult since the player who abandons his agreement and conforms is likely to be the one who escapes all penalty in the resulting conforming imputation $\underline{p}$. On the whole heresies of this type appear likely to be unattractive to at least one of the potential participants; other three person heresies from imputations $\underline{a}^i$ are similar in character.

With two person heresies against these imputations $\underline{a}^i$, it is at least possible for all heretics to come out with a whole skin, though this result is somewhat problematical. For heresies by H = (1, 4) against $\underline{a}^4$ = (8, 8, 8, 0), if any player is penalized it must be 1, since 4 has nothing to lose. The possible policing imputations are (0, 8, 8, 8), which is penalizing, and all six of the imputations $\underline{b}^{ij}$, of which three are penalizing and three are not. From any given $\underline{h}$, the way is always open both to a penalizing and to a non-penalizing imputation. If P = (2, 3, 4) the result is always penalizing, but if P has only two players, then P has no obvious control over whether the result is penalizing or not. Nevertheless there is no obvious way in which 1 can protect himself from a penalizing outcome, and on the whole it would seem that 1 in this situation would tend to let well enough alone and decline to join in such a heresy, especially since even if he is fortunate enough not to be penalized, he winds up no better off than originally. The imputations $\underline{a}^i$, if not strictly strong, seem nevertheless reasonably stable.

Imputations $\underline{b}^{ij}$ are perhaps a little weaker. If we consider b^{34} = (8, 8, 4, 4), heresies by (1, 2, 4) or (1, 2, 3) are fratricidal. Heresies by (1, 3, 4) which leave player 2 with 4 or more are suicidal for 1, the only policing imputation being (0, 8, 8, 8); but on the other hand heresies giving player 3 a share of 8 or more require P = (2, 4) and $\underline{p}$ = (4, 4, 8, 8) or (8, 4, 4, 8) of which the first penalizes 1 and the second is non-penalizing; symmetrical results obtain for $h_4 > 8$. For such heresies as to $\underline{h} = (9, 0, 6, 9)$ or $(9, 0, 9, 6)$ the prospects for penalizing and for non-penalizing policing action are exactly similar if we assume that the history behind the heresy has no influence on the way in which it is policed; the prospects of the heretics coming out with a whole skin are perhaps brighter than in the preceding case, yet there is still no obvious method whereby the heresy $(1, 3, 4)$ can assure 1 against suffering a penalty, so the temptation may still be insufficient for 1.

Heresies against $(8, 8, 4, 4)$ by $H = (3, 4)$ have perhaps a better chance in that the only penalizing imputations are $(8, 8, 8, 0)$ with $P = (1, 2, 3)$ and $(8, 8, 0, 8)$ with $P = (1, 2, 4)$. Indeed, if the heretics are so fortunate as to be able to produce an $\underline{h}$ with either $h_1 > 8$ or $h_2 > 8$, such a heresy will be safe. But while $(8, 6, 5, 5)$, for example, dominates $(8, 8, 4, 4)$ with effective set $H = (3, 4)$, the domination is loose, and even though the set H is still not securing for its own members as much as it could but is, so to speak, keeping something in reserve, it is hard to see how it can use this reserve power to insure that player 1, for example, gets 8. To be sure of this point one way or another one would have to know more about the fine structure of the underlying game than is revealed by the characteristic function; even so, the relation between the amount of such reserve power and the degree of influence that can be exerted on the division of the shares among the non-members of set would seem to be potentially quite complex.

On the other hand, even assuming that H exerts itself fully in the immediate interests of its own members and thus insures that some imputation of the form say $\underline{h}(x) = (x, 12 - x, 6, 6)$ emerges, each member of H may consider that even if a penalizing imputation emerges, the two penalizing imputations are equally likely, and that therefore each member of H has as much chance of emerging with 8 as of emerging with 0 in the policing imputation. The possibility of gain may thus offset the possibility of loss, in such cases, with the gain during the persistence of the heresy as net profit, and the additional possibility that if by good fortune or otherwise a non-penalizing imputation occurs, at least one of the heretics will gain more permanently without the other losing anything. Thus the imputations $\underline{b}^{ij}$ are considerably weaker than the $\underline{a}^i$, but still something short of being fully weak.

Moreover, the weakness of the $\underline{b}^{ij}$ is likely to communicate itself to some extent to the $\underline{a}^i$, since the $\underline{a}^i$ derive much of such strength as they have from the existence of penalizing imputations among the $\underline{b}^{ij}$. If a $\underline{b}^{ij}$, once reached, is itself not likely to last long, the penalty that it imposes on a heretic from an $\underline{a}^i$ may appear relatively mild and thus still less effective than otherwise in inhibiting heresies from the $\underline{a}^i$ to persist, if once reached, and that while imputations $\underline{b}^{ij}$ will occur only rarely if at all, the fact that they enjoy a degree of social approval second if anything only to the $\underline{a}^i$ is still an essential part of the environment that serves to make the imputations $\underline{a}^i$ stable.

REMOTER MOTIVATIONS

In general, where a heresy is subject to policing imputations that are not themselves strong enough to be expected to persist for a considerable "time", the potential heretics may have to look beyond the immediate policing imputations if they are to judge the ultimate consequences of their heresy. In any but very special cases of this sort, it seems likely that if disturbance of an existing imputation is expected to lead to an extensive sequence of moves, only a very general expectation of a sort of average outcome can be determined. *A priori* it seems extremely difficult to determine such a general expectation for a given player by reference to the imputations of the solution themselves and of the most likely heresies, weighted in terms of some sort of frequency of occurrence, particularly as a fairly intricate circularity of cause and effect is involved. Rather one is tempted to refer to some value for each player computable directly from the characteristic function such as the Shapley value, s_i.[13] One would then consider the hypothesis that players would be generally readier to join heresies whenever their share in the existing conforming imputation falls short of this average value, and generally more reluctant when their current share exceeds this value, however it be determined. In the former case the players may perhaps be termed underprivileged, and in the latter, privileged.

Considerations of this nature might call for important reconsideration of the line of argument thus far presented. It might be reasonable to suppose, for example, that a player i who is privileged in imputation $\underline{c}$ might be reluctant to join even in a safe heresy if its policing imputations are in some way sufficiently less stable than $\underline{c}$. In an indeterminate heresy, if policing imputations that do not penalize a privileged player i are less stable than $\underline{c}$, this would also tend to increase this reluctance, thus adding to the stability of $\underline{c}$. Conversely, a player j who is underprivileged in $\underline{c}$ may be willing to engage in a heresy even though it is suicidal for him, provided that the imputations than penalize him are sufficiently less stable than $\underline{c}$, thus detracting from the stability of $\underline{c}$.

---

[13] L. S. Shapley, "A value for n-person games," Annals of Mathematics Study No. 28 (Princeton, 1953), pp. 307-317.

Unfortunately the definition of a stability measure, or even of a stability ranking, for the imputations in a proposed approved pattern in a manner that would take into account the above considerations appears quite difficult, as the stability of one imputation will depend on the stability of others and vice versa. Moreover in some cases it may be hard to justify the use of the Shapley values s_i as the mean expectation of what a player i may expect over the long run subsequent to the heresy if he breaks up the stability of the existing imputation by joining in a heresy. This seems particularly true if, for example, the approved pattern is a solution in which the Shapley value is a lower or upper bound for the shares which a particular individual can get within the solution, as can happen, for example, to the odd player in the discriminatory solutions of the three-person constant-sum game. In principle, one might think of defining a value as an average of the player's shares in the imputations of the pattern with weights determined as a function of the strength of the various imputations in relation to the solution; this introduces a further element of circularity as the strength would in general depend again on these weights. However, it seems a reasonable conjecture that no pattern over which the Shapley values would constitute an unreasonable average would exhibit any important self-policing properties.

There are circumstances under which one can be reasonably sure that the above considerations will not destroy the stability of a solution that is strong in terms of the immediate consequences of the policing of possible heresies. Let $U(\underline{a})$ be the set of players who are not privileged in imputation $\underline{a}$: $U = \{i | a_i \leq s_i\}$. Then if neither U nor any subset of U can obtain for its members more than their shares in $\underline{a}$ (i.e., $v(S) < \Sigma_S a_i$; $S \subseteq U$), then the imputation may be called firm relative to the expectation $\underline{s}$. A "firm" solution consisting only of firm imputations would then seem to be somewhat more stable than those that do not meet this criterion, and in particular if the solution is also strong, its stability would be unlikely to be disturbed by such longer-run considerations. The strong solutions of constant-sum simple games are firm in this sense, relative to any expected value in the interior of the convex hull of the solution. Indeed, in the cases thus far examined, strong and firm solutions appear to coincide, though it seems unlikely that this should invariably be true. Firm imputations may occur in weak solutions. For example, in the non-symmetrical solution to the three-person game with $k > 8$, imputations near the edges of the triangle are firm, despite their being weak.

Another approach would be to insist not only that every heresy be suicidal for some member of the heretic set, but that it be suicidal for a heretic who is privileged. Again, such a requirement does not seem in itself to add much in the way of restraint for the case of simple games, but for other games the question remains open.

DECOMPOSABLE GAMES

While it is not easy to discover strong solutions for non-simple games, they do exist; we have already mentioned the strong solutions of the three-person variable-sum games, and another class of games for which strong solutions are known is the class of games produced by the composition of two or more homogeneous simple games; these games are in general not themselves simple.[14] Strong solutions for the combined games are formed by taking the cartesian product of the strong solutions of the component games; cartesian products of strong with weak and of weak with weak solutions produce weak solutions in the composite game. Additional solutions to such games conforming to the von Neumann-Morgenstern definition do occur that involve transfers from the group of players belonging to one of the component games to other players, but such solutions appear generally to be weak; indeed as the very existence of such solutions depends on an extremely loose kind of domination in which the effective set seems to be very far from having the power to enforce or even very strongly to induce the corresponding move, one is led to wonder whether such solutions have any real significance at all except as the purely mathematical consequences of an arbitrary definition.

For example, in the simplest case, which is the six-person game formed from two three-person constant-sum games of equal weight, we have the following characteristic function, again in the $(0, 24)$ normalization, using a and b to denote variable members of the set of players $(1, 2, 3)$ of the first component game, and f and g to denote variable members of $(4, 5, 6)$: $v((a)) = v((f)) = v((a, f)) = 0$; $v((a, b)) = v((f, g)) = v((a, b, f)) = v((1, 2, 3)) = v((4, 5, 6)) = v((1, 2, 3, f)) = v((a, 4, 5, 6)) = 12$; $v((a, b, f, g)) = v((1, 2, 3, f, g)) = v((a, b, 4, 5, 6)) = 24$. The strong solution consists of the nine $(1, 2, 3) \times (4, 5, 6)$ permutations of $(6, 6, 0, 6, 6, 0)$. But there is also a solution involving a transfer from players $(1, 2, 3)$ to players $(4, 5, 6)$ consisting of all imputations $\underline{a}(y) = (0, 4, 4, 4 - 2y, 6 + y, 6 + y)$ for $0 \leq y \leq 2$, and the eight other $(1, 2, 3) \times (4, 5, 6)$ permutations of these imputations.

If we consider $\underline{h} = (6, 6, 0, 6, 6, 0)$ as a heretical imputation that might occur as a result of action by $H = (1, 2)$ which is loosely effective for $\underline{h}$ against any of the imputations in the solution, then the only policing imputations are those with $p_6 = 4 - 2y$; the only possible policing coalitions are $P = (4, 6)$ or $(5, 6)$ (since neither

[14] TGEB, p. 453, 51.5.2.

1 nor 2 can gain anything in a return to the solution, any subset of (3, 4, 5, 6) that includes both 4 and 5 would be ineffective against $\underline{h}$, and P = (3, 4, 6) and P = (3, 5, 6) would require $p_3 = 4$, and $p_4 + p_6 \leq 8$ or $p_5 + p_6 \leq 8$ if P is to be effective to $\underline{p}$, but this requires y = 2 and $p_6 = 0$, so player 6 has no motive to join P). Either of these policing coalitions can secure for its members the respective amounts 6 + y and 4 - 2y called for by the solution, but if players 1 and 2 stick together, it is hard to see, within the specifications of the game as represented by the characteristic function, how the policing set can compel (1, 2) to abandon their shares of 6 and 6 in the outcome of the game, nor is it easy to see how 1 or 2 could be offered an adequate inducement, within the solution as specified, to abandon the coalition (1, 2). Indeed, the characteristic function, in terms of its very definition, would seem to assert that not even the set (3, 4, 5, 6) has this power. Social sanctions in favor of imputations in such a solution that would thus overcome the patent interest of players 1 and 2 in sticking to their heresy seem too important to be left unexamined behind the scenes.

One could, perhaps, imagine a "district attorney" trying his blandishments on player 1 (or 2): "you can't expect to hold out forever, you know, and when you two do break it up, if your pal is the first to rat, you'll wind up with zero, while if you play ball with us now, we'll fix you up in a coalition with 3, 4 and 5, which will be a strictly effective set and you will be able to be sure of at least 4 instead of zero; we're working on your partner too, you know so you'd better hurry up before he makes up his mind." The trouble with this type of "policing" from a theoretical point of view is that if it is admitted as effective, it does away with the necessity for the direct domination of all heresies by conforming imputations and permits a wide range of patterns to be admitted as self-policing patterns that are not even subsets of solutions. For consider the three-person constant-sum game in (0, 12) normalization: The imputations (6, 6, 0), (6, 0, 6) and (0, 6, 6) are the symmetrical solution, but if the above type of policing be admitted, then the set (5, 5, 2), (5, 2, 5) and (2, 5, 5) is a self-policing pattern, since the heresies that are not directly dominated by conforming imputations, such as $\underline{h}$ = (6, 6, 0) are subject to just such policing.

If such policing is considered less reliable, or if, as seems reasonable, its success requires a higher degree of social pressure than the more direct policing, then patterns that are self-policing only by virtue of such policing can be considered weakly-self-policing. This concept suggests the following postulate, which may be useful in further analysis:

POSTULATE I. In any strongly self-policing pattern, let t_i be the maximum share that player i can get in any imputation in the pattern; then for any set H, $\Sigma_H t_i \geq v(H)$.

If this does not hold, then H will be able to effect a heresy in which each of the members of H gets more than he can within the approved pattern, and from any conforming imputation. Since strict domination is irreflexive, no strict domination by any conforming imputation is possible. Any policing set P that is effective for a conforming imputation $\underline{p}$ cannot intersect H, since the members of H all prefer $\underline{h}$ to any $\underline{p}$, and if the range of imputations $\underline{S}$ determined by the action of P fails to intersect the range $\underline{R}$ of the heresy, then $\underline{S}$ will also fail to intersect the approved pattern.

SUMMARY

Introduction of the idea of self-policing patterns as giving an indication of possible outcomes of games that may warrant special consideration yields, in some cases, a very substantial narrowing of the range of outcomes to be considered, as compared with the notion of solution, but introduces a rather large number of new and subtle difficulties. In many cases the self-policing patterns are also solutions and serve to single out particular solutions for special consideration. Particularly if minimal self-policing patterns are sought, the range of indicated outcomes may be quite narrow, being a unique set of imputations for homogeneous majority games, and for games decomposable into such games. However, for many important games, such as non-homogeneous majority games and four-person constant-sum games near the center, it can be shown that none of the solutions constitute strictly self-policing patterns, and it may be doubted whether in such cases there exist self-policing patterns that are not solutions. Attempts in such cases to find solutions that are self-policing in a less stringent sense lead to rather subtle difficulties that have not yet been adequately explored. Also still to be explored are the possibilities for self-policing patterns in non-simple four-person games outside the central region, in four-person variable-sum games and in non-simple games of five or more players.

William Vickrey

Columbia University

SIMPLE SOLUTIONS*

H. M. Gurk and J. R. Isbell

INTRODUCTION

Main simple solutions of simple n-person games were first introduced by von Neumann and Morgenstern [4] as an especially natural type of solution for such games. In a simple game each coalition is either completely winning or completely losing. Analogously, in these solutions each individual player has only two states, complete success and complete failure, although he does achieve both these states in different elements of the solution. Success for the i-th player in an imputation of such a solution consists in his obtaining a fixed non-negative amount x_i (the game being normalized $0 - 1$) and is achieved by joining with other players to form any of the minimal winning coalitions. If a player fails, he receives nothing.

This paper is a further study of a larger class of solutions introduced in [2] under the name "pseudo-main solutions". However, we now use a more intuitive terminology. A <u>simple solution</u> of any n-person game (normalized, $0 - 1$) is a solution in which the i-th player ($i = 1, \ldots, n$) gets $x_i \geq 0$ or 0 in each imputation of the solution.

We indicate some existence conditions for simple solutions of any n-person game, including a necessary and sufficient condition, and then give some conditions for special classes of games. The first section deals mainly with general conditions and with simple solutions of non-simple games. The other two sections deal mainly with simple solutions of simple games, which we call <u>extreme simple solutions</u>. The second section is concerned with existence conditions and general properties of such solutions corresponding to strong and non-strong simple games. The third section contains a number of results which indicate uniqueness of some extreme

* Part of this material appears in Gurk's doctoral thesis at the University of Pennsylvania; part was presented by Isbell to the American Mathematical Society, April 16, 1955.

simple solutions and offer criteria for determining such solutions for some simple games.

§1. EXISTENCE CONDITIONS

The solutions with which we are concerned are those called pseudo-main in [2]; we now call them simple solutions. We recall the definition as a finite set of imputations α^S "associated one-to-one with certain sets S of players; for each player i there is a (non-negative) constant x_i such that $\alpha_i^S = x_i$ for i in S, $\alpha_i^S = 0$ for i not in S. The set $\{\alpha_S\}$ must form a solution of the game." In view of the possibility that $x_i = 0$ and thus $\alpha^S = \alpha^T$ for $S \neq T$ we drop the phrase "one-to-one".

Given any set of n weights x_i $(i = 1, \ldots, n)$, we may refer to the vector x and will use the following notation:

(a) $x(S) = \Sigma_{i \epsilon S} x_i$;

(b) For a given game G with characteristic function v we define the property $P(S)$ to mean there is a set T, disjoint from S, such that $x(T) = v(T) > 0$ and $x_i > 0$ for $i \epsilon T$.

Now suppose V is a simple solution using weights x_i. Then we call x a simple vector.

Let x be a simple vector for an n-person game G. Then necessarily

I. $x(S) < 1$ implies $P(S)$ for $S \subset N = \{1, \ldots, n\}$.

For, there is an imputation β such that $\beta_i > x_i$ for all i in S. Then β is not in the solution V. Hence β is dominated by some element α^U of V, with effective set T. Clearly $T \cap S = 0$ and $x(T) \leq v(T)$. But we can show

LEMMA 1.1. If x is a simple vector for a game with characteristic function v, then $x(S) \geq v(S)$ for $S \subset N$.

PROOF. Suppose the lemma is false. Then $x(S) < v(S) \leq 1$, and by the same argument as above we obtain T_1, disjoint from S, such that $x(T_1) \leq v(T_1)$. Let $S_1 = S \cup T_1$. Then $1 \geq v(S_1) \geq v(S) + v(T_1) > x(S) + x(T_1) = x(S_1)$. By iteration we exhaust the set N and get $x(N) < 1$. But then V is empty, which is absurd.

Thus we have 1.1 and therefore I.

Essentially the same induction proves

LEMMA 1.2. If x is a simple vector for a game with characteristic function v, then $x(S) = v(S) < 1$ implies that there exists a T disjoint from S such that $x(T) = v(T) = 1 - v(S)$.

Note that 1.1 and 1.2 can be derived from I above, both by the same arguments as above, with $x(N) < 1$ contradicting I.

Also we see that if x is the simple vector corresponding to the solution V then

II. $V = \{\alpha^S \mid x(S) = 1$ and $P(S)$ is false$\}$.

For, if $x(S) = 1$ and $P(S)$ is false, then α^S is an imputation undominated by V and hence is in V. Conversely, if α^S is in V, then $x(S)$ must be 1. If $P(S)$ is true, then there exists an R, disjoint from S, such that $x(R) = v(R) > 0$. If $v(R) = 1$, then we let $R = T$. If not, then by 1.2 there exists a U disjoint from R such that $x(U) = v(U) = 1 - v(R)$. In this case let $T = R \cup U$. Then α^S is dominated by α^T on R. But $P(T)$ is false and hence α^T is in V, a contradiction.

Thus we need only prove the sufficiency part of

THEOREM 1.3. The game with characteristic function v has a simple solution determined by the vector x if and only if I holds. If I holds, then the solution is given by II.

PROOF. It remains to prove that no α^S in V dominates α^T in V, which follows from the fact that $P(T)$ is false (using 1.1), and to prove that every imputation β is either in the solution or in its dominion. For the latter let $S = \{i \mid \beta_i \geq x_i\}$. If $x(S) < 1$ then I and 1.2 imply that β is dominated by V; if $x(S) = 1$, then either $P(S)$, in which case, by 1.2, β is dominated by V, or not $P(S)$, whence $\beta = \alpha^S$ is in the solution.

Note that 1.1 and 1.2, though necessary, are far from sufficient, being realized if $x_i > 0$ for all i, $x(N) = v(N) = 1$ and $v(S) = 0$, $S \subset N$. The problem of satisfying I, and with it 1.1, is a problem of fitting an additive function to a superadditive function. (Cf. [6].) But we have not been able to exploit this observation. Instead we shall concentrate on the simple vector x and investigate the properties of vectors which are necessary or sufficient for them to be simple vectors of appropriate games.

A simple vector is non-negative (since otherwise the α^S would not be imputations). If some $x_i = 0$, one can replace x_i by any positive number and still have a simple vector x'. For, $x'(S) < 1$ implies $x(S \cup \{i\}) < 1$. Another way of putting this is that a game with a simple solution may be restricted to players with non-zero weights and it will still have a simple solution determined by the restriction of x. The extra players need not be dummies [2], but they play no part in the given solution.

In a similar vein (only now we change the characteristic function) we have

THEOREM 1.4. If x is a simple vector in the game with characteristic function v, then for any (normalized) characteristic function $v' \geq v$, x is a simple vector in the game defined by v'.

PROOF. It suffices to show that the sets for which $x(S) = v(S)$ all satisfy $x(S) = v'(S)$. If $v(S) = x(S)$, 1.2 gives us T disjoint from S such that $v(T) = x(T) = 1 - v(S)$. Then v' must agree with v on S and T; hence I still holds.

From 1.4 it immediately follows that every simple vector determines a solution in a strong (i.e., constant-sum) game, since a non-strong game with characteristic function v can be strengthened by appropriately choosing $v' \geq v$.

The above discussion makes no specific mention of simple games and would lead one to believe that there should be many examples of simple solutions for non-simple games. Most of the simple solutions which have appeared in the literature have been such that simple games could be found having the same solutions. However, this is not the case in general, as can be seen by the following example given by Gelbaum [1]. The game is the symmetric five-person game with $v(S) = 1/2$ if S has 2 or 3 members, and hence $v(S) = 1$ if S has 4 or 5 members. Let $x_i = 1/4$ for all i; then I holds, and it is not the winning sets but the two-person sets which support the condition.

However, there is a convenient class of games which has the property that every simple solution of any n-person game is a solution of a member of the class.

Let the *atoms* of a game be the minimal sets S such that $v(S) > 0$. Then an *atomic game* is one such that

(i) Every set S has a disjoint family of atomic subsets S_i such that $v(S) = \Sigma v(S_i)$.

Note that S need not be the union of a disjoint family of maximal atomic subsets.

A <u>maximal atomic game</u> is an atomic game such that

(ii) The sum of the values of the atoms in any maximal disjoint family of atoms is 1.

Now we have

THEOREM 1.5. Every simple solution of an n-person game is also a simple solution of some maximal atomic game.

This is easily verified by choosing as atoms all subsets T of N minimal with respect to the property that $v(T) = x(T) > 0$, and letting $v'(T) = v(T)$. The values of the other coalitions are determined by the maximum sum of values of disjoint atomic subsets.

The converse of 1.5 is obviously not true (i.e., not all maximal atomic games have simple solutions) since all simple games are maximal atomic games, but not all have simple solutions (Section 3 of this paper). Nor is every simple solution of any game also a simple solution of some strong maximal atomic game. This can be shown by a game of 14 players, whose lengthy description we omit. A maximal atomic game need not be strong (see any non-strong simple game); nor need a strong atomic game be maximal. For there are the five-person games whose atoms each consist of two players, with (12), (13), and (23) having the value t, $1/2 < t < 1$, and all other pairs the value $1 - t$.

We may mention an unsolved problem in this connection. For each game G having a simple solution V, 1.5 gives us a maximal atomic game G' with the same solution. Can G' be so chosen as to have all the simple solutions of G? By 1.4, we can choose G' so that it will have no new simple solutions (and in proving 1.5 we have done so). Incidentally, G' is not unique; a related, vague problem is whether reasonable additional conditions can be imposed so as to make G' unique.

We pause here to note the strong resemblance which maximal atomic games bear to simple games, in that the sole objective is the formation of atoms. It follows from (ii) that the players in a disjoint collection of atoms cannot gain by reorganizing, without taking on new partners. It remains that a player may wish to leave an atom for another which will treat him more favorably, but this possibility exists already in simple games. A simple solution may be viewed as a proposed specification of efficient atoms (those such that $x(S) = v(S)$); if it is adopted, then the objective of the i-th player is indeed uncomplicated, it being simply to obtain x_i.

Some of the results we shall obtain in Sections 2 and 3 for simple games can be generalized directly for atomic games. However, since we do not have so complete a set of results for these games as for simple games, we will couch all the theorems in terms of simple games.

The example above, of a simple solution of a non-simple game, can be generalized as follows,[1] to give all simple solutions of symmetric games. For each m and $k \geq 2$, there is a game $G(k,m)$ on $mk - 1$ players, in which each set of k players has value $1/m-1$, and other sets have the least values consistent with these. For $m = 2$ this is the direct majority game on $2k - 1$ players; and the game $G(2,3)$ was described above. Each of these games has a simple solution $V(k,m)$ given by $x_i = 1/mk - k$.

THEOREM 1.6. The only simple solutions of symmetric games are the solutions $V(k, m)$ of the games $G(k, m)$.

PROOF. Suppose first that the solution is symmetric. Let t be the least number such that a set of t players wins (i.e., has value 1). By 1.1, we have $x_i \geq 1/t$. In order for some imputation in V to be effective, there must exist S such that $x(S) \leq v(S)$. By 1.1, $x(S) = v(S)$. By 1.2, there is T containing S such that $x(T) = v(T) = 1$. Since T has at least t elements, $x_i \leqq 1/t$.

Let k be the least number such that the value of a k-player set is k/t. Then the greatest integer $[\frac{n}{k}] \leq t/k$, where n is the number of players in the game. Thus $n - t \leq k - 1$; but by I, $n - t + 1 \geq k$, i.e., t/k is an integer $m - 1$ and $n = mk - 1$. Then the value of a k-player set is $1/m - 1$; there are $m - 1$ disjoint sets of k players and $k - 1$ players left over, and therefore the game is $G(k, m)$. Since the weights x_i determine the solution, it is $V(k, m)$.

For any simple vector x, consider the set M of all players with minimum weight. For any subset S of M, of weight less than 1, I gives us a disjoint set T such that $x(T) = v(T)$. If T were not contained in M then we could replace an element of $T - M$ by an element of S to get a set T' such that $v(T') = v(T) > x(T')$, contradicting 1.1. Thus I holds for the game restricted to M; therefore x restricted to M defines a solution $V(k, m)$ of the restricted game $G(k, m)$. If the set N of all players contained one not in M then we should have $v(N) \geq m/m - 1 > 1$, which is absurd.

Von Neumann and Morgenstern showed [4] that convex composition of games with disjoint sets of players composes solutions. For convenience we give the short proof for simple solutions.

THEOREM 1.7. Let $\Sigma\alpha_i G_i$ be a composition of the games G_i, each of which has a simple vector x^i; then $x = \Sigma\alpha_i x^i$ is a simple vector for the composition.

---

[1] Gelbaum, Kalisch and Nering have found the generalization (unpublished) but not the uniqueness.

PROOF. Let G_i be on N_i and designate $S \cap N_i$ as S_i. If $x(S) < 1$ then $x(S_i) = \alpha_i x^i(S_i) < \alpha_i$ for some i; hence $P(S_i)$ holds in G_i, and hence $P(S)$ holds.

THEOREM 1.8. Every simple solution of a decomposable game is decomposable.

PROOF. We may let $G = \alpha G_1 + (1 - \alpha)G_2$, G_i on N_i, $x = \alpha x^1 + (1 - \alpha)x^2$, where x is a simple vector. By 1.2 (applied to the null set), there are completely winning sets of weight 1; hence, using 1.1, N_1 contains sets of value and weight α (and similarly for N_2). Then any subset of N_1 of weight $< \alpha$ can be extended to a set S of weight < 1 such that $v(N_2 - S) = 0$; and there is $T \subset N_1 - S$ such that $x(T) = v(T)$. Thus x^i is a simple vector in G_i. We have already noted that $P(S_i)$ implies $P(S)$. Therefore, by II, the solution given by x is simply the Cartesian product of the solutions of G_i given by x^i, except that the latter are multiplied by α and $1 - \alpha$ respectively for $i = 1, 2$.

§2. SIMPLE GAMES

A simple vector x, and the corresponding solution, will be called <u>extreme</u> if the following strong form of I holds:

III. If $x(S) < 1$, then there is T disjoint from S such that $x(T) = v(T) = 1$.

An extreme simple solution for v is a solution for the simple game v', defined $v'(S) = 0$ if $v(S) < 1$, $v'(S) = 1$ if $v(S) = 1$. Evidently any simple solution of a simple game is extreme. In fact,

THEOREM 2.1. Let G be a simple game on the set N, $a_i > b_i \geq 0$ for $i \in N$, and V a set of imputations α such that $\alpha_i = a_i$ or b_i for all $\alpha \in V$, $i \in N$. If no element of V is dominated by another element of V, but V dominates a dense subset of its complement, then $b_i = 0$ for all i and a is an extreme simple vector in G.

PROOF. If G is inessential the theorem is trivial. Otherwise (the game being normalized) $\Sigma b_i < 1$, so that there is a non-empty open set of imputations β such that $\beta_i > b_i$ for all i. Since one of these is dominated, there is α in V such that $T = \{i \mid \alpha_i > b_i\}$ is a winning set. For any i we can choose α so that i is not in T; for otherwise, since $\{i\}$ is less than a winning set (and V has at least two elements), there would be an undominated non-empty open set given by $\beta_i > a_i$ and $\beta_j > b_j$ for all j. Then if $b_i > 0$, there is a non-empty open set given by $\beta_i \geq 0$, $\beta_j > \alpha_j$ for $j \neq i$; and since $N - T$ loses, this set is undominated. Hence $b_i = 0$, for all i.

Now if $a(S) < 1$ then there is a non-empty open set of imputations β such that $\beta_i > a_i$ for i in S. One of these is dominated by an element of V, and hence, there is a winning set T disjoint from S such that $a(T) \leqq 1$. If $a(T) < 1$ then we should have a winning set disjoint from T, which is absurd. Hence a satisfies III and the proof is complete.

We cannot conclude that V is a subset of the solution belonging to the simple vector a. For example, in the projective plane game [5] with minimal winning sets 125, 345, 136, 246, 147, 237, 567, the unique simple vector is given by $x_i \equiv 1/3$, and the solution is $\{\alpha^S \mid S$ is minimal winning$\}$. But any six of these points dominate all but five other imputations. Thus the set $\{\alpha^S \mid S = 125, 345, 136, 246, 147, 237, 123, 124, 134, 234\}$ is not contained in the solution but does satisfy 2.1.

Theorem 2.1 is not valid for non-simple games. Consider the symmetric seven-person game with $v(S) = 1/4$ if S has 2 members, 1/2 if S has 3 members, and the least possible for other sets. The 21 permutations of (1/6, 1/6, 1/6, 1/6, 1/6, 1/12, 1/12) dominate all but each other and the 7 permutations of (1/6, 1/6, 1/6, 1/6, 1/6, 1/6, 0). (All 28 together make a solution.) We do not know whether 2.1 holds in atomic games, or whether any game has an actual solution of the indicated character with $b_i > 0$.

In order to clearly formulate the relationships of simple games, simple vectors, and solutions, let us adopt Shapley's convention [7] that all games are on a given infinite set of players, almost all of whom are dummies in any one game. Note that dummy players get 0 in each imputation in any solution [4]. Now in a particular (simple) game G, a simple solution V is called <u>main simple</u> if V can be represented as $\{\alpha^S \mid S$ a minimal winning set$\}$. V is <u>proper simple</u> if it is not main simple but it can be represented as $\{\alpha^S \mid S \in U\}$, where U is a family of minimal winning sets. V is <u>pseudo-main simple</u> if it can be represented as $\{\alpha^S \mid S \in U\}$, where all minimal winning sets are in U, but V is not main simple. V is <u>pseudo-proper simple</u> in the remaining case, i.e., U cannot include all minimal winning coalitions and must include some losing coalitions.

THEOREM 2.2. Every extreme simple solution is a main simple solution of some game; and it is either a main simple or a proper simple solution of some strong simple game.

PROOF. Let V, x^o, G, be given, with some set N of players. Let N_o be the set of all players i such that $\alpha_i > 0$ for some α in V. Let x_i be x_i^o for $i \in N_o$, $x_i = 0$ for all other i. Let U be the set of all $S \subset N_o$ such that $S = \{i \mid \alpha_i > 0\}$ for some $\alpha \in V$. Let U_1 be the set of all $S \in U$ such that S meets all other elements of U; $U_2 = U - U_1$. Perhaps the most natural game to associate with V is the game G_o on N_o with minimal winning sets precisely the elements of U_1. The property III for x^o in G implies the same for x in G_o. Obviously the change from x^o to x is irrelevant. Then let $x(S) < 1$. We have $x(S \cup (N - N_o)) < 1$ and hence $N_o - S$ contains a set T winning in G and such that $x(T) = 1$. Then T must be in U_1.

Generally V is pseudo-main simple in G_o, but it is main simple if U_2 is empty. If U_2 is non-empty, adjoin a player k to N_o and let the minimal winning sets of G_1 be those in U_1 plus $S \cup \{k\}$ for all S in U_2. Then V is main simple in G_1. Finally, the family W of minimal winning sets of G_1 (as well as the family U_1) has the

> <u>Property</u> $\mathcal{S}$: Every pair of elements of W meet, but no proper subset of an element of W meets every element of W.

The first part is obvious. For the second, if we remove a player in N_o we reduce the weight to < 1; if we remove k from an element of U_2, then the required condition holds by definition of U_2. Thus we finish the proof with

> LEMMA 2.3. A family of subsets of a set, which has the property $\mathcal{S}$, can be extended to the family of all minimal winning sets of some strong simple game.

PROOF. Extend to a maximal family W with property $\mathcal{S}$. For each S, either S or $N - S$, say S, meets all elements of W. Let T be a minimal subset of S with the property of meeting all elements of W. Then T is in W. Thus the supersets of elements of W are the winning sets of a strong game, and that W is just the collection of minimal winning sets follows from property $\mathcal{S}$ and maximality.

Thus if G_1 (or G_o) is not strong, we can extend it to a strong game in which the sets of U_1 will still be minimal winning; thus III still holds.

For what games is V a solution? Certainly no element of U_2

and no proper subset of an element of U_1 can be winning sets. But not every element of U_1 need win, as is shown by the projective plane game [5] with 13 players. The natural definition is that each of the 13 (four-person) lines wins; and the solution given by $x_i \equiv 1/4$ has just 13 points. But if any one of these lines is made losing, it remains a blocking set, thus remains in U, and even remains in U_1. Thus the best general condition we can give is III.

Given a solution V in a particular game G, one cannot always represent V as $\{\alpha^S \mid S \in U\}$, where U has property $\mathcal{S}$. This is shown by the 7-person projective plane game with one line struck out, say 567; with all $x_i > 0$ we have α^S both for $S = 567$ and for $S = 123$. However, there is

THEOREM 2.4. For any extreme simple solution V of the game G on the set N, there exist a simple vector x and a family $W = W_1 \cup W_2$ of sets, such that

(1) $V = \{\alpha^S \mid S \in W\}$, where $\alpha_i^S = x_i$ for $i \in S$, $\alpha_i^S = 0$ otherwise;

(2) W_1 is a family of minimal winning sets with property $\mathcal{S}$;

(3) $x_i = 0$ if and only if i is in no element of W_1;

(4) W_2 is a family of not completely winning sets, all contained in the union of W_1.

PROOF. Let N_0 be defined as above, $\{i \mid \alpha_i > 0$ for some $\alpha \in V\}$. Let W_1 be the family of all (necessarily minimal) winning sets $S \subset N_0$ such that α^S is in V; let W_2 be the family of all losing sets having these two properties. Then (1) and (2) are obvious; and we can let $x_i = 0$ for i not in N_0. It remains to show that each i in N_0 is in some element of W_1. By definition of N_0, i is at least in some element S of W_2. Then by II, S meets all elements of W_1; if i is removed, the weight of the remainder is less than 1. Hence, by I, i is in some element of W_1.

Thus a method of hunting extreme simple solutions of a game G is to examine all possible families W_1. In many cases there are no families with property $\mathcal{S}$ and thus we immediately know that there are no simple solutions. If such families W_1 exist, one must solve the equations $x(S) = 1$ for all S in W_1; and for all sets T which meet all elements of W_1, one must have $x(T) \geq 1$. If these conditions are met then one has III. Some x_i, i in the union of W_1, may vanish, but $x_i < 0$ is impossible. For there is S in W_1 not containing i, and $S \cup \{i\}$ would have weight less than 1. However, a better formulation of the method will refer only to minimal sets meeting all elements of W_1; then $x_i \geqq 0$ is an extra condition (for each i).

If one carries out the modified method and finds a solution x, with $x_i < 0$ for i in some non-empty set M, then one has an extreme simple solution, resting on a smaller set $W_1' = \{S \in W_1 \mid S \subset N - M\}$. For let $y_i = \max(x_i, 0)$. If $y(S) < 1$ then $x(T) < 1$ for all $T \subset S \cup M$. Hence there is a set U in W_1, disjoint from S, such that $y(U) = x(U) = 1$.

We may summarize our conclusions, using the von Neumann-Morgenstern notation [3] W^+ for the family of all sets which meet all elements of W.

THEOREM 2.5. The extreme simple solutions of a game G correspond one-to-one with the pairs (W, x) such that

(1) W is a family of minimal winning sets with property $\mathcal{S}$;

(2) x is an additive set function such that $x(S) = 1$ for all S in W and $x(S) \geq 1$ for all (minimal) S in W^+;

(3) $x_i > 0$ for i in the union of W and $x_i = 0$ for all other i.

If (W, x) satisfies (1) and (2) then one can modify W and x canonically to yield a solution of (1)-(3).

If one finds a simple solution of any game G which does not have the symmetry of the game, then obviously the motions of the game will yield other simple solutions. We do not know whether every game having a (extreme or not) simple solution has one possessing the symmetries of the game.

Another unsolved problem is whether $x_i = 0$ in one simple solution of a game implies $x_i = 0$ (by the convention of 2.5) in all simple solutions of that game. For the next result below, we may gain generality by permitting $x_i > 0$ though i is in no element of W. Let $U(x)$ be the family of all sets S such that $x(S) = 1$ and not $P(S)$. Let $P_x(S)$ be $P(S)$ relative to the proposed simple vector x.

THEOREM 2.6. If x is a simple vector for the simple game G and x' is any other vector such that $U(x')$ contains $U(x)$, then there is $p > 0$ such that $qx' + (1 - q)x$ is simple whenever $0 \leqq q \leqq p$.

PROOF. In simple games $P_x(S)$ implies there is T disjoint from S such that $x(T) = v(T) = 1$ and $x_i > 0$ for $i \in T$. Hence, not $P_x(T)$, and $T \in U(x)$. But then $T \in U(x')$ also, and therefore we have $P_{x'}(S)$. Therefore if $x'(S) < 1$ but $P_{x'}(S)$ is false, $P_x(S)$ is false, which can only be because $x(S) > 1$. Let $p(S)$ be the solution of $px'(S) + (1 - p)x(S) = 1$. Then $p = \min p(S)$.

There is a corresponding, complicated result for general simple vectors. 2.6 applies, for example, in the 13-player projective plane. The main simple solution is given by $x_i \equiv 1/4$. For any line L, there is another simple vector $y(L) = y$ given by $y_i = 2/5$ for i in L, $y_i = 1/5$ for i not in L. The families $U(x)$ and $U(y)$ are not comparable; but for each r between 0 and 1, $rx + (1 - r)y = z$ is simple, and $U(x)$ contains $U(z)$. However, neither in this example nor in any other which we have come across have we actually gained new information from 2.6.

We remark that the example above is the first instance of non-uniqueness of simple solutions. One may verify that the set of simple vectors is not convex, for $1/2y(L_1) + 1/2y(L_2)$ is not simple.

Referring back to 2.2, we inquire when an extreme simple solution V is a main simple solution of some strong simple game. If the game G_1 constructed in the proof of 2.2 is strong, then in the reduced game G_o on the set N_o we have

IV. For each S, either S or $N - S$ contains an element of U_1, or both contain elements of U_2.

Conversely, if IV holds in G_o, then G_1 is strong. Now G_o was defined purely in terms of the set V of imputations (on the fixed infinite set of players, almost all coordinates being 0). Thus we have

THEOREM 2.7. The finite set V of imputations is a main simple solution of some strong simple game if and only if IV holds in G_o, and thus if and only if G_1 is strong.

PROOF. All that remains is to observe that if V is main simple in the strong simple game G on N then IV holds in G_o. That is, either S or $N_o - S$ wins, or both $S \cup (N - N_o)$ and $N - S$ win.

We define the <u>committee composition</u>[2] of an n-person simple game G_o and n simple games $G_1, \ldots, G_n$, given abstractly (i.e., no longer on a fixed set of players). Let N_i be the set of players of G_i, and N

[2] This concept is introduced in unpublished lecture notes of L. S. Shapley.

the union of disjoint copies of the several N_i. The set $S \subset N$ is winning in $G_o(G_1, \ldots, G_n)$ if and only if the set of all i such that $S \cap N_i$ wins in G_i, is a winning set in G_o.

THEOREM 2.8. If $x^o, x^1, \ldots, x^n$, are simple vectors in $G_o, G_1, \ldots, G_n$, respectively, then $G_o(G_1, \ldots, G_n)$ has a simple vector $y = (y_{ij})$ defined by $y_{ij} = x_i^o x_j^i$.

We omit the easy verification. The smallest non-trivial committee composition is the five-person weighted majority game (2, 2, 1, 1, 1), where $n = 3$, G_o and G_3 are the 3-person strong simple game, and G_1 and G_2 are the one-person game. A composition of majority games need not be a majority game. If the strong player in (2, 1, 1, 1) is replaced by a three-person majority game, the result is the six-person game described in [4] in the following terms: the minimal winning sets are those three-person sets which contain an even number of the first three players.

Note that the committee composition can be formed if G_o is not simple by giving S the value which $\{i \mid S \cap N_i \text{ wins}\}$ has in G_o. In this situation 2.8 still holds.

§3. UNIQUENESS CONDITIONS

With respect to a strong simple game G, the equations $\Sigma_{i \in S} x_i = 1$, for all minimal winning sets S, are called the equations for a main simple solution. If $x = (x_i)$ solves the equations, then $y_i = \max(x_i, 0)$ defines a simple vector, by 2.5. If all $x_i \geq 0$ then the solution is main simple.

Let us call the corresponding homogeneous system $\Sigma x_i = 0$, the <u>h.m.s.</u> equations for G.

THEOREM 3.1. The h.m.s. equations for a strong simple game with no dummies have no non-zero solutions.

PROOF. Suppose (x_i) is such a solution. Let $P = \{i \mid x_i > 0\}$, $M = \{i \mid x_i < 0\}$. Neither P nor M is a winning set and hence $Z = N - P - M$ is non-empty. In fact, for any subset S of Z, either S or $Z - S$ wins. For $P \cup S$ or $M \cup Z - S$ wins, and contains a minimal winning set, which must be a subset of Z. But now $P \cup M$ consists of dummies and hence is empty.

COROLLARY 3.2. A strong simple game has at most one main simple solution, which hence has the symmetry of the game.

The symmetry appears in the family $\{S \mid \alpha^S \in V\}$ by definition; but it is from 3.1 that we can conclude $x_i = x_j$ when a motion of the game sends i into j.

COROLLARY 3.3. A strong simple n-person game with no dummies has at least n minimal winning sets; if it has exactly n, then its equations for a main simple solution are solvable and yield (by 2.5) a simple solution.

Both 3.2 and 3.3 fail for the (non-strong) modified 13-player plane game whose minimal winning sets are some 12 of the lines. The Corollary 3.3 follows from 3.1, because if the matrix of coefficients for the h.m.s. equations had rank less than n then 3.1 would fail. If the matrix is n by n, of full rank, then it transforms n-space onto itself and hence sends some point to $(1, 1, \ldots, 1)$.

In the last conclusion of 3.3 we cannot say "main simple". At least, it can be shown that solving the equations for a main simple solution need not yield a main simple solution. But the hypothesis of just n minimal winning sets is very strong; the only examples known are certain homogeneous weighted majority games (see [3]) and the 7-player plane games. These, of course, have main simple solutions.

By essentially the same proofs it can be shown that 3.1, 3.2 and 3.3 hold for strong atomic games, where atoms replace minimal winning sets in the definitions.

THEOREM 3.4. Let x be an extreme simple vector in the game G on the set N. If $x(N) < 2$ then G is a strong weighted majority game and x is a homogeneous system of weights; hence the solution is main simple. If $x(N) = 2$ then every minimal winning set, and every blocking set, has weight 1; hence the solution is either main simple in G or main simple in an n+1-person extension of G, which is strong simple.

PROOF. If $x(N) < 2$ then always S or $N - S$ has weight less than 1; hence $N - S$ or S wins, and G is strong. If S wins then $x(N - S) < 1$; hence $x(S) = 1$ if S is minimal winning. Thus x is a homogeneous system of weights and the solution is main simple. If

$x(N) = 2$ then no minimal winning or blocking set has weight greater than 1; and by III none has weight less than 1. The rest is obvious.

That 3.4 is false for non-extreme simple vectors is illustrated by Gelbaum's example described after 1.4. For some of the solutions considered in 3.4 we have the following theorem, for which we again suppose given a fixed infinite set of players.

THEOREM 3.5. If V is a main simple solution of a strong simple game G, $G = G_0(V)$ (as in 2.7), and $x(N_0) \leqq 2$, then V is not an extreme simple solution of any other game.

PROOF. If these conditions hold, consider any game G' solved by V. The set N of players of G' contains N_0, and (we may choose) $x_i = 0$ for i in $N - N_0$. By 3.4 we may suppose V is a main simple solution of G' and G' is strong simple. Then for every subset S of N_0 which is minimal winning in G_0, there is a minimal winning set T of G' such that $\alpha^T = \alpha^S$, i.e., $T \cap N_0 = S$. If $T \neq S$ then $N - S$ wins in G' and there is a minimal winning set $R \subset N - S$ such that $\alpha^R \in V$, which is absurd since G_0 is strong and α^R would dominate α^S. Thus all minimal winning sets in G_0 are also minimal winning in G', and $N - N_0$ consists of dummies.

For uniqueness it is obviously necessary that G_0 be strong simple, as otherwise G_1 is a different game. However, the main simple solution of the 7-person plane game, in which $x(N_0) = 2\ 1/3$, belongs to no other game. For no three-person set other than a line can win, and if not all lines win then there is a three-person blocking set. One can formulate the crucial property in general terms, but the value of doing so is not apparent.

THEOREM 3.6. A simple game having a simple vector x such that $x(N) \leqq 2$ has no other simple solutions.

PROOF. Let x' be another simple vector, M the set of all i such that $x_i' < x_i$, Z the set of all i such that $x_i' = x_i$. Let S be a subset of Z with $x(S) < 1$. Then $x(S \cup M) \leqq 1$. Otherwise there would be a minimal winning set $T \subset S \cup M$ with $x(T) = 1$ and $x'(T) < 1$, which is absurd. But now $x'(S \cup M) < 1$. Hence the complement of $S \cup M$ contains a minimal winning set U such that $x'(U) = 1$. Since $x(U) \geqq 1$, $U \subset Z - S$. Therefore the game played by subsets of Z already has a simple solution determined by the restriction of x. If $x(N - Z) > 0$ then $x(Z) < 2$ and $N - Z$ consists of dummies; otherwise $x(Z) = 2$, M

is empty, and x' differs from x only in giving non-zero weights to certain irrelevant players. In either case x' and x define the same solution.

THEOREM 3.7. If there exists a system $w = (w_i)$ of non-negative weights, all but at most two non-zero, such that every minimal winning set in the simple game G has weight at least $1/2\, w(N)$, then G has at most one simple solution.

PROOF. Let x be simple in G, and choose a pair (S, i) satisfying the following conditions:

(1) S is a minimal winning set, $x(S) = 1$, $i \in S$;
(2) $x_i > 0$;
(3) subject to (1) and (2), w_i is a minimum;
(4) subject to (1), (2), and (3), x_i is a maximum;

Then $x(S - \{i\}) < 1$; hence $T = (N - S) \cup \{i\}$ contains a winning set U such that $x(U) = 1$. We now show $x(T - U) \leq x_i$. Since $w(U) \geq \frac{1}{2} w(N) \geq w(N - S)$, hence $w(T - U) + w(U) + w(S) \geq w(N) + w(T - U)$. But $w(N) + w_i = w(T - U) + w(U) + w(S)$. Therefore $w_i \geq w(T - U)$. Thus by choice of S and i, $T - U$ can only consist of players j failing to satisfy (1), for which we have set $x_j = 0$, of those failing to satisfy (2), and at most one more j, for which $w_j = w_i$ and $x_j \leq x_i$. Then $x(N) = x(S) + x(U) + x(T - U) - x_i \leqq 2$. By 3.6, the proof is complete.

COROLLARY 3.8. A non-homogeneous weighted majority game has no simple solutions.

With 3.4 and 3.7 one can demonstrate that a given game has no extreme simple solutions by (a) exhibiting a suitable system w, and (b) showing that a certain system of linear equations is unsolvable. If the game is strong simple then one uses the equations for a main simple solution; otherwise one has $x(S) = 1$ for all minimal winning and blocking sets S. In either case one has also $x_i = x_j$ whenever a motion of the game sends i into j. A solution of the equations must be non-negative and is the only main vector. By this method one finds

COROLLARY 3.9. Among the 30 strong simple games on 6 or fewer players, 22 have unique simple solutions, which are main simple, and the others have none.

Evidence will be given below.

The non-strong five-person simple games have not been listed nor counted. But if one of them had a simple solution, it would appear as a simple solution of a strong simple six-person game, with $x_6 = 0$ (but 6 not a dummy). Since there are none such, we conclude

COROLLARY 3.10. No non-strong simple game on 5 or fewer players has a simple solution.

The seven strong simple games on 5 or fewer players are described in [4]; all are homogeneous weighted majority games, and hence, by 3.6, their main simple solutions are their only simple solutions. We list the strong simple six-person games, with integral multiples of their simple vectors or weights w as in 3.7. Proofs, including the proof that the list is complete, are simple and are omitted; the completeness proof is sketched in Isbell's "Six-person constnat-sum simple games", Logistics Project, Princeton University.

8 homogeneous weighted majority games

(2,1,1,1,1,1)	(4,2,2,1,1,1)
(3,2,1,1,1,1)	(4,3,3,1,1,1)
(3,3,2,1,1,1)	(5,2,2,2,1,1)
(4,1,1,1,1,1)	(5,3,3,2,1,1)

7 other games with main simple solutions

Minimal winning sets	Simple vector
123, 124, 135, 146, 156, 236, 245, 256, 345, 346	(1,1,1,1,1,1)
123, 145, 146, 156, 245, 246, 256, 345, 346, 356	(1,1,1,1,1,1)
123, 124, 126, 134, 135, 145, 234, 235, 245, 346	(1,1,1,1,1,1)
123, 124, 125, 134, 136, 235, 236, 156, 246, 345	(1,1,1,1,1,1)
123, 124, 125, 134, 136, 146, 234, 236, 246, 345	(1,1,1,1,1,1)
12, 135, 136, 145, 146, 234, 256	(2,2,1,1,1,1)
12, 135, 145, 146 234, 245, 256	(2,2,1,1,1,1)

8 games without simple solutions

(2,2,2,1,1,1)

(3,3,2,2,2,1)

(3,2,2,2,1,1)

(4,3,3,2,2,1)

(4,3,2,2,1,1)

(5,4,3,2,2,1)

123, 124, 125, 126, 135, 136, 145, 146, 234, 256, 3456 } w = (1,1,1,1,1,1)

123, 124, 125, 126, 134, 135, 146, 234, 236, 245, 3456 } w = (1,1,1,1,1,1)

In 3.7 we required that at most two weights vanish. If four vanish, the conclusion can be false. Our example is a strong simple 10-person game, one of whose simple vectors is $1/16\ (8,8,3,3,3,3,2,2,2,2) = x$. A relative weighted majority in the first six players wins. Player 7 decides the ties a34 vs. b56, and 8 decides a35 vs. b46, where a and b are 1 and 2; 9 decides 136 vs. 245, 10 decides 145 vs. 236. This game also has a main simple solution described by $(3,3,1,1,1,1,1,1,1,1) = 6y$. Yet for 3.7 there are the weights $(1,1,p,p,p,p,0,0,0,0)$, $0 \leq p \leq 1/2$. (Any convex combination of x and y is a simple vector. Note that this example shows that a strong simple game can have a main simple solution and also other simple solutions.)

If 3.7 applies except that three weights vanish, there is evidence to suggest that the conclusion may be true. But a much more tortuous argument would have to be used. Consider the seven-person game with minimal winning sets 123, 124, 134, 234, 125, 345, 136, 246, 147, 237. With the weights $(1,1,1,1,0,0,0)$, each winning set has half the total weight or more. The simple vector $x_i = 1/3$ is unique; but it does not satisfy $x(N) \leq 2$. The uniqueness of the simple vector is actually a special case of the more general

THEOREM 3.11. If a seven-person simple game G has a simple vector y given by $y_i = 1/3$, then G has no other simple solution.

We omit the involved argument.

3.11 also applies to the seven-person projective plane game. We may notice the curious relationship between uniqueness of the simple solution and uniqueness of the game. They are not equivalent, for there are non-strong games with unique simple solutions. But we do not know the answers to any of the modified questions which may occur to the reader.

In conclusion we also mention without proof

THEOREM 3.12. If G is a projective plane game with r points in a line, then the main simple solution of G, given by $x_i = \frac{1}{r}$, is unique.

BIBLIOGRAPHY

[1] GELBAUM, B. R., "Symmetric games," Bull. A. M. S. 59 (1953), p. 526 (abstract).

[2] ISBELL, J. R., "A class of game solutions," Proc. A. M. S. 6 (1955), pp. 346-348.

[3] ISBELL, J. R., "A class of majority games," Quart. J. Math. 7 (1956), pp. 183-187.

[4] von NEUMANN, J., and MORGENSTERN, O., Theory of Games and Economic Behavior, Princeton 1944, 3rd ed., 1953.

[5] RICHARDSON, M., "On finite projective games," Proc. A. M. S. 7 (1956), pp. 458-464.

[6] SHAPLEY, L. S., Additive and Non-additive Set Functions, Princeton, 1953 (thesis).

[7] SHAPLEY, L. S., "A value for n-person games," Annals of Mathematics Study No. 28 (Princeton, 1953), pp. 307-317.

H. M. Gurk
J. R. Isbell

University of Pennsylvania
and
The George Washington University

EDGEWORTH MARKET GAMES*

Martin Shubik

§1. ECONOMIC INTRODUCTION

Much mathematical work has been done in the investigation of the von Neumann and Morgenstern solution to an n-person game. However very little economic analysis has been applied to the formulation of the problems studied. In order to connect this particular type of game theory with economic analysis we will examine games whose characteristic functions are constructed in such a manner that they reflect certain economic aspects of a market.

The specific economic problem which is to be examined here is an extension of a bargaining or bilateral monopoly problem studied by Edgeworth[1]. Edgeworth considered a market which can be characterized by a two-person non-zero sum game. Each player was assumed to have a set of preferences which could be represented by a family of convex curves (not necessarily the same for each player) which at least gave an ordering to each player's preference for any combination of the two goods being traded. The player's initial fortunes were assumed to consist of $(a, 0)$ for the first and $(0, b)$ for the second. These state, respectively that the first player has a units of the first good and 0 units of the second good; while the second player has 0 units of the first good and b units of the second.

Edgeworth selected as a solution to this trading problem a set of distributions which satisfy the conditions that at any of these distributions a player is at least as well off as he would be if he failed to trade. No distribution of the set is preferred by both players to any

* Most of the work in this paper was done at The Center for Advanced Study in the Behavioral Sciences. The writer is indebted to Roy Radnor and Howard Raiffa for their conversations on this topic.

1 Edgeworth, F. Y., Mathematical Psychics, London, 1881, pp. 37-42.

other distribution of the set. Any distribution not in the set is dominated by some distribution in the set in the sense that both players can find a distribution of the set which they both prefer to one not contained in the set. This set of distributions usually forms a curve (infinite divisibility of the goods being traded and continuous and convex indifference curves are assumed) and is called the contract curve.

The Edgeworth solution can be demonstrated easily by means of a diagram often used in economics textbooks. In Figure 1 below the initial fortune of the first player is represented by the line Oa. His preference system is represented by the family of indifference curves of which ad is one. Any point on ad for instance, represents a combination of quantities of goods of the first and second type which has the same utility to the first player as does the initial combination of (a, 0).

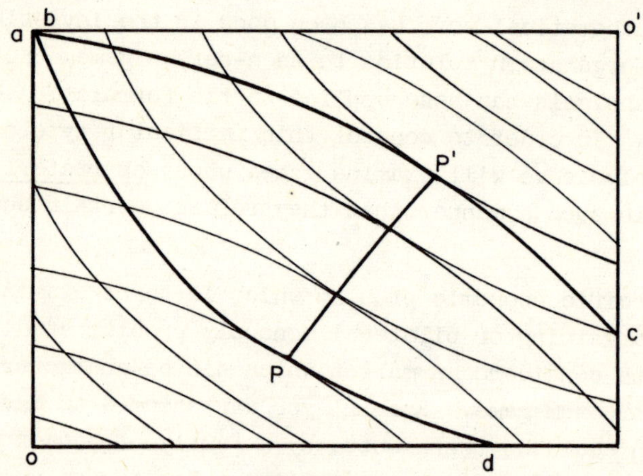

FIGURE 1

The preference system for the second player is portrayed by a second family of curves of which bc is one. The diagram is constructed in such a manner that the point 0 represents the state at which the first player has (0, 0) and the second has (a, b), whereas at 0' the first has (a, b) and the second (0, 0). The Edgeworth solution or contract curve is the locus of the tangent points between the two families of indifference curves, bounded respectively by ad and bc. This is indicated by PP'. It is easy to see that this solution is very similar to that of von Neumann and Morgenstern in that it consists of a set of distributions which do not dominate each other but dominate all other distributions. The solution is not identical with that of von Neumann and Morgenstern in as much as sidepayment in the form of a third commodity or "utils" is not considered. It is not even necessary to assume that the two players have comparable utilities for the goods being traded. However Edgeworth noted that if the players had comparable utilities then there would be a single point which he called the "utilitarian point" at which trade could take place in such a manner that joint utility, i.e., the sum of the utilities, would be maximized.[2] The von Neumann and Morgenstern solution selects this point to determine the trading and then determines a range for the sidepayments. The range of the von Neumann-Morgenstern solution is bounded by the values assigned to the positions of the players prior to trading.

Edgeworth considered bargains with more than two players and concluded that if the players became numerous on both sides of the market (under certain plausible conditions) that the bargaining would be such that the "range of indeterminacy" characterized by the contract curve would narrow and in the limit a single distribution would result. This distribution would correspond to the market price under pure competition.

In this paper we examine the limiting behavior of the solutions and certain other aspects of an n-person game related to the Edgeworth model, as the number of players becomes large.

§2. THE MATHEMATICAL MODEL

An Edgeworth market game consists of a set $I = M \cup N$ of players, consisting of traders in commodity 1 (the set of players, M) and traders in commodity 2 (the set of players, N). Each player has a utility function consisting of a set of strictly convex indifference curves (i.e., an indifference map). The utility function of the ith player is denoted by $\psi_i(x, y)$ where x and y are the quantities of the first and second commodity. It is assumed that:

$$\lim_{x \to \infty} \psi_i(x, y) < \infty \quad \text{and} \quad \lim_{y \to \infty} \psi_i(x, y) < \infty$$

and that the ψ_i are twice differentiable. The ith player in the set M starts with initial resources $(a_i, 0)$ of goods 1 and 2; where the first number is the amount of the first good and the second, the amount of the second. The jth player in set N has initially $(0, b_j)$. The characteristic function for the game is:

$$v(\theta) = 0$$

$$v(S) = \max_{x_i, y_i} \left[\sum_{i \in S \cap M} \psi_i(a_i - x_i, y_i) + \sum_{j \in S \cap N} \psi_j(x_j, b_j - y_j) \right] ,$$

where $\Sigma x_i = \Sigma x_j$, $\Sigma y_i = \Sigma y_j$ and $x_j, y_i, \geq 0$.

In order to examine this game fully we would have to consider many special cases depending upon the shapes of the utility functions, the

[2] Ibid., pp. 42-43.

initial stocks of the traders and the number of traders. We limit ourselves to the simple case where all traders have identical utility functions $\psi_i(x, y) = \psi(x, y)$. Each trader belonging to M initially has (a, o) and each trader belonging to N has (o, b).

Because of the above restrictions we can immediately write down the explicit characteristic function for any game of this type. If there is no joint gain to be obtained from trade, the game is inessential and the value of any coalition is the sum of the individual initial utilities. If the game is essential and trade takes place, then by the symmetry and convexity, at a position of joint maximization for any group of players all players will obtain equal quantities of the same good (at this point the marginal rate of substitution between the two commodities being traded will be the same for all players trading). The characteristic function is:

$$v(\theta) = 0$$

$$v(S) = s\,\psi\left(\frac{as_m}{s}, \frac{bs_n}{s}\right)$$

where $s = |S|$, $s_m = |S \cap M|$ and $s_n = |S \cap N|$. The characteristic function for the two-person game becomes:

$$v(\theta) = 0$$
$$v(\{1\}) = \psi(a, o)$$
$$v(\{2\}) = \psi(o, b)$$
$$v(\{1, 2\}) = 2\psi\left(\frac{a}{2}, \frac{b}{2}\right) .$$

The solution to this game consists of the set of imputations of the form:

$$\left(2p\,\psi\left(\frac{a}{2}, \frac{b}{2}\right) - p\,\psi(o, b) + (1 - p)\,\psi(a, o), \right.$$
$$\left. 2(1 - p)\,\psi\left(\frac{a}{2}, \frac{b}{2}\right) + p\,\psi(o, b) - (1 - p)\psi(a, o) \right),$$
$$0 \leq p \leq 1 .$$

It can be shown that if the number of traders in one commodity is the same as the number of traders in the other, then for any size of the market there is a solution analogous to the solution to the two-person game (although such a solution will no longer be the only solution). There exists a set of imputations which form a solution such that all traders on the same side of the market always obtain an equal amount. The set of imputations is characterized by a parameter p (as in the two-person game) which can be interpreted as a market price. The range of this price remains constant for all games with the same number of players on each side of the market.

> THEOREM 1. An Edgeworth market game denoted by $[n, n]$ for any size n where the number of traders in one commodity is the same as the number of traders in the other commodity and the traders have the same preferences will have a solution consisting of all imputations of the form:

$$\left(2p\,\psi\left(\frac{a}{2}, \frac{b}{2}\right) - p\psi(o, b) + (1-p)\,\psi(a, o), \ldots, 2(1-p)\,\psi\left(\frac{a}{2}, \frac{b}{2}\right) \right.$$

$$\left. + p\psi(o, b) - (1-p)\,\psi(a, o), \ldots \right), \tag{1}$$

$$0 \leq p \leq 1 .$$

PROOF. No imputation of the form of (1) can dominate another imputation of that form. A set to be effective for any imputation requires members from both M and N, i.e., trade must take place; but no imputation of the form (1) dominates another of form (1) for a set containing members from both M and N.

Any imputation not of the form (1) has a pair $\gamma'_i + \gamma''_j < 2\psi(\frac{a}{2}, \frac{b}{2})$ where γ'_i is the amount a player $i \in M$ and γ''_j is the amount a player $j \in N$ obtains from this imputation. If we select a p such that

$$\frac{\gamma'_i - \psi(a,o)}{2\psi\left(\frac{a}{2},\frac{b}{2}\right) - \psi(a,o) - \psi(o,b)} < p < 1 - \frac{\gamma''_j - \psi(o,b)}{2\psi\left(\frac{a}{2},\frac{b}{2}\right) - \psi(a,o) - \psi(o,b)}$$

then the imputation of form (1) for this value of p dominates the imputation γ. This completes the proof.

Although the solutions to the Edgeworth market games do not exhibit any tendency to converge to a single point as the number of players increases, it can be shown that under certain conditions the <u>core</u> of the games approaches a single imputation in the limit. This single imputation has economic meaning either in the theory of monopoly or the theory of pure competition.

The core of a game consists of the set of imputations which are undominated in all solutions. The first class of games to be examined have as an economic anologue a monopolist selling to many customers. We consider the Edgeworth market game with one trader of the first type and examine the behavior of the core as the number of traders of the second type increases. The core approaches a single imputation at which the single trader of the first type acts as a perfectly discriminating monopolist and obtains all of the gain from trading.

THEOREM 2. In an Edgeworth market game denoted by $[1, n]$, for any n the imputation:

$$\eta = \left((n+1)\,\psi\left(\frac{a}{n+1}, \frac{nb}{n+1}\right) - n\psi(o,b),\ \overbrace{\psi(o,b), \psi(o,b), \ldots, \psi(o,b)}^{n} \right)$$

lies in the core. For any ϵ there exists an $n(\epsilon)$ such that for all $n \geq n(\epsilon)$ no imputation ζ with $\zeta_1 = \eta_1 - \epsilon$ lies in the core.

PROOF. No set not containing player 1 is effective for η. Consider a set containing player 1 and k players contained in N:

$$(2)\qquad (n+1)\psi\left(\frac{a}{n+1}, \frac{nb}{n+1}\right) - n\psi(o,b) \geq (n-k+1)\psi\left(\frac{a}{n-k+1}, \frac{(n-k)b}{n-k+1}\right) - (n-k)\psi(o,b)$$

for any integer k, however, in any coalition smaller than I the quantity on the right-hand side of (2) is as much as can be given to player 1 if the others are to obtain at least $\psi(o, b)$ each. Hence no set containing player 1 is effective for the imputation η except I. Given I as the only effective set for η there exists no imputation which dominates η which is therefore in the core.

LEMMA. For any arbitrarily small ϵ there always exists an $n(\epsilon)$ large enough such that

$$|v(I) - v(I - \{j\}) - \psi(o, b)| < \frac{\epsilon}{n} \quad \text{for all} \quad n \geq n(\epsilon)$$

where $\{j\}$ is a player belonging to N.

PROOF. Expand $n\,|\,v(I) - v(I - \{j\}) - \psi(o, b)|$ by Taylor series about (o, b).

$$v(I) = (n+1)\,\psi\left(\frac{a}{n+1}, \frac{nb}{n+1}\right) = (n+1)\psi\left(o + \frac{a}{n+1}, b - \frac{b}{n+1}\right)$$

$$= (n+1)\psi(o, b) + a\psi_1 - b\psi_2 + \frac{1}{2}\left\{\frac{a^2}{n+1}\psi_{1,1} - 2\frac{ab}{n+1}\psi_{1,2} + \frac{b}{n+1}\psi_{2,2}\right\}$$

$$+ R_1 \quad .$$

$$v(I - \{j\}) = n\psi\left(\frac{a}{n}\,\frac{(n-1)b}{n}\right) = n\psi\left(o + \frac{a}{n}, b - \frac{b}{n}\right)$$

$$= n\psi(o, b) + a\psi_1 - b\psi_2 + \frac{1}{2}\left\{\frac{a^2}{n}\psi_{1,1} - 2\frac{ab}{n}\psi_{1,2} + \frac{b^2}{n}\psi_{2,2}\right\}$$

$$+ R_2 .$$

$$n|v(I) - v(I - \{j\}) - \psi(o, b)| \leq \frac{1}{2}\left|\frac{na^2}{n+1}\psi_{1,1} - a^2\psi_{1,1}\right|$$

$$+ \left|\frac{nab}{n+1}\psi_{1,2} - ab\psi_{1,2}\right| + \frac{1}{2}\left|\frac{nb^2}{n+1}\psi_{2,2} - b^2\psi_{2,2}\right| + |R_1 - R_2|$$

as $n \longrightarrow \infty$ all terms on the right can be made arbitrarily small.

Now select an ϵ, then for all games of size n where $n \geq n(\epsilon)$ we consider any imputation of the form:

$$\gamma = \left\{v(I) - n\psi(o, b) - \epsilon, \psi(o, b) + \mu_2, \psi(o, b) + \mu_3, \ldots \psi(o, b) + \mu_{n+1}\right\}$$

where

$$\sum_{j=2}^{n+1} u_j = \epsilon \quad .$$

There will be at least one player $j \in N$ who obtains $\psi(o, b) + \mu_j$ where $\mu_j \geq \epsilon/n$, then by excluding this player, the coalition $I - \{j\}$ will be effective for γ. Say $v(I - \{j\}) - [v(I) - \psi(o, b - \frac{\epsilon}{n}] = \delta$, then the imputation π obtained by adding $\frac{\delta}{n}$ to all components of the imputation γ, except the jth component, dominates γ. Hence γ cannot be in the core.

THEOREM 3. In an Edgeworth market game denoted by $[m, n]$ such that $m = km'$, $n = kn'$ and

$$\psi\left(\frac{m'a}{m'+n'}, \frac{n'b}{m'+n'}\right) = \max \psi\left(\frac{s_m a}{s}, \frac{s_n b}{s}\right)$$

the imputation

$$\eta = \left(\frac{v(I)}{m+n}, \frac{v(I)}{m+n}, \ldots, \frac{v(I)}{m+n}\right)$$

is always in the core; and for any ϵ there exists a $k(\epsilon)$ such that for all $k \geq k(\epsilon)$ no imputation ζ with a component

$$\zeta_i < \frac{v(I)}{m+n} - \epsilon$$

is in the core, $i = 1, 2, \ldots, m + n$.

The proof of this theorem is similar to that of Theorem 2. An interpretation of the condition that

$$\psi\left(\frac{m'a}{m'+n'}, \frac{n'b}{m'+n'}\right) = \max \psi\left(\frac{s_m a}{s}, \frac{s_n b}{s}\right)$$

is obtained by examining the nature of the function

$$s\psi\left(\frac{s_m a}{s}, \frac{s_n b}{s}\right) \quad .$$

It is homogeneous in the two variables, s_m and s_n. If s_m and s_n are increased in a constant ratio the value of the optimum trade increases in a constant manner. In economic terminology there are constant returns to scale. As the ratio of s_m to s_n varies there are different returns. Figure 2 indicates the lines of constant return and

$$\psi^* = \max \psi\left(\frac{s_m a}{s}, \frac{s_n b}{s}\right) \quad .$$

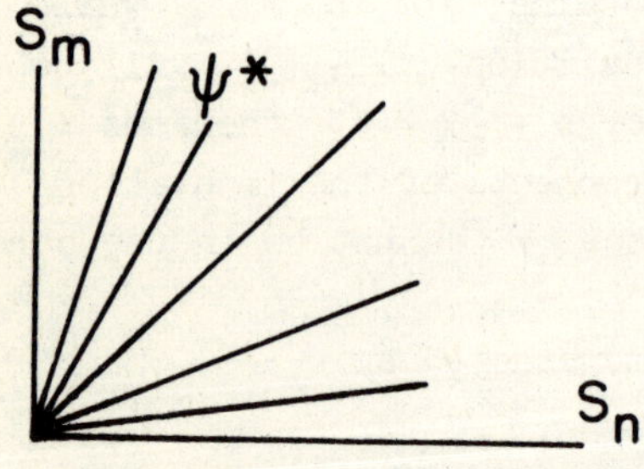

FIGURE 2

The function

$$\psi\left(\frac{s_m a}{s}, \frac{s_n b}{s}\right)$$

represents the average return to all members of S. As the ratios between the types of traders change, the average return fluctuates between the bounds of 0 and ψ*. If we examine the behavior of the core for the limit games whose size increases in any ratio other than that which obtains at ψ* there will always be a subset of players, as soon as the game is sufficiently large, who can obtain an average return of ψ^* in a coalition of this optimal size through which any imputation can be dominated and the core of the limit games will be empty.

Two trivial theorems can be established by considering the meaning of increasing and decreasing returns.

DEFINITION. Increasing returns are present in an economy if for at least two coalitions S and S' where $|S'| = |S| + 1$ and $S' \supset S$

$$\frac{v(S')}{|S'|} > \frac{v(S)}{|S|} .$$

Decreasing returns are present if for S and S', as above:

$$\frac{v(S')}{|S'|} < \frac{v(S)}{|S|} .$$

THEOREM 4. In an n-person economy with increasing returns everywhere, a core will always exist for all size n.

PROOF. The imputation

$$\eta = \left(\frac{v(I)}{n}, \frac{v(I)}{n}, \ldots, \frac{v(I)}{n}\right)$$

must lie in the core. As there are increasing returns everywhere, for any set $S \subset I$

$$\frac{v(S)}{|S|} < \frac{v(I)}{n} \quad \text{or} \quad v(S) < \frac{|S|}{n} v(I)$$

hence there is no effective set for η except I. As no imputation exists which dominates η, η must lie in the core.

THEOREM 5. In an n-person economy with decreasing returns for all coalitions with more than s^* members where $s^* < n$, no core exists.

PROOF. Given decreasing returns

$$\frac{v(I-\{j\})}{n-1} > \frac{v(I)}{n} \quad \text{or} \quad v(I\{j\}) > \frac{n-1}{n} v(I) \quad \text{for all } j.$$

Consider any imputation η; it will always have $n - 1$ components such that:

$$\sum_{i=1}^{n} \eta_i - \eta_j \leq \frac{n-1}{n} v(I) .$$

The coalition $v(I - \{j\})$ is effective for η and we can construct a β which dominates η using $(I - \{j\})$ as the effective set.

§3. ECONOMIC INTERPRETATION OF THE EDGEWORTH MARKET GAMES

The connection between the von Neumann and Morgenstern analysis and Edgeworth's and other economic theories of the market can now be made. The solutions to the n-person games reflect the sociological possibilities available through coalition structures. The core of a game consists of those imputations which are unaffected by any type of coalition structure. Following Edgeworth's terminology we may state that if an imputation in the core existed there would be no motivation for any group to recontract. Certain properties of increasing and decreasing returns to scale appear to be important in these games. If decreasing returns to scale are present then the core of the game is empty, i.e., all imputations are dominated by some coalition. In general this states that in an economy with decreasing returns present everywhere, under an optimal allocation of resources no matter how total product is divided there always will be some group in the economy which can form a coalition that is effective against any imputation that is suggested. This suggests that when decreasing returns are present everywhere the division of total product must depend upon the coalition structure, the social or political features of the economy under all circumstances.

In an economy of "optimum size", i.e., one in which a stage of increasing returns to scale has just given place to constant returns, the core approaches a single point at which every individual is awarded his "marginal value" productivity.[3] In this economy there is only one imputation that is not dominated by some coalition of players.

We prove in Theorem 1 that if the number of players in an Edgeworth Market game are the same on both sides of the market, then a set of imputations analogous to the contract curve[4] will be a solution no matter how many players there are in the market. Even if it happens that

---

3 It may be argued that this is a trading game and there is no production taking place. This makes no real difference however; the players are able to add to joint product by their willingness to trade. In this sense we can assign them a marginal value productivity.

4 Edgeworth, F. Y., op. cit., p. 21.

the core of the game converges to a single point as the players become numerous, this solution will not change. This feature emphasizes the difference between the von Neumann and Morgenstern solution and other economic analysis. The cooperative game analysis takes into account all possible coalition structures, thus even though there may be only one undominated imputation in the economy, there will be sets of mutually stalemating imputations which make up possible solutions given the presence of various coalitions. Edgeworth noted that if coalitions were permitted, the length of the contract curve would not shrink.[5] However, there will be many other von Neumann and Morgenstern solutions as well (only the two person game will, in general, have a single solution).

If there are increasing returns present in an economy (see Theorem 4) there will always be a set of imputations in the core. This merely reflects a basic feature of an expanding economy; the value of extra individuals is sufficiently great that there will be imputations such that no subgroup of the economy can make itself better off by refusing to cooperate.

In Theorem 2 it was demonstrated that as the number of players facing a single monopolist increases, the only undominated imputations approach the single imputation at which the monopolist is totally discriminating and obtains all the gain to be had from trade.

The core of a game is determined by technological conditions. At any imputation in the core all players obtain at least as much as they are able to enforce in any coalition. By means of n-person game analysis and an examination of the core we can give meaning to such vague ideas as "a world of monopolies". In any economy, any imputation of wealth which does not lie in the core implies that some group of individuals is profiting at the expense of another group. As the core will be empty if there are decreasing returns (see Theorem 5), in such an economy there will be no *a priori* "non-discriminatory" division of product.

If the core contains more than one point, then the combination of technological and coalition power is not sufficient to enforce all the possible imputations in a society. There remains an area on which the outcome (as is the case for the two person contract curve) must be determined by other considerations. If there is more than one monopolist in an economy (consider the Edgeworth games [2, n], or if there are three goods and three sets of players M, N and R, [1, 1, r]) and if all the monopolists are essential (i.e., their trade adds value to the economy) the core will not contain a single imputation, but there will be an area of indeterminacy between the shares of the monopolists.

5 *Ibid*., pp. 42-43.

The games considered in Theorems 1, 2 and 3 were all highly symmetric. Little is known about the behavior of solutions for this type of trading game as the number of players increases in a non-symmetric manner. The observations about the behavior of the core, however, are quite general and do not depend on symmetry but only upon the properties of the indivisibilities and complementarities present in the economy; i.e., the nature of the super-additivity of the characteristic function.

Martin Shubik

General Electric Company

A NOTE ON THE ARTICLE "SOME EXPERIMENTAL n-PERSON GAMES"[1]

R. Duncan Luce

§1. INTRODUCTION

The purpose of this note is to present a different, and I feel revealing, analysis of some of the data reported by Kalisch, Milnor, Nash, and Nering in Decision Processes [1]. I shall assume that the reader is familiar with their paper, and so it will be necessary to review only a few aspects of this work. First, while several different experiments are reported, only one of these was devoted to cooperative games with unrestricted side-payments. In that case, two four-person constant-sum games, each in two S-equivalent forms, were each run 16 times -- eight runs per S-equivalent form, one five-person constant-sum game was run three times, and one seven-person constant-sum game was run twice. Only the four-person games were sufficiently free of experimental artifacts, e. g., the effects of the seating arrangements of the subjects, and were repeated sufficiently often for our purposes, so we shall restrict our attention to those cases. Second, the subjects were told what each possible coalition would receive, i.e., they were given the characteristic function directly, and they were left to their own devices for 10 minutes to form coalitions and to agree upon a division of the spoils. Their agreements were supposed to be reported to an umpire, and he enforced them. Evidently, the subjects did not report all agreements on the division of payments, but this did not result in disputes. Third, the authors compared their experimental results with the several equilibrium theories existing at the time and with the basic assumption of n-person theory that S-equivalent games should receive the same strategic considerations. With respect to the major theory of n-person games -- the theory of solutions due to von Neumann and Morgenstern [5] -- they were forced to the conclusion that "It is extremely difficult

[1] This work was supported in part by the Office of Naval Research through the Behavioral Models Project, Bureau of Applied Social Research, Columbia University. Reproduction for any purpose of the United States Government is permitted. It is paper A 199 of the Bureau.

to tell whether or not the observed results corroborate the von Neumann-Morgenstern theory. This is partly so because it is not quite clear what the theory asserts." (p. 313.) The average observed imputation was compared with the Shapley value [6] and they found that "There seems to be a reasonably good fit between the observed data and the Shapley value, considering the small number of plays which are being averaged. There is, however, a tendency for the actual outcome to be more extreme than the Shapley value." (p. 309.) Milnor [4] has offered three different definitions each limiting the admissible outcomes in some more or less intuitively acceptable way. With respect to the data, two appear not to be very relevant, but one is. If $b(i)$ denotes the maximal incremental contribution of player i to the coalitions of which he is a member, Milnor argues that player i's outcome should not exceed $b(i)$. "It appears that $b(i)$ is usually compatible with the outcomes, in particular in the case of the four-person games (a deviation in one of the cases was subsequently related to the statement made by one of the subjects after the experiment was over that he had made a mistake in reasoning). The agreement is less favorable in the case of the five-person game -- some players getting more than their "maximum share" $b(i)$ -- a circumstance which may be related to the fact that the players rushed into coalitions, splitting the payoff evenly, without really studying the strategic possibilities." (p. 315.) Another important comparison, though not directly of an equilibrium nature, was made with the concept of strategic equivalence. They compared the average imputations of the two pairs of S-equivalent games, and they reached the conclusion that S-equivalent games appear not to receive similar treatment. They argue that "the fact that coalitions with high characteristic function are most likely to form, the tendency of members of a coalition to split evenly, and the non-linearity of the utility function all tend to disrupt the concept of strategic equivalence." (p. 313.)

In this note, their data -- fortunately they gave tables of raw data -- will be compared with an equilibrium theory which was not in existence at the time of the experiment. The comparison is interesting in its own right, and it suggests that with regard to the assumption that S-equivalent games shall receive similar strategic considerations a somewhat more subtle interpretation of the data is required than that sketched above. Indeed, one is almost able to argue that the assumption is confirmed.

§2. Ψ-STABILITY

Elsewhere [2, 3], I have offered an equilibrium notion called Ψ-stability. Since it will be assumed that the reader is familiar with at least one of these papers, the central concepts can be recapitulated without presenting any intuitive defense. First, it is assumed that an outcome of a game is not completely characterized by an imputation X, but rather a pair (X, τ) is required, where τ is a partition of the players into coalitions. A partition τ is called a coalition structure. Second, there is assumed to exist a function Ψ with domain the set of coalition structures and range the class of all sets of subsets of the players. It is further assumed that if $T \in \tau$, then $T \in \Psi(\tau)$. S is interpreted as a possible change from the coalition structure τ if and only if $S \in \Psi(\tau)$. Third, if v denotes the 0, 1 normalized characteristic function of a game, then a pair (X, τ) is said to be Ψ-stable if no admissible change insures profit to the participants in the change, i.e.,

i) for every $S \in \Psi(\tau)$, $v(S) \leq \Sigma_{i \in S}\, x_i$,

and any member of a non-trivial coalition receives more than he could insure himself when playing in isolation under the most adverse conditions, i.e.,

ii) $x_i = 0$ implies $\{i\} \in \tau$.

The crux of any application of this definition is the determination of the function Ψ, which summarizes the "sociological" limitations on changes in collusive arrangements. The authors' general discussion of the subjects' behavior suggests a likely candidate for Ψ. "Coalitions of more than two persons seldom formed except by being built up from smaller coalitions. Further coalition forming was usually also a matter of bargaining between two groups rather than more." (p. 306.) While they are not explicit on this point, we may also suppose that if a player demanded so much of a coalition that they would be better off without him, they threatened him with expulsion. Thus, the function Ψ suggested is characterized by admitting all coalitions in τ, the union of any two coalitions in τ, and the expulsion of any single player from a coalition in τ, i.e.,

$$\Psi(\tau) = [S \mid \text{either there is an } i \text{ such that } S \cup \{i\} \in \tau \text{ or there exist } T \text{ and } T' \in \tau \text{ such that } S = T \cup T'].$$

§3. COMPARISON WITH THE DATA

Tables I and II present the data reported by Kalisch et al in slightly modified form: first, they have been regrouped according to the equilibrium coalition structure, and second, they have all been reduced to 0, 1 normalization. The characteristic functions of the two games are given in the captions by giving their values for three two-element coalitions -- this is sufficient because the games are constant-sum and normalized. The exact form in which they were presented to the subjects may be found in the report of the experiment. In the statement of the

TABLE I

Comparison between data from RAND experiment and Ψ-stability theory. Symmetric 4-person constant-sum game in 0, 1 normalization: $v(\{1, 2\}) = v(\{1, 3\}) = v(\{1, 4\}) = 1/2$. Data are reduced to 0, 1 normalization; the round-off error is 0.01. The function Ψ is described in text. In all cases the lower limit on x_i is omitted as it is always confirmed.

Coalition Structure and corresponding Ψ-stable imputations	Game No.	Run No.	Observed Imputation Players 1	2	3	4	Incompatibilities between theory and data
$[\{1\}, \{2, 3, 4\}]$	2	7	.00	.40	.51	.09	None
$x_2 + x_3 \geq .50$	2	8	.00	.30	.43	.28	None
$x_2 + x_4 \geq .50$	3	2	.00	.26	.36	.38	None
$x_3 + x_4 \geq .50$	3	6	.00	.36	.28	.36	None
$x_1 = .00$							
$[\{3\}, \{1, 2, 4\}]$							
$x_1 + x_2 \geq .50$	3	4	.38	.36	.00	.26	None
$x_1 + x_4 \geq .50$							
$x_2 + x_4 \geq .50$							
$x_3 = .00$							
$[\{1, 2, 3\}, \{4\}]$							
$x_1 + x_2 \geq .50$	2	2	.48	.20	.33	.00	None
$x_1 + x_3 \geq .50$	2	4	.25	.31	.44	.00	None
$x_2 + x_3 \geq .50$	3	3	.34	.33	.34	.00	None
$x_4 = .00$							
$[\{1, 4\}, \{2, 3\}]$	2	1	.45	.13	.38	.05	None
$x_1 + x_4 = .50$							
$x_2 + x_3 = .50$	2	3	.19	.19	.31	.31	None
	2	5	.21	.19	.31	.29	None
	2	6	.28	.19	.31	.23	None
$[\{1, 2\}, \{3, 4\}]$	3	1	.25	.25	.25	.25	None
$x_1 + x_2 = .50$	3	7	.25	.25	.25	.25	None
$x_3 + x_4 = .50$							
$[\{1, 3\}, \{2, 4\}]$							
$x_1 + x_3 = .50$	3	5	.25	.25	.25	.25	None
$x_2 + x_4 = .50$							
$[\{1, 2, 3, 4\}]$	3	8	.25	.25	.25	.25	Total (see text)
Not stable							

TABLE II

Comparison between data from RAND experiment and Ψ-stability theory. Non-symmetric 4-person constant-sum game in 0, 1 normalization: $v(\{1, 2\}) = 3/4$, $v(\{1, 3\}) = 1/2$, $v(\{1, 4\}) = 1/4$. Data are reduced to 0, 1 normalization; the round-off error is 0.01. The function Ψ is described in text. In all cases the lower limit on x_i is omitted as it is always confirmed.

Coalition Structure and corresponding Ψ-stable imputations	Game No.	Run No.	Observed Imputation Players 1	2	3	4	Incompatibilities between theory and data
[{1}, {2, 3, 4}]	1	1	.00	.40	.30	.30	$x_2 + x_3 = .70 < .75$
$x_2 + x_3 \geq .75$	1	2	.00	.43	.43	.15	None
$x_2 + x_4 \geq .50$	4	2	.00	.42	.42	.17	None
$x_3 + x_4 \geq .25$							
$x_1 = 0$							
[{2}, {1, 3, 4}]	4	1	.38	.00	.25	.38	None
$x_1 + x_3 \geq .50$							
$x_1 + x_4 \geq .25$	4	3	.29	.00	.46	.25	None
$x_3 + x_4 \geq .25$	4	8	.29	.00	.42	.29	None
$x_2 = 0$							
[{3}, {1, 2, 4}]	1	6	.43	.43	.00	.15	None
$x_1 + x_2 \geq .75$	1	8	.44	.44	.00	.11	None
$x_1 + x_4 \geq .25$	4	4	.38	.54	.00	.08	None
$x_2 + x_4 \geq .50$	4	5	.37	.53	.00	.10	None
$x_3 = 0$	4	7	.38	.54	.00	.08	None
[{1, 2, 3}, {4}]	1	4	.13	.44	.44	.00	Incompatible
$x_1 = .25$							
$x_2 = .50$	1	7	.19	.44	.38	.00	Incompatible
$x_3 = .25$							
$x_4 = .00$							
[{1, 4}, {2, 3}]	1	3	.13	.38	.38	.13	None
$x_1 + x_4 = .25$	4	6	.13	.38	.38	.13	None
$x_2 + x_3 = .75$							
[{1, 2}, {3, 4}]	1	5	.25	.50	.13	.13	None
$x_1 + x_2 = .75$							
$x_3 + x_4 = .25$							

conditions implied by Ψ-stability theory -- the first column -- condition ii has been omitted in all cases because it was always satisfied. In the last column it is indicated whether the data are compatible with the prediction. To be compatible, the data were only required to be within 0.01 of the predicted value since the reduction to normalized form introduced a round-off error. In the 32 cases there are four discrepancies; it is worth examining these in some detail.

In Table I, run 8 of game 3, resulted in the set of all players forming a single coalition and an equal split of the proceeds. For the given function Ψ, stability theory states that there is no stable pair with the set of all players as a coalition. On the other hand, the theory does admit the pair $(\|1/4, 1/4, 1/4, 1/4\|, [\{1\}, \{2\}, \{3\}, \{4\}])$ as stable; however, experimentally, this case could not arise since there was no possible way for this coalition structure to collect a payment different from 0. So, had the players not wished to cooperate, they would have been forced by the experimental design to call themselves a single coalition in order to collect. Can it be that this explains this anomalous case?

The other three exceptions are in Table II, and they all occur in game 1 -- this was the S-equivalent form of the normalized non-symmetric game. In run 1, the error is five percentage points in a total predicted value of 75; by some standards this would not be considered a grevious error. In runs 4 and 7, the equilibrium coalition structure leads to a prediction of a unique imputation, and there can be no doubt that the subjects missed it. To be sure, there is some relation between the observed imputations and the predicted one, but there is no denying that they could not have calculated as the theory assumes they should.

§4. CONCLUSIONS

Excluding the peculiar case of the four-person coalition, the data are compatible with the predictions when the subjects received the characteristic function in normal form. The same is true for the S-equivalent form of the symmetric game, but not for the non-symmetric one. The incompatibilities in the last case tend to support the view that the participants did not fully appreciate the logic of the situation for a presentation of the characteristic function in a non-normalized form. I would argue, however, that by examining the data as we have done the evidence for non-equivalent treatment of S-equivalent games appears to be not nearly so strong as when average imputations are compared. One can be encouraged to hope that if the experiment were replicated and the 10 minute time limit eliminated, no exceptions would occur.

These remarks raise the question as to what a comparison of

average imputations can mean; can such a comparison ever reject the assumption of S-equivalence? I would think not for the following reason: Any equilibrium theory so far constructed leads to multiple equilibrium imputations, and presumably the "correct" theory will do so also. However, such theories neither do nor can be expected to indicate the probability distribution over these equilibrium imputations. Presumably, a dynamic theory of coalition formation is necessary to predict the likelihood that a particular equilibrium state will be achieved. The data strongly suggest that these probabilities are dependent upon the form of presentation of the characteristic function, but it is much less certain that the form of presentation influences the states which are in equilibrium. This is to say, while two S-equivalent games may have corresponding equilibrium states, they need not have the same dynamic characteristics and, therefore, not the same probability distribution over equilibrium states. If this is so, then there is no reason at all to expect the average imputations of S-equivalent games to be the same, or even similar. This analysis encourages one not yet to reject the argument that equilibrium concepts should be invariant under S-equivalence; more data will have to be amassed before that is necessary.

Finally, the relative success of this analysis suggests that it may be neither completely foolish to postulate the existence of the "sociological" functions Ψ nor impossible to estimate them for certain experimental and existing situations.

BIBLIOGRAPHY

[1] KALISCH, G., MILNOR, J. W., NASH, J., and NERING, E. D., "Some experimental n-person games," Decision Processes, New York, John Wiley and Sons, 1954, pp. 301-327.

[2] LUCE, R. D., "A definition of stability for n-person games," Annals of Mathematics 59, (1954), pp. 357-366.

[3] LUCE, R. D., "Ψ-stability: a new equilibrium concept for n-person game theory," Mathematical Models of Human Behavior, Stamford, Conn., Dunlap and Associates, 1955, pp. 32-44.

[4] MILNOR, J. W., "Reasonable outcomes for n-person games," The RAND Corporation Memorandum RM-916, Santa Monica, 1952.

[5] von NEUMANN, J., and MORGENSTERN, O., The Theory of Games and Economic Behavior, Princeton, 1944, 2nd ed., 1947.

[6] SHAPLEY, L. S., "A value for n-person games," Annals of Mathematics Study No. 28 (Princeton, 1953), pp. 307-317.

Harvard University R. Duncan Luce

ACCEPTABLE POINTS IN GENERAL COOPERATIVE n-PERSON GAMES

by

Robert J. Aumann

§1. INTRODUCTION

In this paper, we introduce a new theory for n-person games. Like many previous theories for n-person games, it is based on the search for a reasonable "steady state" in the "supergame",[1] i.e., the game each play of which consists of a long sequence of plays of the original game. We use two approaches: The usual approach, in which a single play is studied, and an attempt is made to find an n-tuple of strategies that intuitively speaking might be acceptable for a steady state in the supergame (Sections 4 and 5); and an approach based on a mathematical analysis of the supergame itself (Sections 6 and 7). The results we obtain by the two approaches turn out to be essentially equivalent (Sections 9 and 10).

We will consider here only cooperative games, i.e., games in which coalitions are permitted; side payments will be forbidden.[2] However, the ideas introduced here also apply to non-cooperative games. Subsequent papers will consider non-cooperative as well as cooperative games.

We will not explicitly consider games involving chance; however, everything we say here carries over with no essential change to games involving chance.

Section 2 describes the notation we will use throughout. The reader may omit this section at first and use it only for reference. Section 8 is devoted to the proof of lemmas needed in the subsequent sections; it makes strong use of the approachability-excludability theory of Blackwell [4]. Sections 11, 12, and 13 are devoted to some miscellaneous remarks and counter-examples.

---

1 The name is due to Luce and Raiffa [6].

2 Of course it is known (see [1]) that the cooperative game with side payments is a special case of the cooperative game without side payments.

§2. NOTATION

Throughout the paper, we will be concerned with an n-person game G. N will denote the set of all players; the individual players will be denoted by the positive integers 1, ..., n. If x^i is an object defined for each $i \in N$, and if $B \subset N$, then x^B will denote the set of x^i where $i \in B$; it will be called a B-vector. When $B = N$, we will often omit the superscript and the prefix; that is, we will write x instead of x^N, and will refer to x simply as a vector rather than as an N-vector. If X is a set of vectors, we denote by X^B the set of all x^B where $x \in X$. When we speak of a subset B of N, i.e., of a $B \subset N$, we will mean a non-null subset, unless we specifically include the null subset. All B-vector equalities or inequalities will be taken to hold term by term.

We will be concerned only with random variables taking a fixed finite number of values, and possibly with infinite sequences of such random variables. If x is a random variable ranging over a finite set Z, then the probability distribution of x may be considered as a function y from Z to the reals satisfying the relations:

$$y(z) \geq 0, \quad z \in Z \tag{2.1}$$

and

$$\sum_{z \in Z} y(z) = 1 \ . \tag{2.2}$$

We can then write

$$y = \sum_{z \in Z}^{*} y(z)z \tag{2.3}$$

where the * indicates the sum is merely symbolic and is not meant to be an ordinary sum. $y(z)$ is simply the probability that a random variable distributed according to y will take the value z.

The set of all probability distributions on Z will be denoted by C(Z). The set of all pure strategies p^i for player i will be denoted by P^i. P is then the space of pure strategy vectors p. C(P) is the space of <u>correlated</u> strategy vectors. A correlated strategy vector is a probability distribution on the set of all pure strategy <u>vectors</u>, just as a mixed strategy is a probability distribution on individual strategies. In general, to make use of a correlated strategy vector, the players have to agree to consult the <u>same</u> random device in choosing the pure strategies they will play. Correlated strategy vectors are thus not usable in non-cooperative games. We will usually abbreviate C(P) to C. The prefix c- in front of a word will stand for "correlated".

Let X^B be a space of B-vectors, and let $B = \cup_{j=1}^{k} B_j$ be a partition of B into subsets. If for $j = 1, \ldots, k$, we have

$$y^{B_j} \in C(X^{B_j}) \quad ,$$

then $(y^{B_1}, \ldots, y^{B_k})$ denotes the joint probability distribution of $y^{B_1}, \ldots, y^{B_k}$.

If Z_1 and Z_2 are finite sets and $y \in C(Z_1 \times Z_2)$, then $y|Z_1$ (sometimes also written $y|C(Z_1)$) will be the distribution given by

$$y|Z_1 = \sum_{z_1 \in Z_1}^{*} z_1 \sum_{z_2 \in Z_2} y(z_1, z_2) \quad .$$

If $B_1 \subset B_2$ and $y^{B_2} \in C(X^{B_2})$, then we will simply write y^{B_1} instead of $y^{B_2}|X^{B_1}$. This applies to all symbols except γ. Thus γ^{B_1} is not the same as $\gamma^{B_2}|P^{B_1}$. (γ^B will be defined on $C^B(P)$).

Let Z_1 and Z_2 be finite sets, and suppose we have a function f that takes Z_1 into Z_2. If nothing is said to the contrary, we will assume that the definition of f is extended linearly; that is, f is a function from $C(Z_1)$ to $C(Z_2)$ defined by

$$f(y) = \sum_{z_1 \in Z_1}^{*} f(z_1) y(z_1) \quad .$$

If for each $i \in N$, we have a function f^i defined on X^i, then we may define a vector function f on X by means of

$$f(x) = (f^1(x^1), \ldots, f^n(x^n)) \tag{2.4}$$

for all $x \in X$. It is important to remember, though, that the inverse process does not in general work. A vector function f defined on X does not yield a unique definition of a function f^i defined on X^i.

In general, when we have defined f^i on X^i, we will use the function f on X in the above sense.

For $p \in P$, the payoff to the $i^{\underline{th}}$ player will be denoted by $E^i(p)$. If $c \in C$, $H(c)$ will be the mean of the distribution $E(c)$. $E(c)$ is called the payoff distribution when c is played, $H(c)$ the expected payoff when c is played. If $p \in P$, we have $H(p) = E(p)$.

The null set will be denoted by $\emptyset$.

We will be concerned with partitions of N. Let $\mathcal{B}_N$ denote the set of all subsets of N. Formally, a partition of N is a vector d whose components are members of $\mathcal{B}_N$, and which satisfies the condition

$$j \in d^i \Longleftrightarrow d^i = d^j, \quad i, j \in N .$$

The set of all partitions of N will be denoted by D. The <u>constant</u> partition d_N is defined by

$$d_N^i = N, \quad i \in N . \tag{2.5}$$

The <u>identity</u> partition d_e is defined by

$$d_e^i = \{i\}, \quad i \in N .$$

Denote by R the set of all vectors r, each of whose components r^i is a member of $\mathcal{B}_N$ containing i. Clearly $D \subset R$, and $D^i = R^i$. For a given d, the number of distinct d^i will be called $n(d)$. We choose a set of representatives $i_1, \ldots, i_{n(d)}$, one out of each distinct d^i. If c^B is a c-strategy B-vector, define

$$e(c^B) = B .$$

(The e stands for exponent; B is called the <u>exponent</u> of c^B.) Denote by T the set of all vectors t, each of whose components is some correlated strategy B-vector (not necessarily the same B for each member of N), for which

$$i \in e(t^i), \quad i \in N .$$

Let

$$d^i(t) = \{j \mid t^j = t^i\}, \quad i \in N, \quad t \in T . \tag{2.6}$$

Then $d(t)$ is a partition of N, and we have

$$d^i(t) \subset e(t^i), \quad i \in N .$$

Define

$$c(t) = \left((t^{i_1})^{d^{i_1}(t)}, \ldots, (t^{i_{n(d(t))}})^{d^{i_{n(d(t))}}(t)} \right) \tag{2.7}$$

where $i_1, \ldots, i_{n(d(t))}$ are a set of representatives for $d(t)$.

In discussing games in extensive form, we will call the vertices of the game tree that are not moves by the name "terminal" rather than "play". This is necessary to avoid confusion, as we will be dealing with repeated plays of the same game, in the more common sense of the word.

As a final remark, we mention that everything we have stated probabilistically can be restated, where necessary, in the language of measure theory, and rigorously justified in that language. All sets we use can be proved to be measurable.

§3. GENERAL COOPERATIVE GAMES

We use the word "general" in the title of the paper and of this section in order to emphasize that, unlike the cooperative games of von Neumann-Morgenstern, the games to be considered here exclude side payments. On the other hand, full and free cooperation is permitted between the players, in the sense that they may communicate with each other freely prior to each play and form any coalitions they may consider convenient. The agreements under which the coalitions are chosen are to be considered enforceable. (However, see Section 13 for further discussion of this point.) We will usually omit the word "general" and refer simply to a cooperative game. This is in accordance with the usage introduced by Nash [1].

In applications, the prohibition on side payments may be the result either of a physical barrier to side payments (in economic applications this might take the form of certain provisions in an anti-trust law) or of the lack of a common unit of measurement for the payoff. As has been pointed out by Raiffa [8], if there is a common unit of measurement for the payoff but there exists some physical barrier to the making of side payments, then the theory need only be invariant under linear transformations of the payoff function that do not destroy the common unit of measurement; i.e., under transformations of the form

$$H' = aH + b \tag{3.1}$$

where both a and b are scalar multiples of the vector $(1, \ldots, 1)$ (a by a positive factor, b by an arbitrary factor). If, however, there is no common unit of measurement for the payoff, then the theory should be invariant under independent linear transformations of the payoff function; i.e., under transformations of the form (3.1), where a is an arbitrary positive vector and b is an arbitrary vector. In fact, the theory presented in this paper is invariant under the wider class of transformations, so that the results may be used in either kind of application.

Mathematically, the cooperative game is given by a vector payoff function that is assumed to be linear on the space C of correlated strategy vectors. The cooperative game differs from the non-cooperative game only in that the use of correlated strategy vectors is permitted in the former, but not in the latter.

This points up a reason for studying cooperative games even if our a priori interest lies in the field of non-cooperative games only. Consider a long sequence of plays (we call this a <u>superplay</u>) of the game

	1	2
1	(4, 0)	(0, 0)
2	(0, 0)	(0, 4)

One of the most "natural" ways for the players to proceed in such a superplay would be to alternate between playing $(1, 1)$ and $(2, 2)$. This would tend to happen even in the non-cooperative case, by a sort of "silent gentlemen's agreement". However, the resulting payoff, $(2, 2)$, is obtainable only by a correlated strategy vector, not by mixed strategies. Thus if we wish to arrive at results that are valid for the supergame by considering a single play, we must permit ourselves the use of correlated strategy vectors. This amounts to considering cooperative games.

§4. ACCEPTABLE POINTS

Let c be a correlated strategy vector in G. Obviously, c will be a candidate for a possible steady state in the supergame unless some subset B of the players can, by concerted action, increase their payoff. Here it is not sufficient that the players of B be able to increase their payoffs by changing their strategies while the players of $N - B$ maintain their strategies as they are. For the players of $N - B$ will not in general maintain their strategies fixed in the face of a change on the part of B, but will also change <u>their</u> strategies in order to meet the new conditions; thus B's glory would be short-lived indeed. In order to rule out c as a candidate for a possible steady state, we have to be sure that the players of $N - B$ cannot prevent the players of B from improving their payoff by concerted action. Of course, we must require that <u>all</u> the members of B obtain more at the new point than they do at c, for they must all have an incentive to cooperate in the venture.

DEFINITION 4.1. Let $c_o \in C$. c_o is said to be c-acceptable if there is no $B \subset N$ such that for each $c^{N-B} \in C^{N-B}$, there is a $c^B \in C^B$ for which

$$H^B(c^B, c^{N-B}) > H^B(c_o) .$$

We write "c-acceptable" instead of "acceptable" because in subsequent papers of this series, we will be defining a slightly different kind of acceptability, and we will use prefixes to distinguish them. When in heuristic discussion in the sequel we use the word "acceptable" without a prefix, then it is to be taken to mean "acceptable" with an arbitrary prefix (but one that is fixed throughout the discussion). For the purposes of this paper alone, "acceptable" can be taken to mean the same thing as "c-acceptable".

The set of all c-acceptable c-strategy vectors is denoted by A_c. Note that the notion of acceptability is a "global" one. In other words, whether or not a point is acceptable depends only on the payoff at that point, not on the point itself. This is also easy to accept from the intuitive standpoint; as long as he is getting the same payoff, it makes little difference to a player which strategy he is playing.

DEFINITION 4.2. A payoff vector h is said to be c-acceptable, if for some $c \in A_c$, we have

$$H(c) = h .$$

It will be convenient in the sequel to have available the following trivial restatement of 4.2.

LEMMA 4.3. A payoff vector h is c-acceptable if and only if for each $B \subset N$, there is a $c^{N-B} \in C^{N-B}$, such that for all $c^B \in C^B$, there is an $i \in B$ for which

$$H^i(c^B, c^{N-B}) \geq h^i .$$

§5. TWO-PERSON, ZERO-SUM GAMES

A sine qua non of theories of n-person games is that for two-person, zero-sum games, they reduce to the von Neumann theory. Because of the global nature of our theory, we can only expect that every acceptable payoff vector yield to each player the value of the game to that player. This is indeed the case.

THEOREM 1. In a two-person, zero-sum game G, a payoff vector h is c-acceptable if and only if

$$h^1 = v \quad , \tag{5.1}$$

where v is the value of G to player 1.

PROOF. Let h be c-acceptable. From Lemma 4.3 applied to $B = \{1\}$, we deduce that there is a $c_o^2 \in C^2$, such that for all $c^1 \in C^1$, we have

$$H^1(c^1, c_o^2) \leq h^1 \quad .$$

Hence

$$\max_{c^1 \in C^1} H^1(c^1, c_o^2) \leq h^1 \quad .$$

Hence

$$\min_{c^2 \in C^2} \max_{c^1 \in C^1} H^1(c^1, c^2) \leq h^1 \quad . \tag{5.2}$$

C^1 and C^2 are the mixed strategy spaces for players 1 and 2 respectively. Hence the left side of (5.2) is v, and we deduce

$$v \leq h^1 \quad . \tag{5.3}$$

The value of G to player 2 is $-v$. Proceeding as in the proof of (5.3), we obtain

$$-v \leq h^2 \quad . \tag{5.4}$$

But

$$h^2 = -h^1 \quad . \tag{5.5}$$

Combining (5.4) and (5.5), we obtain

$$v \geq h^1 \quad . \tag{5.6}$$

Combining (5.3) and (5.6), we obtain (5.1).

Conversely, assume (5.1). Then (5.3) and (5.2) follow at once. But for $B = \{1\}$, (5.2) is precisely the condition for c-acceptability given by Lemma 4.3. Similarly we establish the condition for $B = \{2\}$. To establish it for $B = \{1, 2\}$, note that for <u>any</u> (c^1, c^2), we have

$$H^1(c^1, c^2) + H^2(c^1, c^2) = 0 = h^1 + h^2 \quad ,$$

whence we must have either

$$H^1(c^1, c^2) \leq h^1$$

or

$$H^2(c^1, c^2) \leq h^2 \quad .$$

Thus h is c-acceptable.

§6. STRATEGIES AND PAYOFFS IN THE SUPERGAME

In the notion of supergame that will be used in this paper, each superplay consists of an infinite number of plays of the original game G. On the face of it, this would seem to be unrealistic, but actually it is more realistic than the notion in which each superplay consists of a fixed finite (large) number of plays of G. In the latter notion, the fact that the players know when they have arrived at the last play becomes the decisive factor in the analysis, overshadowing all other considerations.[3] This is unnatural, because in the usual case, the players can always anticipate that there will still take place in the future an indefinite number of plays. This is especially true in political and economic applications, and even holds true for the neighborhood poker "club". Of course when looked at in the large, nobody really expects an infinite number of plays to take place; on the other hand, after each play we do expect that there will be more. A. W. Tucker has pointed out that this condition is mathematically equivalent to an infinite sequence of plays, so that is what our notion of supergame will consist of.

Roughly speaking, we will be interested in "behavior" strategies in the supergame; that is, we will be interested in strategy vectors that determine the c-strategy vector to be played on each play as a function of the information possessed by the players about the outcomes of the previous plays. Before we pass to the formal definition of a supergame c-strategy vector, let us examine in detail the procedure followed during the play of a cooperative supergame. The first question to be answered is: What information about the outcome of each previous play does each player i actually possess? The most he can know is exactly what pure strategy vector was actually played; the least he can be sure of knowing, in every game, is the strategy he himself played and the payoff he himself received.

---

3 See, for example, the excellent analysis of the supergame of the game known as the "Prisoner's Dilemna" that appears in 5.5 of [6].

In between is a whole range of possibilities, depending on the rules of G. The situation can be concisely described by means of a function u^i that takes P onto some identification space of P. We define u^i so that for $p_1, p_2 \in P$, $u^i(p_1) = u^i(p_2)$ if and only if player i cannot tell after a play of G, whether the vector p_1 or the vector p_2 has been played. The minimum knowledge that each player i has, in accordance with our remarks above, insures us that if

$$u^i(p_1) = u^i(p_2),$$

then

$$p_1^i = p_2^i$$

and

$$H^i(p_1) = H^i(p_2) \quad .$$

Every game G has associated with it an information vector u satisfying these conditions. In particular, these conditions assure us of the existence of functions ϕ^i and ψ^i which map $u^i(P)$ onto P^i and $H^i(P)$ respectively. Some relations connecting the vector functions H, ϕ, ψ, and u are the following:

$$\phi \circ u = \text{identity}$$

$$\psi \circ u = H$$

$$\psi = H \circ \phi \quad .$$

The information vector just defined is not included in the classical formulation of the extensive form of a game, as given, say, in [3]. That is because the state of information at the <u>end</u> of a game has heretofore been of little interest, as the payoff in no way depends on it. In a number of consecutive plays of a game, however, it is seen to be of vital importance. Incidentally, inclusion of this concept in the extensive structure of a game necessitates a revision in the notion of equivalence as formulated in [5]. We hope to treat the subject in more detail in a subsequent paper. For the present, we content ourselves with the remark that the information vector may be included in the extensive description of the game by associating with each player i, an information partition on the set of all terminals. In a game of perfect information, this partition would be the identity partition for each i.

All that we have said in this section up to now is quite general, applying to the non-cooperative as well as to the cooperative case. In the latter case, the information function does not exhaust the information that a player has to work with. Prior to each play, consultation takes place among the various players to determine the constitution of the coalitions and the correlated strategy B-vector that each coalition B will use. Formally, each player designates the coalition B to which he wishes to belong, and the correlated strategy B-vector he wishes B to play. Those players who can agree among each other form coalitions and play the agreed upon c-strategy B-vector.[4]

Naturally, although each player knows nothing about the coalitions in which the other players participate, he does remember the coalition B of which he himself was a member on previous plays.

The formal definition is as follows:

DEFINITION 6.1. A supergame c-strategy f^i for player i is an infinite sequence of functions

$$f_0^i, f_1^i, \ldots, f_k^i, \ldots$$

where f_k^i is a function from

$$((U_1^i \times R_1^i) \times \ldots \times (U_k^i \times R_k^i))$$

to T^i; here U_j^i is a copy of $u^i(P)$ and R_j^i is a copy of R^i, for $j = 1, \ldots, k$.

The vector f is called a supergame c-strategy vector.

In the sequel, we will denote $U_j^i \times R_j^i$ by J_j^i. J_j^i represents the information about the jth play available.

To define the payoff to a supergame c-strategy vector, we must first define the payoff to a given super<u>play</u>, as a function of its individual constituent plays. For this purpose, we use the arithmetic mean of the payoffs (in the limiting case, this amounts to the first Cesàro sum); Blackwell [4] also used the arithmetic mean under similar circumstances. Heuristically, one can argue both for and against this method; in my mind the heuristic advantages outweigh the disadvantages. Other methods have been proposed (see [6]) for obtaining the superplay payoff from the payoffs to its constituent plays; the most prominent seems to be the discounting of future payoffs back to the present, using a fixed rate of

---

[4] If $B_1 \subset B_2$ and every member of B_1 proposed (B_2, c^{B_2}) while the members of $B_2 - B_1$ proposed something else, then the members of B_1 form a coalition and play c^{B_1}.

interest. This has the disadvantage of putting an unnaturally heavy stress on the beginning of the supergame. Nevertheless, it would be interesting to see an analysis based on this payoff notion, especially for certain kinds of economic applications.

A more perplexing problem in defining the payoff to a supergame c-strategy vector f is that of somehow combining the various possible superplays that might occur if f is played. The first impulse is to use some notion of expected payoff. We shall indeed define such a notion, but it is not an appropriate criterion for the players in choosing their supergame c-strategy. The chief reason for our being interested in expected payoffs for individual plays is that in a long sequence of plays, the law of large numbers turns an expected payoff into an actual payoff. A payoff for the whole superplay that is only expected, but not eventually approached, will not satisfy most players. Thus if a player has joined a coalition on the strength of a high expectation and is being disappointed in this expectation, he will soon express his disappointment by resigning from the coalition, regardless of his originally high expectations. For a clear example of the pitfalls of expected payoffs, see Section 12. What is required is some kind of criterion based on a law of large numbers. We give a restricted definition of this kind in this section, but reasoning of this type in general will dominate the remainder of the paper.

Let us first return to the question of expected payoff. For each $t \in T$, define

$$s(t) = (u(c(t)), \quad d(t)) \quad . \tag{6.2}$$

Let f be a supergame c-strategy vector. For each $k \geq 1$, we may define a member $z_k(f) = z_k$ of $C(J_1 \times \ldots \times J_k)$ recursively as follows:

$$z_1 = s(f_0) \tag{6.3}$$

$$z_k = \sum_{y \in J_1 \times \ldots \times J_{k-1}}^{*} z_{k-1}(y)\ (y,\ s(f_{k-1}(y))), \quad k > 1 \quad . \tag{6.4}$$

z_k is the probability distribution of possible outcomes of plays up to the kth. Define a member $x_k(f) = x_k$ of $C(J_k)$ as follows:

$$x_k = s(f_{k-1}(z_{k-1})), \quad k \geq 1 \quad . \tag{6.5}$$

x_k is the probability distribution of outcomes of the kth play. Finally, define $E_k = E_k(f)$ and $H_k = H_k(f)$ by

$$E_k = E(c(f_{k-1}(z_{k-1}))) \tag{6.6}$$

and

$$H_k = H(c(f_{k-1}(z_{k-1}))) \quad . \tag{6.7}$$

E_k represents the probability distribution of possible payoff vectors on the kth play; H_k is the expected payoff for the kth play.

If

$$H(f) = \lim_{k \to \infty} \frac{1}{k} \sum_{r=1}^{k} H_r(f) \tag{6.8}$$

exists, then f is said to be <u>summable in the mean</u>.

Let $v = (v_1, \ldots, v_k, \ldots)$ be a sequence of random variables in which v_k is distributed according to the distribution $x_k(f)$. These random variables are not in general independent. Instead of saying that the random variables v_k are distributed according to $x_k(f)$, we will sometimes say that the random variable v is distributed according to f. The probability of a statement concerning a random variable v distributed according to f will usually be denoted Prob_f.

DEFINITION 6.9. A supergame c-strategy vector f is said to be <u>summable</u> if it is summable in the mean and if a sequence of random variables distributed according to $E_k(f)$ obeys the strong law of large numbers.[5]

If v is an infinite sequence $(v_1, \ldots, v_k, \ldots)$ where $v_k \in J_k$ for all k, define

$$H_k(v) = \psi(v_k | U_k) \; (= H(\phi(v_k | U_k))) \tag{6.10}$$

and

$$S_k(v) = \frac{1}{k} \sum_{j=1}^{k} H_j(v) \quad . \tag{6.11}$$

---

[5] See [2], p. 207. The law is there stated for one dimensional random variables; the extension to n dimensions is straightforward.

Note that if v is a random variable distributed according to f, then $H_k(v)$ is a random variable distributed according to $E_k(f)$.

LEMMA 6.12. A necessary and sufficient condition that a supergame c-strategy vector f be summable is that there exist a vector $H(f)$ such that for each vector $\epsilon > 0$, we have

$$\lim_{k \to \infty} \mathrm{Prob}_f \left(|S_r(v) - H(f)| \geq \epsilon \text{ for some } r \geq k \right) = 0 \quad .$$

Furthermore, this $H(f)$ is then identical with the $H(f)$ defined by (6.8).

The proof, which is straightforward, will be omitted.

§7. STRONG EQUILIBRIUM POINTS

When we come to investigate what supergame c-strategy vectors f are liable actually to occur, we are faced with a problem somewhat different from the one we discussed in Section 4. There one could always look forward to future plays in order to punish deviations by the other players. In the supergame, however, there is only one superplay. Thus the players will have to make sure the first time that deviations won't pay. This amounts to saying that no coalition B can increase the payoff of all its members by substituting a different supergame c-strategy B-vector g^B while $N - B$ maintains the supergame c-strategy (N-B)-vector f^{N-B}. It is clear that such an f would be very strongly stable, since no coalition could have any incentive to move away from it, especially if we demand that B cannot even increase its payoff with positive probability, no matter how small. This is a much stronger version of what corresponds in the supergame to an ordinary equilibrium point.

The formal definition is as follows:

DEFINITION. Let f be a summable supergame c-strategy vector. f is a strong equilibrium c-point if there is no $B \subset N$ for which there is a supergame c-strategy vector g satisfying

7.1) $$g^{N-B} = f^{N-B}$$

and

7.2) there is a B-vector $\epsilon^B > 0$ for which

$$\lim_{k \to \infty} \text{Prob}_g \, (S_r^B(v) \geq H^B(f) + \epsilon^B \text{ for some } r \geq k) > 0 \quad .$$

(The limit exists because the sequence involved is monotone decreasing.)

The set of all strong equilibrium c-points is denoted by S_c.

It is necessary to use the form $S_r^B(v) \geq H^B(f) + \epsilon^B$ rather than $S_r^B(v) > H^B(f)$ because substitution of the latter statement would make condition 7.2 true even for $f = g$. We wish to include only those cases in which substitution of g for f will yield B an advantage that at least does not tend to the vanishing point, doubtful and rare as it may be.

Condition 7.2 is very weak, which means that the concept of strong equilibrium c-point is a very strong one (i.e., comparatively few supergame c-strategy vectors are strong equilibrium c-points). On the other end of the spectrum, we could replace 7.2 by

7.3) There is a B-vector $\epsilon^B > 0$ for which

$$\lim_{k \to \infty} \text{Prob}_g \, (S_r^B(v) \geq H^B(f) + \epsilon^B \text{ for all } r \geq k) = 1 \quad .$$

Denote by $\tilde{S}_c$ the set of supergame c-strategy vectors obtained when 7.3 is substituted for 7.2 in the definition of strong equilibrium c-point.

It is easy to see that

$$7.3 \Longrightarrow 7.2 \quad ,$$

whence it follows that

$$S_c \subset \tilde{S}_c$$

and

$$H(S_c) \subset H(\tilde{S}_c) \quad . \tag{7.4}$$

We will demonstrate the opposite inequality as well. This shows that the notion of strong equilibrium point does not really depend on the strength of the condition 7.2.

§8. APPLICATION OF APPROACHABILITY AND EXCLUDABILITY THEORY

For two-person, zero-sum games with vector payoffs, Blackwell [4] has defined the notion of "approachability" or "excludability" of a set in the payoff space with a supergame strategy employed by one or the other of the players. Blackwell does not use the information vector u which we introduced in Section 6, but apart from that, our Definition 6.1 applied to two-person, zero-sum games reduces to his definition of supergame strategy (for games not involving chance). Most of his definitions and results can be used here without any changes being necessary due to our use of the information vector.

Let us consider the two-person, zero-sum game with vector payoffs that is obtained from our game G by considering the coalition B to be the first player and the coalition $N - B$ the second player, the payoff to the first player being given by H^B.

LEMMA 8.1. Let $\Lambda(f, \epsilon^B)$ denote the closed convex set consisting of all B-vectors λ^B for which

$$\lambda^B \geq H^B(f) + \epsilon^B \quad .$$

Then if there is an $\epsilon^B > 0$ such that $\Lambda(f, \epsilon^B)$ is approachable with g^B, then 7.3 holds.

PROOF. This lemma follows at once from the definition of approachability; see Section 1 of [4]. Our games do not involve chance, so the elements of the game matrix should be considered as B-vectors rather than probability distributions on B-vectors. We have abbreviated "approachable in the game matrix" to "approachable"; this will be done throughout the sequel.

LEMMA 8.2. If there is an $\epsilon^B > 0$ such that, for each $c^{N-B} \epsilon C^{N-B}$, there is a $c^B \epsilon C^B$ for which

$$H^B(c^B, c^{N-B}) \geq H^B(f) + \epsilon^B \quad ,$$

then there is a g^B for which 7.3 holds.

PROOF. The condition given in the lemma is equivalent to the statement that for each $c^{N-B} \epsilon C^{N-B}$, $H^B(C^B, c^{N-B})$ intersects $\Lambda(f, \epsilon^B)$. In accordance with Theorem 3 of [4], this in turn is equivalent to the statement that there is a g^B with which $\Lambda(f, \epsilon^B)$ is approachable. Applying 8.1, we obtain the conclusion.

The information situation in [4] is slightly different from ours. In our application, the requirements of [4] may be stated as follows: Both B and N - B know the payoffs to B on each previous play; nothing else is known. In our situation, B knows its own previous payoffs, but N - B need not know what the payoffs to B were; on the other hand, both B and N - B may have information given by the information function, that they do not have in [4]. An analysis of the proof of Theorem 3 of [4] shows that these differences do not affect the truth of the theorem, so that we may still apply it here.

To see this intuitively, note that the only difference that could interfere with the approachability of $\Lambda(f, \epsilon^B)$ with g^B would be the additional information that N - B has in our case that it does not have in [4]. Examination of the proof of Theorem 3 of [4] indicates that g_k^B depends only on the previous payoffs, and can be defined for <u>any</u> set of previous payoffs without affecting the truth of the theorem; thus no amount of additional information can possibly help N - B in preventing B from approaching $\Lambda(f, \epsilon^B)$ with g^B.

LEMMA 8.3. If there is a $c^{N-B} \in C^{N-B}$ (called γ^{N-B}) such that for each $c^B \in C^B$, there is an $i \in B$ for which

$$H^i(c^B, \gamma^{N-B}) \leq H^i(f) \quad ,$$

then for every B-vector $\epsilon^B > 0$, and every supergame c-strategy vector β for which for all $(v_1, \ldots, v_k)$, we have

$$\beta_k^i(v_1, \ldots, v_k) = \gamma^{N-B}, \quad i \in N - B, \quad k \geq 0 \quad , \tag{8.4}$$

we have

$$\lim_{k \to \infty} \operatorname{Prob}_\beta (S_r^B(v) \geq H^B(f) + \epsilon^B \text{ for some } r \geq k) = 0$$

PROOF. The hypothesis of the lemma is equivalent to the assertion that for each $\epsilon^B > 0$, $\Lambda(f, \epsilon^B)$ fails to intersect $H(C^B, \gamma^{N-B})$. Applying Theorem 3 of [4], we conclude that for each $\epsilon^B > 0$, $\Lambda(f, \epsilon^B)$ is excludable with the supergame c-strategy (N-B)-vector given by (8.4). From the definition of excludability (Section 1 of [4]) it then follows that there is a number $\delta > 0$ such that

$$\lim_{k \to \infty} \text{Prob}_\beta \text{ (For all } r \geq k, \text{ there is an } i \in B \text{ such that}$$

$$S_r^i(v) < H^i(f) + \epsilon^i - \delta) = 1 \quad ,$$

whence

$$\lim_{k \to \infty} \text{Prob}_\beta \, (S_r^B(v) \geq H^B(f) + \epsilon^B \text{ for some } r \geq k) = 0 \quad .$$

§9. THE MAIN THEOREM: FIRST HALF

We will now show that our two approaches culminating in the definition of c-acceptability (4.1) and strong equilibrium (7.1, 7.2, and 7.3) actually yield results that are essentially the same. More precisely, for every c-acceptable strategy vector c, there is a strong equilibrium c-point (according to either 7.2 or 7.3) in supergame c-strategies which has the same payoff as c; and conversely. Thus c-acceptable points for a single play are seen to correspond exactly to strong equilibrium c-points in the supergame. The main theorem can be concisely stated by means of the equation

$$H(A_c) = H(S_c) \; (= H(\tilde{S}_c)) \quad .$$

This section is devoted to the proof of the relation

$$H(\tilde{S}_c) \subset H(A_c) \quad .$$

The proof rests heavily on Lemma 8.2. In addition to 8.2, the main fact needed is that if B cannot be prevented from obtaining more than h^B, then they cannot be prevented from obtaining a fixed amount more than h^B either. This is shown in Lemma 9.1.

LEMMA 9.1. Let h be a vector and B be a subset of N. If for each $c^{N-B} \in C^{N-B}$, there is a $c^B \in C^B$ for which

$$H^B(c^B, c^{N-B}) > h^B \quad ,$$

then there is a positive B-vector ϵ^B such that for each $c^{N-B} \in C^{N-B}$, there is a $c^B \in C^B$ for which

$$H^B(c^B, c^{N-B}) \geq h^B + \epsilon^B \quad .$$

PROOF. We first remark that if $G(x, y)$ is a continuous real-valued function of two variables x and y ranging over compact sets X and Y respectively, then the function defined by

$$F(y) = \max_{x \in X} G(x, y)$$

is continuous on Y.

Now define a function F on C^{N-B} by

$$F(c^{N-B}) = \max_{c^B \in C^B} \min_{i \in B} (H^i(c^B, c^{N-B}) - h^i) \quad .$$

By the above remark, F is continuous. Since C^{N-B} is compact, F takes on its minimum at a point c_0^{N-B} in C^{N-B}. By the hypothesis of the lemma,

$$F(c_0^{N-B}) > 0 \quad .$$

Setting

$$\epsilon^i = F(c_0^{N-B})$$

for all $i \in B$, we obtain the conclusion of the lemma.

THEOREM 2. $H(\tilde{S}_c) \subset H(A_c)$.

PROOF. Let f be a summable supergame c-strategy vector; set

$$h = H(f) \quad .$$

Suppose

$$h \notin H(A_c) \quad , \tag{9.2}$$

i.e., suppose there is a B for which the hypothesis of Lemma 9.1 holds. Then the conclusion of Lemma 9.1, which is the same as the hypothesis of Lemma 8.2, holds. Hence we deduce the conclusion of Lemma 8.2, i.e., there is a g^B for which 7.3 holds. Setting

$$g^{N-B} = f^{N-B} \quad ,$$

we obtain a B and a g obeying 7.1 and 7.3, whence

$$f \notin \tilde{S}_c \quad . \tag{9.3}$$

As we can go through the same argument for any f satisfying $h = H(f)$, we may deduce from (9.3) that

$$h \notin H(\tilde{S}_c) \quad . \tag{9.4}$$

We have shown that (9.2) $\Longrightarrow$ (9.4), whence we can deduce that

$$H(\tilde{S}_c) \subset H(A_c) \quad ,$$

which is what we set out to prove.

§10. THE MAIN THEOREM: SECOND HALF

This section is devoted to the proof of the relation

$$H(A_c) \subset H(S_c) \quad ;$$

together with Theorem 2, this yields the main theorem. The train of thought of the proof is somewhat as follows:

Suppose h to be a c-acceptable payoff. It is easy to set up a supergame c-strategy vector f whose payoff is h. But we must also incorporate in f a foolproof system for the punishment of any deviators; f must make sure that crime does not pay. The machinery for accomplishing this is at hand; by the definition of c-acceptability, for each B there is a c-strategy (N-B)-vector γ^{N-B} whose use prevents B from obtaining more than h^B. It remains only to <u>find</u> the culprits. To do this, we note that any players who at one time or another deviated from f, were at the time of their deviation certainly not included in the same coalition as the "orthodox" players. Since each player knows who was in his coalition on all previous plays, the deviators are thus easily spotted. The set B_k of players who deviated on some play up to the (k+1)st is monotone increasing with k. Every time it grows, i.e., every time another player joins the ranks of the deviators, the remaining players revise their strategy to punish at least one of the <u>new</u> set of deviators. Since N is finite, there must be a set B that includes all the deviators and that is actually attained by B_k after a finite number of plays. The use of the strategy γ^{N-B} will then make certain that crime does not pay for at least one of the deviators. The proof is complicated by the fact that B_k and B usually depend on which pure strategies were chosen from among those considered by the correlated strategy vectors used on previous plays.

The strong equilibrium point f just described is one of

"unrelenting ferocity"[6] against offenders. It exhibits a zeal for meting out justice that is entirely oblivious to the sometimes dire consequences to oneself or to the other faithful - i.e., those that have not deviated. There are other, more reasonable strong equilibrium points, which give deviating players a chance to return to the fold. These are important in connection with the general non-cooperative game; I will return to them in a subsequent paper on that topic. Here it is simpler and just as efficacious to use the strong equilibrium point of unrelenting ferocity.

THEOREM 3. $H(A_c) \subset H(S_c)$.

PROOF. Let $h \in H(A_c)$, and suppose $\gamma^N \in A_c$ is such that

$$H(\gamma^N) = h \quad . \tag{10.1}$$

Then by 4.3, for each $B \subset N$ there is a $c^{N-B} \in C^{N-B}$, which we will call γ^{N-B}, such that for each $c^B \in C^B$, there is an $i \in B$ for which

$$H^i(c^B, \gamma^{N-B}) \leq h^i \quad . \tag{10.2}$$

Now define a supergame c-strategy vector f as follows:

$$f_o^i = \gamma^N, \qquad i \in N$$

$$f_k^i(v_1^i, \ldots, v_k^i) = \gamma^{v_k^i|R^i}, \quad k > 0, \quad i \in N, \quad v_j^i \in J_j^i, \quad j \leq k. \tag{10.3}$$

We must prove that the payoff to f is h and that f is a strong equilibrium c-point. It is convenient to break the proof into a number of lemmas. However, these lemmas will depend on previous assumptions and formulae, and are valid only within the context of this proof. They will be numbered like ordinary formulae used within the proof.

LEMMA 10.4. For $k > 0$ and $i \in N$, we have

$$f_k^i(z_k) = \gamma^N \quad .$$

PROOF. The proof is by induction on k. For $k = 0$, (10.4) follows at once from (10.3). Assume (10.4) for a given k. Applying (2.6) and (2.7), we obtain

---

6 The term is due to Luce and Raiffa [6].

$$c(f_k(z_k)) = \gamma^N$$

and

$$d(f_k(z_k)) = d_N \quad .$$

Hence, by (6.2),

$$s(f_k(z_k)) = (u(\gamma^N),\ d_N) \quad . \tag{10.5}$$

Now by (6.4) and the linearity of f_k^i we have

$$f_{k+1}^i(z_{k+1}) = \sum_{y \in J_1 \times \ldots \times J_k}^{*} z_k(y) f_{k+1}^i(y,\ s(f_k(z_k)))$$

$$= \sum_{y \in J_1 \times \ldots \times J_k}^{*} z_k(y) f_{k+1}^i(y,\ (u(\gamma^N),\ d_N)) \qquad \text{(by (10.5))}$$

$$= \sum_{y \in J_1 \times \ldots \times J_k}^{*} z_k(y) f_{k+1}^i(y^i,\ (u^i(\gamma^N),\ N)) \qquad \text{(by (2.4) and (2.5))}$$

$$= \sum_{y \in J_1 \times \ldots \times J_k}^{*} z_k(y) \gamma^N \qquad \text{(by (10.3))}$$

$$= \gamma^N \sum_{y \in J_1 \times \ldots \times J_k} z_k(y)$$

$$= \gamma^N \quad . \qquad \text{(by (2.2))}$$

This completes the induction and the proof of the lemma.

LEMMA 10.6. For $k, j \geq 0$ and $k \neq j$, $c(f_k(z_k))$ is statistically independent of $c(f_j(z_j))$.

PROOF. By (10.3), the choice of strategies after each play depends only on the coalitions previously chosen. Thus the sequence of partitions of N into coalitions is strictly determined. For a given play, the choice of c-strategy B-vectors depends only on this sequence (which is, in fact, a sequence of constant partitions), and <u>not</u> on the strategies previously chosen. This completes the proof.

LEMMA 10.7. f is summable, and $H(f) = h$.

PROOF. Applying (2.7) to (10.4), we obtain

$$c(f_k(z_k)) = \gamma^N \quad ,$$

whence by (6.6)

$$E_k = E(c(f_k(z_k))) = E(\gamma^N) \quad . \tag{10.8}$$

By Lemma 10.6, the strong law of large numbers[7] applies to each component of $c(f_k(z_k))$, hence to $c(f_k(z_k))$ itself, and hence also to E_k. By (10.8) and (6.7), we have

$$\begin{aligned} H_k &= H(\gamma^N) \\ &= h \quad . && \text{(by (10.1))} \end{aligned}$$

Hence

$$\begin{aligned} H(f) &= \lim_{k \to \infty} \frac{1}{k} \sum_{r=1}^{k} H_r && \text{(by (6.8))} \\ &= \lim_{k \to \infty} \frac{1}{k} kh \\ &= \lim_{k \to \infty} h \\ &= h \quad . \end{aligned}$$

By Definition 6.9, the proof is complete.

LEMMA 10.9. If $v_k|R$ is a partition and $k > 0$, then

$$j \in e(f_k^i(v_1, \ldots, v_k)) \Longrightarrow f_k^j(v_1, \ldots, v_k) = f_k^i(v_1, \ldots, v_k) \quad .$$

PROOF. From (10.3) and the left side of the conclusion, we deduce that $j \in v_k^i|R^i$. Hence since $v_k|R$ is a partition, it follows that

$$v_k^i|R^i = v_k^j|R^j \quad ,$$

7 See [2], p. 207, <u>The Kolmogoroff Criterion</u>, or p. 208, the Theorem.

and then the right side of the conclusion follows at once from (10.3).

In order to show that f is a strong equilibrium c-point, we will first assume that it is not.

ASSUMPTION 10.10. There is a $B \subset N$ (which we will call B^*) for which there is a supergame c-strategy vector g for which

$$g^{N-B^*} = f^{N-B^*} \tag{10.11}$$

and there is a B^*-vector $\epsilon^{B^*} > 0$ for which

$$\lim_{k \to \infty} \text{Prob}_g(S_r^{B^*}(v) \geq H^{B^*}(f) + \epsilon^{B^*} \text{ for some } r \geq k) > 0. \tag{10.12}$$

It will be our task to show that (10.10) leads to an absurdity.

Let $v = (v_1, \ldots, v_k, \ldots)$ be an arbitrary sequence for which $v_k \in J_k$, $k \geq 1$. Let $B(v)$ denote the set of i for which there is a $k_i \geq 0$ such that

$$g^i_{k_i}(v_1, \ldots, v_{k_i}) \neq f^i_{k_i}(v_1, \ldots, v_{k_i}) \; . \tag{10.13}$$

It follows from (10.11) and (10.13) that

$$B(v) \subset B^* \; . \tag{10.14}$$

DEFINITION 10.18. For each v and for $k \geq 0$, let $B_k(v)$ denote the set of i for which there is a k_i such that

$$0 \leq k_i \leq k$$

and such that (10.13) holds. Define

$$B_{-1}(v) = \emptyset \; . \tag{10.19}$$

For each v, there is clearly a $k(v)$ such that

$$B_k(v) = B(v), \quad k \geq k(v) \; . \tag{10.20}$$

Furthermore, it is not difficult to see that (10.21): For each v, $B_k(v)$ is monotone increasing with k. We will in the sequel restrict use of the symbol $k(v)$ to the <u>first</u> $k(v)$ satisfying (10.20).

DEFINITION 10.22. We say that v occurs with <u>positive probability</u> if for each $k > 0$,

$$z_k(g)(v_1, \ldots, v_k) > 0 \quad ;$$

that is, if for each $k > 0$, and under the assumption that the supergame c-strategy vector g is played, $(v_1, \ldots, v_k)$ occurs with positive probability.

LEMMA 10.23. If v occurs with positive probability, then for each $k > 0$, $v_k|R$ is a partition.

PROOF. By (6.4) and (10.22), $(v_1, \ldots, v_k)$ occurs with positive coefficient in

$$(v_1, \ldots, v_{k-1}, s(g_{k-1}(v_1, \ldots, v_{k-1}))).$$

Hence v_k occurs with positive coefficient in $s(g_{k-1}(v_1, \ldots, v_{k-1}))$. Applying (6.2), we conclude that $v_k|R$ occurs with positive coefficient in $d(g_{k-1}(v_1, \ldots, v_{k-1}))$. Since by definition, $d(g_{k-1}(v_1, \ldots, v_{k-1}))$ is a probability combination of a number of partitions of N, it follows that $v_k|R$ must be a partition, which completes the proof.

LEMMA 10.24. Let v occur with positive probability, and assume that (10.10) holds. For $k \geq 0$ and $i \in N\text{-}B_k(v)$, we have

$$g_k^i(v_1, \ldots, v_k) = \gamma^{N-B_{k-1}(v)} \quad .$$

PROOF. We will write B instead of $B(v)$, and for each k, B_k instead of $B_k(v)$. The proof is by induction on k. First let $k = 0$. If $i \in N - B_o$, then it follows from Definition 10.18 that there is no k_i satisfying $k_i = 0$ and (10.13). In other words, for $i \in N - B_o$, we have

$$g_o^i = f_o^i \quad .$$

Hence by (10.3), we have for $i \in N - B_o$,

$$\begin{aligned} g_o^i &= \gamma^N \\ &= \gamma^{N-B_{-1}} \qquad \text{(by (10.19))}, \end{aligned}$$

which establishes (10.24) for $k = 0$.

Next suppose we have established (10.24) for some $k \geq 0$ and all preceding k. For $i \in B_k - B_{k-1} = (N - B_{k-1}) \cap B_k$, we must have, by Definition 10.18, that (10.13) holds for $k_i = k$. Hence

$$g_k^i(v_1, \ldots, v_k) \neq f_k^i(v_1, \ldots, v_k), \quad i \in B_k - B_{k-1} \quad . \tag{10.25}$$

Now if $k = 0$, it follows from (10.19), (10.25), and (10.3) that

$$g_0^i \neq \gamma^{N-B_{-1}}, \quad i \in B_0 - B_{-1} \quad . \tag{10.26}$$

If $k > 0$, we have for $j \in B_k - B_{k-1}$ that

$$j \in N - B_{k-1} = e(g_k^i(v_1, \ldots, v_k)), \quad i \in N - B_k \quad \text{(by induction hypothesis)}$$

$$= e(f_k^i(v_1, \ldots, v_k)), \quad i \in N - B_k \quad \text{(by (10.18))}.$$

Applying Lemma 10.9 and Lemma 10.23, we obtain

$$f_k^i(v_1, \ldots, v_k) = f_k^j(v_1, \ldots, v_k),\ k > 0,\ i \in N - B_k,\ j \in B_k - B_{k-1}. \tag{10.27}$$

From (10.18) and the induction hypothesis, we obtain

$$f_k^i(v_1, \ldots, v_k) = g_k^i(v_1, \ldots, v_k) = \gamma^{N-B_{k-1}}, \quad i \in N - B_k, \quad k > 0 \quad .$$

Combining this with (10.27), we obtain

$$f_k^j(v_1, \ldots, v_k) = \gamma^{N-B_{k-1}}, \quad j \in B_k - B_{k-1}, \quad k > 0 \quad ,$$

i.e.,

$$f_k^i(v_1, \ldots, v_k) = \gamma^{N-B_{k-1}}, \quad i \in B_k - B_{k-1}, \quad k > 0 \quad .$$

From this and (10.25) we deduce, for $k > 0$, that

$$g_k^i(v_1, \ldots, v_k) \neq \gamma^{N-B_{k-1}}, \quad i \in B_k - B_{k-1} \quad ; \tag{10.28}$$

the same result for $k = 0$ is given by (10.26). Hence (10.28) holds for $k \geq 0$.

We certainly have

$$g_k^i(v_1, \ldots, v_k) \neq \gamma^{N-B_{k-1}}, \quad i \in B_{k-1} \quad ,$$

for otherwise $i \notin e(g_k^i(v_1, \ldots, v_k))$, contradicting the membership of g_k^i in T^i. Combining this with the induction hypothesis and (10.28), we obtain

$$g_k^i(v_1, \ldots, v_k) \begin{cases} = \gamma^{N-B_{k-1}}, & i \in N - B_k \\ \neq \gamma^{N-B_{k-1}}, & i \in B_k . \end{cases}$$

It follows that

$$d^i(g_k(v_1, \ldots, v_k)) = N - B_k, \quad i \in N - B_k \quad ,$$

and in particular, since $N - B_{k+1} \subset N - B_k$, we have

$$d^i(g_k(v_1, \ldots, v_k)) = N - B_k, \quad i \in N - B_{k+1} . \tag{10.29}$$

Now for $i \in N - B_{k+1}$, we have by Definition 10.18 that

$$g_{k+1}^i(v_1, \ldots, v_{k+1}) = f_{k+1}^i(v_1, \ldots, v_{k+1}) = \gamma^{v_{k+1}^i|R^i} . \tag{10.30}$$

By reasoning similar to that used in the proof of (10.23), we obtain that $v_{k+1}^i|R^i$ occurs with positive coefficient in $d^i(g_k(v_1, \ldots, v_k))$. Combining this with (10.29) and (10.30), we obtain that for $i \in N - B_{k+1}$

$$g_{k+1}^i(v_1, \ldots, v_{k+1}) = \gamma^{N-B_k} .$$

This completes the induction and establishes (10.24).

From (10.24) and (10.20) we deduce that if v occurs with positive probability, then

$$g_k^i(v_1, \ldots, v_k) = \gamma^{N-B(v)}, \quad i \in N - B(v), \quad k > k(v) \quad ,$$

whence

$$c(g_k(v_1, \ldots, v_k)) = (\gamma^{N-B(v)}, c^{B(v)}(g_k(v_1, \ldots, v_k))), \quad k > k(v) \quad .$$

Setting

$$c^B(g_k(v_1, \ldots, v_k)) = \gamma_k^B , \; B \subset N, \quad k > k(v) \quad ,$$

we obtain

$$c(g_k(v_1, \ldots, v_k)) = (\gamma_k^{B(v)}, \gamma^{N-B(v)}), \quad k > k(v) \quad . \tag{10.31}$$

LEMMA 10.32. For each $m \geq 0$, there is a positive integer k, such that for all v, r, and $h_1, \ldots, h_m$ for which

$$r \geq k \tag{10.33}$$

and

$$h_j \in H(P), \quad j = 1, \ldots, m \tag{10.34}$$

and

$$S_r^{B*}(v) \geq H^{B*}(f) + \epsilon^{B*} \tag{10.35}$$

we have

$$r > m \tag{10.36}$$

and

$$\frac{1}{r}\left(\sum_{j=1}^{m} h_j^{B*} + \sum_{j=m+1}^{r} H_j^{B*}(v)\right) \geq H^{B*}(f) + \frac{1}{3}\epsilon^{B*} \tag{10.37}$$

PROOF. First of all, we may choose

$$k > m \quad ,$$

so that (10.36) is satisfied. Next, choose

$$k \geq \max_{i \in B*} \frac{3m}{\epsilon^i} \max_{p \in P} H^i(p) \quad . \tag{10.38}$$

We then have

$$\left| S_r^{B*}(v) - \frac{1}{r}\left(\sum_{j=1}^{m} h_j^{B*} + \sum_{j=m+1}^{r} H_j^{B*}(v)\right) \right|$$

$$= \left| \frac{1}{r}\left(\sum_{j=1}^{r} H_j^{B*}(v)\right) - \frac{1}{r}\left(\sum_{j=1}^{m} h_j^{B*} + \sum_{j=m+1}^{r} H_j^{B*}(v)\right) \right| \qquad \text{(by (6.11))}$$

$$= \frac{1}{r}\left| \sum_{j=1}^{m} (H_j^{B*}(v) - h_j^{B*}) \right|$$

$$\leq \frac{1}{k} \sum_{j=1}^{m} \left| H_j^{B*}(v) - h_j^{B*} \right| \qquad \text{(by (10.33))}$$

$$\leq \frac{1}{k} \sum_{j=1}^{m} 2 \max_{p \in P} H^{B*}(p) \qquad \text{(by (10.34))}$$

$$= \frac{2m}{k} \max_{p \in P} H^{B*}(p)$$

$$\leq \frac{2}{3} \epsilon^{B*} \qquad \text{(by (10.38))}$$

Combining this with (10.35), we obtain (10.37). This completes the proof.

COROLLARY 10.39. For each $m \geq 0$, the following holds: For all sufficiently large k, we have that for all v, r, and $h_1, \ldots, h_m$ for which (10.33), (10.34), and (10.35) hold, (10.36) and (10.37) follow.

PROOF. Because if (10.32) is true for a given k, it is certainly true for all larger k.

LEMMA 10.40. For each $m \geq 0$ and $B \subset N$, including the null set, we have

$$\lim_{k \to \infty} \text{Prob}_g((S_r^{B*}(v) \geq H^{B*}(f) + \epsilon^{B*}$$

$$\text{for some } r \geq k) \text{ and } k(v) = m \text{ and } B(v) = B) = 0 \quad .$$

PROOF. First let $B = \emptyset$. Then for the v we are considering (inside Prob_g), we have

$$B(v) = \emptyset \quad . \tag{10.41}$$

Since $B_k(v)$ is monotone increasing with k and always contained in $B(v)$, it follows from (10.41) that for all k,

$$B_k(v) = \emptyset \quad .$$

Applying (10.24), we obtain for $k \geq 0$ and $i \in N$,

$$g_k^i(v_1, \ldots, v_k) = \gamma^N \quad . \tag{10.42}$$

Let θ be a supergame c-strategy vector given for all v by

$$\theta_k^i(v_1, \ldots, v_k) = \gamma^N, \quad k \geq 0, \; i \in N \quad . \qquad (10.43)$$

From (10.42) and (10.43), we obtain

$$\lim_{k \to \infty} \text{Prob}_g((S_r^{B*}(v) \geq H^{B*}(f) + \epsilon^{B*} \text{ for some } r \geq k) \text{ and } k(v) = m \text{ and } B(v) = B) \; .$$

$$= \lim_{k \to \infty} \text{Prob}_\theta((S_r^{B*}(v) \geq H^{B*}(f) + \epsilon^{B*} \text{ for some } r \geq k) \text{ and } k(v) = m \text{ and } B(v) = B)$$

$$\leq \lim_{k \to \infty} \text{Prob}_\theta(S_r^{B*}(v) \geq H^{B*}(f) + \epsilon^{B*} \text{ for some } r \geq k)$$

$$\leq \lim_{k \to \infty} \text{Prob}_\theta(S_r^i(v) \geq H^i(f) + \epsilon^i \text{ for some } r \geq k)$$

$$= \lim_{k \to \infty} (1 - \text{Prob}_\theta(S_r^i(v) < H^i(f) + \epsilon^i \text{ for all } r \geq k))$$

$$= 1 - \lim_{k \to \infty} \text{Prob}_\theta(S_r^i(v) < H^i(f) + \epsilon^i \text{ for all } r \geq k)$$

$$\leq 1 - \lim_{k \to \infty} \text{Prob}_\theta(|S_r^i(v) - H^i(f)| < \epsilon^i \text{ for all } r \geq k) , \qquad (10.44)$$

where i is an arbitrary member of $B*$.

Now when v is distributed according to θ, we deduce from (10.43) and (6.10) that the $H_j^i(v)$ are independent random variables whose mean is $H^i(\gamma^N)$. Applying (6.11), (10.1), (10.7), and the strong law of large numbers, we obtain that the second term of the right side of (10.44) is 1, whence we deduce that the left side vanishes. This completes the proof in the case $B = \emptyset$.

Next, suppose $B \neq \emptyset$. From (10.14) we obtain that $B(v) \subset B*$. Thus if $B \not\subset B*$, then $\text{Prob}_g(B(v) = B) = 0$, and our lemma is already proved. Thus we may assume without loss of generality that

$$B \subset B* \; . \qquad (10.45)$$

For $j = 1, \ldots, m$, suppose the random variable w_j^B to be distributed according to an arbitrary but fixed member of $C^B(J_j^B)$, which we will call γ_j^B, while w_j^{N-B} is distributed according to γ^{N-B}. Set

$$h_j = H(w_j), \qquad j = 1, \ldots, m \quad . \tag{10.46}$$

Applying Corollary 10.39, we obtain

$$\begin{aligned}&\lim_{k\to\infty} \mathrm{Prob}_g((S_r^{B*}(v) \geq H^{B*}(f) + \epsilon^{B*} \quad \text{for some } r \geq k) \\ &\qquad\qquad \text{and } k(v) = m \\ &\qquad\qquad \text{and } B(v) = B) \\ &\leq \lim_{k\to\infty} \mathrm{Prob}\left(\left(\frac{1}{r}\left(\sum_{j=1}^{m} h_j^{B*} + \sum_{j=m+1}^{r} H_j^{B*}(v)\right) \geq H^{B*}(f) + \frac{1}{3}\epsilon^{B*} \right.\right. \\ &\qquad\qquad \left.\text{for some } r \geq k\right) \\ &\qquad\qquad \text{and } k(v) = m \\ &\qquad\qquad \left.\text{and } B(v) = B\right) ,\end{aligned} \tag{10.47}$$

where $v_{m+1}, \ldots, v_r$ are distributed according to $x_{m+1}(g), \ldots, x_r(g)$, while h_j is given by (10.46). If for $j = k+1, \ldots, r$, we will let w_j be a random variable whose distribution is $(\gamma_j^B, \gamma^{N-B})$, where the w_j^B may be dependent on w_ℓ with $\ell < j$ but the w_j^{N-B} must be independent of all w_ℓ with $\ell \neq j$, then we may conclude from (10.31) and (6.10) that the right side of (10.47) is

$$\begin{aligned}&\leq \lim_{k\to\infty} \mathrm{Prob}\left(\frac{1}{r}\sum_{j=1}^{r} H^{B*}(w_j) \geq H^{B*}(f) + \frac{1}{3}\epsilon^{B*}\right. \\ &\qquad\qquad \left.\text{for some } r \geq k\right) \\ &\leq \lim_{k\to\infty} \mathrm{Prob}\left(\frac{1}{r}\sum_{j=1}^{r} H^{B}(w_j) \geq H^{B}(f) + \frac{1}{3}\epsilon^{B}\right. \\ &\qquad\qquad \left.\text{for some } r \geq k\right)\end{aligned} \tag{10.48}$$

by (10.45). Now there is certainly a supergame c-strategy vector β for which

$$c(\beta_{j-1}(z_{j-1}(\beta))) = (\gamma_j^B, \gamma^{N-B}) \quad , \quad j \geq 1 \quad ,$$

whence the right side of (10.48)

$$= \lim_{k\to\infty} \mathrm{Prob}_\beta(S_r^B(v) \geq H^B(f) + \frac{1}{3}\epsilon^B \quad \text{for some } r \geq k) \quad . \tag{10.49}$$

Applying (10.2), (10.7) and Lemma 8.3, we obtain that the right side of (10.49) vanishes. Combining this with (10.48) and (10.47), we obtain the lemma.

LEMMA 10.50.

$$\lim_{k \to \infty} \text{Prob}_g(S_r^{B^*}(v) \geq H^{B^*}(f) + \epsilon^{B^*} \quad \text{for some} \quad r \geq k) = 0 \quad .$$

PROOF. Letting $\cup$ stand for mutually exclusive disjunction, we have

$$1 = \text{Prob}_g\left(\bigcup_{m=o}^{\infty} k(v) = m \right)$$

$$= \sum_{m=o}^{\infty} \text{Prob}_g(k(v) = m)$$

$$= \sum_{m=o}^{\infty} \text{Prob}_g(k(v) = m \quad \text{and} \quad \bigcup_{B \subset N} (B(v) = B))$$

$$= \sum_{m=o}^{\infty} \sum_{B \subset N} \text{Prob}_g(k(v) = m \quad \text{and} \quad B(v) = B) \quad . \tag{10.51}$$

Let $Q(v, k)$ stand for the expression

$$S_r^{B^*}(v) \geq H^{B^*}(f) + \epsilon^{B^*} \quad \text{for some} \quad r \geq k \quad .$$

Then

$$\lim_{k \to \infty} \text{Prob}_g(Q(v, k))$$

$$= \lim_{k \to \infty} \text{Prob}_g(Q(v, k) \quad \text{and} \quad \bigcup_{m=o}^{\infty} (k(v) = m) \quad \text{and} \quad \bigcup_{B \subset N} (B(v) = B))$$

$$= \lim_{k \to \infty} \sum_{m=o}^{\infty} \sum_{B \subset N} \text{Prob}_g(Q(v, k) \quad \text{and} \quad k(v) = m \quad \text{and} \quad B(v) = B)$$

$$= \sum_{m=o}^{\infty} \sum_{B \subset N} \lim_{k \to \infty} \text{Prob}_g(Q(v, k) \quad \text{and} \quad k(v) = m \quad \text{and} \quad B(v) = B)$$

(since the series is dominated by the series on the right side of (10.51))

$$= \sum_{m=0}^{\infty} \sum_{B \subset N} 0 \qquad \text{(by Lemma 10.40)}$$

$$= 0 \quad .$$

This completes the proof of Lemma 10.50.

Lemma 10.50 contradicts (10.12) and thus establishes the truth of Theorem 3.

COROLLARY 4. $H(A_c) = H(S_c) = H(\tilde{S}_c)$.

PROOF. Combine (7.4), Theorem 2 and Theorem 3.

§11. EXISTENCE OF ACCEPTABLE POINTS

All two-person games, zero-sum or not, have acceptable points. The proof, which is not difficult, will be given in a subsequent paper of this series devoted exclusively to the application of this theory to two-person games. Unfortunately, though, when we go beyond two-person games, we find games in which there are no acceptable points. It is instructive to examine an example of such a game.

G is a three-person, zero-sum game. The three persons form a community and at regular intervals hold elections for mayor and vice-mayor. The mayor draws a salary of 2; the vice-mayor draws a salary of 1; the remaining player pays the salaries.

Let h be an arbitrary payoff vector. Since G is zero-sum, not all the players can have a positive payoff. Without loss of generality, let

$$h^1 \leq 0 \quad . \tag{11.1}$$

Now the minimum payoff to 1, even in pure strategies, is -3; hence since G is zero-sum, it follows that

$$h^2 + h^3 \leq 3 \quad . \tag{11.2}$$

From (11.2) it follows that at least one of 2 and 3 gets at most $1\frac{1}{2}$; so we may write without loss of generality

$$h^2 \leq 1\frac{1}{2} \quad . \tag{11.3}$$

Now by agreeing to vote for 2 for mayor and for 1 for vice-mayor, 1 and 2 can both increase their payoffs; and there is nothing that 3 can do to prevent this.

Actually, the game G just described does not deserve an acceptable point. In a long sequence of plays of G, a steady state will never be reached; no sooner do we start settling down to one, than two of the players will see that it is relatively disadvantageous for them, and will certainly not be willing to agree to it for the remainder of the superplay. The game is inherently unstable. The same argument holds for all games not possessing acceptable points. Thus in seeking a steady state for the supergame, it would seem that we may restrict ourselves to games possessing acceptable points. These are called stable games.

§12. THE EXPECTED PAYOFF

In order to see why the notion defined by (6.8) is essentially misleading, let us consider the three-person game G of perfect information which is given in extensive form in Figure 1.

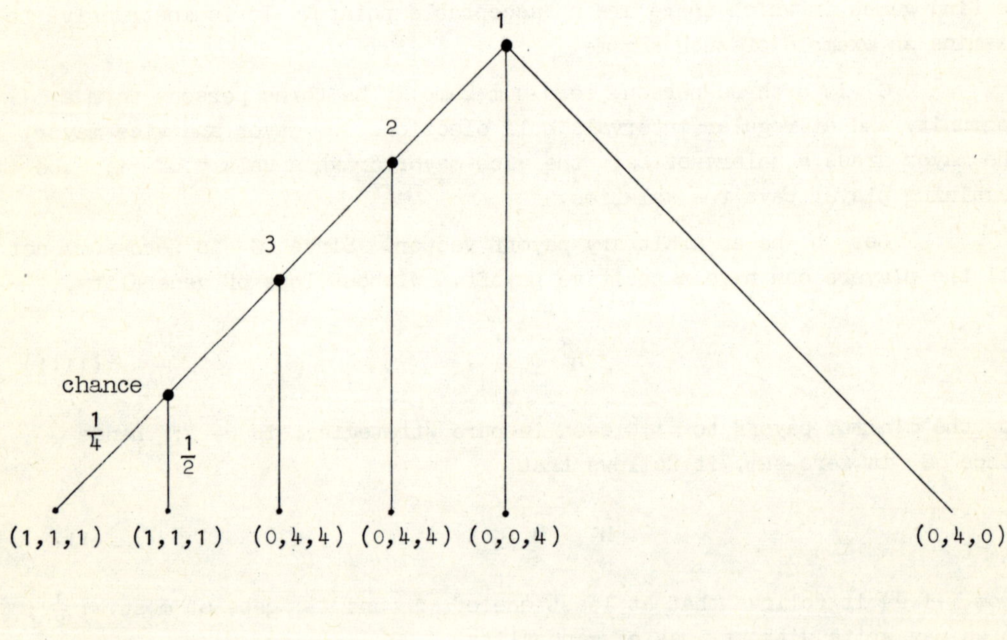

FIGURE 1. The Payoff Vector at each Terminal gives the Payoff to Players 1, 2, and 3 in that Order

The payoff vector at each terminal gives the payoff to players 1, 2, and 3 in that order.

Player 1 has three strategies, ℓ^1, c^1, and r^1 (left, center, and right). The other players have two strategies each; they will be denoted by ℓ^j and r^j, where $j = 0, 2,$ and 3. Chance[8] is denoted by 0.

The point (ℓ^1, ℓ^2, ℓ^3), yielding a payoff of (1, 1, 1), is acceptable. No player or group of players can obtain a higher payoff. However, if 1 plays the supergame c-strategy given by (10.3), there is a supergame c-strategy (2,3)-vector that will yield an expected payoff of (0, 2, 2) rather than (1, 1, 1). This supergame c-strategy (2,3)-vector, which we will call the "plan", may be described as follows: On the first play, both players make the "orthodox" choice, i.e., they concur with player 1 in choosing (ℓ^1, ℓ^2, ℓ^3). On all subsequent plays, we distinguish two cases:

1) If chance played ℓ^0 on the first play, then player 2 makes the "orthodox" choice, i.e., that given by (10.3), while player 3 plays r^3.

2) If chance played r^0 on the first play, then player 3 makes the "orthodox" choice, while player 2 plays r^2.

According to (10.3), player 1 will respond to this treachery with a wounded roar and proceed to wreak dire vengeance. In case 1 he will finish off player 3 by playing r^1; in case 2 he will play c^1, thus avenging himself on the treacherous player 2. In each case, he is convinced that the orthodox player has been loyal all along, and therefore will not feel too unhappy about his getting 4 while player 1 himself gets nothing. The expected payoff is (0, 2, 2), an improvement for both 2 and 3. For 2 and 3, this sounds like a good set-up, at least on paper. Who wouldn't want an expected payoff of 2 instead of 1? Let's see how it works out in practice.

Imagine that chance has played ℓ^0 on the first play. For the second play, player 1, all unsuspecting, suggests (ℓ^1, ℓ^2, ℓ^3), and player 2 concurs, according to plan. Player 3 is now faced with the choice of going along with the plan and playing r^3, thus bringing down

---

8 It may be objected that we have excluded chance from consideration in this paper. Actually, chance is used in this example only to provide, as conveniently and cheaply as possible, a random device on which players 2 and 3 will be able to peg their choices. We could eliminate chance at the cost of complicating the example considerably. In order to exhibit the principle involved as clearly as possible, we have retained chance; but it is important to remember that it does not form an essential part of the example.

upon himself the eternal wrath of player 1, or of concurring in the choice of (ℓ^1, ℓ^2, ℓ^3), thus possibly making player 2 angry at him, but still assuring himself of a steady income of 1. It won't take him long to make his choice. He has nothing to gain and everything to lose by going along with the plan; he will get nothing but 0 for the rest of eternity for his pains, while player 2 enjoys a steady payoff of 4. Player 3 will not now be consoled by the fact that if chance had played the other way, he would have been getting 4 while player 2 would have been getting 0; in no sense does this constitute an incentive for him to follow the plan as things stand now. The point is that in a single play of a game, it may be worthwhile for each of the parties to plan to co-operate, even if it hurts, because in future plays, he may be the beneficiary. But in a superplay, there are no future superplays; this is it; by going along with the plan, player 3 will be ruining himself not just for one play, but for always. It just isn't worthwhile, and no player would do it. Similar remarks apply to player 2. The plan does yield a higher expected payoff, but will never be carried out.

It is possible to define strong equilibrium c-points using an appropriate formula involving expected payoffs in place of 7.2. If this is done, the first half of the main theorem, the analogue of Theorem 2, remains true. However, the second half, the analogue of Theorem 3, fails. The game G that we have been discussing provides the counterexample. If we start out with $h = (1, 1, 1)$, then not only does the f given by (10.3) fail, as we have shown, but it can be shown that there is then no strong equilibrium c-point (in the sense of expected payoffs) whose payoff is h.

§13. ENFORCEABILITY OF AGREEMENTS

There are two kinds of agreements involved in a cooperative supergame. One is the formal agreement that is contracted prior to a given play by the members of a coalition B, who agree to play a certain correlated strategy B-vector on that play. This agreement expires immediately after the play in question is completed; its validity does not extend to any subsequent plays. The other is the informal "gentlemen's agreement" in which the players undertake informally to make certain choices or join certain coalitions on each of a number of plays, or even for the entire superplay; they do not form a part of the formal mathematical structure introduced in Section 6. These are the agreements that are involved in the supergame strategy vectors that we have used in our proofs. When in heuristic discussions above we referred to "orthodoxy" or "deviationism", it was the second kind of agreement to which we were referring. To distinguish between the two kinds of agreements, we will call the first formal, the second informal.

As mentioned in Section 3, together with the formal agreement there comes a formal enforcement mechanism. Put completely precisely, if for the $k^{\underline{th}}$ play the players choose the vector $t \in T$, then each $i \in N$ must become a member of $d^i(t)$, and the coalition $d^i(t)$ must then play $(t^i)^{d^i(t)}$. If the reader wishes, he may consider that once the choice of t has been made, the remainder of the play is no longer in the hands of the individual players, but in the hands of an umpire. Such an extreme interpretation is unnecessary for practical purposes; we refer to it only to make clear the intuitive meaning of formal enforcement.

On the other hand, the informal agreement can be enforced in no such formal way. Violation of such an agreement can be "prevented" only by the threat of implied threat of retaliation on subsequent plays. The exact nature of this "pseudo-enforceability" is of course the crux of what we have been investigating in the foregoing sections.

In the light of these remarks, it is an interesting fact that if in the supergame we abandon formal enforceability for the formal agreements, the main theorem remains true. The possibility of retaliation is sufficient to enforce even the formal agreements, without any necessity for formal enforcement apparatus. We have not proved this statement in the foregoing; the structure introduced in Section 6 was constructed with formal enforceability of formal agreements in mind. A structure permitting the violation of formal agreements would be somewhat more complicated.[9] However, upon a little reflection, the reader will be able to convince himself of the truth of the statement, at least intuitively.

Of course, when considering just a single play rather than the supergame, as in the definition of acceptability, then there are only formal agreements and these are formally enforceable.

---

[9] Roughly speaking, each play would be described by a choice of a $t \in T$ as in Section 6, followed by a chance choice of a pure strategy vector in accordance with $c(t)$, followed by some move permitting the players to renege on their commitments if they wish.

BIBLIOGRAPHY

[1] NASH, J. F., "Non-cooperative games," Annals of Mathematics 54 (1951), pp. 286-295.

[2] FELLER, W., An Introduction to Probability Theory and its Applications, John Wiley and Sons, 1950.

[3] KUHN, H. W., "Extensive games and the problems of information," Annals of Mathematics Study No. 28 (Princeton, 1953), pp. 193-216.

[4] BLACKWELL, D., "An analog of the minimax theorem for vector payoffs," Pacific J. Math. 6 (1956), pp. 1-8.

[5] THOMPSON, F. B., "Equivalence of games in extensive form," The RAND Corporation, Research Memorandum RM-759, 1952.

[6] LUCE, R. D., and RAIFFA, H., Games and Decisions, John Wiley and Sons, 1957.

[7] von NEUMANN, J., and MORGENSTERN, O., Theory of Games and Economic Behavior, Princeton, 1944, 3rd ed., 1953.

[8] RAIFFA, H., "Arbitration schemes for generalized two-person games," Annals of Mathematics Study No. 28 (Princeton, 1953), pp. 361-387.

The Hebrew University Robert J. Aumann

A BARGAINING MODEL FOR THE COOPERATIVE n-PERSON GAME*

by

John C. Harsanyi

§1. INTRODUCTION

One of the important achievements of the von Neumann-Morgenstern theory of games[1] has been the development, for the finite two-person constant-sum game, of a determinate solution,[2] which is based on a few very natural postulates of rational behaviour, and which defines a unique value (payoff) for the game to each player. But the theory has not offered a determinate solution in this sense for what are called essential games, i.e., all two-person non-constant-sum games and n-person games with the exception of some degenerate cases. This fact seriously limits the use of the theory both for analytical (explanatory and predictive) and for policy-making purposes. If we want to set up an analytical model for a certain social situation, we need a theory that can make determinate predictions (at least in a probabilistic sense) on people's behaviour in this situation[3]

* The author is greatly indebted to Professor Kenneth J. Arrow, of Stanford University, for constant help, criticism and encouragement. He also wishes to thank Dr. E. S. Barnes of Sydney University, Professor S. Karlin of Stanford University, Professor J. Marschak of Cowles Foundation at Yale University, Professor T. C. Schelling of Yale University, and the Editorial Referee of this paper, for helpful comments. After earlier preliminary work this research project has been completed during the author's Rockefeller Fellowship at Stanford University.

1 J. von Neumann and O. Morgenstern, The Theory of Games and Economic Behavior, 3rd edition, Princeton, 1953.

2 We use the term 'determinate solution' to distinguish solutions which consist in a unique imputation (unique payoff vector), from the von Neumann-Morgenstern concept of a 'solution', which is defined in general by a set of alternative imputations.

3 The only type of indeterminacy which seems to be admissible in models based on rationality postulates is leaving open the choice between 'equivalent' strategies, which yield the same payoffs - though it may be often preferable to avoid even this type of indeterminacy by introducing additional postulates.

- preferably predictions that can be confirmed or refuted by observation and experiment. In the same way, if we want to advise one of the participants on what strategy to adopt, we again need a theory that can make determinate recommendations on what is the best strategy, or the best probability mixture of strategies, in this situation.

A theory of n-person game that yields a determinate solution can make important contributions to the analysis of a number of economic, political and other social phenomena, such as oligopoly with more than two oligopolists; many-cornered bargaining situations (e.g., bargaining on the labour market among several employer and employee organizations and possibly even government agencies); the distribution of income and power in a given social organization or in a whole society; the distribution of territory and spheres of influence among several countries in international politics; and in general the balance of power among several individuals or social groups.

§2. COMPARISON WITH OTHER THEORIES

The need for defining determinate solutions (values) also for essential games has been first pointed out by Dr. Nash.[4] He also actually developed a bargaining model yielding a determinate solution for the cooperative two-person non-zero-sum game[5] both for the case without, and the case with, a choice among alternative threats by the players. His theory is, moreover, independent of the restrictive assumption of transferable utility,[6] needed in the von Neumann-Morgenstern theory.

In an earlier paper[7] I have tried to show that Nash's solution is mathematically equivalent (in the case of unique threats) to Professor Zeuthen's theory of bargaining,[8] going back to the pre-game theory period.

---

[4] J. Nash, "The bargaining problem," Econometrica 18 (1950), pp. 155-162, esp. p. 157.

[5] Op. cit. and also his "Two-person cooperative games," Econometrica 21 (1953), pp. 128-140.

[6] Transferable utility means that side payments between players are allowed and have no effect on the sum of all players' utilities, the payee's utility gain being always exactly equal to the payer's utility loss. This requires that money transfers should not be subject to costs or subsidies, and that the marginal utility of money be constant to all players over the relevant range.

[7] "Approaches to the bargaining problem before and after the Theory of Games: a critical discussion of Zeuthen's, Hicks', and Nash's theories," Econometrica 24 (1956), pp. 144-157.

[8] F. Zeuthen, Problems of Monopoly and Economic Warfare, London, 1930, Chap. IV.

I have argued that Zeuthen's model for the actual bargaining process gives important additional support to Nash's solution. The bargaining model of this paper is essentially a generalization of Nash's theory of two-person bargaining, for the n-person case.

A determinate solution for the n-person game, restricted to the case of transferable utility, has already been suggested by Dr. Shapley.[9] However, his solution is meant to define the value to each player of the prospect of having to play a given game, while our solution is meant to define the actual payoffs that the players of the game, if they act in accordance with certain rationality postulates, tend to agree upon in bargaining.

Moreover, as Shapley's theory depends on the use of the von Neumann-Morgenstern characteristic function of the game, it involves the rather unsatisfactory assumption that in the event of a conflict between two player coalitions each side would use a maximin strategy against the other side, i.e., would act on the expectation that the other side would always try to inflict the highest possible damage on them - irrespective of the costs of this policy to themselves. This interpretation of rational behaviour in conflict situations is quite natural in the case of constant-sum games, where one side's loss is necessarily the other side's gain. But in the case of non-constant-sum games it is more natural to assume that in a conflict each side would avoid, and would also expect the other side to avoid, strategies disproportionately expensive in relation to the damage caused to the other side.

Furthermore, Shapley uses a bargaining model (though this model does not play an essential role in his theory) which is based on the assumption that any player joining a coalition is able to obtain the whole increment in the value (joint profit) of this coalition that results from his joining, whereas most of us would rather predict some sort of profit-sharing agreement in this situation between the new member of the coalition and the old ones.

Our own model is based, with respect to behaviour in conflict and bargaining situations, on assumptions derived from Nash's theory, which avoid the objections outlined above. From Nash's theory it also inherits the independence of any restrictive assumption as to transferable utility or even as to the possibility of side payments. If side payments are allowed they simply become moves incorporated in the strategies available to the players. But, notwithstanding the mentioned differences in our basic assumptions, our solution turns out to be mathematically very closely

---

9 L. S. Shapley, "A value for n-person games," Annals of Mathematics Study No. 28 (Princeton, 1953), pp. 307-317.

related to Shapley's, so that it may be regarded not only as a generalization of the Nash solution but also as a generalization of a - somewhat modified - Shapley solution.

§3. THE n-PERSON BARGAINING GAME WITH GIVEN DISAGREEMENT PAYOFFS

We shall first consider the special case of a cooperative game where the n players by joint action can achieve any payoff vector $u = (u_1, \ldots, u_n)$ within a given compact and convex prospect space P while if even one player fails to cooperate the players obtain only a given disagreement-payoff vector $d = (d_1, \ldots, d_n) \in P$.

To this simple case we can extend Nash's theory as follows. A given payoff vector $\bar{u}$ will represent the equilibrium outcome of multilateral bargaining among the n players only if no pair of players i and j has the incentive to redistribute their payoffs between them so long as the payoffs of all other players $k \neq i, j$ remain fixed. But according to Nash's theory of bilateral bargaining this will be the case only if the payoffs $u_i = \bar{u}_i$ and $u_j = \bar{u}_j$ of the two players maximize the product[10]

$$\pi_{ij} = (u_i - d_i) \cdot (u_j - d_j) \tag{1}$$

subject to the conditions:

$$\begin{aligned} & u \in P \\ & u_i \geqq d_i \\ & u_j \geqq d_j \\ & u_k = \bar{u}_k = \text{constant for all } k \neq i, j \\ & d_i, d_j = \text{constant} \quad . \end{aligned}$$

In view of the convexity of the prospect space P, the sufficient and necessary condition for this is[11]

---

[10] To denote the components of a vector we shall add subscripts to the symbol denoting the vector itself.

[11] To simplify our argument, we shall assume that all relevant partial derivatives exist at the point $\bar{u}$. No generality is lost by this assumption because our results can be trivially extended to games where these derivatives do not exist as we can always approximate these games, as closely as desired, by games for which these derivatives do exist.

$$(2) \qquad \left(\frac{\delta \pi_{ij}}{\delta u_i} \right)_{H(u_1, \ldots, u_n) = 0} = 0$$

where $H(u_1, \ldots, u_n) = 0$ is the equation of the upper boundary hypersurface H of the prospect space P. This can also be written as

$$(3) \qquad \frac{u_i - d_i}{u_j - d_j} = - \left(\frac{\delta u_i}{\delta u_j} \right)_{H = 0} .$$

As we can form all together $n(n-1)/2$ pairs out of the n players, we obtain $n(n-1)/2$ equations of this form, out of which, however, only $(n-1)$ equations are independent. These $(n-1)$ equations together are equivalent to the requirement of choosing the payoff vector $\bar{u}$ in such a way as to maximize the product

$$(4) \qquad \pi = \prod_{i=1}^{n} (u_i - d_i)$$

subject to the conditions:

$$u \in P$$

$$u \geqq d$$

$$d = \text{constant} \quad .$$

In view of the compactness and convexity of P, such a vector $\bar{u}$ always exists and is always unique - with the exception of the degenerate case where $u_i - d_i = \text{constant}$ identically in the whole prospect space for one or more of the players. (Since this implies that this player or these players cannot derive any benefit at all from cooperation with the other players, this degenerate case does not represent a cooperative game in the full sense of the word and will not be discussed in this paper.)

Alternatively, we can extend Nash's theory, and obtain requirement (4), by directly extending his axioms to the n-person case. We then need the following five axioms.

AXIOM 1. Efficiency. If $\bar{u}$ is a solution (solution payoff vector) for the game Γ with the prospect space P, then there is no point u in P such that $u_i > \bar{u}_i$ for every player $i = 1, \ldots, n$. In other words, $\bar{u}$

must lie on the upper boundary of P.

AXIOM 2. Symmetry. Any symmetric game Γ has a solution $\bar{u}$ with $\bar{u}_1 = \bar{u}_2 = \dots = \bar{u}_n$. That is, every symmetric game has a symmetric solution.

AXIOM 3. Linear invariance. If Γ' is a game that results from Γ when we subject the payoff function (utility function) of some player i to an order-preserving linear transformation but leave all other players' payoff functions unchanged, and if $\bar{u}$ is a solution for Γ, then the image of $\bar{u}$ under this transformation will be a solution for Γ'. In other words, if we change the zero point and/or the unit of measurement for a given player's utility function, this will make no difference to this player's utility payoff in a 'real' sense but will change only the way how it is measured - while the other players' payoffs will be left completely unaffected.

AXIOM 4. Independence of irrelevant alternatives (to borrow a term introduced by Dr. Arrow in a different context for a similar idea). If $\bar{u}$ is a solution for the game Γ with prospect space P and disagreement payoff vector d, and if Γ' is a game that results from Γ if we restrict the prospect space to $P' \subset P$ but so that both $\bar{u}$ and d still remain in the new prospect space, then $\bar{u}$ will be a solution also for Γ'. In other words, exclusion from the prospect space of possible payoff vectors other than the solution and the disagreement payoff vector themselves, does not change the solution. Axiom 4 essentially means that $\bar{u}$, which by Axiom 1 must lie on the upper boundary of the game's prospect space, depends on the shape of this boundary only in the neighborhood of $\bar{u}$ but is independent of the more distant parts of this boundary.[12]

[12] The obvious similarity between this last axiom and Arrow's irrelevant-alternative postulate has led some expositors of Nash's theory to regard this axiom, also, as a rationality postulate. But I think that Nash's own interpretation is preferable, according to which this axiom expresses a structural property of the actual bargaining process rather than a normative postulate: it expresses the fact that bargaining essentially consists in progressively narrowing down the range of alternative deals considered by the bargaining parties, so that in the end any proposed deal has to compete only with very close alternatives but does not have to compete any more with more distant alternatives already excluded from consideration at earlier stages.

AXIOM 5. Completeness. A game has no other solutions than those defined by Axioms 1, 2, 3 and 4. (Instead of postulating that the solution is unique, as Nash does, we make only the weaker assumption of completeness, which will allow us at a later stage of our argument to use a solution concept that lacks uniqueness - though the solution concept we shall eventually arrive at will be unique.)

In deriving requirement (4) from these axioms we can fairly closely follow Nash's original argument for the two-person case. We subject all players' utility functions to a linear transformation $\mathcal{T}$ by changing their zero points and their units so as to carry the disagreement payoff vector d into $d' = (0, \ldots, 0)$ and to carry the point $\bar{u}$, at which the product $\pi = \Pi(u_i - d_i)$ takes its maximum value π_o over the set P, into the point $\bar{u}' = (1, \ldots, 1)$. Let us denote the image of P by P', and the image of H, the upper boundary hypersurface of P, by H'. P' will be, like P itself, a compact and convex set. Over the set P', the product $\pi' = \Pi(u_i - d_i') = \Pi u_i$ will take its maximum value $\pi_o' = 1$ at $\bar{u}'$. Hence $\bar{u}'$ will be the point of tangency between H' and the hyperboloid J whose equation is $\Pi u_i = 1$, and the two hypersurfaces will have a common tangent hyperplane K at this point, with the equation $\Sigma u_i = n$ where n is the number of players. Let K^- denote the half-space $\Sigma u_i \leqq n$. Owing to the convexity of P' we shall have $P' \subset K^-$. Consequently if we construct a half-hypersphere P'' in K^- around the point $\bar{u}'$ with a radius ρ sufficiently large we shall have $P' \subset P''$.

Now consider the symmetric game Γ'' whose prospect space is P'' and whose disagreement payoff vector is d'. By Axioms 1 and 2, the point $\bar{u}'$ will be a solution for Γ''. Next consider the game Γ' whose prospect space is P' and whose disagreement payoff vector is d'. By Axiom 4, $\bar{u}'$ will be a solution also for Γ'. Finally again consider the original game Γ whose prospect space is P and whose disagreement payoff vector is d. By Axiom 3, the point $\mathcal{T}^{-1}\bar{u}' = \bar{u}$ will be a solution for Γ. On the other hand, by Axiom 5, $\bar{u}$ will be the only solution for Γ. This establishes requirement (4).[13]

§4. A BARGAINING MODEL FOR THE GENERAL CASE

In the general case we cannot assume that the disagreement payoffs

---

[13] A similar extension of Nash's solution to the special case of n-person games in which no coalitions can be formed, has been suggested also by R. Duncan Luce and Howard Raiffa, Games and Decisions, New York, 1957, pp. 349-350.

which the players would receive in the absence of an all-player agreement are given. Rather, these payoffs will depend on the pattern of cooperation and/or conflict that would emerge among different groups of players should a general agreement fail, so that these disagreement payoffs themselves become unknowns to be determined by our theory.

Like in the special case discussed in Section 3, in the general case, also, our main aim is to define a set of payoffs that represent the stable equilibrium outcome of multilateral bargaining among rational players. The model we shall use is not necessarily meant to predict the actual bargaining moves of the different players. Rather, it is meant primarily to exhibit the equilibrium conditions that stable agreements among rational players have to satisfy - whatever the actual details of the preceding bargaining process. These equilibrium conditions will be used, then, to define the equilibrium payoffs of the game.

Intuitively, our fundamental assumption is that in general every possible subset of the players (we shall call these subsets 'syndicates') will agree to cooperate in protecting their common interests against the rest of the players. The set N of all n players will be called the all-player syndicate; all proper subsets S of N will be called sectional syndicates. In contrast to the von Neumann-Morgenstern model, where each player can be a member of only one coalition, in our model each player will be a member of 2^{n-1} different syndicates. (We have adopted the term 'syndicate', instead of the term 'coalition' of the von Neumann-Morgenstern theory, as a reminder of this difference in the assumptions made.) The total number of syndicates in the game, including the all-player syndicate, will be $(2^n - 1)$.

Thus, in the general case, the outcome of the game will be determined, not by one unique agreement among the n players, but by a whole network of $(2^n - 1)$ agreements concluded by the members of the different syndicates. The terms that a given player obtains in any of these agreements will influence his bargaining position with respect to all other agreements in which he participates. Hence in equilibrium all these agreements concluded by the members of the various syndicates must satisfy certain equilibrium conditions of mutual consistency and interdependence. The basic task of our analysis will be to define these equilibrium conditions.

More specifically, we shall assume that the members of each syndicate S publicly guarantee certain payments, called 'dividends', to one another and agree to cooperate in securing these dividends to all members of the syndicate. These dividends will be assumed to be additional to the dividends that each member of the syndicate may receive from other syndicates. A syndicate may also decide to pay negative dividends, i.e., to reduce the payoffs of its members by certain amounts. Any decision of a syndicate is supposed to require the consent of all members. The dividend of player i from syndicate S will be denoted by w_i^S. To simplify notation we adopt the convention that for any syndicate S of which i is not a member, $w_i^S = 0$.

We shall assume that these dividend-guaranteeing agreements are backed by each sectional syndicate's announcing a threat strategy ϑ^S, i.e., a joint strategy[14] they would follow if no agreement could be reached on on the final payoffs of the game with the players not in S. The threat strategies ϑ^S and $\vartheta^{\bar{S}}$ announced by a given syndicate S and by the complementary syndicate $\bar{S}$ fully determine the conflict payoffs c_i^S and $c_j^{\bar{S}}$ that each member i of S and each member j of $\bar{S}$ (i.e., all players) would receive in the event of a conflict between S and $\bar{S}$. (No meaning will be attached to the symbol c_i^S when player i is not a member of syndicate S.)

We want to exclude dividend-guaranteeing agreements that are in themselves unrealistic or are inconsistent with other commitments of the participants. In particular, we shall require that the dividends guaranteed to any player i by a given syndicate S and all its subsets taken together should not exceed the conflict payoff c_i^S, to which he would be confined in the case of a conflict between S and $\bar{S}$. That is, we require that

$$\sum_{R \subseteq S} w_i^R \leq c_i^S \quad . \tag{5}$$

When all sectional syndicates have announced their threat strategies and their dividends, the n players may agree to distribute a further set of dividends w_i^N among themselves, which may be regarded as the dividends of the all-player syndicate N. Thus the final payoff of the game to player i will be

$$\bar{u}_i = \sum_{R \subseteq N} w_i^R \quad . \tag{6}$$

Obviously we have to require that these final payoffs agreed upon in this way should represent a set of possible payoffs, i.e., the vector formed of these payoffs must be a point in the prospect space of the game. That is

---

[14] Of course, any joint strategy formed out of the pure strategies available to the members of S, or any jointly randomized mixed strategy available to them, may serve as a threat strategy ϑ^S.

(7) $$\left(\sum_{R \subseteq N} w^R \right) \in P \ .$$

But it may happen that if we restrict summation to the proper subsets of N we obtain a vector outside of P, that is, we may have

(8) $$\left(\sum_{R \subset N} w^R \right) \notin P \ .$$

This is not inconsistent with requirement (7) but only means that the all-player syndicate N has to pay negative dividends at least to some players. Similarly, it may happen that for some sectional syndicates S with respect to certain players i we have

(9) $$\sum_{R \subset S} w_i^R > c_i^S \ .$$

Again, this is not inconsistent with requirement (5) but only means that syndicate S has to pay a negative dividend to i. However, cases (8) and (9) do imply that the syndicates R corresponding to subsets of N or S respectively cannot guarantee their dividends unconditionally to their members, but can guarantee these dividends only _under the condition_ that the larger syndicate N or S will be able to agree on a set of negative dividends satisfying (7) and (5). In such cases we shall speak of '_conditional dividends_'.

To facilitate compliance with requirements (5) and (7) we shall assume that the game is played in three stages. In stage I all sectional syndicates announce their threat strategies. These threat strategies automatically determine all conflict payoffs, and thereby determine the upper bounds to permissible dividends according to requirement (5). In stage II all sectional syndicates announce their dividends in compliance with these restraints. Finally, in stage III the all-player syndicate announces its own dividends in accordance with requirement (7), and by this means determines the final payoffs of the game.

To complete the description of our model, we add the following _penalty rule_. If the members of any set $S \subset N$ of players are unable to agree on a threat strategy ϑ^S and/or on a dividend vector w^S, then their dividends from S and from all supersets T of S will be zero. The intuitive meaning of this rule is that effective cooperation within a given set T of players is possible only if there is agreement and cooperation within all subsets S of T.

§5. ANALYSIS OF THE MODEL IN THE CASE WHERE THERE ARE NO CONDITIONAL DIVIDENDS

We shall first make the simplifying assumption that conditions (8) and (9) do not apply to any syndicate of the game, which means that all syndicates are in the position to give unconditional guarantees for their dividends.

If all dividends are unconditional we can reduce the game of our model to a set of bargaining games with given disagreement payoffs in the sense of Section 3.

First consider only stage III of the game. The players have already agreed on their dividends from all sectional syndicates and are sure to receive at least the sum of these dividends. Thus player i is sure to receive at least the payoff

$$d_i^N = \sum_{R \subset N} w_i^R \quad . \tag{10}$$

Therefore with respect to the bargaining on the final payoffs, the payoffs $d_1^N, \ldots, d_n^N$ represent the disagreement payoffs of the players. Hence we can conclude that the players will agree on such a final payoff vector $u = \bar{u}$ which maximizes the product

$$\pi^N = \prod_{i=1}^{n} (u_i - d_i^N) \tag{11}$$

subject to the conditions:

$$u \in P$$
$$u \geqq d^N$$
$$d^N = \text{constant} \quad .$$

This means in view of (6) that each player i will receive from N the dividend

$$w_i^N = \bar{u}_i - \sum_{R \subset N} w_i^R = \bar{u}_i - d_i^N \quad . \tag{12}$$

In order to extend our analysis to the complete game, including stages I and II, we subject the game to a type of <u>normalization</u>. That is, we shall assume that each syndicate $S \subseteq N$ at the beginning of the game agrees on a comprehensive strategy σ^S which will govern the policy of the syndicate during the whole course of the game. By the von Neumann-Morgenstern normalization theorem this assumption can be made without loss of generality.

For the all-player syndicate N this comprehensive strategy σ^N must specify what final payoff vector $\bar{u}$ will be selected in stage III of the game, given any possible combination of threat strategies ϑ^S and dividend vectors w^S that may be chosen by the sectional syndicates S. This essentially amounts to choosing a function U which maps the product space of the sets of all possible threat strategies ϑ^S, and of the sets of all possible dividend vectors w^S, for all $S \subset N$ into the set P of all possible final payoff vectors $\bar{u}$. Thus we can write

$$U \equiv \{\bar{u} - U(\vartheta^S, \vartheta^{S'}, \ldots; w^S, w^{S'}, \ldots) : S, S', \ldots \subset N\}$$

and

$$\sigma^N \equiv \{U\} \quad .$$

For each sectional syndicate S this comprehensive strategy σ^S must specify:

(A) a unique threat strategy ϑ^S that the syndicate will announce in stage I of the game; and

(B) a set of alternative dividend vectors w^S that the syndicate will announce in stage II of the game, depending on the threat strategies $\vartheta^{S'}, \vartheta^{S''}, \ldots$ selected by the other sectional syndicates $S', S'', \ldots \neq S$ and $\subset N$. This again amounts to choosing a function

$$W^S \equiv \{w^S = W^S(\vartheta^{S'}, \vartheta^{S''}, \ldots)\}$$

which maps the product space of the sets of possible threat strategies for each sectional syndicate other than S into the space of possible dividend vectors w^S.

Thus we can define:

$$\sigma^S \equiv \{\vartheta^S; W^S\} \quad .$$

Under these assumptions, of course, the final payoff vector $\bar{u}$ of the game becomes a single-valued function of all comprehensive strategies taken together so that we can write

$$\bar{u} = V(\sigma^N, \sigma^S, \sigma^{S'}, \ldots)$$

(13)

$$\bar{u}_i = V_i(\sigma^N, \sigma^S, \sigma^{S'}, \ldots)$$

Using these notations we can reformulate requirement (11) as follows. If the comprehensive strategies of all sectional syndicates are kept constant, the all-player syndicate N will choose a comprehensive strategy σ^N such that maximizes the product

(14) $$\pi^N = \prod_{i=1}^{n} (u_i - d_i^N) = \prod_{i=1}^{n} \left[V_i(\sigma^N, \sigma^S, \sigma^{S'}, \ldots) - d_i^N\right]$$

subject to the conditions:

$$\sigma^N \in \Sigma^N$$

$$\sigma^S, \sigma^{S'}, \ldots = \text{constant for all } S, S', \ldots \subset N$$

$$u \geq d^N$$

$$d^N = \text{constant},$$

where Σ^N represents the set of all comprehensive strategies available to N [all of which, of course, must satisfy requirement (7)].

On the other hand, if we keep constant the comprehensive strategies $\sigma^N, \sigma^{S'}, \sigma^{S''}, \ldots$ of all syndicates other than some given sectional syndicate S, then the choice of a comprehensive strategy σ^S by the members of S can also be regarded as a bargaining game with given disagreement payoffs. For, according to our penalty rule, if the members of S cannot agree on the choice of a threat strategy and a dividend vector - which amounts essentially to their being unable to agree on the choice of a comprehensive strategy σ^S - each player i in S will receive only the disagreement payoff

(15) $$d_i^S = \sum_{\substack{R \supset S \\ R \subset N}} w_i^R \quad .$$

Therefore, according to the theory of Section 3, the members of S will agree on such a comprehensive strategy σ^S which maximizes the product

(16) $$\pi^S = \prod_{i \in S} (\bar{u}_i - d_i^S) = \prod_{i \in S} \left[V_i(\sigma^N, \sigma^S, \sigma^{S'}, \ldots) - d_i^S\right]$$

subject to the conditions:

$$\sigma^S \in \Sigma^S$$

$$\sigma^N, \sigma^{S'}, \sigma^{S''}, \ldots, = \text{constant for all } N, S', S'', \ldots \neq S \text{ and } \subseteq N$$

$$\bar{u} \geq d^S$$

$$d^S = \text{constant},$$

where Σ^S denotes the set of all comprehensive strategies available to the syndicate S [all of these strategies, of course, must satisfy requirement (5)].

If the comprehensive strategies of all syndicates satisfy requirements (14) [or (11)] and (16), we shall say that they are mutually optimal in relation to one another.

In the following Section we shall extend this concept of mutual optimality to the general case where conditional dividends are also allowed. In Section 7 we shall show that, under this more general definition, a set of mutually optimal comprehensive strategies always exists.

§6. THE CASE OF CONDITIONAL DIVIDENDS

We now drop the simplifying assumption that all dividends are unconditional, i.e., that conditions (8) and (9) do not apply with respect to any syndicate.

If there are conditional dividends we cannot reduce the game to a set of subgames with given disagreement payoffs, because it will not be true any more that the players can count upon the payoffs d_i^N and d_i^S [defined by (10) and (15)] irrespective of whether they reach an agreement or not, as by assumption now some of the dividends will be conditional upon an agreement being reached. But even though we cannot now use the theory of Section 3, the five axioms of that Section with the addition of two further axioms will enable us to define a solution for the game.

For the sake of uniform notation we shall still write

(17) $$d_i^N = \sum_{R \subseteq N} w_i^R \quad \text{as in (10).}$$

(18) $$d_i^S = \sum_{\substack{R \supseteq S \\ R \subseteq N}} w_i^R \quad \text{as in (15).}$$

but, to avoid confusion, we shall now call these quantities '<u>claims</u>' (rather than 'disagreement payoffs') since they represent the total claim of a given player arising from unconditional or conditional dividends, prior to an agreement on dividends from the syndicate N or S.

The additional axioms we shall use are as follows:

AXIOM 6. <u>Positive relation of payoffs to claims</u>. Suppose that games Γ and Γ' have the same prospect space P and that all players other than player i have the same claims in both games while the claim of player i is larger (smaller) in game Γ' than in game Γ. Then, the game Γ' has a solution $\bar{u}'$ that assigns to player i a payoff $\bar{u}_i'$ no smaller (no larger) than is the payoff $\bar{u}_i$ assigned to him by any solution $\bar{u}$ of the game Γ. That is, other things being equal, a larger claim tends to improve, and a smaller claim tends to worsen, a player's bargaining position.

In the absence of conditional dividends, the property asserted by Axiom 6 is a direct implication of our model since, as is easy to show, requirement (11) makes each final payoff $\bar{u}_i$ a monotone non-decreasing function of the disagreement payoff d_i^N if the prospect space P of the game and all other disagreement payoffs d_j^N $(j \neq i)$ are kept constant. But in the general case we have to make it an axiom by its own right. Its intuitive meaning is that in stage III of the game each player's bargaining position will depend on his total claim, i.e., on the sum of all his unconditional and conditional dividends from all sectional syndicates, his bargaining position being stronger the larger his total claim. Conditional dividends will be no less effective than unconditional dividends in strengthening a player's bargaining position because he will in the same way have the complete support of the other members of the relevant syndicates when he insists on receiving full credit for these dividends in determining the final payoffs.

DEFINITION. Let $\bar{u}$, $\bar{u}^*$, and $\bar{u}^{**}$ be payoff vectors on the upper boundary H of the prospect space. Then $\bar{u}$ is called a point <u>intermediate</u> between $\bar{u}^*$ and $\bar{u}^{**}$ if $\bar{u}_i \geqq \min(\bar{u}_i^*, \bar{u}_i^{**})$ for all players i.

AXIOM 7. <u>Admissibility of intermediate solutions</u>. If $\bar{u}^*$ and $\bar{u}^{**}$ are both solutions of the game Γ then all points $\bar{u}$ of H such that are intermediate between

$\bar{u}^*$ and $\bar{u}^{**}$ are also solutions of Γ.

COROLLARY A OF AXIOM 7. The set of solutions of a given game Γ is a simply connected compact region of H, cut off by n hyperplanes each of which is perpendicular to one of the axes.

COROLLARY B OF AXIOMS 6 AND 7. If two games Γ and Γ' have the same prospect space P and the same claim vector (disagreement payoff vector) d^N, then they have the same solution or solutions.

We also change Axiom 5 so as to include Axioms 6 and 7 among the axioms referred to in Axiom 5.

We shall first again restrict our analysis to stage III of the game.

The definition of unconditional dividends implies that if a game has only unconditional dividends then the claim vector (disagreement payoff vector) d^N must lie on or below the upper boundary H of the prospect space P, i.e., d^N must either be a point of H or a point dominated by some points of H. In the general case, where conditional dividends are also admitted, this restriction no longer applies. In this general case the solutions of the game can be characterized by the following four propositions.

PROPOSITION (a). If the claim vector d^N lies on the upper boundary H of the prospect space P then d^N itself is a solution of the game.

PROPOSITION (b). A point $\bar{u}$ on H which either dominates the claim vector d^N or is itself dominated by the claim vector d^N is a solution if the hyperplane $K(\bar{u})$ whose points u satisfy the equation

$$\sum_{i=1}^{n} \frac{u_i - \bar{u}_i}{d_i^N - \bar{u}_i} = 0 .$$

is a supporting (or tangent) hyperplane of the prospect space P.

As $K(\bar{u})$ is nothing else but the tangent hyperplane drawn at the point $\bar{u}$ to the hyperboloid $J(\bar{u})$ whose points satisfy the equation

$$\prod_{i=1}^{n} (u_i - d_i^N) = \prod_{i=1}^{n} (\bar{u}_i - d_i^N) = \text{constant}$$

we can also say that $\bar{u}$ is a solution if either $\bar{u} \geq d^N$ or $\bar{u} \leq d^N$ and if $\bar{u}$ is a point of tangency between the hypersurface H and the hyperboloid $J(\bar{u})$.

PROOF OF PROPOSITIONS (a) and (b). If the point d^N lies on or below the hypersurface H then the two propositions are equivalent to requirement (11). So it is sufficient to prove proposition (b) in the case where d^N lies above H. The proof is very similar to the one used to derive requirement (11) in Section 3.

Consider the game Γ where the point $\bar{u}$ is dominated by the claim vector d^N and satisfies the requirement that $K(\bar{u})$ is a supporting hyperplane of the prospect space P. As $\bar{u}$ is a point on H, this is only possible if P lies in the lower half-space $K^-(\bar{u})$ defined by $K(\bar{u})$. Consequently all points u of P must satisfy the inequality

$$\sum_{i=1}^{n} \frac{u_i - \bar{u}_i}{d_i^N - \bar{u}_i} \leqq 0 \quad .$$

Now subject all players' utility functions to the linear transformation $\mathcal{V}$ by changing their zero points and units so as to carry the point d^N into $d' = (0, \ldots, 0)$ and to carry the point $\bar{u}$ into $\bar{u}' = (-1, \ldots, -1)$. Let us call Γ' the new game resulting from Γ under the transformation $\mathcal{V}$. Let P' and H' denote the images of P and H under $\mathcal{V}$. Then all points u of P' will satisfy the inequality

(19)
$$\sum_{i=1}^{n} \frac{u_i - \bar{u}_i'}{d_i' - \bar{u}_i'} \leqq 0$$

As $\bar{u}_i' = -1$ and $d_i' = 0$ we can also write

(20)
$$\sum_{i=1}^{n} u_i \leqq -n$$

where n is the number of players.

Let us rotate P' around the straight line connecting d'

and $\bar{u}'$ and call Q the body traced out by the points of P'. Moreover, let P" be the convex hull of Q. As all points of P' satisfy inequality (20), this inequality will also be satisfied by all points of Q and of P".

Now consider the symmetric game Γ'' whose prospect space is P" and whose claim vector is d'. By Axioms 1 and 2, the point $\bar{u}'$ will be a solution of Γ''. Next consider the game Γ' which has been derived from our original game Γ by means of the transformation $\mathcal{V}$. Γ' of course has the prospect space P' and the claim vector d'. By Axiom 4 and Corollary B, the point $\bar{u}'$ will also be a solution of Γ'. Finally consider our original game Γ with the prospect space P and the claim vector d^N. By Axiom 3, the point $\bar{u} = \mathcal{V}^{-1}\bar{u}'$ will be a solution of Γ. This completes the proof.

In the case where the claim vector d^N lies on or below the hypersurface H we have found that a given prospect space P and a given claim vector d^N determine a unique solution for the game in terms of requirement (11) (corresponding to the unique point of tangency between the hypersurface H convex to above, and a hyperboloid concave to above in the relevant region). But in the case where d^N lies above H there may be several solution points belonging to a given prospect space P and claim vector d^N because the hypersurface H and the family of hyperboloids J (or even one particular hyperboloid of this family) may have several points of tangency as both H and J are convex to above in the relevant region. In addition, Propositions (c) and (d) below may also define further solutions for the same game.

> PROPOSITION (c). Let $\hat{u}_i$ denote the highest payoff and $\check{u}_i$ the lowest payoff to the player i, associated with any point of the upper boundary H of the prospect space P of the game Γ. If a given player i has a claim $d_i^N > \hat{u}_i$ then the game Γ has a solution $\bar{u}$ such that $\bar{u}_i = \hat{u}_i$ for player i and $\bar{u}_j = \check{u}_j$ for all players $j \neq i$. That is, if the claim of a given player i exceeds the highest payoff that he can obtain according to the rules of the game, then there is a solution under which he obtains the highest possible payoff, while all other players obtain the lowest payoffs consistent with the efficiency axiom.

PROOF. Consider the one-parameter family of the games $\Gamma^*(\mu)$, which have the same prospect space P as Γ has and which have a claim vector $d^*(\mu)$ assigning the same payoff $d_i^* = d_i^N$ as assigned by the claim vector d^N of Γ to player i, but assigning the payoff $d_j^* = \mu$ to all other players j, with $\mu > d_j^N$ and $\mu > \hat{u}_j$ for all $j \neq i$. By Proposition (b), each game $\Gamma^*(\mu)$ will have a solution at the point (or points) $u = u^*(\mu)$ at which the product $\dot{\pi} = \Pi(d_k^* - u_k)$, $k = 1, \ldots, i, \ldots, n$, takes its minimum value $\dot{\pi} = \dot{\pi}_o$ for all $u \in P$. This is so because all points of the prospect space P lie on or below the hyperboloid $J(u^*)$ whose points u satisfy the equation $\Pi(d_k^* - u_k) = \dot{\pi}_o$, and _a fortiori_ lie on or below the hyperplane $K(u^*)$ which is tangential to $J(u^*)$ at the point u^* and whose points u satisfy the equation $\Sigma(u_k - u_k^*)/(d_k^* - u_k^*) = 0$. Clearly, the points $u^* = u^*(\mu)$ depend continuously on the parameter μ.

As μ goes to positive infinity, the hyperplane $K(u^*)$ becomes perpendicular to the u_i axis in the limit. For, the equation of $K(u^*)$ can also be written as $(u_i - u_i^*)/(d_i^N - u_i^*) + \Sigma(u_j - u_j^*)/(\mu - u_j^*) = 0$. As μ tends to infinity, all terms, except for the first, tend to zero, and in the limit we obtain the equation $u_i = u_i^*$ defining a hyperplane perpendicular to the u_i axis. Consequently, as μ goes to infinity, the point $u^*(\mu)$ at which the hyperplane $K(u^*)$ touches the upper boundary H of the prospect space P must tend to $\bar{u}$, defined by $\bar{u}_i = \hat{u}_i$ and $\bar{u}_j = \check{u}_j$ for all $i \neq j$, since, by the convexity of P, $\bar{u}$ is the only point of H where H can have a supporting hyperplane perpendicular to the u_i axis. Hence, with respect to the whole set of points $u^*(\mu)$ corresponding to all possible values of μ, we have $\sup u_i^* = \bar{u}_i$ and $\inf u_j^* = \bar{u}_j$. But, by Axiom 6, our original game Γ must have a solution u^{**} such that $u_i^{**} \geqq \sup u_i^* = \bar{u}_i$ and $u_j^{**} \leqq \inf u_j^* = \bar{u}_j$. The only point of the prospect space P that satisfies these conditions is $u^{**} = \bar{u}$ itself. Therefore $\bar{u}$ must be a solution of Γ.

PROPOSITION (d). All points on the upper boundary hypersurface H which are intermediate among the solutions defined by Propositions (a), (b) and (c) are also solutions of the game.

This proposition follows from Axiom 7.

THEOREM 1. For any given prospect space P and any given claim vector d^N there exists at least one solution.

PROOF. If the claim vector d^N is such that $d_i^N > \hat{u}_i$ for at least one player i then Proposition (c) defines at least one solution. In contrast, if $d_i^N \leqq \hat{u}_i$ for all players i then d^N must either (1) lie on the hypersurface H or (2) dominate some points of H or (3) be dominated by some points of H. In Case 1 Proposition (a) defines the solution. In Cases 2 and 3 Proposition (b) will define at least one solution: in Case 2 that point of H where the product $\Pi(d_i^N - u_i)$ takes its maximum value will always be a solution; in Case 3 that point of H where the product $\Pi(u_i - d_i^N)$ takes its maximum value will always be a solution [in accordance with requirement (11)].

> COROLLARY C. To any prospect space P and any claim vector d^N there corresponds a non-empty set $Y(P, d^N)$ of solutions $\bar{u}$.

To extend our analysis from stage III of the game to the complete game, we shall once more consider the game normalized in terms of comprehensive strategies for all syndicates.

Given the comprehensive strategies $\sigma^S, \sigma^{S'}, \sigma^{S''}, \ldots$ of all sectional syndicates $S, S', S'', \ldots \subset N$, the claim vector d^N is also given, and so is the set $Y(P, d^N)$ of all final payoff vectors (solutions) $\bar{u}$ corresponding to P and d^N. Thus the all-player syndicate N will have to choose some comprehensive strategy σ^N such that yields a final payoff vector $\bar{u} \in Y(P, d^N)$. Any comprehensive strategy σ^N which satisfies this requirement will be called <u>optimal</u> with respect to these comprehensive strategies $\sigma^S, \sigma^{S'}, \sigma^{S''}, \ldots$ of the sectional syndicates.

On the other hand, given the comprehensive strategies $\sigma^N, \sigma^{S'}, \sigma^{S''}, \ldots$ of all syndicates other than some particular sectional syndicate S, the final payoff vector $\bar{u}$ will depend only on the comprehensive strategy σ^S chosen by the syndicate S. Thus the choice of a comprehensive strategy σ^S by the members of S can be regarded as a subgame Γ^S, to which Propositions (a), (b), (c) and (d) apply. Any final payoff vector $\bar{u} = (\bar{u}_1, \ldots, \bar{u}_n)$ can be partitioned into a payoff vector $\bar{u}^S$ consisting of the payoffs of the members of the syndicate S, and a payoff vector $\bar{u}^{\bar{S}}$ consisting of the payoffs of the other players (who are members of the complementary syndicate $\bar{S}$). Let P^S be the set of all possible payoff vectors $\bar{u}^S$ for the members of S, such that can result from any alternative comprehensive strategy σ^S available to the syndicate S, given the comprehensive strategies $\sigma^N, \sigma^{S'}, \sigma^{S''}, \ldots$ of all other syndicates. P^S may be regarded as the prospect space P^S of the subgame Γ^S. On the other hand, the claim vector d^S defined by equation (18) may be regarded as the claim vector of this subgame. Thus the members of the syndicate S will have to agree on such a comprehensive strategy σ^S which yields to them a final payoff vector $\bar{u}^S \in Y(P^S, d^S)$. Any comprehensive strategy σ^S which satisfies this requirement will be called <u>optimal</u> with respect to the comprehensive strategies $\sigma^N, \sigma^{S'}, \sigma^{S''}, \ldots$ of the other syndicates.

A set of comprehensive strategies containing exactly one strategy for all syndicates $S \subseteq N$ will be called <u>mutually optimal</u> if each element of the set is optimal with respect to all other elements of the set. The threat strategies ϑ^S and the dividend vectors w^S corresponding to a set of mutually optimal comprehensive strategies will also be called mutually optimal. Finally, any final payoff vector $\bar{u}$ determined by a set of mutually optimal comprehensive strategies will be called a <u>solution</u> of the game.

§7. THE EXISTENCE OF A SOLUTION

We now propose to show that a solution $\bar{u}$, corresponding to a set of mutually optimal comprehensive strategies on the part of all syndicates, always exists for every cooperative game. We have already seen that there always exists a solution corresponding to a given prospect space P and a given claim vector d^N (Theorem 1). But, while the prospect space P is a datum for our theory, the claim vector d^N is a dependent variable, and so it has to be shown that a set of mutually optimal comprehensive strategies - which will define an admissible claim vector for the game - always exists.

Clearly, the choice of an optimal comprehensive strategy by any particular syndicate $S \subseteq N$ can also be described in terms of maximizing a certain function. For instance, we may define the functions Φ^N and Φ^S as follows.

$$(21) \qquad \Phi^N(\sigma^N; \sigma^S, \sigma^{S'}, \ldots,) = \begin{cases} 0 \text{ if } \bar{u} \in Y(P, d^N) \\ -D[\bar{u}, Y(P, d^N)] \text{ if } \bar{u} \notin Y(P, d^N) \end{cases}$$

where $\bar{u} = V(\sigma^N, \sigma^S, \sigma^{S'}, \ldots)$ denotes the final payoff vector corresponding to the comprehensive strategies $\sigma^N, \sigma^S, \sigma^{S'}, \ldots$; and where $D[\bar{u}, Y(P, d^N)]$ denotes the distance between $\bar{u}$ and the point of the set $Y(P, d^N)$ nearest to $\bar{u}$.

$$(22) \qquad \Phi^S(\sigma^S; \sigma^N, \sigma^{S'}, \ldots) = \begin{cases} 0 \text{ if } \bar{u}^S \in Y(P^S, d^S) \\ -D[\bar{u}^S, Y(P^S, d^S)] \text{ if } \bar{u}^S \notin Y(P^S, d^S) \end{cases}$$

where $\bar{u} = V^S(\sigma^S, \sigma^N, \sigma^{S'}, \ldots)$ denotes the payoff vector for the members of S corresponding to $\sigma^S, \sigma^N, \sigma^{S'}, \ldots$; and where D is again a distance function.

Then, any comprehensive strategy σ^N optimal with respect to the comprehensive strategies of the sectional syndicates can be defined by the requirement of maximizing the function

$$\Phi^N = \Phi^N(\sigma^N;\ \sigma^S,\ \sigma^{S'},\ \ldots) \tag{23}$$

subject to the conditions:

$$\sigma^N \in \Sigma^N$$
$$\sigma^S,\ \sigma^{S'},\ \ldots = \text{constant for all } S, S', \ldots \subset N,$$

where Σ^N denotes the set of all comprehensive strategies available to N.

Similarly, any comprehensive strategy σ^S optimal with respect to the comprehensive strategies of the other syndicates can be defined by the requirement of maximizing the function

$$\Phi^S = \Phi^S(\sigma^S;\ \sigma^N,\ \sigma^{S'},\ \ldots) \tag{24}$$

subject to the conditions:

$$\sigma^S \in \Sigma^S$$
$$\sigma^N,\ \sigma^{S'},\ \ldots = \text{constant for all } N, S', \ldots \neq S \text{ but } \subseteq N,$$

where Σ^S denotes the set of all comprehensive strategies available to S.

We now are in the position to state:

THEOREM 2. Every cooperative game has at least one solution, i.e., at least one final payoff vector corresponding to a set of mutually optimal comprehensive strategies on the part of all syndicates.

PROOF. Suppose that each subset S of N, including $S = N$, at the beginning of the game appoints a trustee with the instruction to select a comprehensive strategy σ^S for the syndicate S in such a way as to maximize the function Φ^S in accordance with requirement (23) or (24). This will give rise to a (2^n-1)-person non-cooperative game $\bar{\Gamma}$ among the (2^n-1) trustees. According to Nash's equilibrium point theorem as generalized by Debreu,[15] this non-cooperative game will have at least one equilibrium point, formed by pure strategies for each trustee. It is easy to see that the functions Φ^N and Φ^S satisfy the continuity and

[15] See J. Nash, "Non-cooperative games," Annals of Mathematics 54 (1951), pp. 286-295; and G. Debreu, "A social equilibrium point existence theorem," Proc. Nat. Acad. Sci. 38 (1952), pp. 886-895.

contractibility requirements of Debreu's theorem.

Any equilibrium point in pure strategies of this (2^n-1)-person non-cooperative game $\bar{\Gamma}$ will determine a set of mutually optimal comprehensive strategies for the syndicates of our original n-person cooperative game Γ. Any final payoff vector $\bar{u}$ corresponding to such a set of comprehensive strategies will be a solution of Γ.

§8. MAKING THE SOLUTION UNIQUE

We have already seen that in the general case there may be several admissible solutions for a game with a given prospect space P and a given claim vector d^N. Actually, there may even be several admissible claim vectors d^N for a given game Γ, corresponding to different alternative sets of mutually optimal comprehensive strategies, defined by different alternative equilibrium points of the non-cooperative game $\bar{\Gamma}$.

However, if a game had several equally admissible solutions, none of them could represent a really stable equilibrium point in bargaining among rational players.[16] For, each player would naturally prefer that particular solution that would yield him the highest payoff among all admissible solutions; and obviously no agreement could be reached on this basis.

Once more, the only way of overcoming this deadlock is further bargaining among the players. It is natural to assume that this bargaining will involve two steps. First the players will agree that each of them should receive at least the <u>lowest</u> payoff that would be assigned to him by any of the solutions corresponding to an equilibrium point of the non-cooperative game $\bar{\Gamma}$. Let this non-controversial minimum payoff to player i be denoted by $\tilde{u}_i$.

The second step will be a final round of bargaining on the distribution of further payoffs. This may now be regarded as a bargaining game with the given disagreement payoffs $\tilde{u}_1, \ldots, \tilde{u}_n$. According to the theory of Section 3, the players will agree on such a final payoff vector $u = \bar{\bar{u}}$ which maximizes the product

$$\bar{\bar{\pi}} = \prod_{i=1}^{n} (u_i - \tilde{u}_i) \tag{25}$$

---

[16] In other words, the equilibrium conditions derived in Sections 5 and 6, and used to define the "solutions" of the game, will in general only represent necessary conditions, but will not also represent sufficient conditions, for stable bargaining equilibrium.

subject to the conditions:

$$u \in P$$

$$u_i \geq \tilde{u}_i \quad \text{for all players } i$$

$$\tilde{u} = \text{constant}.$$

The final payoff vector $\bar{\bar{u}}$ so defined will be called the stable solution or the bargaining equilibrium point of the game.

In view of the convexity of P, this definition makes the vector $\bar{\bar{u}}$ always unique. Obviously, in the case where the game has only one solution $\bar{u}$, this solution itself becomes the stable solution of the game because then $\bar{u} = \tilde{u} = \bar{\bar{u}}$. (As an example, we refer to the case of games with transferable utility: as will be shown in the Appendix, such games always have only one unique solution.)

APPENDIX

GAMES WITH TRANSFERABLE UTILITY: RELATION OF OUR SOLUTION TO THE SHAPLEY VALUES OF THE GAME

1. In the special case of games with transferable utility, the mathematical properties of our model become much simpler. We shall base our analysis on the original ('extensive') version of our model rather than on its normalized form in terms of comprehensive strategies.

We shall again start with analysis of stage III of the game. In the case of transferable utility, the upper boundary H of the prospect space P is always a hyperplane with an equation of the form[17]

$$\sum_{i=1}^{n} u_i = \alpha^N = \text{constant} \quad . \tag{26}$$

Given the claim vector d^N, there will be a unique solution $\bar{u}$ defined by Proposition (a) or (b). In both cases the solution will satisfy the equation

[17] As in the case of games with transferable utility the prospect space includes the whole hyperplane H, which of course extends to infinity, we have to relax the assumption (made only for mathematical convenience in Section 3, but in fact unnecessary for our further argument) that P is a compact set, though we still have to assume that it is a closed set. Or alternatively we may retain the compactness assumption but make utility freely transferable only in a certain - sufficiently large - region of the prospect space, as our conclusions in this Appendix do not in fact require unlimited utility transferability.

$$\bar{u}_i - d_i^N = \bar{u}_j - d_j^N \quad \text{for all} \quad i, j \in N \quad . \tag{27}$$

In the case where Proposition (a) applies this is obvious as then $\bar{u} = d^N$. In the case where Proposition (b) applies, the hyperplane H must coincide with the hyperplane $K(\bar{u})$ whose equation is

$$\sum_{i=1}^{n} \frac{u_i - \bar{u}_i}{d_i^N - \bar{u}_i} = 0 \quad . \tag{28}$$

As $\bar{u}$ is a point of H, it must also satisfy equation (26) so that

$$\sum_{i=1}^{n} \bar{u}_i = \alpha^N \tag{29}$$

which in conjunction with (26) itself gives

$$\sum_{i=1}^{n} (\bar{u}_i - u_i) = 0$$

for all points u of H. As these points must also satisfy (28), equation (27) follows.

In view of (27) we can write

$$\sum_{j=1}^{n} (\bar{u}_j - d_j^N) = n(\bar{u}_i - d_i^N) \quad .$$

Considering (29) this implies

$$\bar{u}_i = \frac{1}{n} \alpha^N + d_i^N - \frac{1}{n} \sum_{j \in N} d_j^N \quad . \tag{30}$$

By equation (17), this is equivalent to

$$\bar{u}_i = \frac{1}{n} \alpha^N + \sum_{R \subset N} w_i^R - \frac{1}{n} \sum_{j \in N} \sum_{R \subset N} w_j^R \tag{31}$$

2. Next consider stages I and II of the game, where each syndicate S has to choose a threat strategy ϑ^S and a dividend vector w^S.

The assumption of transferable utility means that, in case of a conflict between the syndicate S and the complementary syndicate $\bar{S}$, the members of syndicate S could freely redistribute their conflict payoffs c_i^S among themselves without affecting the <u>sum</u> Σc_i^S of their own conflict payoffs and without affecting the conflict payoffs of the players in the complementary syndicate $\bar{S}$. In other words, the conflict strategy ϑ^S of the syndicate S can be regarded as consisting of two components which can be varied independently, viz.

(i) a '<u>fighting strategy</u>' $\hat{\vartheta}^S$, which, together with the fighting strategy $\hat{\vartheta}^{\bar{S}}$ of the complementary syndicate $\bar{S}$, would in the case of a conflict between S and $\bar{S}$ determine the joint payoffs

$$(32) \qquad \alpha^S = \sum_{i \in S} c_i^S \qquad \text{and} \qquad \alpha^{\bar{S}} = \sum_{j \in \bar{S}} c_j^{\bar{S}}$$

of the syndicates S and $\bar{S}$ so that we can write

$$(33) \qquad \alpha^S = \alpha^S(\hat{\vartheta}^S, \hat{\vartheta}^{\bar{S}}) \qquad \alpha^{\bar{S}} = \alpha^{\bar{S}}(\hat{\vartheta}^{\bar{S}}, \hat{\vartheta}^S)$$

and

(ii) a '<u>distribution strategy</u>' $\check{\vartheta}^S$, which would in case of a conflict between S and $\bar{S}$ determine the distribution of the joint conflict payoff α^S among the members of S.[18]

---

[18] More formally the fighting strategy $\hat{\vartheta}^S$ can be defined as an equivalence class of joint strategies for the syndicate S, viz., the set of joint strategies ϑ^S such that yield the same joint payoff α^S to the members of S against a given joint strategy $\vartheta^{\bar{S}}$ of the complementary syndicate $\bar{S}$. It can then be shown that the partition of the joint strategies ϑ^S of the syndicate S into these equivalence classes $\hat{\vartheta}^S$ is independent of the joint strategy $\vartheta^{\bar{S}}$ used by $\bar{S}$.

The distribution strategy $\check{\vartheta}^S$ in turn can be formally identified with the vector

$$\left(\frac{c_i^S}{\alpha^S}, \frac{c_{i'}^S}{\alpha^S}, \frac{c_{i''}^S}{\alpha^S}, \ldots \right)$$

where $i, i', i'', \ldots$ are the members of S.

In order to find the optimal distribution strategy $\check{\vartheta}^S$ and the optimal dividend vector w^S for a given syndicate S, we may keep the fighting strategy $\hat{\vartheta}^S$ of this syndicate constant, as well as keep constant the threat strategies (including both the fighting and the distribution strategies) and the dividend vectors of all other syndicates. Then the choice of $\check{\vartheta}^S$ and of w^S by the members of S can be regarded as a subgame $\check{\Gamma}^S$, to which the conclusions of Section 6 apply. The members of S have to agree on such $\check{\vartheta}^S$ and w^S that the resulting vector $\bar{u}^S$ of their final payoffs should be an admissible solution of their subgame $\check{\Gamma}^S$. As the dividends of all syndicates other than S are assumed to be given, by equation (31) the final payoffs of the members of S (and of course the final payoffs of the other players also) are single-valued functions of the dividend vector w^S chosen by syndicate S. On the other hand, the dividend w_i^S of each player i in S is subject to restraint (5), i.e.,

(34) $$\sum_{R \subseteq S} w_i^R \leqq c_i^S \quad .$$

By (32) this implies

(35) $$\sum_{i \in S} w_i^S \leqq \alpha^S - \sum_{i \in S} \sum_{R \subset S} w_i^R \quad .$$

By (31), for the final payoffs $\bar{u}_i$ of the members of S this gives the restriction

(36) $$\sum_{i \in S} \bar{u}_i \leqq \frac{n-s}{n} \alpha^S + \frac{n-s}{n} \sum_{i \in S} \sum_{\substack{R \not\subseteq S \\ R \subseteq N}} w_i^R + \frac{s}{n} \alpha^N - \frac{s}{n} \sum_{i \notin S} \sum_{R \subseteq N} w_i^R$$

where s denotes the number of players in S. Inequality (36) thus defines the prospect space of the subgame $\check{\Gamma}^S$ among the members of S. By Axiom 1, the solution of this subgame must lie on the upper boundary of this prospect space, so that in (36) the sign $\leqq$ can be replaced by the sign $=$. Consequently the signs $\leqq$ can be replaced by signs $=$ also in (35) and in (34), and we can write

(37) $$\sum_{R \subseteq S} w_i^R = c_i^S \quad .$$

A similar equation will apply to the conflict payoffs c_i^T that are assigned to player i by all the $(2^{s-1} - 1)$ syndicates T which are proper subsets of S and contain player i as member. If we solve this set of linear equations for the variables w_i^S we obtain the result (which can be verified by substitution) that

$$(38) \qquad w_i^S = \sum_{R \subseteq S} (-1)^{s-r} c_i^R$$

where s and r are the number of players in S and R. By (37) this in turn implies that

$$(39) \qquad \sum_{i \in S} w_i^S = \sum_{R \subseteq S} (-1)^{s-r} \alpha^R \quad .$$

Similarly in view of (6) and (31) we can write

$$(40) \qquad \sum_{i=1}^{n} w_i^N = \sum_{R \subseteq N} (-1)^{n-r} \alpha^R \quad .$$

Going back to the subgame $\check{\Gamma}^S$, it will have a unique solution $\bar{u}^S$ defined by Proposition (a) or (b). By an argument similar to the one we used above in connection with equation (27), it can be shown that in either case the solution must satisfy the condition

$$(41) \qquad \bar{u}_i - d_i^S = \bar{u}_j - d_j^S \text{ for each } i, j \in S \quad .$$

By equations (6) and (18) this can also be written as

$$(42) \qquad \sum_{R \supseteq S} w_i^R = \sum_{R \supseteq S} w_j^R \text{ for any } i, j \in S \quad .$$

By equations (6) and (27) we have

$$(43) \qquad w_i^N = w_j^N \text{ for any } i, j \in N \quad .$$

There are all together $(n - 1)(2^n - 2)$ independent equations of form (42) and $(n - 1)$ independent equations of form (43). If we add to these the $(2^n - 1)$ equations of forms (39) and (40), we obtain a set of $n(2^n - 1)$ linear equations, which can be solved for the variables $w_i^S : S \subseteq N$. The solution is, as can be verified by substitution, as follows:

$$(44) \qquad w_i^S = \frac{1}{s} \sum_{R \subseteq S} (-1)^{s-r} \alpha^R \quad .$$

In view of equation (6) this gives the expression for the final payoffs of the game

$$\bar{u}_i = \sum_{\substack{S \subseteq N \\ S \ni i}} \frac{(s-1)!\,(n-s)!}{n!}\,\alpha^S - \sum_{\substack{S \subseteq N \\ S \not\ni i}} \frac{s!\,(n-s-1)!}{n!}\,\alpha^S \quad . \tag{45}$$

3. Our last task is to define the optimal fighting strategies $\hat{\vartheta}^S$. If the fighting strategies of all other syndicates are kept constant, the final payoffs of the members of S (and of the other players as well) become, in view of (45) and (33), single-valued functions of the fighting strategy $\hat{\vartheta}^S$ chosen by the syndicate S. Now on the right-hand side of equation (45) the only terms dependent on $\hat{\vartheta}^S$ are the terms containing α^S and $\alpha^{\bar{S}}$. The former has the coefficient

$$\frac{(s-1)!\,(n-s)!}{n!}$$

while the latter has the coefficient

$$-\frac{\bar{s}!\,(n-\bar{s}-1)!}{n!} = -\frac{(s-1)!\,(n-s)!}{n!}$$

where $\bar{s} = n - s$ denotes the number of players in $\bar{S}$. Hence the payoff of each member of the syndicate S will be maximized if the difference $(\alpha^S - \alpha^{\bar{S}})$ is maximized.

Consequently the optimal fighting strategy $\hat{\vartheta}^S$ of the syndicate S will be such that maximizes the difference $(\alpha^S - \alpha^{\bar{S}})$ against the fighting strategy $\hat{\vartheta}^{\bar{S}}$ chosen by the complementary syndicate $\bar{S}$. On the other hand, the optimal fighting strategy $\hat{\vartheta}^{\bar{S}}$ of $\bar{S}$ will be such that maximizes the difference $(\alpha^{\bar{S}} - \alpha^S)$, i.e., minimizes $(\alpha^S - \alpha^{\bar{S}})$ if $\hat{\vartheta}^S$ is kept constant. Thus the optimal fighting strategies of two complementary syndicates in a game with transferable utility have maximin and minimax properties. The choice of such optimal fighting strategies by the two complementary syndicates represents a two-person zero-sum subgame, which has a unique solution. Though there may be several optimal fighting strategies for each syndicate, all of these optimal strategies will correspond to the same joint conflict payoffs α^S and the same final payoff vector $\bar{u}$ of the game. Hence in the case of transferable utility a game has only one <u>unique</u> solution, which accordingly represents a stable solution or bargaining equilibrium point.

Moreover, in the case of transferable utility our model of the cooperative game reduces, not to one unique (2^n-1)-person non-cooperative game, but to a set of $(2^{n-1} - 1)$ mutually independent two-person zero-sum subgames.[19]

4. Inspection of equation (45) will show that our solution for games with transferable utility is defined by an expression formally identical with the Shapley value,[20] except that in our formula the von Neumann-Morgenstern characteristic function $v(S)$ is replaced by a modified characteristic function that we have denoted by α^S. In our notation $v(S)$ would be defined as

$$v(S) = \max_{\vartheta^S} \min_{\vartheta^{\bar{S}}} \alpha^S(\vartheta^S, \vartheta^{\bar{S}})$$

while α^S is defined as

$$\alpha^S = \alpha^S(\vartheta_o^S, \vartheta_o^{\bar{S}})$$

where ϑ_o^S and $\vartheta_o^{\bar{S}}$ satisfy the requirement

$$\alpha^S(\vartheta_o^S, \vartheta_o^{\bar{S}}) - \alpha^{\bar{S}}(\vartheta_o^{\bar{S}}, \vartheta_o^S) = \max_{\vartheta^S} \min_{\vartheta^{\bar{S}}} [\alpha^S(\vartheta^S, \vartheta^{\bar{S}}) - \alpha^{\bar{S}}(\vartheta^{\bar{S}}, \vartheta^S)] \quad .$$

That is, $v(S)$ corresponds to a joint strategy on the part of S such that is associated with the maximin value of the joint payoff itself of S, while α^S corresponds to a joint strategy that is associated with the maximin value of the difference between the joint payoffs of S and $\bar{S}$.

Quite apart from the bargaining model proposed in this paper, it seems preferable to use this modified characteristic function in the expression defining the Shapley value. For, as has been shown above, if the payoffs of the players are in fact determined by this expression, each member of a given syndicate (coalition) S of players will maximize his own payoff by selecting a joint strategy ϑ^S with a view to maximizing

[19] For the case of the cooperative two-person game, Howard Raiffa has suggested a similar solution based on minimaxing the difference between the two players' payoffs, subject to the overriding requirement of efficiency (Pareto-optimality). Though, the motivation of Raiffa's solution is somewhat different as it is based on interpersonal utility comparisons and on the idea of arbitration, rather than bargaining. See Howard Raiffa, "Arbitration schemes for generalized two-person games," Annals of Mathematics Study No. 28 (Princeton, 1953), pp. 361-387. Cf. also R. Duncan Luce and Howard Raiffa, Games and Decisions, New York, 1957, pp. 143-145.

[20] Shapley, op. cit., equation (11) on p. 311.

the difference $(\alpha^S - \alpha^{\bar{S}})$ rather than maximizing α^S itself.[21]

As many cooperative games _without_ transferable utility can be reasonably well approximated by a game _with_ transferable utility, the _modified_ Shapley value of the latter game will often be found a convenient approximation of the exact bargaining equilibrium point of the original game.

[21] This modification also overcomes the objection raised by Luce and Raiffa, Games and Decisions, pp. 137-138, to the Shapley value in its original form, viz., that it is insufficiently sensitive to the players' threat possibilities.

John C. Harsanyi

Stanford University
and
Cowles Foundation at Yale University

ABSOLUTE GAMES[1]

J. R. Isbell

INTRODUCTION

This paper introduces a theory of cooperative games in which we do not assume, like von Neumann and Morgenstern [7], linear transferability of utility between the players. It is not a generalization of the Neumann-Morgenstern theory because it rests upon a different theory of games of two players. That theory rests in turn upon a utility theory which is a modification of that of von Neumann and Morgenstern. Thus the entire development is parallel and heavily indebted to that introduced in [7].

There are thus three stages in the argument. The first is the introduction of <u>absolute utility</u>, with an interpretation supporting the further work. We assume that a person's utility space, with the metric structure induced by the Neumann-Morgenstern axioms on probability combinations, is bounded. Then we have an absolute scale of utilities from 0 to 1 and a basis for interpersonal comparison of utility. We go on to adopt an arbitration scheme which keeps ratios of absolute utilities invariant. This is supported by some plausible arguments in connection with which we are indebted to T. Carrington, but essentially it amounts to a specialization.

The second stage is the theory of two-player games. This is little more than a modification of the work of Raiffa [8], with an interpretation which is modified from Nash [5]. The modification is necessary because Nash and Raiffa deal with games of two <u>persons</u>, where a person has a utility function, but we deal with games of two <u>players</u>, where a player is a coalition of persons. Thus Nash and Raiffa can assume that the given game A is bilinear, and derive a modified game A' which has a solution; in our case A' is bilinear, and therefore A'' has a solution.

---

[1] Supported (in part) by the Office of Naval Research under Contract N7onr41904. Based on the author's doctoral thesis, Princeton University. Presented to the American Mathematical Society, September 2, 1955.

The third stage is the description of the general game in terms of its two-player quotient games. Here we have not sought any innovation but worked out as faithful as possible a copy of the Neumann-Morgenstern theory. Nevertheless, there are both technical and conceptual difficulties which we have overcome only using the invariant ratio scheme; and even so, there are certain indications that we have not found the right definitions.

Before giving a more critical introduction we explain more fully the relation of this work to the previous literature. Our dissatisfaction with the hypotheses of von Neumann and Morgenstern was aroused by a criticism of McKinsey [4]; we present this criticism in Section 2 and use it as a lead to a sketch of the work to be done in the next four sections. In seeking an axiomatic basis for McKinsey's argument we became convinced that utility spaces are bounded. Savage has remarked [9] that this seems more reasonable than the contrary assumption, but he does not use it, and we find no other reference to the question. Now the use of a person's base utility before a game, or some number, as a measure of his bargaining power in the game, has been touched on by Shapley [10]; but Shapley raises the point only after putting a game in the Neumann-Morgenstern normal form, and he obtains results which seem odd and are not paralleled in our theory. We use base utilities in getting a general game into normal form, specifically in solving two-player games; but once we have the normal form, the rest of our treatment is independent of base utilities.

In connection with two-player games account should be taken of Braithwaite's work. Braithwaite's [1], Nash's [5], and Section 5 of this paper, represent three theories of two-player games each of which is essentially implicit in Raiffa's general theory [8]. It should be noted that Nash's paper is the earliest. Also, Nash earlier [6] indicated a possible general theory, of which [5] seems to be the first installment. The present work throws little light on how Nash's program might be carried out, for Nash rejects interpersonal comparison of utility, which we find vital for the definition of a coalition.

In games of more than two players we find no previous work without linear transferability. The work with linear transferability has all rested on the Neumann-Morgenstern normal form. Here we are concerned mainly with getting the normal form; then any of the various competing ideas for dealing with the normal form of a linear game (which is a _characteristic function_ [7]) should be applicable to our normal form, which is a _projective form_. This does not seem to be true in our present formulation. We are able to do some work with the Neumann-Morgenstern solution concept, which is a _stable set_. We are not able to generalize the Shapley _value_ [12], but we include some remarks on what such a generalization would require. There are other ideas available, none of them well worked out even for characteristic functions. But the suggestions of Gillies [2] and Luce [3] should be noted. Gillies' cores pop up in an example below, though they should be much less frequent in this theory than in that of [7]; Luce's k-stability is hard to verify, and we have not tried.

Specifically, Section 1 introduces bounded utilities, with brief supporting argument, and goes through some methodological discussion to the ratio scheme. The best positive argument we have for the ratio scheme is a "derivation" using the old idea of equal subjective differences and the Weber-Fechner Law. Tucker Carrington is not responsible for this argument, but he pointed out the possibility. We feel that the scheme will stand comparison with those of Nash and Braithwaite, but generalization would be desirable. Section 2 takes up McKinsey's criticism and example, formulates the program of the next four sections, and carries it through for the example. Sections 3 and 4 do the required preliminary work in utility spaces, all the proofs being gathered in Section 4. (Thus the first three sections contain no proofs.) We represent payoffs at the end of a game by end games, a generalization of bargaining games.

Section 5 deals with two-player games, using any arbitration scheme from a certain class introduced by Raiffa [8], which we characterize. In Section 6 we set up the projective form of a general game, using the ratio scheme. Here two main difficulties appear. The first is the restriction to the ratio scheme. The reason for this is that a coalition must have a sufficiently definite constitutional agreement to be able to act as one man in a two-player game, and we have not been able to formulate such agreements in general. Using the ratio scheme, we require agreement on the ratios of the utilities of the members of the coalition. The second main difficulty is that the projective form cannot take the place of an end game, so that we have no theoretical basis for constructing sequences of games. The reason for this is that the projective form is projective, in the sense of having no structure of convex combination. In terms of games, the projective form of a convex combination is not determined by the projective forms of the terms. This is true already for the values of two-player games; therefore the Neumann-Morgenstern idea of describing the general game in terms of the values of two-player quotient games is not sufficient in general to preserve the convex structure (and a fortiori the sequence structure) of games. In Section 7 we show by examples that the situation is worse than that; and we have no suggestions on how to get a manageable normal form which is more than projective. The end games have a convex structure, but they too are inadequate for two-player games. The problem will evidently have to be attacked on the two-player level.

Section 7 deals with three-player games including several examples, and Section 8 concludes the paper by readmitting the hypothesis of linear transferability. The resulting theory is, of course, closely similar to the original theory of von Neumann and Morgenstern.

§1. AXIOMS FOR UTILITY THEORY

We begin with the axiomatic utility theory of von Neumann and Morgenstern [7]. Their axioms are not categorical but have precisely eleven different realizations (counting the empty set). We select one of these, partly for technical convenience. As von Neumann and Morgenstern showed [7], their axioms imply that the utility space of a person is isomorphic with a subset of the real numbers. The structure consists of a linear order and an algebra of convex combination. The group of this structure is the group of all order-preserving linear transformations. One may verify that the convex subsets of the real line fall into eleven equivalence classes under this group. We select the equivalence class of the interval $[0, 1)$. We regard the actual zero and the ideal 1 as fixed, so that our utility space has no proper motions.

Whether 0 and 1 are in or out of this space makes little difference. What is vital is that the space should be bounded. This point has been almost unanimously ignored in the literature. Savage [9] has observed that utility spaces should be bounded, noting that the St. Petersburg paradox arises in any unbounded utility space. That is, if a person has unbounded finite utilities, then we may construct a prospect for which he has infinite utility; a coin will be tossed until heads turns up, and if this does not happen till the n-th toss then he will receive utility 2^n. But quite aside from this, introspection convinces the present author firmly that, e.g., there is no prospect so desirable as to be worth an even bet of his life. He who claims a utility space unbounded above must in principle stand ready to bet his life at any odds provided the price is right.

Then we accept bounded utility spaces. The desired half-open interval may be characterized by three axioms added to those of [7]: (a) there is a least utility, 0; (b) there is no greatest utility; (c) for each utility $\epsilon > 0$ there is a probability $\alpha > 0$ such that for all utilities u, $\epsilon > \alpha u + (1 - \alpha)0$.

At the referee's suggestion we emphasize that Axiom (c) does not make the utility space non-archimedean, though it may seem to. In fact, being bounded, it is more or less uniformly archimedean; for each large u and small ϵ not only is there α so small that $\epsilon > \alpha u$, but α can be chosen independently of u. At the suggestion of Lloyd Shapley we emphasize that (a), (b), and (c) are easy consequences of (hence equivalent to) the hypothesis that the utility space is equivalent to $[0, 1)$.

Having bounded utilities, we are not automatically better off. In fact it has been customary, while tacitly permitting unbounded utilities, to deal only with problems in which the possible outcomes are bounded (though there are a few significant exceptions, which, on the basis of the above comments, the author regards as mistakes). We propose to use bounded utility as a justification for interpersonal comparison of utility. The formal basis for comparison is obvious; we mean to treat utility u on the absolute scale [0, 1), for one person, as the same utility as u on anyone else's scale. Now this declaration, whatever emotions it may excite, has no meaning until we specify the operations involved in the treatment. But we choose to state a sweeping generalization concerning method at this point; then when we criticize the comparison procedure below, we shall criticize only the mechanics, not calling the possibility of comparison into question. The generalization is this; when two objects have identical descriptions which are presumably categorical (in a given context), then those objects "presumably" have identical properties. The two occurrences of "presumably" are meant to be equivalent; we say that if an objection to the identification of the two objects is to be heard, it must allege that the description is not categorical.

This matter deserves further elucidation. In the following section we level a criticism of precisely the kind just contemplated at a symmetry in the Neumann-Morgenstern theory which we contend is unjustifiable. We base our contention on an allegation that the relevant identical descriptions are not categorical for the context; and we specify the non-symmetric considerations which we think should be included in the description. Now we anticipate similar criticism of the theory we shall propose; as we said in the Introduction, we think it may be supported by better arguments than we have found, but it certainly has weaknesses. We hope to stimulate constructive criticism, particularly any which identifies measurable attributes of a game which we fail to recognize. But we demand that any criticism be to the point in that it either refutes our presumably categorical description or refutes the principle we have stated above.

Another general concept, besides that of interpersonal comparison, usable in the description of the operation we propose, is that of <u>equal difference</u> of utilities. Before von Neumann and Morgenstern introduced their axioms which put a metric structure on utilities by reference to probability combinations, the idea had been considered of creating a metric structure by axiomatizing the concept of equal difference. (See [7].) This is of course Euclid's way of putting a metric on the line. But nobody ever seemed to have an operational criterion for equal differences. Our proposal may be viewed as a step toward an operational criterion for equal differences. The view is difficult to criticize in the absence of ideas on how to make the remaining step which is necessary: the operational interpretation of the "solution" of a game.

We consider a pure bargaining problem, which we describe in the two-person case for simplicity. Person I has utility a and II has utility b. Alone, neither can improve his position. By joint action the persons may improve their utilities, I to x and II to y, where (x, y) is to be chosen from a set which we must assume is closed. (Once into game theory we shall also assume the set is convex, with the usual justification.) The choice of (x, y) is up to I and II jointly; if they cannot agree then they get (a, b), and in effect both of them lose. We say that the solution is that point (x, y) such that x = ta, y = tb, and t is a maximum. This saying may be regarded as an assumption, a pronouncement, or the definition of a special class of bargaining problems; whatever it is, it is used throughout the paper. Whatever it is, it is not an operational statement; for nobody knows what a solution is.

We have said roughly that cooperation will take place not, as suggested by the Shapley value [12], with an even division of the unearned increment, but with each player jealously defending his relative position as measured by the ratio a:b. Another way of putting it is that I and II receive equal increments of utility in going from (a, b) to (ta, tb); this involves a definition of equal increment or equal difference, together with the assertion that equal differences determine the solution of a bargaining problem. Note that if a > 0 and b = 0 then the utility of II will remain zero. We do not regard this as a defect; if zero utility exists (we have said it does, but only for convenience), a person with zero utility is not playing games. If a = b = 0 we have defined no solution and shall define none.

The arbitration scheme which assigns solutions to bargaining problems and games in the manner indicated has considerable advantages in comparison with alternative arbitration schemes, which will appear in Section 5. The only positive argument we have for this scheme is a weak one; it was worked out in a squadroom discussion with the chemist Tucker Carrington. We must refer to the equal difference idea and draw an analogy. The magnitude of a star, which is supposedly a subjectively linear measure of brightness, is proportional to the logarithm of the energy of radiation received; similarly the decibel scale of sound intensity is logarithmic in energy. In fact there is a shaky generalization of nineteenth century psychology, known as the Weber-Fechner Law, which states that equal differences in sensation always correspond to equal ratios in stimulus strength, or that the subjective scale is logarithmic in the objective scale. We have such a relationship between the proposed utility-difference scale and the Neumann-Morgenstern probability-calibrated scale.

§2. SKETCH OF THE THEORY

An important motivation for this study is the following example of McKinsey [4]. A worker, A, has two alternatives; he can work for the boss B, giving B a profit of 10 and A nothing, or he can cut off his arm, giving 0 to B and - 1000 to A. In the von Neumann theory, one reasons thus; the maximum which A can assure to A by his own efforts is 0, the maximum which B can assure to B is 0, and the maximum which {A, B} can assure by cooperation is 10. The characteristic function, v, of the game, is defined accordingly, $v(A) = v(B) = 0$, $v(AB) = 10$; and the further theory depends only on v. McKinsey suggests that such a treatment ignores some significant considerations.

Clearly in any theory based on v alone, the example is a symmetric game. For example, its Shapley value is (5, 5) [12]. Now Nash's uncompleted program of treating cooperative games by applying non-cooperative theory to models including certain moves which represent steps toward cooperation [5, 6], does cover this game; and it prescribes the solution (0, 10). With McKinsey, we prefer this result. However, Nash's rationale quite essentially demands incomparable utilities for different persons, so that we must make some changes.

Given an n-person game, we use absolute utilities and thus only part of the positive orthant of E_n. Determine the convex hypersurface M consisting of all maximal n-tuples of utilities, $(x_1, \ldots, x_n)$, which the set of all players can obtain for its respective members. Consider each partition of the set N of all players into two opposing subsets, $S = \{1, 2, \ldots, s\}$, $N - S = \{s + 1, \ldots, n\}$. Suppose that within S, and within N - S, agreement has been reached on the ratios of the utilities of the members; thus we have an s-vector $y = (y_1, \ldots, y_s)$, and an (n-s)-vector z, both defined only up to a scalar multiple. Accordingly we restrict to the subset of M consisting of all $(ay_1, \ldots, ay_s, bz_{s+1}, \ldots, bz_n)$. For simplicity we let S and N - S choose agents, i, j, such that $y_i \neq 0 \neq z_j$, and project the above typical vector to (ay_i, bz_j). Now i, commanding all the strategic resources of S, aims to maximize the ratio $ay_i : bz_j$ (which j similarly aims to minimize). Since internal utility ratios in S and N - S are fixed, we may expect the result to be independent of the choice of agents.

The minimax ratio is obtained by use of Nash's model [5]. The players i, j, have strategy spaces X_i, X_j (representing the entire coalitions, S, N - S); the vector objective function (ay_i, bz_j) is continuous but not generally bilinear on $X_i \times X_j$. But we demand that i and j choose strategies, s, t, called <u>threats</u>; they do not use these threats, but compute the ratio ay_i/bz_j which would result if s and t were played, and then adopt a more cooperative procedure leading to the element of M determined by y, z, and the minimax ratio. It turns out that we do not get a solution unless we iterate the construction; i and j consider probability mixtures of threats as strategies in a new game, threaten particular mixtures, and arbitrate the result. But two stages suffice, and the value is uniquely determined by optimal threats to threaten to act.

So far we have one imputation in M for each choice of S, y, N - S, z. The collection of all these points gives us $2^{n-1} - 1$ (n-2)-dimensional boundaries in the (n-1)-dimensional surface M, representing frontiers of the domains of effectiveness of sets S, N - S. This figure is a straightforward generalization of the normal form of von Neumann and Morgenstern [7], and there is no serious difficulty in carrying over the definition of a stable set solution.

Let us return to McKinsey's example. The utility-1000 is inadmissible, but we may keep the given numbers and ask where the absolute zeros of A and B are located. If they are at all nearly symmetric then we get the solution on which Nash and McKinsey agree. If A's absolute zero is at - 2000 on the given scale, and B's at - 5, then A has his choice of base points (- 1000, 0) and (0, 10), which represent increments of (1000, 5), (2000, 15) respectively over absolute zero. He will prefer the first, which gives him a better ratio for bargaining. The solution divides the possible increment of 2015 in the ratio 1000:5, which is $2005 - \epsilon$: $10 + \epsilon$; or in the given scale, A will not work unless he is paid nearly half the net profit, $5 - \epsilon$. This seems reasonable, since the possible gain means so much to the boss and so little to the worker. In effect by cutting off his arm he can injure the boss as much as himself, which is of course an unlikely situation which McKinsey evidently meant to exclude.

We have supposed that the 10 units of possible profit are linearly transferable to A. In our theory it hardly matters. Except in the unlikely case just examined, the solution is (0, 10) whatever transfer is possible. In that extreme case, with the base point (- 2000, - 5), B will get 5 units of the profit and almost all of the rest will go to A - supposing transfer is possible at a rate much less than 200 to 1. If the base point is (- 2000, - 5) and transfer is at some such rate as 200 to 1 then all the implications of the numbers originally given have disappeared.

McKinsey's treatment of the example is simply based on common sense; he says nothing about absolute zeros or exchange rates. Nash's treatment is based entirely on the exchange rate. Now the rate at which profits exchange between two players is likely to give some indication of the ratio of their absolute utilities; but we cannot endorse Nash's theory, for absolute utilities seem to be a more basic element of bargaining strength.

We shall see in Section 7 that exchange rates play a significant part in the present theory, in games of three or more players. It is worth noting that when two persons are in a bargaining situation they often have in mind the possibility of later collaboration in a many-person problem. Thus exchange rates could creep in very easily.

§3. UTILITY SPACES

For any finite set N, the concrete, absolute utility space I_N is the Cartesian product over N of intervals $[0, 1)$. An element of a utility space is an imputation; for convenience I_N will sometimes be regarded as embedded in a Euclidean space E_N, imputations becoming vectors. The indexing set N is always finite; generically its power is designated n, and its elements, $1, 2, \ldots, n$.

We distinguish between "person" and "player". The coordinate indices of a utility space are persons; and any person may be agent for a (unique) set of persons, which set is a player. A player must have only one agent in any given situation, so far as utility spaces are concerned.

The only motions of a utility space are permutations of coordinates. There are endomorphisms, multiplications by scalars between 0 and 1; and our theory will be invariant under these. We remark that it is not evident on heuristic grounds that an appropriate theory should be invariant under these transformations.

Imputations carry three orderings: $x < y$ means $x_i < y_i$ for all i, $x \leq y$ means $x_i \leqq y_i$ for all i and $x_i < y_i$ for some i, $x \leqq y$ means $x \leq y$ or $x = y$. Properly, only $\leqq$ is a partial ordering; every non-void subset of I_N has a greatest lower bound (base), in the sense of $\leqq$. An imputation x in a set A will be called maximal (weakly maximal) if $x \leq y$ $(x < y)$ is false for y in A.

The natural projection of I_N on I_S, $S \subset N$, will be designated P^N_S. $S \cup T$ in super- and subscripts will be written $S + T$. We shall also wish to use central projection along rays. A ray $0x$ is the set of all (necessarily positive) multiples of x in I_N, and similarly for rays with origin not zero; we restrict the term to positively directed rays, i.e., rays xy with $x \leq y$. Observe that the rays through a point in I_T form a $(t-1)$-dimensional simplex, in the natural (quotient) linear structure. Unfortunately this structure is only projective and does not have convex combinations.

Given a situation involving a set N of persons, we wish to describe the set of weakly maximal imputations they can achieve. We suppose (informally) that convex combinations of possibilities are possible, by randomizing with suitable probabilities. We also admit limits, for obvious reasons. In order to keep the limits in utility space, we consider only sets of points whose coordinates are bounded below 1; and we use the term bounded in this sense.

DEFINITION. The top of a bounded set S in I_N is the set of all maximal elements of the convex closure (convex hull of the closure) of S. The weak top of S over x, $W(S, x)$, is formed as follows:

(i) Let $T(S, x)$ be the set of all y in the top of S such that $y \geqq x$;

(ii) Let $W^*(S, x)$ be the set of all z such that

(a) $z \geqq x$;

(b) there is at least one y in $T(S, x)$ such that $z \leq y$;

(c) there is no y' in $T(S, x)$ such that $z < y'$.

Then $W(S, x) = T(S, x) \cup W^*(S, x)$.

3.1. If the set S is the (weak) top of a bounded set (over a point) then it is the (weak) top of itself (over its base; except that the null set has no base).

We postpone proofs, since our assertions are geometrically nearly evident. We illustrate in two dimensions in Figure 1. The curve ab, or any closed connected part of it, is a top. The weak top of ab over x is the curve pq. The weak top of ab over a, however, would consist of a alone; over p, it would be the segment pa.

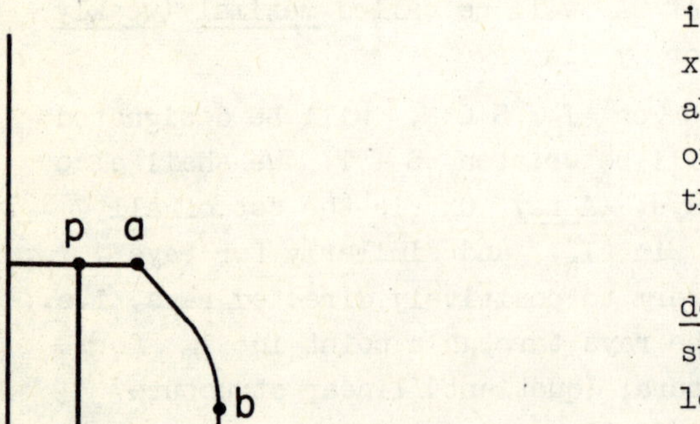

FIGURE 1

Call a weak top W over x degenerate if there is an index i such that for all y in W, $y_i = x_i$; let $Z(W)$ be the set of all such i. Let $J = J(W)$ be the complementary set of $Z(W)$. If $Z(W)$ is empty, call W general.

3.2. A weak top meets every ray through its base. A weak top is degenerate if and only if it contains its base. If W is a general weak top over x then W meets each ray R through x in just one point, $f(R)$; and f is a homeomorphism. If W is degenerate then P^N_J, restricted to W, is an order-isomorphism; and $P^N_J(W)$ is a convex set which is the convex hull of its weak top and its base. Thus a degenerate weak top may be described in complete generality as follows: Take a weak top V over c in I_J. Adjoin some non-empty set of persons, Z, to J; and for each i in Z prescribe a coordinate b_i. Then W consists of all points p in I_N, $N = Z \cup J$, such that $p_i = b_i$ for i in Z and $P^N_J(p)$ is in the convex hull of $V \cup \{c\}$.

Convex (probability) combinations of utilities must be generalized to tops. A convex combination of sets, $\Sigma\alpha_i A_i$, is the set of all points $\Sigma\alpha_i a^i$, a^i in A_i. The <u>top combination</u> of weak tops, $T\Sigma\alpha_i W_i$, $\alpha_i \geqq 0$, $\Sigma\alpha_i = 1$, is the weak top of their set combination over the combination of their bases, with these coefficients. Weak tops with a fixed base are partially ordered by $V \geqq W$ if for no v in V, w in W, is $v \leq w$. A bounded family of weak tops with the same base has a least upper bound, namely the weak top of their union over their base. Now a weak top-valued function φ on a set Z with convex structure is <u>concave</u> if (i) the point-valued function Ψ, where $\Psi(z)$ is the base of $\varphi(z)$, is linear, and (ii) $\varphi(\Sigma\alpha_i z_i) \geqq T\Sigma\alpha_i\varphi(z_i)$, for all convex combinations.

Heuristically speaking, for a set of persons, not agents, a probability combination will be represented by the corresponding top combination; but given a situation involving general players, all we can say is that the relevant functions should be concave. Note further that the set of all quotients of pairs of concave real-valued functions (denominator non-zero) on a compact convex set is precisely the set of all continuous functions. Therefore we shall not have bilinearity or concavity of payoff functions in two-player games.

DEFINITION. A bargaining game $\mathfrak{G} = \mathfrak{G}_N$ on the set N of players is an ordered pair (x, W), where W is a weak top in I_N with base x. The game is <u>free</u> if W is a top.

To play a bargaining game, the players bargain, trying to agree on some point in W. If there is not unanimous agreement, they get x, i.e., the i-th player gets x_i. We propose a point in W, which may be regarded as a value, solution, or arbitrated solution, derived as we have indicated from the principle of preserving absolute utility ratios. There are special difficulties if W is degenerate or if $x = 0$.

DEFINITION. The <u>ratio scheme</u> is the function R defined on all bargaining games $\mathfrak{G} = (x, W)$, whose value, for $x \neq 0$, is the greatest imputation common to the ray Ox and W; $R((0, W)) = W$.

The <u>free ratio scheme</u> is the function F defined on all bargaining games $\mathfrak{G}_N = (x, W)$ as follows:

(i) If $x = 0$ then $F(\mathfrak{G}) = R(\mathfrak{G}) = W$.

(ii) Let $R(\mathfrak{G}) = y^1 \in W$. Inductively, having y^p, let $S_p \subseteqq N$ be the set of those indices i for which there is no z in W such that $y^p \leq z$ and $y^p_i < z_i$. Let $u^p = P^N_{S_p}(y^p)$. If $y^p = u^p$, then $F(\mathfrak{G}) = y^p$. Otherwise let y^{p+1} be the greatest point common to $u^p y^p$ and W.

3.3. The free ratio scheme is well-defined and coincides on free bargaining games with the ratio scheme.

Note that the treatment of degenerate cases is not homogeneous under F or R, and F is not lower semicontinuous. We shall not attempt to settle the degenerate cases.

DEFINITION. A <u>coalition</u> C is an ordered triple (S, Ox, i), where S is a non-void finite set, Ox is a ray in I_S, and $i \in S$. Here S is the <u>ground set</u> (but may be loosely called the coalition), Ox is the <u>ratio</u>, and i is the <u>agent</u>.

The <u>section</u> W^x of a weak top W in I_N by a ray Ox in I_S, $S \subset N$, is the set of all y in W such that $P^N_S(y) \in Ox$. A coalition (S, Ox, i) is <u>admissible</u> relative to a weak top W if W^x is non-void and $x_i > 0$.

The <u>quotient set</u> of N by $C = (S, Ox, i)$, $S \subset N$, is the set $U = (N - S) \cup \{i\}$. The <u>quotient</u> of a weak top W in I_N by C, admissible relative to W, is the weak top W^C of $P^N_U(W^x)$ over its base.

It is contemplated that the collection S agrees, as sketched above, to divide their proceeds in the proportion given by x, and leaves i to negotiate with $N - S$ on behalf of S. We have admitted coalitions which may section a (degenerate) weak top only in its base. We must do so, because these weak tops may appear as part of the apparatus of a game whose other structure makes these coalitions desirable or significant. In fact, to assure admissibility, we shall take all the weak tops in a game with base 0 (except in one-person utility spaces). Then by the first sentence of 3.2, every coalition is admissible.

If W is a degenerate weak top over 0, then sectioning by a coalition contained in $Z(W)$ is a vacuous operation; for all points x in W have $x_i = 0$ for i in $Z(W)$, and thus lie in each such section. Sectioning by a coalition S which meets $J(W)$ and has a non-zero ratio for some person in $Z(W)$ reduces all players in S to zero utility, as is easily seen; thus it amounts to enlarging the degenerate set. We define a <u>general</u> coalition as one which is in neither of these cases.

3.4. Let W be a weak top over 0 in I_N. Let $C = (S, 0x, i)$ be a general admissible coalition; let U be the quotient of N by C. Then $P_U^N(W^x)$ is a weak top over 0, order-isomorphic under P_U^N to W^x.

The conclusion of 3.4 is evidently true also for the former of the two excepted cases; it is always false for the latter. The essential commutativity conclusions of 3.5 are easily seen to be true in all cases.

3.5. If W is a weak top in I_N, $C_1 = (S, 0x, i)$ and $C_2 = (S, 0x, j)$ are general admissible coalitions, U_1 and U_2 the quotient sets of N by C_1, C_2, respectively, $W_1 = P_{U_1}^N(W^x)$+ $W_2 = P_{U_2}^N(W^x)$, then W_1 and W_2 are order-isomorphic to W^x and to each other. If $C_3 = (T, 0y, k)$ is admissible, and $S \cap T$ is void or consists of the one player $i = k$, then the quotient of W_1 by C_3 is identical with the quotient by C_1 of the quotient W_3 of W by C_3.

DEFINITION. An <u>end game</u> $\ni = \ni_N$ on the set N of players is a function on non-void subsets S of N to weak tops W_S in I_S such that

(i) if $s > 1$, W_S is a weak top over 0;

(ii) for disjoint S and T, $x \in W_S$, $y \in W_T$,

there exists z in W_{S+T} such that
$P_S^{S+T}(z) \geqq x$ and $P_T^{S+T}(z) \geqq y$;

(iii) all $W_{\{i\}}$, $i \in N$, are > 0.

The <u>quotient</u> of the end game $\ni = \ni_N$ by the coalition $C = (S, 0x, i)$ is the end game $E = E_U$ defined as follows:

(a) U is the quotient set of N by C;
(b) if $i \notin T \subset U$, $E(T) = \ni(T)$;
(c) if $i \in T \subseteqq U$, $E(T)$ is the quotient of $\ni(S \cup T)$ by C.

Condition (ii) of the definition is the translation into the present context of superadditivity; if S can attain $P_S^{S+T}(z')$ and T can attain $P_T^{S+T}(z')$, the join can do as well by independent action and possibly better. One must show the persistence of the condition in quotients.

3.6. The quotient of an end game by an admissible coalition is an end game.

The weak tops appearing in end games are obviously non-degenerate $(s > 1)$; for none contains its base, 0.

3.7. The subset W of I_S is a non-degenerate weak top over 0 if and only if

(1) W does not contain 0,
(2) W meets each ray through 0 in just one point,
(3) no convex combination of elements of W is strictly greater than an element of W.

An end game-valued function $\ni : x \longrightarrow \ni_x$ is <u>concave</u> if each of its coordinates $\ni^S : x \longrightarrow \ni_x(S)$ is concave.

The role of end games in the theory is to serve as (concave) payoffs to strategies. We do not propose any solution concept special for end games.

§4. PROOFS FOR SECTION 3

PROOF of 3.1. Let T be the top of A', A the convex closure of A', and B the convex closure of T. We have T the top of A and $B \subseteqq A$, by definitions. Hence every element of T is maximal in B. On the other hand, by Zorn's lemma, every element of $A - T$ is less than some element of T; for in a compact set every chain of comparable elements has an upper bound. In particular, no element of $B - T$ is maximal in B. Hence T is the top of T.

Now label the figure as follows: W is the weak top of S over x. Then $W = W(T, x)$, where T is the top of S. We may suppose $T = T(S, x)$. Then T is the top of the convex closure B of W. In fact, an element of B represented as $\Sigma\alpha_k p^k$, $\alpha_k > 0$, $\Sigma\alpha_k = 1$, and each $p^k \in W$, is certainly non-maximal if $p^1 \in W^*(T, x)$. For $p^1 \leq y^1$ for some y^1 in T such that $p^k \leqq y^k$; then $\Sigma\alpha_k p^k \leqq \Sigma\alpha_k y^k$, and the latter expression is $\leqq$ an element of T. Hence if we had $y \leqq \Sigma\alpha_k p^k$ for some y in T, we should have the contradiction $y \leq y^*$ in T.

Further, by definition x is a lower bound of W. For any $x' \geq x$, there is a coordinate i such that $x'_i > x_i$. Now in one dimension $W = T$, because (b) and (c) of (ii) in the definition conflict; then we are done. In $n > 1$ dimensions, take a coordinate index $j \neq i$. Consider the ray R of all points z such that $z_k = x_k$ for $k \neq j$, $z_j \geqq x_j$. The set of all points $z \in R$ such that there is no y in T such that $z < y$ is a non-void set, since R is unbounded; it is closed by definition. Accordingly there is a least point z^o with this property; and then there is $y \in T$ such that $z^o_j \leqq y_j$. With the definition of R, either $z = y \in T$ or $z \leq y$ and $z \in W$. Hence $W = W(B, x)$, and the proof is complete.

PROOF of 3.2. A weak top W meets each ray R through the base of W in the least element of R not less than any element of W, by the same argument as in the preceding proof. If W is general then by 3.1 W contains an element $y > x$, x the base of W, and hence $x \notin W$. If W is degenerate then its base satisfies the defining conditions.

If a ray through x met the general weak top W over x in two points, $z \leq z'$, then for some $\epsilon > 0$, $(1 - \epsilon)z' + \epsilon y$ (y chosen as above) is $> z$, a contradiction. Now since x is not in W, the mapping f^{-1} is the restriction to W of a projection which is continuous on the whole space, less x; it remains to show f is continuous. In fact all such functions are equicontinuous in the interior of the cone. If the ray R be given by direction cosines, $(\lambda_1, \ldots, \lambda_n) = \lambda$, where $f(R) = k\lambda$ has least component $d = k\lambda_1 > 0$, consider R' given by $\lambda' = (\lambda'_1, \ldots, \lambda'_n)$, $f(R') = k'\lambda'$. If $\Sigma|\lambda_j - \lambda'_j| < \epsilon\, \lambda_i$, then $k' \leqq \frac{k}{1 - \epsilon}$, for otherwise $f(R') > f(R)$; also $k' \geqq \frac{k}{1 + \epsilon}$, for otherwise $f(R') < f(R)$. Finally if a sequence of rays R^i in the interior converges to a boundary ray R, then $f(R)$ cannot be less in R than the limit of any subsequence $\{f(R^{u_i})\}$, since it would then be less than some $f(R^{u_i})$ in I_N. Neither can $f(R)$ be greater in R than the limit of such a subsequence, for then a convex combination of $f(R)$ with some $y > x$ in W would be greater in I_N than almost all terms in the subsequence. Hence $\{f(R^i)\}$ converges and has the limit $f(R)$.

Finally, if W is degenerate one readily verifies that it must be precisely the set described below the theorem.

PROOF OF 3.3. It is impossible to have $S_p = 0$. For then for each i there would exist z^i such that $z^i_i > y^p_i$ and for all j, $z^i_j \geqq y^p_j$; the center of gravity of such points would be $z > y^p$, and there would be $t \in W$ with $t \geqq z > y^p$, which is absurd. Further, $S_{p+1} \supseteqq S_p$ by definition, and unless the inclusion is proper, $u^{p+1} = y^{p+1}$. Hence there can be at most as many different y^j as there are players. Finally, if (x, W) is free, i.e., W is a top, then the set S_1 of the definition is N, so that $u^1 = y^1 = R((x, W))$.

PROOF OF 3.4. There is p in W^x such that $P^N_{N-S}(p) = 0$. For let x^* in I_N satisfy $P^N_S(x^*) = x$, $P^N_{N-S}(x^*) = 0$. By 3.1, $0x^*$ meets W in a point p which has the required property. With a symmetric argument, 0 is the base of $P^N_U(W^x)$.

That P^N_U on W^x is an order-isomorphism follows from routine considerations on $S \cap Z(W)$ and the three other Boolean conjunctions of these sets. It follows that $P^N_U(W^x)$ is contained in its weak top over 0 and contains its top, T. If $y \in I_U$ is $\leqq$ some element t of T, and not $<$ any element of T, then the inverse image y^* is $\leqq$ the inverse image t^* and not $<$ any element of W^x. Hence the theorem.

PROOF OF 3.5. The first assertion was established in the proof of 3.4. For the second, each iterated quotient is identical with an object definable in one stage. If S and T are disjoint, let $V = (N - S - T) \cup \{ik\}$, $Z = P^N_V(W^x \cap W^y)$. By the definitions the quotient of W_1 by C_3 is Z. In particular, Z, and hence W^y, is non-void. The condition for z to be an element of the quotient of W_3 by C_1 is precisely that it be an element of Z. For the remaining case, the above argument holds; it happens that $\{ik\} = \{i\}$.

PROOF OF 3.6. By 3.2 if $i \in T$ and $t > 1$, there is p in $\ni(S \cup T)$ such that $P^{S+T}_S(p) \in 0x$ and $P^{S+T}_{T-i}(p) = 0$; and there is q in $\ni(S \cup T)$ such that $P^{S+T}_S(q) = 0$. If $i \notin T$, or $t = 1$, condition (i) persists by definition.

For superadditivity, we may suppose V and W disjoint, both in U, $i \in W$. For $y \in E(V)$ and $z \in E(W)$, we have $y \in \ni(V)$ and $z' \in \ni(W \cup S)$, with $P^{W+S}_W(z') = z$, $P^{W+S}_S(z') \in 0x$. There is an imputation h in $\ni(V \cup W \cup S)$ such that $P^{V+W+S}_V(h) \geqq y$ and $P^{V+W+S}_{W+S}(h) \geqq z'$. Consider the point g in I_{V+W+S} whose projections are y and z';

$h \geqq g$. There is a maximal point f common to Og and $\ni(V \cup W \cup S)$; necessarily $f \geqq g$, and $f \in (\ni(V \cup W \cup S))^x$. Hence $P_{V+W}^{V+W+S}(f) = e \in E(V \cup W)$, and $P_V^{V+W}(e) \geqq Y$ and $P_W^{V+W}(e) \geqq z$.

PROOF OF 3.7. Condition (3) is obviously necessary, and the others are necessary by 3.2. Conversely, if these conditions hold, then each w in W either is in the top of W or $w \leqq x$ for some x in the top of W. Then $w \geqq 0$ and by (3), there is no y in $T(W, 0)$ such that $w < y$; hence w is in $W^*(W, 0)$. Hence W is a subset of its weak top over 0, and by (2) W is all.

§5. TWO-PLAYER ABSOLUTE GAMES

A <u>strategy space</u> is a convex subset of sequential Hilbert space, compact in the weak topology.

DEFINITION. An <u>absolute game</u> $A = A_N$ consists of the following:

(i) a set N of players;
(ii) for each $i \in N$, a strategy space X_i;
(iii) a simplex Q of <u>outcomes</u>;
(iv) a multilinear function F, the <u>outcome function</u>, from $\times_{i \in N} X_i$ to Q;
(v) a concave end-game-valued function G on Q. G is the <u>objective</u>; its values are end games on N.

The composition GF is the <u>payoff</u> E. If $E(x) = \ni_x$, $\ni_x(S)$ may be designated $E(x, S)$. Thus there is a concave weak-top-valued function, also labeled E and also called the <u>payoff</u>, on $X \times 2^N$. The least upper bound W of the set of all $E(x, N)$ is the <u>weak top of</u> A. A is <u>free</u> if W is a top.

The first and main step in the playing of a general absolute game is that the players organize, considering especially the possible quotient games by coalitions on complementary sets of players, S, $N - S$. This is the key idea in the Neumann-Morgenstern theory [7], and represents a sharp contrast with the Nash program [6]. But in treatment of the two-player games we follow Nash [5], and more closely, we follow Raiffa [8], which is in sharp contrast with the treatment of von Neumann and Morgenstern.

The rationale is fairly involved. First suppose the players, 1, 2, choose strategies x^1 from X_1, x^2 from X_2, respectively. The pair $x = (x^1, x^2)$ determines an end game $E(x)$. This end game assigns to 1 a utility $e_1(x)$ and to 2 a utility $e_2(x)$; it also determines a weak top $E_{12}(x)$ in I_{12}. If the players actually play the strategies x^1, x^2, they will arrive at a situation in which they have the respective utilities $e_1(x)$, $e_2(x)$, and by exchange or other cooperative behavior they can reach various imputations in $E_{12}(X)$. These imputations are generally better than $(e_1(x), e_2(x))$; but in general the players could do even better by employing quite different strategies. Accordingly we propose that the players simply threaten to use the strategies x^1, x^2. Once the threats are made, the players are to agree that the ratio of their utilities shall be $e_1(x): e_2(x)$, and to proceed cooperatively to obtain the (maximal) imputation in the weak top of the game, which is determined by this ratio.

Thus we have described the pure threat game A' corresponding to the given game A. We shall not go into a study of the formal properties of pure threat games here. Instead, consider the players to be presented with the game A and the model A'. The latter involves the strategy spaces X_1, X_2, and a payoff function f, defined on $X_1 \times X_2$ to the weak top of A. The value $f(x^1, x^2)$ is the point in I_{12} common to the ray Oe, where $e = (e_1(x), e_2(x))$, and the weak top of A.

The pure threat game A' is closely analogous to Nash's threat game in [5]. However, Nash's method of going from e to f permits him to conclude that the threat game has a saddle point in pure threats. This is not true here. Hence we shall have to use mixed threats, i.e., probability measures on the strategy spaces. To play a probability measure μ^1 on X_1, the player employs a random device to select a threat x^1 in X_1, which he then announces (the announcement being his only move in the game A'). (The same applies for μ^2 on X_2.) Naturally there is a saddle point for A' in mixed threat strategies. But the corresponding value for A' need not be an element of the weak top of A. Therefore we shall have to construct a second-order threat game A''. If A'' has a solution in pure strategies, then the value will have to be in the weak top of A.

A move of the game A'' consists of choosing a mixed strategy in A'. Each player makes such a choice, and the players announce their choices, μ^1, μ^2. The payoff $(f_1(\mu^1, \mu^2), f_2(\mu^1, \mu^2)) = f$, which is the expectation for the players if they use the strategies μ^i in A', is computed; then the payoff in A'' to the strategy pair (μ^1, μ^2) is the point common to the weak top of A and the ray Of.

We can solve a two-player game in the same way relative to arbitration schemes different from the ratio scheme. We have not been able to generalize the definition of quotient games suitably, to obtain a theory of many-person games not tied to the ratio scheme; but we give what general results we have, and conclude by pointing out a special class of games which may be easier to handle.

We leave it to the reader to verify

5.1. In two dimensions, a non-degenerate weak top W over 0 is a topological arc meeting both axes, in points $(a, 0)$ and $(0, b)$. There are constants, c, d, such that (a, x) is in W if and only if $0 \leqq x \leqq c$, and (y, b) is in W if and only if $0 \leqq y \leqq d$. For any (x, y) in W for which $x \neq a$ and $y \neq b$, $x > d$ and $y > c$; and all such (x, y) are strongly maximal in W.

We consider a family of arbitration schemes $\mathcal{T}$ introduced by Raiffa [8]. $\mathcal{T}$ is a family of functions $T_\alpha : R_\alpha \longrightarrow W_\alpha$. For any non-degenerate weak top W_α over 0 in the plane, T_α maps the set $R_\alpha = \{p \mid 0 \leqq p \leqq q$ for some q in $W_\alpha\}$ into W_α in such a manner that

(a) T_α is continuous;

(b) $T_\alpha(e) \geqq e$;

(c) if $T_\alpha(e) = T_\alpha(e')$ then for all r between 0 and 1, $T_\alpha(re + (1 - r)e') = T_\alpha(e)$.

The region R_α is convex and, for any one game, compact; however, it is convenient to think of the "universal" region which is the convex hull of W_α and 0, less the point 0. Let us drop the subscript, and refer to the sets $T^{-1}(x)$ as <u>flow lines</u>.

Each flow line $T^{-1}(x)$ is a compact convex set containing x, except that 0 may be an omitted limit point. Each is contained in the lower left quadrant from x. For x strongly maximal in W, $T(x) = x$; but for x in the end intervals described in 5.1, $T(x)$ may be a greater point in the same interval. If $T^{-1}(x)$ contains neither of the end points of W, then $T^{-1}(x)$ separates R; thus no flow line consists of a single point unless it is an end point of W. For x not an end point, if $T(e) = x$ and $e' = (1 + r)e - rx$ is in R, then $T(e') = x$; i.e., the flow line is the intersection with R of a convex cone with vertex x. Suppose first that $T^{-1}(x)$ separates R, but our assertion is violated by x, e, and e'. There is $f \neq x$ in $T^{-1}(x)$ and in the boundary of R, since the boundary is not separated by one point; and f can be chosen not in W, for otherwise x is in an end interval and the convex set $T^{-1}(x)$ does not cut. Label this conclusion (*): such an f exists. Now the ray xe cuts R, and f is on one side of it, say the upper left side. Let q be the last point in the segment fe' such that $T(q) = x$; we know $q \neq e'$. By continuity there is p in the segment qe' such that the ray pe meets W in a point ℓ below x. Then $T(p) = y$ is below ℓ, as otherwise the segment py would meet $T^{-1}(x)$. Now $T^{-1}(\ell)$

separates R, since ℓ is certainly not an end point; hence $T^{-1}(\ell)$ satisfies (*). But a segment from ℓ to a boundary point must meet either ex or py, which is absurd.

In the other case x is not an end point, by hypothesis; but $T^{-1}(x)$ contains an end point z, since it does not separate R. Replace the segment fe' in the above argument with ze'; one has the same conclusion.

Thus the only exception to our rule is a flow line $T^{-1}(x)$ meeting W only in the end point x. By (b), $T^{-1}(x)$ is contained in the coordinate axis on which x lies. One can easily see that the exception is realized for some T, i.e., $T^{-1}(x)$ can be a proper subsegment of that part of the axis which bounds R.

In summary, (a), (b) and (c) imply that the flow lines are line segments, except that countably many may be wedges bounded by two line segments and a piece of the boundary of R. Thus $T^{-1}(x)$, x not an end point, is determined either by its slope or by the slopes of its upper and lower boundaries. The slope function is right- and left-continuous and satisfies certain inequalities expressing the condition that two flow lines cannot meet in R.

The most obvious choice for $\mathcal{T}$ is a family of lines with constant (non-negative) slope. The horizontal family and the vertical family define the two minorant games, which are of fundamental importance in the Neumann-Morgenstern theory [7]. Another possibility is the family of all lines passing through some point p less than every point of R; with p = 0, this is the ratio scheme. A third possibility is the choice of the (upper/lower) slope of $T^{-1}(x)$ as the negative of the slope of W at x. Nash has shown that this scheme is invariant under independent order-preserving linear transformations of the utilities of the two players, and that this is the only such scheme which satisfies certain other natural restrictions [5].

Let us notice what seems a serious defect in the Nash scheme, not remarked either by Nash or by Raiffa (who does examine the scheme at some length [8]). It is possible to leave the non-cooperative machinery, i.e., the function H, alone, but to enlarge the set of possible exchanges, in such a way as to injure one player. Examples are readily constructed; for T(p) assumes a certain value x if only W contains x and satisfies a certain <u>local</u> condition at x.

The most natural way to exclude this possibility is to define the flow lines once and for all throughout the space I_{12} — as we have done with the ratio scheme. But it is not completely clear that the possibility ought to be excluded. If a single scheme is desired, and if it is to satisfy Nash's Axioms I-VII [5], then it must be the Nash scheme. If, e.g., the scheme is to satisfy (a) - (c) and flow lines are to be invariant under multiplication by scalars between 0 and 1, then (obviously) it must be the ratio scheme. The question indeed can only be settled by constructing theories and comparing their predictions with empirical observations.

To solve the game, A with payoff function H and weak top W_α by means of the scheme $\{T_\alpha\}$, one passes to the game A' with payoff $T_\alpha \cdot H$ in the weak top W_α. Raiffa actually requires W to be a top (his schemes map every point in R to a point in the top of W, so that this is not a special assumption); then he shows that if H is continuous and bilinear, and T_α satisfies (a) and (c) above, then A' has a solution [8]. The sense of the term "solution" is that there is an imputation (V_1, V_2) such that player i has a strategy guaranteeing that he obtains at least V_i, and (hence) that his opponent j obtains at most V_j. Letting W be a weak top, we can draw the same conclusion except for the parenthetic "hence". For A' has an equilibrium point, by Raiffa's argument. If there is an equilibrium point V not weakly comparable with any other element of W, then still any strategy guaranteeing i at least V_i must restrict the opponent j to at most V_j. But if V is weakly comparable with some element V' of W, then either V_1 or V_2 is a maximum by 5.1. Say V_1 is a maximum. Then there is a real number $s > 0$ such that (V_1, t) is in W if and only if $0 \leqq t \leqq s$. Consider the two-player zero-sum game B with the given strategy spaces, X_1, X_2, and payoff $K(x) = (-H_2(x), H_2(x))$. B has a solution and a value t; and $0 \geqq t \geqq -s$ (for player 1 has a strategy assuring that he gets V_1 and hence that 2 gets at most s; but 2 always gets at least 0). Then any good strategies x_1, x_2, in B, satisfy $H(x_1, x_2) = (V_1, -t)$; $H_1(x_1, t_2) \geqq V_1$ for all y_2 in X_2; $H_2(x_1, y_2) \leqq -t$ for all y_2 in X_2; $H_1(y_1, x_2) \leqq V_1$ (trivially); and $H_2(y_1, x_2) \geqq -t$, for all y_1 in X_1.

To state the conclusion neatly, let us call the two-player absolute game A <u>bilinear</u> if its objective function, evaluated for the two one-player sets, is linear; thus the mapping here called H is bilinear continuous. Let us call the scheme $\mathcal{T}$ <u>admissible</u> if it satisfies (a) - (c). Replace the notation A' above with $A'(\mathcal{T})$. Finally

DEFINITION. A two-player game A has a <u>solution</u> if there exist an imputation $v = (V_1, v_2)$ and strategies x_1, x_2 such that

(1) v is in the weak top of A;

(2) for any y_2 in X_2, $H_1(x_1, y_2) \geqq v_1$ and $H_2(x_1, y_2) \leqq v_2$;

(3) for any y_1 in X_1, $H_2(y_1, x_2) \geqq v_2$ and $H_1(y_1, x_2) \leqq v_1$.

5.2. For any bilinear two-player absolute game A and any admissible arbitration scheme $\mathcal{T}$, $A'(\mathcal{T})$ has a solution.

If A is not bilinear, then A' need not have a solution. As indicated above, we introduce mixed strategies in A' which are probability measures over X_1, X_2. Now the values of the payoff are not necessarily in W; but A' becomes a bilinear game. Hence we conclude

5.3. For every two-player absolute game A, and every admissible arbitration scheme $\mathcal{T}$, $A''(\mathcal{T})$ has a solution.

Defining a value for A as a value for A'', we obtain the following result.

5.4. Every two-player absolute game has a value with respect to each admissible arbitration scheme.

Our definition of solutions involves the prediction that a player will seek to minimize his opponent's utility even when he can gain no further profit by doing so. This is clearly the safest assumption for the opponent. It may be defended also on the view that the player's knowledge of the utilities involved is somewhat uncertain; for maintaining pressure on the opponent cannot be a losing policy and may be a winning one.

We conclude with a few words on a special class of two-player games. Let us provisionally call such a game <u>independent</u> if the threat point, i.e., the payoff in A to a pair of strategies good in A'', is independent of the scheme $\mathcal{T}$ (provided $\mathcal{T}$ is admissible). Obviously (a) the threat point and the weak top of an independent game constitute a quite full description of the game, (b) a pure bargaining game is independent, (c) a game having a solution is independent. In fact, conditions (2) and (3) in the definition of a solution suffice.

These conditions are not necessary, as is shown by the 2×2 bilinear game with $G(x_1^1, x_2^1) = (3/4, 3/4)$, $G(x_1^2, x_2^1) = G(x_1^1, x_2^2) = (1/2, 1/2)$, $G(x_1^2, x_2^2) = (1/4, 1/4)$. Either player can secure $1/2$ or hold his opponent to $1/2$, but he cannot do both at once. Nevertheless for any line through $(1/2, 1/2)$ with positive slope, each player has a (pure) strategy assuring that the payoff is on the side of that segment which is favorable to him.

Precising the definition, we have

DEFINITION. The two-player game A is <u>independent</u>

> if there exists a point (t_1, t_2) such that for every a between 0 and 1 there exist strategies x_1, x_2, such that $aG_1(x_1, y_2) - (1 - a)G_2(x_1, y_2) \geqq at_1 - (1 - a)t_2 \geqq aG_1(y_1, x_2) - (1 - a)G_2(y_1, x_2)$, for all y_1 in X_1 and all y_2 in X_2.

In view of our characterization of admissible schemes, this implies that for each $\mathcal{T}$ there are strategies good in $A''(\mathcal{T})$ (namely those given for a, where $(1 - a)/a$ is the slope of the flow line through (t_1, t_2)), which lead to the threat point (t_1, t_2).

Perhaps we should remark that a convex combination of games, $\Sigma\alpha_i A_i$, is played by beginning with a chance move, which selects the game to be played next, A_i having probability α_i.

> 5.5. A convex combination of independent games is independent.

For each a we can use the given strategies; for the required inequalities are preserved under convex combination.

§6. THE PROJECTIVE FORM

There remains one detail preliminary to representation of a general absolute game in the projective form. A coalition has somewhat greater strategic resources than do its members acting independently; the strategies of the several players can be correlated. In the case of finite multilinear games, the simplex of correlated pure strategies for a coalition of players having $m_1, \ldots, m_n$ pure strategies, respectively, has $\Pi_{i=1}^{i=n} m_i$ vertices, one for each choice of vertex for each player in the coalition. This is a tensor product (precisely, the convex hull of the image of the Cartesian product of the individual strategy spaces in the familiar tensor product of any linear spaces containing them).

> DEFINITION. The <u>tensor product</u> M of a finite set of strategy spaces, $S_1, \ldots, S_n$, is the convex closure of the range of the transformation T defined herewith. Let coordinates in S_i be x^i_j; let A be the collection of all functions α on $\{1, 2, \ldots, n\}$ to the union of the natural numbers and zero. For each α in A, define T_α on $\times S_i$ by $T_\alpha(x^1, \ldots, x^n) = \Pi_{i=1}^{i=n} f^i_\alpha(x^i)$, where if $\alpha(i) \neq 0$, $f^i_\alpha(x^i) = x^i_{\alpha(i)}$, but $f^i_\alpha \equiv 1$ otherwise. Then $T(x^1, \ldots, x^n)$ is the point, in the sequential Hilbert space with coordinates indexed by A, whose α-th coordinate is $T_\alpha(x^1, \ldots, x^n)$.

6.1. The tensor product M of a finite set of strategy spaces S_i is a strategy space. Every multilinear function F on $\times S_i$ induces a linear function F^* on M; where $p \in M$ is a convex combination $\Sigma c_i T((x_i^k))$, the value $F^*(p) = \Sigma c_i F((x_i^k))$ is independent of the representation. F^* is continuous if and only if F is continuous.

PROOF. Since a subset of sequential Hilbert space is conditionally compact if and only if it is bounded, M is a strategy space. The arbitrary multilinear function F is simply a linear combination of T_α, $F = \Sigma d_\alpha T_\alpha$. If $\Sigma c_i T((x_i^k)) = \Sigma e_j T((y_j^k))$, then in particular $\Sigma c_i T_\alpha((x_i^k)) = \Sigma e_j T_\alpha((y_j^k))$. The expressions $\Sigma_i (c_i F((x_i^k)) = \Sigma_\alpha d_\alpha \Sigma_i c_i T_\alpha((x_i^k))$ and $\Sigma_j e_j F((y_j^k))$ are therefore equal. The equivalent condition for continuity of F or of F^* can be expressed in terms of $F = \Sigma\, d_\alpha\, T_\alpha$ as follows. Let k_α be the diameter of the range of T_α. Then $\Sigma k_\alpha |d_\alpha| < \infty$.

DEFINITION. The quotient of the game A_N by the coalition $C = (S, 0x, i_o)$, $S \subseteqq N$, consists of

(i) the quotient of N by C;
(ii) for each i in $N - S$, the strategy space X_i from A_N; for i_o, the tensor product of the strategy spaces of the members of S;
(iii) the same outcome space as in A_N;
(iv) the multilinear function constructed after 6.1 from the outcome function of A_N;
(v) the function whose values are the quotients by C of the corresponding values of the objective of A_N.

The extension of the weak top of the quotient game over N is the set of all imputations z in the weak top of A_N such that $P_S^N(z) \in 0x$.

A partition of the game A_N is an iterated quotient game of A_N having two players.

Observe that the threat models for two-player games may be regarded as games played to determine what one-player quotient game will be formed. By 3.5 (if agents have non-zero coordinates in the ratios of their coalitions), the extension of the weak top W of a quotient game is order-isomorphic to W, and the order of operations in forming iterated quotients is immaterial. Especially, the quotients of a free game are free. We define the value of a two-player game as the value with respect to the ratio scheme (5.4).

DEFINITION. The <u>projective form</u> of the free absolute game A_N is a function D on non-void subsets of N to subsets of the weak top W of A_N, defined as follows: $D(N) = W$; for S a proper subset of N, D(S) is the set of all imputations z in W such that there is a partition P of A_N by coalitions with ground sets S and N - S, both agents having non-zero components of the ratios, such that z is the point in the extension of the weak top of P which corresponds, under the natural isomorphism, to the value of P.

By definition $D(S) = D(N - S)$. The values of D are connected sets generally homeomorphic with the simplex of dimension $n - 2$. The proof follows. Consider the point p in $D(S) = D(N - S)$, having projections x and y in I_S, I_{N-S}, respectively. Suppose S changes from the ratio Ox to the ratio Ou. Every strategy available to S in one partition determines a strategy in the other as follows: Let σ assure S of the imputation θx; then carrying out the operations prescribed by σ, and then having certain members throw away utility to reach the ray Ou, assures S of an imputation τu not $< \theta x$. From this we can conclude by continuity that the imputation q in D(S) whose projections lie in the rays Ou, Oy, is not strictly less than p. From this and the symmetric inequality for N - S one concludes by the same sort of argument as in 3.2 that D(S) is arcwise connected. If there is one point $p > 0$ in D(S) then an analogous argument shows that D(S) is homeomorphic with the Cartesian product of the simplexes of dimensions $s - 1$ and $n - s - 1$.

There is a weak superadditivity condition, which we present in connection with the concept of <u>domination</u>. In the additive case of von Neumann [7], an imputation x is said to dominate y with effective set S if $P^N_S(x) > P^N_S(y)$ and S can attain $P^N_S(x)$. Here what S can attain depends on what N - S is doing.

DEFINITION. Relative to a projective form D, the imputation x is <u>attainable by</u> S if there exists z in D(S) such that $P^N_S(z) \geqq P^N_S(x)$ and $P^N_{N-S}(z)$ and $P^N_{N-S}(x)$ are collinear with the origin. Then x <u>dominates</u> y with <u>effective</u> set S if x is attainable by S and $P^N_S(x) > P^N_S(y)$.

6.2. If x is attainable by the two disjoint sets S, T, then x is attainable by $S \cup T$.

PROOF. If $x = 0$, or $S \cup T = N$, the fact is evident. Otherwise x is attainable by S if and only if the partition by S and $N - S$ on ratios taken from x gives S at least $P^N_S(x)$; thus we reduce to a three-player game. Let $1 \in S$, $2 \in T$, $3 \in N - S - T$, and form the quotient by these three coalitions with the indicated agents, ratios the projections of x, and the agents so chosen that their components in the ratios are positive.

Now player 1 must be able to assure the ratio $x_1/(x_2 + x_3)$ if 2 and 3 restrict themselves to the subspace of their joint strategy space which consists of uncorrelated strategies. Similarly for player 2; and it follows that 1 and 2 can attain x without troubling to correlate their strategies.

Relative to a projective form (characteristic complex) von Neumann and Morgenstern define a class of admissible imputations [7]. (They use the term imputation with this sense and no other.) We generalize their remark, a point not in $D(N)$ cannot dominate a point in $D(N)$, and if $x \leq y$, y in $D(N)$, then any z dominating y also dominates x. Thus we lose no generality by restricting admissible imputations to $D(N)$. We restrict further: An imputation x in $D(N)$ is admissible if for each i in N, x is attainable by $N - \{i\}$. If this condition were applied to the weak (non-constant-sum) games of Neumann-Morgenstern, it would be absurd; but in our formulation there are no weak games and the conditions of admissibility form a direct generalization of those of Neumann-Morgenstern. For a criticism of this last restriction, see [11].

Let the class of admissible imputations relative to A be designated $I(A)$. Then a stable set [7; there called a solution] is a subset V of $I(A)$ such that (1) no element of V dominates another element of V, but (2) each element of $I(A) - V$ is dominated by an element of V. (The term stable set is due to Shapley [11]. A stable set is closed, by the same argument as in [7]. It is not known whether stable sets always exist even in additive games.)

6.3. A free two-player game has a unique stable set consisting of its value.

The proof: $I(A)$ consists of the value.

§7. THREE-PLAYER ABSOLUTE GAMES

Free absolute games of three players have essentially the same stable sets as do constant-sum three-person games in the classical theory. Note that the projective form consists of a two-dimensional surface together with three curves, each associated with a player from whose point of view his curve is convex. The problem of finding a main solution is the problem of inscribing a triangle in the skew triangle formed by these curves, with sides parallel to the coordinate planes.

> DEFINITION. A main solution of the free three-player game A on $N = \{1, 2, 3\}$ is a set of three imputations, x^1, x^2, x^3, in the weak top of A, such that
>
> (i) $x_1^2 = x_1^3$, $x_2^1 = x_2^3$, $x_3^1 = x_3^2$;
>
> (ii) each x^i is in $D(i)$.

For convenience we call a single point a main solution if it lies in all $D(i)$. Note that if two of the three points coincide, then all three must coincide.

> 7.1. A main solution of a free three-player game is a stable set.

PROOF. There is no domination within the solution; never does more than one player prefer a change, and since the point he dislikes is attainable by his opponents, he is ineffective. Conversely, for any y in $I(A)$ consider the set of those players i for which $y_i \geqq x_i^k$, $k \neq i$. If there are not two such players then y is dominated by an appropriate x^k. But if, say, $y_1 \geqq x_1^3$ and $y_2 \geqq x_2^3$, then $y_3 \leqq x_3^3$ since A is free; hence $y = x^3$, because x^3 is in $D(3)$ and y in $I(A)$.

> 7.2. A free three-player game has at most one main solution.

The proof is omitted.

We give below a construction for a main solution which is feasible if only the weak top of A is non-degenerate; but neither 7.1 nor 7.2 need hold in this generality. Project each $D(\{i, j\})$ into the ij-plane, obtaining sets E_{ij} as images. Now each E_{ij} is a continuous curve and in fact a weak top over 0. In the 23-plane, let F_{23} be the set of all those points (x_2, x_3) such that there is x_1 such that (x_1, x_2) is in E_{12} and (x_1, x_3) in E_{13}. Then F_{23} is a continuum any two points of which are weakly comparable. One easily shows by examining the ends of F_{23} that they lie on opposite sides of E_{23}, whence E_{23} and F_{23} meet.

However, there is no reason for the intersection to be a single point — unless the game is free. Note also that even if the intersection is a single point, the three imputations corresponding need not form a stable set. A later example will show this.

We have indicated the proof of

7.3. A free three-player game has a main solution (possibly a single point).

Now the arguments of von Neumann [7] on the family of all stable sets go through without essential change. The essential point (his 32:5) is that if two imputations, x, y, differ in all coordinates, then one dominates the other. For x or y is greater in two coordinates, and the two-player sets can attain every point in $I(A)$.

Referring to the main solution $\{x^i\}$, let $b_i = x_i^i$, $s_i = x_i^j$ for $j \neq i$. Referring to the vertices $z^1 = D(2) \cap D(3)$, and z^2, z^3, similarly defined, let $c_i = \max_{j \neq i} z_i^j$. Observe that the condition $c_i \leqq c \leqq z_i^i$ is necessary and sufficient for the plane $y_i = c$ to meet both sets $D(j) \cap I(A)$, $j \neq i$. Further, if $I(A)$ is not a single point (which we suppose for the following discussion) then for each i, $b_i \leqq c_i < s_i < z_i^i$.

A stable set V must contain two points, x, y, and these must have a common coordinate, say $x_1 = y_1$. Either for all z in V, $z_1 = x_1$, or not. In the latter case von Neumann's argument, pp. 285-287, shows that V must be the main solution. In the former case the entire line $z_1 = x_1$ (in $I(A)$) must be in V, since no two points on the line dominate each other. The argument of pp. 287-288 (with modifications and a slight addition apropos of c_1) shows that the necessary and sufficient condition for stability is $c_1 \leqq x_1 < s_1$. In summary, we have

7.4. For any free three-player absolute game A, either $I(A)$ consists of one point which is the unique stable set, or $b_i \leqq c_i < s_i$ and the stable sets are precisely (a) the main solution and (b) the lines $y_i = c$, $c_i \leqq c < s_i$.

Here is the example for the remark preceding 7.3. The players 1, 2, 3, are each to choose a partner; if no pair chooses each other, their utilities for the result are (.4, .7, .7). If 1 and 2 choose each other the result is (.5, .8, .7); if 1 and 3 (.5, .7, .8); if 2 and 3, (.4, .75, .75). No transfer of utility is possible; the weak top W of the game is the weak top of the set of these three imputations over 0. The formal conditions for a main solution can be shown to be met uniquely by $x^1 = (.4, .75, .75)$, $x^2 = (.5, .7, .75)$, $x^3 = (.5, .75, .7)$. Clearly $(.5, .8, .7)$ dominates x^1 and is dominated by no point whatsoever. In fact, each point $(.5, .7 + t, .8 - t)$, $0 \leq t \leq .1$, has this property, and the set of all those points is the unique stable set.

The stable set in the above example is a <u>core</u> [2]. But it is easy to show that free absolute games, in our theory, never have cores unless I(A) consists of one point; and the existence of cores is contrary to the spirit of this theory. Note also that the example has the characteristics to be expected in the classical theory if one adds a dummy to a weak two-player game. But addition of a dummy to such a game in our theory yields various results, none of which is as above. We shall illustrate this point. We take leave of the preceding example with the note that its explanation appears to be a serious problem which must be disposed of before any extensive class of non-free games can be successfully treated.

Consider now the game in which 1 and 2 have utilities .5 each and can attain by cooperation $(.9, .7)$, $(.7, .9)$, or any point in the weak top of this pair over 0. Evidently the value (hence the unique stable set) is $(.8, .8)$. If a player 3 is adjoined to the game, having no part in the rules, with no transfer of utility possible from him to 1 or 2, then evidently he does not affect the game. We state the results for three cases of transferable utility; the computations are quite simple and are omitted. For convenience let 3 begin with utility .5.

If 3 can transfer and receive utility at a rate of one-for-one, within bounds (say $\pm .1$), then $s_1 = s_2 = .8$, $s_3 = .6$; $b_1 = b_2 = .7$, $b_3 = .5$. Thus in effect 3, as a source of backing for 1 or 2 in bargaining (with 2 or 1), has freed one-third of the disputed gross of .6; .2 must go to 1 (in addition to his basic .5), and .2 to 2, but the distribution of the remaining .2 depends on the coalition pattern.

The result varies monotonically and continuously with the exchange rate. If 3 must sacrifice 8k of utility in order to increase the utility of 1 or 2 by k, but the take (the .6) is of equal value to 1, 2, or 3, then the result is approximately $s_1 = s_2 = .8$, $s_3 = .53$; $b_1 = b_2 = .77$, $b_3 = .5$. Thus in this case the availability of 3 makes indeterminate the disposition of only ten per cent of the gross profit.

If 1 finds 3 more useful than does 2 then 1's position in the game improves. Suppose 3 can transfer utility to 1 at a one-for-one rate; 3's base utility cannot be transferred to 2 at all; but the profit is still equally valuable to all. Then approximately $s_1 = .83$, $s_2 = .77$, $s_3 = .57$; $b_1 = .77$, $b_2 = .70$, $b_3 = .50$.

In all cases $b_3 = .5$, as it should, for 3 has no possible part in the game except as a member of a coalition, and b_3 is his utility when the coalition $\{1, 2\}$ forms against him. In all cases the "freed" part of the gross profit (.20, .06, .13, in the three cases) is equally divided between the members of the victorious coalition. In no case is there such irresolvable indeterminacy as in the degenerate example previously considered. The present examples are not precisely free games; but their weak tops are "free in the region of $I(A)$", so that their solutions are naturally like the solutions of free games. In fact, each could be replaced by a more complicated free game having the same set $I(A)$.

Having granted that the first example below 7.4 represents a serious difficulty we may observe the supporting evidence for the present theory obtained from the next examples, provided our solutions of them be regarded as reasonable. It seems that such solutions cannot be derived unless some sense is provided in which a two-person player has more weight than a one-person player. Our choice of the ratio principle is merely the most obvious device permitting this; in particular, it is an advantage to be able to form coalitions with linear sections and projections. Another particular: To attribute a fixed weight to each player, independent of his position or threatened position in the game, as in [10], would not serve. In fact it would introduce effects of 3's presence in the above family of games even in the case that he cannot fulfill his only proper function of offering backing to 1 or 2, i.e., in the case that his base utility is totally non-transferable. This observation seems to have scotched our hope of building a simpler approximate theory, in which the fluctuations of the game are considered to be negligible in comparison with the base utilities of the players.

We adjoin a tentative list of further impossibilities. One cannot replace end games, in the values of the objective function, by their projective forms. One cannot even form probability combinations of projective forms of two-player bargaining games, because the value of the combination depends on the base points.

One can single out a class of strong games, as in the classical theory. But the most pressing reason for studying such a class is the hope of extending the Shapley value [12]. This cannot be done by Shapley's methods, because a convex combination of strong games need not be strong. To extend the value to all absolute games at one swoop seems infeasible. It might be done in two, if the first swoop involved quite remarkable results on games in functional form. But it appears more promising to extend piecemeal, beginning with a symmetric value for symmetric games.

A reading of chapter titles in [7] suggests the study of four-player games, simple games, and composition of games. All these appear to be promising avenues for research. Note that the above examples show that composition with side payments yields a game whose projective form is not determined by the projective forms of the parts.

A final caution concerning reductions: One cannot even reduce players' strategy spaces to essential strategies. For let player 1 decide whether the outcome is (.3, .5) or (.6, .9). Clearly the first alternative is useless as a threat (assuming the ratio principle) and can be ignored in playing this game. But consider the game which consists of playing this game with probability 1/2, or otherwise (with probability 1/2) receiving (.5, .5). If some transferability exists then it will be advantageous for 1 to threaten (.3, .5); for then the expected result is (.4, .5), a more favorable ratio for 1 than the alternative expectation (.55, .7). Thus the projective form is no more than projective, and in particular it loses convex structure. But as the example shows, this is not a defect peculiar to the form; it is inherent in the ratio scheme. The example shows that the value of a two-player game is not a sufficient description of the game, for combinations; therefore a fortiori the values of two-player quotient games cannot give a sufficient description of the general game.

We conclude with a solution of a problem posed by Shapley in lectures at Princeton in 1953.

Three players, I, II, III, engage in a statutory duel. We present the game diagram in a modified, transparent form (Figure 2). There is a chance move selecting the player to fire next at his choice of the other two players. Each player has equal chance, but the accuracies, i.e., probabilities of causing death to the target, are respectively 4/5, 3/5, 2/5. A superficial oddity is that if each player single-mindedly maximizes the effect of his shot, the one with the highest accuracy comes out worst. The unique non-cooperative solution, in fact, has I firing at II and both II and III firing at I, with the result (.297, .333, .370). (For convenience we value survival at 1.)

Referring to the definition of an absolute game, we know the set of players. The strategy spaces are large, but it fairly obviously suffices to consider stationary strategies. There are three final outcomes, of the form "i survives"; but we may cut off those parts of the game tree in which only two players survive, since there are no personal moves in those subgames. The values of the two-player subgames are easily seen to be the accuracy ratios of the respective pairs; thus in Figure 2 the point P may as well be labeled (4/7, 3/7, 0). These values are used in computing the outcome function below. In the present case the objective function is trivial. (The principal purpose of the objective function is to describe the conditions governing side payments.)

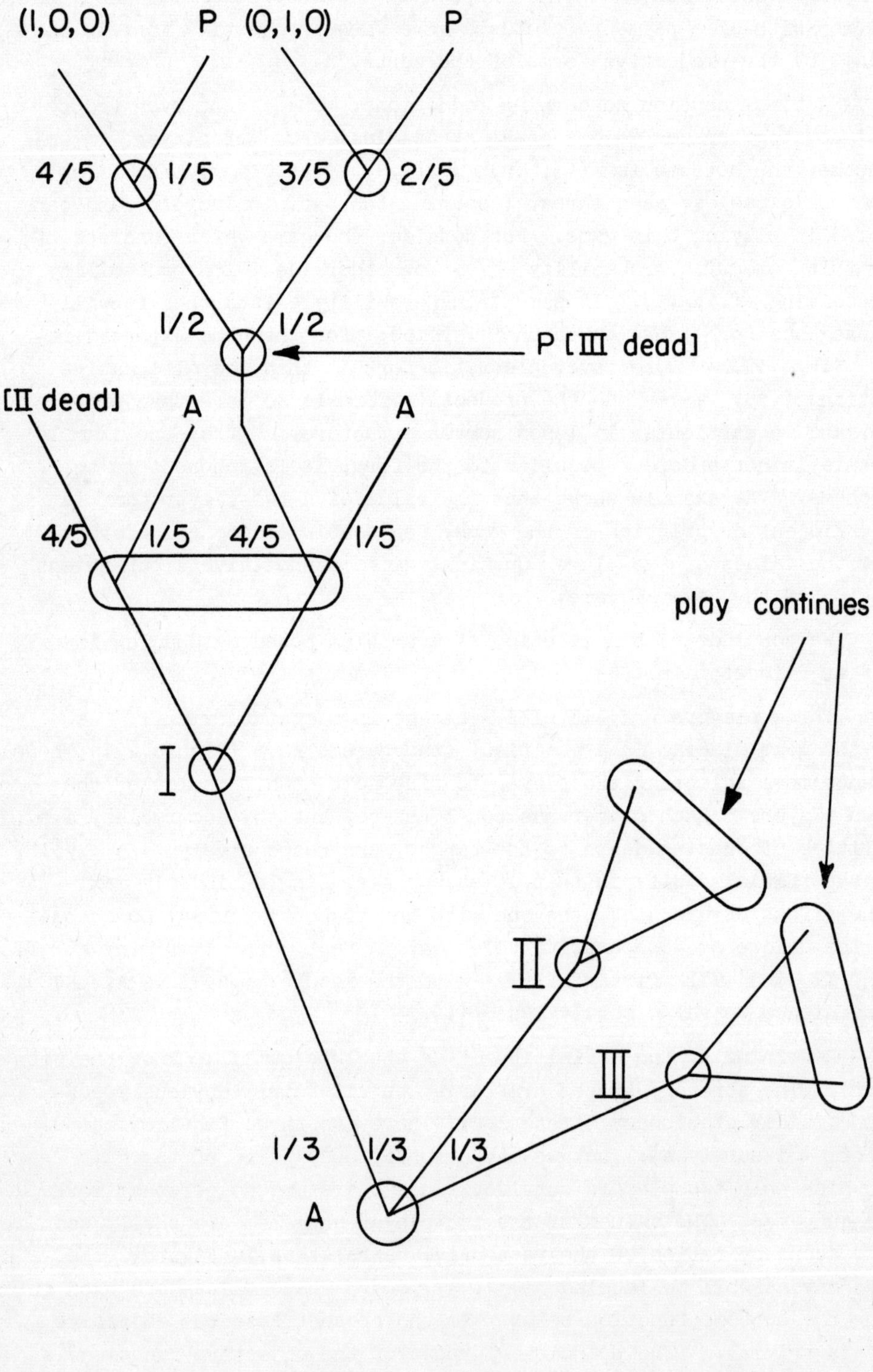

(1,0,0)
P
(0,1,0)
P
4/5
1/5
3/5
2/5
1/2
1/2
P [III dead]
[II dead]
A
A
4/5
1/5
4/5
1/5
play continues
I
II
III
1/3
1/3
1/3
A

FIGURE 2

There are eight triples of pure strategies. Using Arabic numerals to cite the targets in order, we find these values:

211, (.297, .333, .370); 212, (.444, .200, .356);
231, (.487, .276, .237); 232, (.635, .143, .222);
311, (.254, .524, .222); 312, (.402, .390, .208);
331, (.444, .467, .089); 332, (.593, .333, .074).

Now since the accuracy of each player is independent of the target, the values depend linearly on strategy mixtures. Thus the weak top of the game is the weak top of this eight-point set over 0. The curves D(i) are broken lines connecting the following points:

Coalition of I and II

x_1 =	0	.254	.444	.607	.635	.635
x_2 =	.524	.524	.467	.270	.143	0
x_3 =	.222	.222	.089	.123	.222	.222

Coalition of I and III

x_1 =	0	.297	.444	.606	.606
x_2 =	.333	.333	.200	.274	.274
x_3 =	.370	.370	.356	.120	0

Coalition of II and III

x_1 =	.297	.297	.424	.424
x_2 =	0	.333	.473	.473
x_3 =	.370	.370	.103	0

The fourth point under "I and II", for example, is a linear combination of 232 and 332. If I and II agreed on the ratio 607:270, they would threaten III with an appropriate point (.444, x, .089), which they can force upon him. The ratio 607:123 is the same as 444:89, to the present accuracy. Observe that each curve is convex toward the odd player.

Using the method of the proof of 7.3, we observe that I and II get .606, .473, respectively in coalitions with III on a 1:0 ratio. Continuing, we construct the broken line F_{12}, given below with the parameters x_3.

Offers by III to I and II

x_1 =	.606	.606	.606	.444	.297	0
x_2 =	.473	.473	.464	.340	.333	0
x_3 =	0	.103	.120	.356	.370	.370

The intersection of F_{12} and E_{12} is (.508, .389), to three places. Hence the main solution is $\{x^1, x^2, x^3\}$, where

$$x^1 = (.348, .389, .263)$$
$$x^2 = (.508, .229, .263)$$
$$x^3 = (.508, .389, .103) \ .$$

We recapitulate the arguments supporting this solution. To begin with, the players are not supposed to play non-cooperatively; if they were, that would be a different physical situation.

Given that the players wish to make deals, and no special requirements on these deals being physically specified, we impose none. It is not generally to be expected that a problem in cooperative behavior will have a unique solution. We seek rather a stable set of imputations. "No imputation y ... conforming with the 'accepted standard of behavior' — can be upset — i.e., dominated — by another imputation x of the same kind. On the other hand ... the 'standard of behavior' can be used to discredit every non-conforming procedure..." This part of the argument is due to von Neumann and Morgenstern ([7]; quotation from page 41).

The entire range of possibilities must be explored for stable sets — for which, we recall, neither existence nor uniqueness is known to hold. However, the special advantages of a main solution are apparent. In the present theory it is not clear whether main solutions will generalize significantly beyond three-player games.

For the interpretation of "possibility" we are indebted to Nash [5]. Precisely, probability combinations are older, and throwing away utility is an innovation in this paper. Nash's argument applies when the battle lines are drawn; we say that S and N - S will agree on the imputation x, chosen from the optimal possibilities, if they have threats (strategically best) such that they will receive the same relative injuries if they carry out their threats.

The general strategic problem is unsettled. However, in the present example it is clear that the threats involved are possible and best possible. For example, the point x^3 of the main solution is reached by a probability combination of the strategy triples 331, 232, and 332; this requires cooperation from the victim, III. But if I and II both fire at III, the result must be a probability combination of (.444, .467, .089) and (.593, .333, .074). Then II is engaged to throw away utility so that I and II emerge with the ratio 508:389. There is no relative loss, and there is an absolute gain, to III if he agrees to x^3 as proposed. (It cannot be denied that we require a high degree of sophistication of the players in our games.)

Finally, the interpretation of equal injury in terms of ratios of absolute utilities is essential to the justification. We conclude with a pat mythical explanation of this view. Take the closures of the individual utility spaces, $[0, 1]$, or [Hell, Heaven]. The utility u is now equivalent to a lottery ticket giving probability u of Heaven and $1 - u$ of Hell. If several persons lose utility in the same proportion, then they are victims of a uniform devaluation of Heaven. The injuries would appear to be equal.

§8. ADDITIVE GAMES

At this point the theory appears to have gone far enough to invite comparison with the previous theory in the restricted domain of the latter. The assumption which is called for is essentially that all end games involved are functions to a set of parallel hyperplanes. More descriptively, suppose that the i-th player has a utility x_i below which he cannot be depressed in the game under study, that he has a large supply of money, and that in the relevant range his utility for money is linear, at the rate of λ_i per dollar. Then the goods with which the game deals need not be money; if an object is worth p_i to the i-th player, it is really worth $\lambda_i \max_j (p_j/\lambda_j)$, since he can sell it to any other player. (A sale may be another game; but the object is worth the same regardless of the conditions of the sale.) Accordingly we transform from absolute utilities, dividing the i-th player's scale by λ_i, so that the numbers appearing represent money freely transferable at unit exchange rate between players, and for a single player the ratio of the adjusted utilities of two outcomes is the same as the ratio of the corresponding absolute utilities.

We do not define large supplies and relevant ranges, but simply set up another form of game. Instead of a projective form there will be a characteristic function, which also takes the place of an end game. This means that the outcomes of a game can be other games, as in the non-cooperative theory.

DEFINITION. A <u>characteristic function</u> v on the set N of players is a non-negative real-valued function on the set of all subsets of N satisfying

(a) $v(N) > 0$;

(b) if S and T are disjoint,
$v(S \cup T) \geqq v(S) + v(T)$;

(c) $v(S) + v(N - S) = v(N)$.

An <u>additive game</u> $G = G_N$ consists of the following:

(i) a set N of players;

(ii) for each $i \in N$, a strategy space X_i;

(iii) a simplex Q of outcomes;

(iv) a multilinear function F from $\times_{i \in N} X_i$ to Q;

(v) a linear characteristic-function-valued function H on Q.

The composition HF is the <u>payoff</u> E.

The <u>quotient</u> of the characteristic function v on N by the coalition $C = (S, 0x, i)$ is the function v^* on subsets of the quotient set of N by C defined by $v^*(T) = v(T)$ if $i \notin T$, $v^*(T) = v(S \cup T)$ if $i \in T$.

Quotients and partitions of additive games are defined just as for absolute games, using quotient characteristic functions instead of quotient end games. Partitions can be evaluated just as before with $E(x, \{i\})$. (The old symbol has a new meaning.) 5.2 applies directly, since the partitions are already bilinear. Thus there is a value (v_1, v_2) for this game and there are single strategies, x^1, x^2, assuring that the ratio of the payoffs to the players in the original game is at least v_1/v_2, at most v_1/v_2, respectively.

DEFINITION. A <u>value</u> of a two-player additive game is a vector $(v_1, v_2) = v$ maximizing $v_1 + v_2$, such that player 1 has a strategy x^1 assuring that his share of the payoff is at least $v_1/(v_1 + v_2)$, and player 2 a strategy x^2 assuring the reverse.

8.1. A two-player game has a unique value.

The proof has been indicated; a value is unique on the face of the definition. By the definitions, the value of a partition is independent of ratios and agents. If (v_1, v_2) is the value of a game G, call v_1 the value of 1 in G, v_2 the value of 2.

DEFINITION. The <u>characteristic function</u> v_G <u>of the</u> (additive) <u>game</u> G is defined as follows: $v_G(0) = 0$; $v_G(N)$ is the maximum of $v(x, N)$, v a value of the payoff of G; for S a proper subset of N, $v_G(S)$ is the value of S in the partition of G into S, $N - S$.

8.2. The characteristic function of an additive game is a characteristic function.

The proof is omitted.

The principal solution concepts for characteristic-function games are stable sets, values, and cores. We have no more to say on stable sets, and the present (constant-sum) characteristic functions have no cores. We shall just show that the axiomatic justification of the Shapley value holds when restricted to constant-sum functions.

DEFINITION. (Shapley). The value of a characteristic function v is a function ϕ on players. If v is a characteristic function on N,

$$\phi(i) = \sum_{S \subseteqq N} \frac{(s-1)!(n-s)!}{n!} [v(S) - v(S - i)] \quad .$$

The value of an additive game is the value of its characteristic function.

Clearly this definition agrees with our definition for two-player games.

In Shapley's theory characteristic functions are not required to satisfy our condition (c), the constant-sum condition. Accordingly he has a large linear space for which he exhibits a special basis [10]. If R is a set of players, let e_R be the function on sets S of players defined by $e_R(S) = 1$ if $R \subseteqq S$, otherwise $e_R(S) = 0$. Every characteristic function has a unique linear representation in terms of the functions e_R. Accordingly if the value ϕ of e_R has value $1/r$ on elements of R and 0 on other players, and ϕ is to be a linear operator, all values of ϕ are determined and one obtains Shapley's formula. This is reasonable, since e_R is the characteristic function of a pure bargaining game.

In the present theory e_R is not a characteristic function if $r > 1$. However, all characteristic functions in the present sense are linear combinations of the characteristic functions of direct majority voting games on an odd number of players.

Precisely, let F_N be the set of all real-valued functions on subsets of N vanishing on the empty set; F_N is a vector space of dimension $2^n - 1$. The set G_N of all superadditive functions in F_N is a convex cone of full dimension, since it contains the basis consisting of all e_R, $R \subseteqq N$. Let C_N be the set of all characteristic functions in G_N. For each $R \subseteqq N$ with r odd let m_R be the element of C_N which has value 1 on those sets containing a majority of the members of R and 0 in all other sets.

8.3. The set C_N is a convex cone of dimension 2^{n-1} linearly generated by the functions m_R.

PROOF. C_N is a convex cone with vertex 0, i.e., if f and g are in C_N and t and u are positive real numbers, then tf + ug is in C_N. The equations (c) of the definition include 2^{n-1} linearly independent ones; thus the dimension of C_N is at most 2^{n-1}.

There are 2^{n-1} subsets of N of odd power. There are $\binom{n}{s}$ of power s, $\Sigma_{s \leqq n} \binom{n}{s} = 2^n$ and $\Sigma_{s \leqq n} (-1)^2 \binom{n}{s} = 0$. Now each m_R can be represented in terms of the functions e_R by Shapley's formula [10]

$$m_R = \sum_{\substack{S \subseteqq R \\ s \geqq \frac{r+1}{2}}} e_S \sum_{\substack{T \subseteqq S \\ t \geqq \frac{r+1}{2}}} (-1)^{s-t} .$$

This expression involves those e_S for which S is a proper subset of R, and, with non-zero weight, e_R itself. In fact,

$$\sum_{\substack{T \subseteqq R \\ t \geqq \frac{r+1}{2}}} (-1)^{r-t} = \sum_{\frac{r+1}{2} \leqq t \leqq r} (-1)^{r-t} \binom{r}{t} ,$$

which is

$$\sum_{t \leqq \frac{r-1}{2}} (-1)^t \binom{r}{t} ;$$

this is positive if (r - 1)/2 is even and negative if it is odd, by grouping terms in the expansion of $(1 - 1)^r$. Then consider the set B of $2^n - 1$ functions, m_R for r odd and e_S for s even (not zero). Each e_R can be represented as a linear combination of m_R and the collection of all e_S, S a proper subset of R. Hence by induction B forms a basis, and a fortiori the m_R are linearly independent. Hence the theorem.

Hence we conclude

8.4. The Shapley value ϕ is the unique linear symmetric transform of characteristic functions satisfying (i) $\Sigma_{i \in N} \phi(i) = v(N)$ and (ii) $\phi(i) = 0$ whenever $v(S) \equiv v(S - i)$.

PROOF. We refer to Shapley's proof [12] that ϕ has these and other desirable properties. Now ϕ is determined by its values on the basis consisting of all m_R, r odd. These values assign 0 to players in $N - R$ by (ii), and some constant to players in R by symmetry; with (i), ϕ is determined.

REMARKS (added December 1957).

In the recent book "Games and Decisions: A Critical Survey", by R. Duncan Luce and Howard Raiffa (New York, 1957), there is an argument to the effect that no such arbitration scheme as the ratio scheme can exist. The argument is on pages 133-134 and turns on three related bargaining games. The difference between their viewpoint and mine is essentially that I prefer to say no such games can exist.

Luce and Raiffa argue as follows, concerning any arbitration scheme $\mathcal{T}$. $\mathcal{T}$ is to yield solutions to two-player games G defined by the strategies, payoffs, and the closed convex set $R(G)$ of jointly feasible outcomes in a two-dimensional utility space. One might wish to demand that $\mathcal{T}$ satisfy

> AXIOM A. Let G and G' be two games having the same strategy spaces and individual payoffs, but such that $R(G)$ contains $R(G')$. Then $\mathcal{T}(G) \geq \mathcal{T}(G')$.

Axiom A has considerable persuasive force. (The ratio scheme satisfies it, evidently. None of the particular schemes advocated by Nash, Raiffa, or Braithwaite satisfies Axiom A.) However, say Luce and Raiffa, Axiom A is inconsistent with the requirement of Pareto optimality, which they have formulated as follows, for $\mathcal{T}(G) = (t_1, t_2)$:

ASSUMPTION 2. (Pareto optimality). ... iii. There is no (u_1, u_2) in $R(G)$, different from (t_1, t_2), such that $u_1 \geq t_1$ and $u_2 \geq t_2$.

The proof is very simple (see Figure 9, p. 134, op. cit.).

In fact the ratio scheme does not satisfy this strong form of the Pareto optimality principle. This difficulty is noted (though not clearly formulated) in Section 3 above; then it is essentially dismissed on the ground that it applies only to non-free games. The alternative scheme F, the free ratio scheme, satisfies strong Pareto optimality but does not satisfy Axiom A.

In the context of this paper, the ratio scheme satisfies a weak Pareto optimality condition for all games: $\mathcal{T}(G)$ is always an element of the weak top of G, or

There is no (u_1, u_2) in $R(G)$ such that $u_1 > t_1$ and

$u_2 > t_2$.

However, even this is not true if the scheme is to be applied to Luce-Raiffa's Figure 9 (p. 134, op. cit.). For two of the three games there represented, R(G) fails to contain its weak top over its base. I do not believe it is possible to describe a two-person conflict situation which may reasonably be treated as a game of this geometric form; and it is for that reason that no such games are admitted in this paper.

In summary, Axiom A and strong Pareto optimality (2, iii, of Luce-Raiffa) are in conflict, unless every game is free. One should either sacrifice one of these conditions or produce an argument to show that every socially meaningful game is free. Axiom A conflicts even with weak Pareto optimality if Figure 9 of Luce-Raiffa corresponds to any real possibility. I do not believe it does, but have no cogent argument to offer here. However, the following caution should not be lost sight of. Pareto optimality in any form cannot be expected in all contexts, a point which is made in several places in the book of Luce and Raiffa. In the context of arbitration, it certainly seems indispensable.

BIBLIOGRAPHY

[1] BRAITHWAITE, T., Theory of Games as a Tool for the Moral Philosopher, Cambridge, 1955.

[2] GILLIES, D., "Some theorems on n-person games," Thesis, Princeton, 1953.

[3] LUCE, R. D., "Ψ-stability: a new equilibrium concept for n-person game theory," Mathematical Models of Human Behavior, Stamford, Conn., Dunlap and Associates, 1955, pp. 32-44.

[4] McKINSEY, J., Introduction to the Theory of Games, New York, 1952.

[5] NASH, J., "Cooperative two-person games," Econometrica, 21 (1953), pp. 128-140.

[6] NASH, J., "Non-cooperative games," Annals of Mathematics (2), 54 (1951), pp. 286-295.

[7] von NEUMANN, J., and MORGENSTERN, O., Theory of Games and Economic Behavior, 2nd edition, Princeton, 1947.

[8] RAIFFA, H., "Arbitration schemes for generalized two-person games," Annals of Mathematics Study 28 (Princeton 1953), pp. 361-387.

[9] SAVAGE, L., The Foundations of Statistics, New York, 1954.

[10] SHAPLEY, L., "Additive and non-additive set functions," Thesis, Princeton, 1953.

[11] SHAPLEY, L., "Notes on the n-person game, III," RAND Research Memorandum RM-817, April, 1952.

[12] SHAPLEY, L., "A value for n-person games," Annals of Mathematics Study 28 (Princeton 1953), pp. 307-318.

J. R. Isbell

The George Washington University

A NEW APPROACH TO n-PERSON GAMES

John G. Kemeny*

§1. PURPOSE OF A NEW THEORY

The games discussed in this paper will be zero-sum games with at least three players. It will be assumed that players have an opportunity to bargain. Hence the paper belongs to the theory of cooperative n-person games.

There is an essential difference between zero-sum games with two players and with more players. In the former a value is assigned to the game, while in the latter a set of imputations constitutes a solution. The assignment of a value can be thought of as a single imputation, giving v to the first player and $-v$ to the second. For games with more than two players the theory allows us to say only that one of several things occurs. And the number of possibilities in a set of imputations is either infinite, or consists of a number of widely differing imputations.

A numerical example will help to clarify this picture. Consider a 3-person game with the following values for coalitions: $v(\{1, 2\}) = 100$, $v(\{1, 3\}) = 10$, $v(\{2, 3\}) = 1$. There is one so-called "non-discriminating" solution, with the imputations $(54.5, 45.5, -100)$, $(54.5, -10, -44.5)$, and $(-1, 45.5, -44.5)$ forming the solution. There is also an infinity of "discriminating" solutions, in which one player is assigned a payoff between the limits of the above imputations, and the others divide the remainder in some manner. Whichever type of solution we select, it tells us very little about what actually happens. And there is no way of choosing one set of imputations in preference to another.

von Neumann and Morgenstern present reasons why it is not possible to arrive at a more specific solution. While I agree that with the information usually given for n-person games no more can be said, it

* The preparation of this manuscript was aided by the National Science Foundation through a grant to the Dartmouth Mathematics Project.

seems to me that we must ask for more information. The solution of two-person games is so satisfactory because the value is unique and there is (in all but exceptional games) a unique optimal strategy for each player. It is the purpose of this paper to set up a frame-work in which additional information can be supplied. The aim is to add enough new information to enable us to arrive at a single optimal strategy and a single imputation - for all but exceptional games.

What type of information do we need? The original theory uses as its key concept the value of a coalition. This is computed by assuming that the players not in the coalition form a competing coalition, and then computing the value of the resulting two-person game. This procedure seems reasonable. It may be assumed that the aim of a many-person game is to negotiate until a winning coalition is formed, and then the players left out will try to minimize the gain of the coalition. Since the value of a coalition is determined by the theory of two-person games, I will accept this as a satisfactory concept.

We may then think of the basic problem as the determination of which winning coalition is formed, and the division of spoils among the victors. But the answer to the latter question must be known before the former can be settled. When a player decides whether to enter a proposed coalition, he will want to know how much he personally stands to gain. As a matter of fact, if we know how each coalition would divide gains, we could construct a reasonable method for predicting which coalition will actually be formed.

Given the method of division for each coalition, we can assign a value to each coalition for each player. Thus we could well-order the preferences of each player. (I will ignore exceptional games through-out this paper. E.g., here I ignore the possibility of a "tie".) We can then think of the various winning coalitions as participating in an election using a "preferential ballot", and find out which coalition is "elected". This scheme will be worked out in Section 5. Here we merely want to indicate that once the method of division is known, the solution of the game is not far behind.

Thus we see that our missing information is knowledge of how various possible coalitions would split their gains. The original theory supplies us with some bounds on possible imputations, but these leave much to be settled. This is not surprising since there is nothing in the theory to tell us about the psychology of the players. We have to know about their bargaining ability and also about the aggressiveness of their demands.

I will propose a frame-work in which such psychological information can be supplied. It will enable us to decide how each coalition splits its gains, and hence will lead to a theory in which a game normally has a unique imputation associated with it.

§2. METHOD OF DIVISION

A coalition of k players gains G. How will the k players divide this amount? Since we are interested only in winning coalitions, we may assume that $G > 0$. If we have no other information, and we are forced to make a guess, we have to guess G/k for each player. We know that this is not what usually happens. A player's strength in the game, his bargaining ability, his psychology, etc. have an effect on the outcome.

It is hopeless to try to get a complete list of all factors, and to evaluate each one. Instead I will introduce a single factor which is to be a composite of all of these. New information is to be introduced by means of a utility function for each player. Given the functions, we will assume that the coalition divides evenly relative to these. A player who is able to hold out for a large share will have a utility function according to which a proposed share always looks small. A weak player or one unable to bargain successfully will be assigned a utility function that exaggerates his share.

Player number i, P_i, is assigned the function f_i. f_i is a strictly monotone increasing function, defined for non-negative arguments. If P_i is assigned a share G_i of the total G, this will appear to him to be a fraction $g_i = [f_i(G_i) - f_i(0)]/[f_i(G) - f_i(0)]$. We will assume that the division is such that the shares represent the <u>same apparent proportion g_i for each player</u>.

It must, of course, be shown that this can always be achieved. Let $g_i(x) = [f_i(x) - f_i(0)]/[f_i(G) - f_i(0)]$. g_i is strictly monotone increasing, and $g_i(0) = 0$, $g_i(G) = 1$. Let h_i be the inverse function of g_i. This too is strictly monotone increasing, and $h_i(0) = 0$, $h_i(1) = G$. Let $s(x)$ be the sum of the $h_i(x)$. This is a function which, on the unit interval, takes on all values between 0 and kG once. Hence there is a unique α on the unit interval such that $s(\alpha) = G$. We assign a share $G_i = h_i(\alpha)$ to P_i. The shares total G, as required, and the apparent proportion is $g_i(h_i(\alpha)) = \alpha$ for each player.

Since utility functions are normally determined only within a linear transformation, it is important to note that the apparent proportion g_i is not affected by a linear transformation on f_i. It is often convenient to choose $f_i(0)$ as 0, so that $g_i(x)$ simplifies to $f_i(x)/f_i(G)$.

It will be convenient to consider a simple class of utility functions for illustrations. The functions of the form x^t lead to a very simple method of division. $g_i(x) = (x/G)^{t_i}$ and $h_i(x) = G \cdot x^{1/t_i}$. The equation to be solved simplifies to $\Sigma x^{1/t_i} = 1$, which is independent of G. If α is the positive root of this equation, then P_i gets a fraction α^{1/t_i} of whatever the gain may be. This type of utility function leads to a division according to fixed ratios.

The purpose of this section was to show that these utility functions actually accomplish the purpose for which they were introduced. In this sense they provide the frame-work for a new theory. It is the intention that the utility functions must be given as part of the description of the game. The exact interpretation of these functions can be given only in terms of the entire theory: The functions measure those factors that determine how well the player succeeds in bargaining. It can be objected that the utility functions must be hard to determine. But in this they do not differ essentially from the entries in the game-matrix. Once we get outside the realm of games in the ordinary sense, the rules are not clearly stated, and pay-offs must be estimated. The usual justification for applying game-theory to such "games" is that the result will be qualitatively the same for a reasonable range of the parameters. This will also be true for the utility functions. Often the result will depend only on the general shape of the utility curves. It will be important to study the manner in which the solution depends on the utility functions.

§3. THREE-PERSON GAMES

In a 3-person zero-sum game the only question is which coalition of two people is formed. The rules determine how much a given coalition can expect to gain. And the method of division determines how much of this gain goes to each player. Thus a coalition is worth a definite amount to each member of it. We will say that P_i <u>prefers</u> P_j to P_k if the coalition $\{P_i, P_j\}$ is worth more to him than the coalition $\{P_i, P_k\}$. The game is determinate if there is a pair of players who prefer each other. In this case these two players will form a coalition, and hence the result is completely known. The game is indeterminate if no such pair exists. Clearly, this happens only when 1 prefers 2 and 2 prefers 3 and 3 prefers 1, or when 1 prefers 3 and 3 prefers 2 and 2 prefers 1. I will show that for any set of utility functions there exist both determinate and indeterminate games, but that the latter are rare and always of only one of the two possible kinds.

Let us first solve the case where the utility functions are of the form x^t. Then players divide according to fixed ratios. Say that

P_i gets a fraction b_{ij} of the gain of the coalition $\{P_i, P_j\}$. And let us use G_{ij} to denote the total gain of this coalition. Then P_1 will prefer P_2 if: $b_{12}G_{12} > b_{13}G_{13}$. And we get two similar conditions for the other preferences.

It is simplest to solve the game geometrically. I will introduce barycentric coordinates: Let $S = G_{12} + G_{13} + G_{23}$, and $B_{ij} = G_{ij}/S$. In the resulting triangle the condition $b_{12}G_{12} = b_{13}G_{13}$ is represented by a straight line going through the corner (0, 0, 1). [We take B_{12} as first coordinate, B_{13} second, and B_{23} third.] The half triangle towards (1, 0, 0) is an area in which P_1 prefers P_2, while in the other half he prefers P_3. Similarly we get lines for the other two conditions. The resulting picture must be one of the following two.

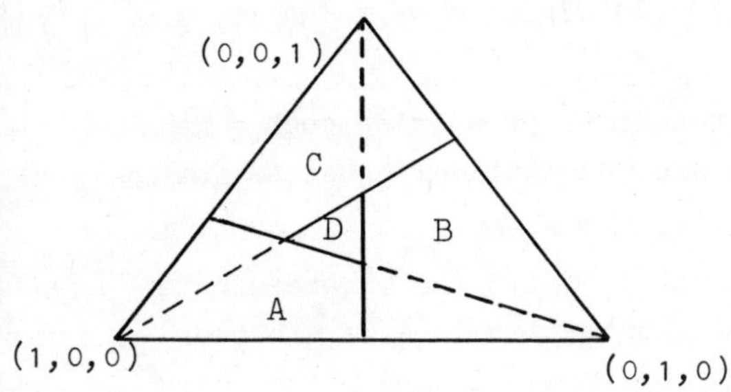

FIGURE 1

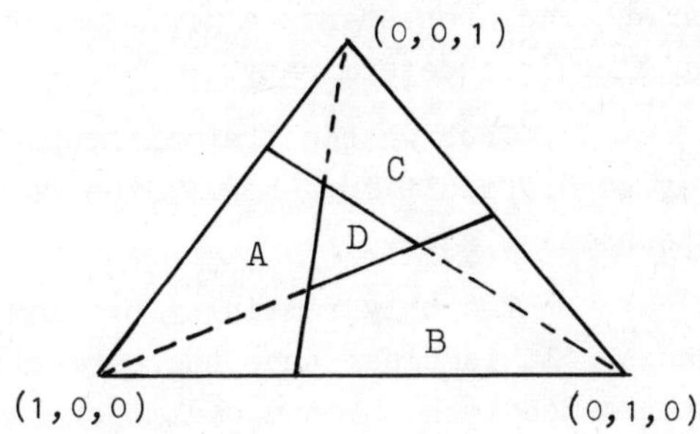

FIGURE 2

In each figure A is the area in which P_1 and P_2 form a coalition, B where P_1 and P_3 combine, C where P_2 and P_3 unite, and D is the area of indeterminacy. The difference between the two figures is the difference between the two kinds of indeterminacy.

Thus we see that for given utility functions of this type most games are determinate, but one kind of indeterminacy is always possible. [I have intentionally ignored the exceptional games falling on a borderline in which one player has a choice of two coalitions, and the exceptional set of functions for which the three lines meet -- in which case there is no possible indeterminacy.] Let us now consider three arbitrary functions.

We can still write down the conditions on the preferences of the players. But we can no longer represent them in barycentric coordinates because such representation assumes that the conditions depend only on the ratios of the G_{ij}, which although true in the previous case is not true in general. We must therefore use the first octant of three-dimensional space. We can construct the surface representing games in which P_1 has no preference. In the planes $G_{23} = 0$ we get a line having positive slope everywhere and going through the origin. Since the preference of P_1 does not depend on G_{23}, the desired surface is the cylindrical surface generated by parallel displacement of this curve. Thus we get three cylindrical surfaces, each containing one axis. The octant is divided into 7 volumes. For each axis there are two of these volumes bordering on the axis (separated by the surface containing the axis). These two volumes represent the formation of a definite coalition. E.g., the volumes bordering on the G_{23} axis have (relatively) low G_{12} and G_{13} values, and hence represent the formation of the $\{P_2, P_3\}$ coalition. The two volumes are separated by the surface discussed above, and the fact that this separation is immaterial results from the fact that in this case it does not matter what P_1's preference is. Thus six volumes represent determinate games. The seventh volume, not bordering on an axis, represents a (relatively) even distribution of gains resulting in indeterminacy.

Thus we see that although the technique of solving such a game may be difficult in general, the results are qualitatively just as for the special case.

The only remaining problem is what to do in the indeterminate cases. It is clear that the formation of a coalition is always better than a deadlock. Yet a coalition can only be formed if one player is willing to settle for an outcome less than optimal for him. I suggest that in this case the player who has least to lose by not following his preference agrees to a coalition under the less favorable terms. Thus we have to subdivide the seventh volume into regions as to who loses least. We do this by drawing in surfaces (within the seventh volume) where two players lose equal amounts. The intersection of two of these surfaces is also the intersection of the third surface with them, since when two pairs of players lose equal amounts, then all three players lose equal amounts. Hence the volume is subdivided into six regions, with a pair of regions representing least loss for a player. Taking pairs of regions we now have three new volumes which we must add to the three volumes that have already been assigned to the three determinate cases, always adjoining a new region to its neighboring old region. We then arrive at a solution where the entire space is divided into three volumes, representing the three possible coalitions. Thus all games (except those on the borders) are uniquely determined.

I will conclude by illustrating the convention in the case represented in Figure 1. In the special case equal losses are represented by straight line segments, and the result is as follows:

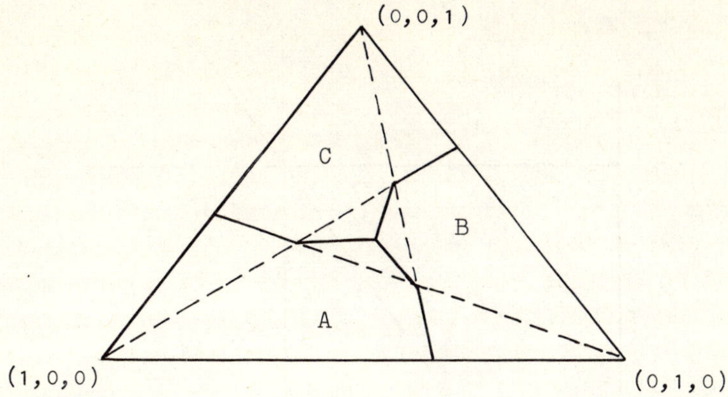

FIGURE 3

§4. AN ILLUSTRATION

Let $f_1(x) = x^{1/2}$, $f_2(x) = x$, $f_3(x) = x^2$. Division is by fixed ratios. To find the ratios for $\{1, 2\}$ we must solve $x^2 + x = 1$. The root of this is $\alpha = .618$. The first player gets a fraction $\alpha^2 = .382$, while the second players gets a fraction α. Solving the other two equations also, we find the values: $b_{12} = .382$, $b_{21} = .618$, $b_{13} = .275$, $b_{31} = .725$, $b_{23} = .382$, $b_{32} = .618$. P_3 drives the hardest bargain; P_1 is the easiest to please.

The conditions for preferences are as follows:

1 prefers 2 as his partner if $.382\ G_{12} > .275\ G_{13}$
2 prefers 3 as his partner if $.382\ G_{23} > .618\ G_{12}$
3 prefers 1 as his partner if $.725\ G_{13} > .618\ G_{23}$

The units for payoffs have not been fixed, so we may let $G_{23} = 1$. The conditions for preference are shown in Figure 4, on the following page.

For sufficiently small values of G_{12} and G_{13}, $G_{12} < .618$ and $G_{13} < .853$, P_1 is not a very useful partner, and hence P_2 and P_3 form a coalition. But since P_1 is easy to please, he can find a partner before G_{12} or G_{13} reaches 1. When G_{12} exceeds .618 or G_{13} exceeds .853, P_1 will be in on the winning coalition. If both values are large, then P_1 can choose his partner, his choice depending on the relative value of G_{12} and G_{13}. Since P_2 is less demanding than P_3, a smaller value of G_{12} will suffice.

It is interesting to note how small the area of indeterminacy is. Actually it is exaggerated in Figure 4, since it is really too small to show. The indeterminacy holds when all three of the inequalities given above hold. A necessary condition for this is that G_{12} is in the interval $.615 < G_{12} < .618$. G_{12} can take on any positive value, but it

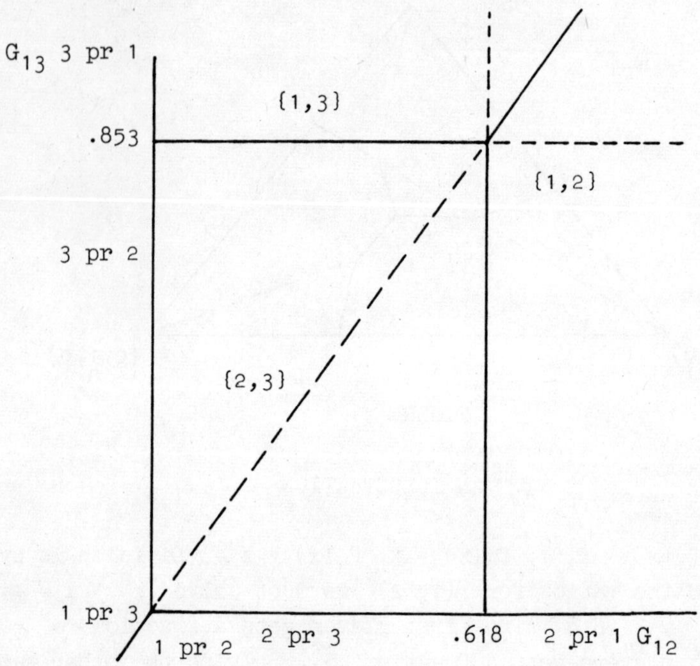

FIGURE 4

must be within an interval of length .003 for there to be a possibility of an indeterminacy. The bounds on G_{13} are of the same order of magnitude.

§5. SKETCH OF A GENERAL THEORY

We are given an n-person game, and a utility function for each player. We consider all possible coalitions, and compute a value for each one. This is done by assuming that the players not in the coalition co-operate to minimize the gain of the coalition, and we solve the resulting two-person game. We consider only winning coalitions, i.e., coalitions with a positive value. Our problem is to find which one of these is formed.

By our method of division we know how much of the gain of a coalition each member gets. Thus we have a value of each coalition for each player. This leads to a well-ordering for each player of all the coalitions in which he is a member. If any winning coalition is a first choice for all its members, then this coalition is formed. If not, then some player must settle for a lower choice, and we need a systematic method for arriving at the coalition that will be formed. I propose the following method: For each winning coalition we compute the <u>sacrifice</u>

each member must make to agree to the coalition. This number is the difference between his share of his first-choice coalition and his share of the given coalition. The coalition actually formed is the one for which the largest necessary sacrifice is a minimum.

There are n players, and N possible winning coalitions. v_{ij} is the share to P_i of the jth winning coalition. ($i = 1, \ldots, n$; $j = 1, \ldots, N$; and $v_{ij} = 0$ if P_i is not a member of the jth coalition.) Let $V_i = \max_j v_{ij}$, $s_{ij} = V_i - v_{ij}$, $S_j = \max_i s_{ij}$ (where P_i is a member of the jth coalition). Then coalition j_0 is formed, where $S_{j_0} = \min_j S_j$. (Ties are again ignored.)

The imputation is then determined as follows: For members of the coalition j_0 we assign v_{ij}. We then consider the remaining players as a single coalition, that loses what our winning coalition gains. This loss is apportioned among the losing players. I will not go into the problem of apportioning losses.

Let us consider a concrete example. There are four political parties. Each has a certain number of seats on the council, P_1 having most, P_4 least. No party alone has enough votes to control the council, but P_1 has enough votes so that together with any of the others it has control. The winning coalitions are the four 3-party coalitions, and the three two-party coalitions having P_1 as a member. We must assign gains to these seven coalitions. $\{P_2, P_3, P_4\}$ has a bare majority, $v = 1$. $\{P_1, P_2, P_3\}$ has an overwhelming majority, $v = 3$. The other coalitions are comfortably off, $v = 2$.

We must also know what the utility functions of the four parties are. $f_1(x) = x^2$, $f_2(x) = f_3(x) = x$, $f_4(x) = x^{1/2}$. These are chosen so that the strongest party expects most, while the weakest party expects least. The relevant facts are summarized in Figure 5.

Coalition	v_{1j}	v_{2j}	v_{3j}	v_{4j}	s_{1j}	s_{2j}	s_{3j}	s_{4j}	s_j
{1,2}	1.236	.764			.264	0			.264
{1,3}	1.236		.764		.264		0		.264
{1,4}	1.450			.550	.050			0	.050
{1,2,3}	1.500	.750	.750		0	.014	.014		.014
{1,2,4}	1.140	.650		.210	.360	.114		.340	.360
{1,3,4}	1.140		.650	.210	.360		.114	.340	.360
{2,3,4}		.414	.414	.712		.350	.350	.378	.378
V_i	1.500	.764	.764	.550					

FIGURE 5

We see from the figure that {1, 2, 3} is formed. It must, however, be pointed out that the coalition {1, 4} is not far inferior. The gains 1, 2, 3 for the various coalitions were somewhat arbitrarily chosen, and a slightly different choice might lead to a different result. The result is clear, however, qualitatively. The strongest player can choose the coalition. He can either form a very strong majority, or a weaker majority that he dominates completely.

It must be pointed out that the present method is but one of several similar procedures. It is not necessary that we should minimize the maximum sacrifice. We could, e.g., minimize the maximum percentage sacrifice. But I hope that I have at least demonstrated that it is possible to build the framework of a theory leading normally to a unique imputation for n-person games.

BIBLIOGRAPHY

[1] von NEUMANN, J. and MORGENSTERN, O., Theory of Games and Economic Behavior, Princeton, 3rd edition, Princetion (1953).

John G. Kemeny

Dartmouth College

A BIBLIOGRAPHY OF GAME THEORY

Dorothea M. Thompson and Gerald L. Thompson

The present bibliography is a preliminary version of a complete subject bibliography that will appear at a later date. It is being prepared at the request of Professor Oskar Morgenstern, and is supported by the Office of Naval Research through the Econometric Research Program at Princeton University.

The following rules were adhered to in compiling this bibliography: Only articles that make substantial use of game theory are included. Derivative articles based on ideas quite far removed from the subject, even though influenced by it, are omitted. Abstracts of papers and talks are generally not included. Papers appearing in dittoed or mimeographed form are usually omitted, exceptions being made only when we felt that it was possible for anyone to obtain copies of these by writing to the source or by going to a sufficiently large library. The articles that appear with a star in the left margin are those that are obviously pertinent but could not be found in the libraries consulted. The other papers have been seen by the compilers.

In order to make the final version of the bibliography as useful as possible, readers of the present bibliography can help by sending corrections or additions to Professor Oskar Morgenstern, Econometric Research Program, 9-11 Lower Pyne, Princeton University, Princeton, New Jersey. Such aid will be greatly appreciated by the compilers.

D.M.T.
G.L.T.

1. ABLOW, C. M. and GEORGES BRIGHAM. "An analog solution of programming problems" J. Oper. Res. Soc. 3 (1955) 388-394.

ACKOFF, RUSSELL L. See Churchman, C. West.

ADAMS, ERNEST W. See Luce, R. Duncan.

2. *AERTS, L. "Input-output analyse en lineair programming in het kader van de economische wetenschap" Economica documentatie (Leuven) (1954) 145-

3. AGGARWAL, OM P. "Some minimax invariant procedures for estimating a cumulative distribution function" Ann. Math. Stat. 26 (1955) 450-463.

4. AGMON, SHMUEL. "The relaxation method for linear inequalities" Canad. J. Math. 6 (1954) 382-392.

AKAIKE, HIROTUGU. See Matusita, Kameo.

5. ALBERT, G. E. "Accurate sequential tests on the mean of an exponential distribution" Ann. Math. Stat. 27 (1956) 460-470.

6. ALBERT, G. E. "On the computation of the sampling characteristics of a general class of sequential decision problems" Ann. Math. Stat. 25 (1954) 340-356.

7. ALCHIAN, ARMEN. "The meaning of utility measurement" Amer. Econ. Rev. 43 (1953) 26-50.

8. ALLAIS, M. "Le comportement de l'homme rationnel devant le risque; critique des postulats et axiomes de l'école americaine" Econometrica 21 (1953) 503-546.

9. ALLAIS, M. "La psychologie de l'homme rationnel devant le risque: La théorie et l'expérience" J. Soc. Stat. Paris 94 (1953) 47-72.

10. ALLEN, R. G. D. Mathematical Economics. London, MacMillan and Co., Ltd.; New York, St. Martins Press, 1956. See chapters 15 and 16.

11. ALLEN, S. G. Jr. "A class of minimax tests for one-sided composite hypotheses" Ann. Math. Stat. 24 (1953) 295-298.

ALLISON, J. R. See Hutton, R. F.

12. AMERICAN MANAGEMENT ASSOCIATION. Top Management Decision Gaming, May 2, 1957. New York, American Management Association, Inc., 1957.

13. *ANDERSON, OSKAR. "Theorie der Glucks-spiele und ökonomisches Verhalten" Schweiz. Z. Volkswirtschaft und Statistik 85 (1949) 46-53.

14. ANDERSON, T. W. "Classification by multivariate analysis" Psychometrika 16 (1951) 31-50.

15. ANTOSIEWICZ, H. A. "A theorem on alternatives for pairs of matrices" Pacific J. Math. 5 (1955) 641-642.

16. ANTOSIEWICZ, H. and A. HOFFMAN. "A remark on the smoothing problem" Management Sci. 1 (1954) 92-95.

17. ANTOSIEWICZ, H. A., ed. Proceedings of the Second Symposium in Linear Programming. Directorate of Management Analysis, DCS/Comptroller, Headquarters U. S. Air Force, Washington, D. C., 1955. 2 vols.

ARNOFF, E. LEONARD. See Churchman, C. West.

18. ARROW, KENNETH J. "Alternative approaches to the theory of choice in risk-taking situations" Econometrica 19 (1951) 404-437.

19. ARROW, KENNETH J. "Alternative proof of the substitution theorem for Leontief models in the general case." In reference 552, pp. 155-164.

20. ARROW, KENNETH J. "Decision theory and operations research" J. Oper. Res. 5 (1957) 765-774.

21. ARROW, KENNETH. "Le role des valeurs boursières pour la répartition la meilleure des risques" Econométrie, Colloques Internationaux du Centre National de la Recherche Scientifique, Paris, Centre National de la Recherche Scientifique, 1953, pp. 41-47.

22. ARROW, KENNETH J. Social Choice and Individual Values. New York, John Wiley and Sons, Inc., 1951, xi, 99.

23. ARROW, K. J., E. W. BARANKIN, and D. BLACKWELL. "Admissible points of convex sets." In reference 575, pp. 87-91.

24. ARROW, K. J., D. BLACKWELL, and M. A. GIRSHICK. "Bayes and minimax solutions of sequential decision problems" Econometrica 17 (1949) 213-244.

25. ARROW, KENNETH J. and GERARD DEBREU. "Existence of an equilibrium for a competitive economy" Econometrica 22 (1954) 265-290.

26. ARROW, KENNETH J., THEODORE HARRIS, and JACOB MARSCHAK. "Optimal inventory policy" Econometrica 19 (1951) 250-272.

27. ARROW, KENNETH and LEONID HURWICZ. "Gradient methods for constrained maxima" J. Oper. Res. 5 (1957) 258-265.

28. ARROW, KENNETH J. and LEONID HURWICZ. "Reduction of constrained maxima to saddle-point problems" Proc. of the Third Berkeley Symposium on Mathematical Statistics and Probability, Vol. 5, Berkeley and Los Angeles, University of California Press, 1956, 1-20.

29. AUMANN, ROBERT. "Acceptable points in general cooperative n-person games." This Study.

30. BABBAR, M. M. "Distributions of solutions of a set of linear equations (with an application to linear programming)" J. Amer. Stat. Assoc. 50 (1955) 854-869.

31. BABBAR, M. M. "A note on aspects of linear programming technique" J. Farm. Econ. 38 (1956) 607-612.

32. BABBAR, M. M., GERHARD TINTNER, and EARL HEADY. "Programming with consideration of variations in input coefficients" J. Farm Econ. 37 (1955) 333-341. Errata, 38 (1956) 612-613.

33. BAHADUR, R. R. "A characterization of sufficiency" Ann. Math. Stat. 26 (1955) 286-293.

34. BAHADUR, R. R. "Sufficiency and statistical decision functions" Ann. Math. Stat. 25 (1954) 423-462.

35. BAHADUR, RAGHU RAJ and LEO GOODMAN. "Impartial decision rules and sufficient statistics" Ann. Math. Stat. 23 (1952) 553-562.

36. BAHADUR, R. R. and E. L. LEHMANN. "Two comments on 'sufficiency and statistical decision functions'" Ann. Math. Stat. 26 (1955) 139-142.

37. BAHADUR, RAGHU RAJ and HERBERT ROBBINS. "The problem of the greater mean" Ann. Math. Stat. 21 (1950) 469-487; 22 (1951) 310.

38. BAHADUR, R. R. and LEONARD SAVAGE. "The nonexistence of certain statistical procedures in non parametric problems" Ann. Math. Stat. 27 (1956) 1115-1122.

39. BAILEY, M. J. "A generalized comparative statics in linear programming" Rev. Econ. Studies 23 (1955-56) 236-240.

40. BALDWIN, ROGER R., WILBERT CANTEY, HERBERT MAISEL, and JAMES McDERMOTT. "The optimum strategy in blackjack" J. Amer. Stat. Assoc. 51 (1956) 429-439.

41. BARANKIN, EDWARD. "On systems of linear equations, with applications to linear programming and the theory of tests of statistical hypotheses" Univ. California Publ. Stat. 1 (1949-1953) 161-214.

42. BARANKIN, E. W. "Some investigations on linear programming." In reference 401, pp. 68-73.

BARANKIN, E. W. See Arrow, K. J.

43. *BARNARD, G. A. "Sampling inspection and statistical decisions" J. Roy. Stat. Soc. Ser. B. 16 (1954) 151-165; discussion 166-174.

44. BARNARD, G. A. "Simplified decision functions" Biometrika 41 (1954) 241-251.

45. BARTLETT, T. E. and A. CHARNES. "Cyclic scheduling and combinatorial topology assignment and routing of motive power to meet scheduling and maintenance requirements. Part II. Generalization and analysis" Naval Res. Log. Q. 4 (1957) 207-220.

46. *BASU, D. "On the minimax approach to the problem of estimation" Proc. Nat. Inst. Sci. India 18 (1952) 287-299.

47. BATCHELOR, JAMES H. "A commercial use of linear programming." In reference 17, pp. 103-116.

48. BATES, JAMES. "A model for the science of decision" Philos. of Sci. 21 (1954) 326-339.

49. BAUMOL, WILLIAM J. "The Neumann-Morgenstern utility index - an ordinalist view" J. Political Econ. 59 (1951) 61-66.

50. *BEALE, E. M. L. "An alternative method for linear programming" Proc. Cambridge Philos. Soc. 50 (1954) 512-523.

51. BEALE, E. M. L. "Cycling in the dual simplex algorithm" Naval Res. Log. Q. 2 (1955) 269-275.

52. BEALE, E. M. L. "Linear programming by the method of leading variables" Conference on Linear Programming (arranged by Ferranti, Ltd.) pp. 56-71.

BEALE, E. M. L. See Vajda, S.

53. BEAUMONT, R. A. and R. P. PETERSON. "Set-transitive permutation groups" Canad. J. Math. 7 (1955) 35-42.

54. BECKMANN, MARTIN. "Aktivitätsanalyse der Produktion und des Wirtschaftens" Z. Gesamte Staatswiss. 109 (1953) 629-644.

55. BECKMANN, M. J. "Comparative statics in linear programming and the Giffen Paradox" Rev. Econ. Studies 23 (1955-56) 232-235.

56. BECKMANN, MARTIN J. "Grundbigriffe der Produktions theorie vom Standpunkt der Aktivitätsanalyse" Weltwirtschaftliches Arkiv 75 (1955) 33-58.

57. BECKMANN, MARTIN J. "On the division of labor in teams" Metroeconomica 8 (1956) 163-168.

58. BECKMANN, MARTIN J. and JACK LADERMAN. "A bound on the use of inefficient indivisible units" Naval Res. Log. Q. 3 (1956) 245-252.

59. BECKMANN, MARTIN and THOMAS MARSCHAK. "An activity analysis approach to location theory" Kyklos 8 (1955) 125-143.

60. BECKMANN, M. and T. MARSCHAK. "An activity analysis approach to location theory." In reference 17, pp. 331-379.

BECKMAN, MARTIN. See Koopmans, Tjalling C.

61. BEJAR, JUAN. "Regresion en mediana y la programacion lineal" Trabajos de Estadistica 7 (1956) 141-158.

62. BELLMAN, RICHARD. "Comment on Dantzig's paper on discrete variable extremum problems" J. Oper. Res. 5 (1957) 723-724.

63. BELLMAN, RICHARD. "Decision making in the face of uncertainty. II" Naval Res. Log. Q. 1 (1954) 327-332.

64. BELLMAN, RICHARD. Dynamic Programming (RAND Corporation Research Study) Princeton, Princeton Univ. Press, 1957, xxv, 342.

65. BELLMAN, RICHARD. "Dynamic programming and multi-stage decision processes of stochastic type." In reference 17, pp. 229-249.

66. BELLMAN, RICHARD. "Functional equations in the theory of dynamic programming. III" Rend. Circ. Mat. Palermo 5 (1956) 297-319.

67. BELLMAN, R. "Mathematical aspects of scheduling theory" J. Soc. Ind. and Appl. Math. 4 (1956) 168-205.

68. BELLMAN, RICHARD. "On a dynamic programming approach to the caterer problem. I" Management Sci. 3 (1956-57) 270-279.

69. BELLMAN, RICHARD. "On an iterative procedure for obtaining the Perron root of a positive matrix" Proc. Amer. Math. Soc. 6 (1955) 719-725.

70. BELLMAN, RICHARD. "On games involving bluffing" Rend. Circ. Mat. Palermo (2) 1 (1952) 139-156.

71. BELLMAN, RICHARD. "On the computational solution of linear programming problems involving almost-block-diagonal matrices" Management Sci. 3 (1956-57) 403-406.

72. BELLMAN, RICHARD. "Some functional equations in the theory of dynamic programming" Proc. Nat. Acad. Sci. U.S.A. 39 (1953) 1077-1082.

73. BELLMAN, R. "Some functional equations in the theory of dynamic programming. I. Functions of points and point transformations" Trans. Amer. Math. Soc. 80 (1955) 51-71.

74. BELLMAN, RICHARD. "The theory of dynamic programming" Bull. Amer. Math. Soc. 60 (1954) 503-515.

75. BELLMAN, RICHARD and DAVID BLACKWELL. "Some two-person games involving bluffing" Proc. Nat. Acad. Sci. U.S.A. 35 (1949) 600-605.

76. BELLMAN, RICHARD, CHARLES E. CLARK, DONALD G. MALCOLM, CLIFFORD J. CRAFT, and FRANC M. RICCIARDI. "On the construction of a multi-stage multi-person business game" J. Oper. Res. 5 (1957) 469-503.

77. BELLMAN, RICHARD, IRVING GLICKSBERG, and OLIVER GROSS. "On some variational problems occurring in the theory of dynamic programming" Proc. Nat. Acad. Sci. U.S.A. 39 (1953) 298-301.

78. BELLMAN, RICHARD, IRVING GLICKSBERG, and OLIVER GROSS. "Some nonclassical problems in the calculus of variations" Proc. Amer. Math. Soc. 7 (1956) 87-94.

79. BELLMAN, RICHARD and OLIVER GROSS. "Some combinatorial problems arising in the theory of multi-stage processes" J. Soc. Indust. Appl. Math. 2 (1954) 175-183 (1955).

80. BELLMAN, RICHARD and SHERMAN LEHMAN. "On the continuous gold-mining equation" Proc. Nat. Acad. Sci. U.S.A. 40 (1954) 115-119.

81. *BENNION, E. G. "Capital budgeting and game theory" Harvard Business Review 34 (1956) 115-123.

82. BERESFORD, R. S. and M. H. PESTON. "A mixed strategy in action" Operational Res. Q. 6 (1955) 173-175.

83. BERGE, CLAUDE. "Le problème du gain dans la théorie généralisée des jeux sans informations" Bull. Soc. Math. France 81 (1953) 1-8.

84. BERGE, CLAUDE. "Sur les ensembles purs et les ultrafiltres" C. R. Acad. Sci. Paris 238 (1954) 2136-2137.

85. BERGE, CLAUDE. "Sur l'inversion des transformateurs" C. R. Acad. Sci. Paris 232 (1951) 134-136.

86. BERGE, CLAUDE. "Sur l'isovalence et la régularité des transformateurs" C. R. Acad. Sci. Paris 231 (1950) 1404-1406.

87. BERGE, CLAUDE. "Sur un théorème de la convexité régulière non linéaire" C. R. Acad. Sci. Paris 238 (1954) 2485-2486.

88. BERGE, CLAUDE. "Sur une convexité régulière non linéaire et ses applications à la théorie des jeux" Bull. Soc. Math. de France 82 (1954) 301-315.

89. BERGE, CLAUDE. "Sur une théorie ensembliste des jeux alternatifs" C. R. Acad. Sci. Paris 232 (1951) 294-296.

90. BERGE, CLAUDE. "Sur une généralisation du théorème de Zermelo - von Neumann" C. R. Acad. Sci. Paris 241 (1955) 455-457.

91. BERGE, CLAUDE. "Sur une théorie ensembliste des jeux alternatifs" J. Math. Pures Appl. 32 (1953) 129-184.

92. BERGE, CLAUDE. Théorie Générale des Jeux à n Personnes. Mémorial des Sciences Mathématiques 138. Paris: Gauthier-Villars, 1957, 114 pp.

93. BERGE, CLAUDE. "The theory of games, a new branch of mathematics" United States Line. Paris Review Vanves, Kapp et l'Imprimerie Illustrations Bobigny, n.d.

94. BERGE, CLAUDE. "Topological games with perfect information." In reference 265, pp. 165-178.

BERGENDAL, GUNNAR. See Welinder, Carsten.

95. BERKOVITZ, L. D. and W. H. FLEMING. "On differential games with integral payoff." In reference 265, pp. 413-435.

96. BERNARD, JESSIE. "The theory of games of strategy as a modern sociology of conflict" Amer. J. Soc. LIX (1954) 411-424.

97. BERNOULLI, DANIEL. "Exposition of a new theory on the measurement of risk" Econometrica 22 (1954) 23-36. Translated from Latin into English by Dr. Louise Sommer, from "Specimen Theoriae novae de mensura sortis" Commentarii Academiae Scientiarum Imperialis Petropolitanae, Tomus V (Papers of The Imperial Academy of Sciences in Petersburg) Vol. V (1738) 175-192. Professor Karl Menger wrote footnotes 4, 9, 10, and 15.

98. BERNSTEIN, C. "Continuous programming." In Economic Activity Analysis, Edited by Oskar Morgenstern. New York, John Wiley and Sons, Inc., 1954, pp. 383-390.

99. BHARUCEHA-REID, A. T. "On the stochastic theory of epidemics." In reference 728, Vol. IV, pp. 111-119.

100. BIRCH, B. J. "On games with almost complete information" Proc. Cambridge Philos. Soc. 51 (1955) 275-287.

101. BIRNBAUM, ALLAN. "Characterizations of complete classes of tests of some multiparametric hypotheses, with applications to likelihood ratio tests" Ann. Math. Stat. 26 (1955) 21-51.

102. BISHOP, CHARLES E. "Programming farm-nonfarm allocation of farm family resources" J. Farm Econ. 38 (1956) 396-407.

BISHOP, CHARLES E. See King, R. A.

103. BISHOP, GEORGE T. "On a problem of production scheduling" J. Oper. Res. 5 (1957) 97-103.

104. BLACKETT, D. W. "Convex programming." In reference 401, pp. 183-185.

105. BLACKETT, D. W. "Some Blotto games" Naval Res. Log. Q. 1 (1954) 55-60.

106. BLACKWELL, DAVID. "An analog of the minimax theorem for vector payoffs" Pacific J. Math. 6 (1956) 1-8.

107. *BLACKWELL, DAVID. "Game theory." In Operations Research for Management, edited by J. F. McCloskey and F. N. Trefethen, Baltimore, Johns Hopkins Press, 1954, pp. 238-253.

108. BLACKWELL, DAVID. "Equivalent comparisons of experiments" Ann. Math. Stat. 24 (1953) 265-272.

109. BLACKWELL, DAVID. "On multi-component attrition games" Naval Res. Log. Q. 1 (1954) 210-216.

110. BLACKWELL, DAVID. "On randomization in statistical games with k terminal actions." In reference 575, pp. 183-187.

111. BLACKWELL, DAVID. "On the translation parameter problem for discrete variables" Ann. Math. Stat. 22 (1951) 393-399.

112. BLACKWELL, DAVID and M. A. GIRSHICK. Theory of Games and Statistical Decisions. New York, John Wiley and Sons, Inc., 1954, xi, 355.

113. BLACKWELL, DAVID and J. L. HODGES Jr. "Design for the control of selection bias" Ann. Math. Stat. 38 (1957) 449-460.

BLACKWELL, DAVID. See Arrow, K. J.

BLACKWELL, DAVID. See Bellman, Richard.

114. BLASBALG, H. "Transformation of the fundamental relationships in sequential analysis" Ann. Math. Stat. 28 (1957) 1024-1028.

115. BLUMENTHAL, L. M. "Metric methods in linear inequalities" Duke Math. J. 15 (1948) 955-966.

116. BLUMENTHAL, L. M. "Two existence theorems for systems of linear inequalities" Pacific J. Math. 2 (1952) 523-530.

117. BLYTHE, COLIN R. "On minimax statistical decision procedures and their admissibility" Ann. Math. Stat. 22 (1951) 22-42.

118. BOHNENBLUST, H. F. and SAMUEL KARLIN. "On a theorem of Ville." In reference 574, pp. 155-160.

119. BOHNENBLUST, H. F., SAMUEL KARLIN, and LLOYD S. SHAPLEY. "Games with continuous, convex pay-off." In reference 574, pp. 181-192.

120. BOHNENBLUST, H. F., SAMUEL KARLIN, and LLOYD S. SHAPLEY. "Solutions of discrete, two-person games." In reference 574, pp. 41-72.

121. BOLDYREFF, ALEXANDER W. "Determination of the maximal steady state flow of traffic through a railroad network" J. Oper. Res. Soc. 3 (1955) 443-465.

122. BOLES, JAMES N. "Linear programming and farm management analysis" J. Farm Econ. 37 (1955) 1-24.

123. BOLES, JAMES N. "Shortcuts in programming computations" J. Farm Econ. 38 (1956) 981-990.

BOLES, JAMES N. See McCorkle, Chester O. Jr.

124. *BOREL, ÉMILE. Le Jeu, La Chance et Les Théories Scientifiques Modernes. Gallimard, 1941.

125. BOREL, ÉMILE. "On games that involve chance and the skill of the players" Econometrica 21 (1953) 101-115. Translated from Eléments de la Théorie des Probabilités, 3e ed., Paris, J. Hermann, 1924, pp. 204-221.

126. BOREL, ÉMILE. "On systems of linear forms of skew symmetric determinant and the general theory of play" Econometrica 21 (1953) 116-117. Translated from C. R. Acad. Sci. Paris 184 (1927) 52-54.

127. *BOREL, ÉMILE. "Sur les jeux où interviennent l'hasard et l'habileté des joueurs" Eléments de la Théorie des Probabilités, 3e ed. Paris: Librairie Scientifique, J. Hermann, 1924, pp. 204-224.

128. BOREL, ÉMILE. "Sur les systèmes de formes linéaires à determinant symétrique gauche et la théorie générale du jeu" C. R. Acad. Sci. Paris 184 (1927) 52-54.

129. BOREL, ÉMILE. "La théorie du jeu et les équations intégrales à noyau symétrique" C. R. Acad. Sci. Paris 173 (1921) 1304-1308.

130. BOREL, ÉMILE. "The theory of play and integral equations with skew symmetric kernels" Econometrica 21 (1953) 97-100.

131. BOREL, ÉMILE. "Un théoreme sur les systèmes de formes linéaires à determinant symétrique gauche" C. R. Acad. Sci. Paris 183 (1926) 925-927.

132. *BOREL, ÉMILE et ANDRÉ CHÉRON. Theorie Mathématique du Bridge à la Portée de Tous. Paris, Gauthier-Villars, 1940. 2 ième éd., 1955.

133. BOTT, RAOUL. "Symmetric solutions to majority games." In reference 575, pp. 319-323.

134. BOWKER, ALBERT H. "Continuous sampling plans." In reference 728, pp. 75-85.

135. BOWMAN, EDWARD H. "Production scheduling by the transportation method of linear programming" J. Oper. Res. 4 (1956) 100-103.

136. BRADT, RUSSELL N. and SAMUEL KARLIN. "On the design and comparison of certain dichotomous experiments" Ann. Math. Stat. 27 (1956) 390-409.

137. BRAITHWAITE, RICHARD B. Scientific Explanation. A Study of the Function of Theory, Probability and Law in Science. Cambridge, Cambridge University Press, 1953, xii, 376.

138. BRAITHWAITE, RICHARD B. Theory of Games as a Tool for the Moral Philosopher. Cambridge, Cambridge University Press, 1955, 76.

139. *BRAMBILLA, FRANCESCO. "Caso, incertezza, organizzazione" Tecnica ed Org. 4 (1953) 3-7 (no. 8); also, no. 9 (1953) 10-15.

140. *BREMS, HANS. "Some notes on the structure of the duopoly problem" Nordisk Tidskrift for Teknisk Økonomi 37 (1948) 50-51.

BRIGHAM, GEORGES. See Ablow, C. M.

141. BROSS, IRWIN D. J. Design for Decision. New York, The Macmillan Co., 1953, viii, 276.

142. BROWN, GEORGE W. "Iterative solution of games by fictitious play." In reference 552, pp. 374-376.

143. BROWN, GEORGE W. "Notes on the solution of linear systems in solving inequalities" Proc. of the 2nd Symp. on Large-Scale Digital Calc. Mach., Cambridge, Harvard U. Press, 1951, 137-140.

144. BROWN, GEORGE W. and TJALLING C. KOOPMANS. "Computational suggestions for maximizing a linear function subject to linear inequalities." In reference 552, pp. 377-380.

145. BROWN, GEORGE W. and JOHN von NEUMANN. "Solutions of games by differential equations." In reference 574, pp. 73-79.

146. BROWN, J. A. C. "An experiment in demand analysis. The computation of a diet problem on the Manchester computer." In Conference on Linear Programming (arranged by Ferranti, Ltd.) pp. 41-53.

147. BROWN, RICHARD H. "The solution of a certain two-person zero-sum game" J. Oper. Res. 5 (1957) 63-67.

BROWN, W. G. See Nelson, Michael.

148. *de BRUIJN, N. G. and D. van DANTZIG. "Inequalities concerning determinants and systems of linear equations" Proc. Kon. Nedrl. Akad. Wetensch. Amsterdam Ser. A 55 (1952) 315-321.

149. BURGER, E. "On extrema with side conditions" Econometrica 23 (1955) 451-452.

150. BURGER, E. "Spieltheoretische Behandlung eines Reklameproblems. (Variante eines Spielmodells von Gillman)" Mitteilungsbl. Math. Stat. 6 (1954) 39-52.

151. BURGER, E. "Über homogene lineare Ungleichungssysteme" Z. Angew. Math. Mech. 36 (1956) 135-139.

152. BURGER, E. "Zur theorie der kooperativen zweipersonenspiele" Arch. Math. 7 (1956) 143-147.

BURROWS, GLENN L. See Waugh, Frederick.

153. *CAHN, A. S. "The warehouse problem" Bull. Amer. Math. Soc. 54 (1948) 1073 (abstract).

154. CANDLER, WILFRED. "A modified simplex solution for linear programming with variable capital restrictions" J. Farm Econ. 38 (1956) 940-955.

155. CANDLER, WILFRED. "A modified simplex solution for linear programming with variable prices" J. Farm Econ. 39 (1957) 409-428.

156. CANSADO, ENRIQUE. "Programación lineal un instrumento mathemático al servicio de los empresarios" Trabajos de Estadistica 7 (1956) 305-333.

CANTEY, WILBERT E. See Baldwin, Roger T.

157. CARVER, W. B. "Systems of linear inequalities" Ann. of Math. 23 (1922-23) 212-220.

158. *CASTAÑEDA, JOSÉ. "Introducción a la programación lineal" Rev. de Ciencias Aplicada 8 (1954) 203-217.

159. CASTAÑEDA, JOSÉ. "La programación lineal y la teoria ecónomica. Aplicaciones de la programación lineal" Trabajos de Estadistica 7 (1956) 97-122.

160. CASTIGLIONI, PIETRO. "Planning the mechanization of the bank" Banca nazionale del Lavoro: Quarterly Review, Rome 34 (1955) 156-162.

CASTLE, E. N. See Nelson, Michael.

161. CAYWOOD, T. E. and C. J. THOMAS. "Applications of game theory in fighter versus bomber combat" J. Oper. Res. Soc. 3 (1955) 402-411.

162. *CERNIKOV, S. N. "Systems of linear inequalities" Uspehi Mat. Nauk. N. S. 8 (1953) 7-73.

163. *CERNIKOV, S. N. "A generalization of Kronecker-Capelli's theorem on systems of linear equations" Mat. Sb. N. S. 25 (1944).

164. CHACKO, GEORGE. "Certain game situations in regional economic development" Indian J. Econ. 37 (1956) 167-175.

165. CHACKO, GEORGE K. "The concept of utility: a different approach to fundamentals" Indian J. Econ. 32 (1951) 99-113.

166. CHACKO, GEORGE K. "Economic behaviour - a new theory" Indian J. Econ. 30 (1950) 349-365.

167. CHAMPERNOWNE, D. G. "Nota sull'articlo di J. von Neumann 'un modello di equilibro economico'" Industria (Milano) (1952) 14-28.

168. CHAMPERNOWNE, D. G. "A note on J. von Neumann's article" Rev. Econ. Stud. 13 (1945-46) 10-18.

169. CHAMPERNOWNE, D. G. "A note on Mr. Farrell's model" Econometrica 22 (1954) 303-309.

170. CHAND, M. "On the theory of games: an expository analysis" Indian J. Econ. 36 (1956) 409-424.

171. CHARNES, A. "Constrained games and linear programming" Proc. Nat. Acad. Sci. U.S.A. 39 (1953) 639-641.

172. CHARNES, A. "Optimality and degeneracy in linear programming" Econometrica 20 (1952) 160-170.

173. CHARNES, ABRAHAM and WILLIAM W. COOPER. "Una facile esposizione attorno ai fini della programmazione lineare" L'Industria (1954) 421-427.

174. CHARNES, ABRAHAM and WILLIAM W. COOPER. "Generalizations of the warehousing model" Operational Res. Q. 6 (1955) 131-172.

175. *CHARNES, ABRAHAM and WILLIAM W. COOPER. "Linear programming as a tool for the solution of management problems" Chemical Eng. Prog. 53 (1957) 170-173.

176. CHARNES, ABRAHAM and WILLIAM W. COOPER. "Management models and industrial applications of linear programming" Management Sci. 4 (1957-58) 38-91.

177. CHARNES, ABRAHAM and WILLIAM W. COOPER. "Nonlinear power of adjacent extreme point methods in linear programming" Econometrica 25 (1957) 132-153.

178. CHARNES, ABRAHAM and WILLIAM W. COOPER. "The stepping stone method of explaining linear programming calculations in transportation problems" Management Sci. 1 (1954) 49-69.

179. *CHARNES, ABRAHAM, WILLIAM W. COOPER, and D. FARR and STAFF. "Linear programming and profit preference scheduling for a manufacturing firm" J. Oper. Res. Soc. 1 (1953) 114-129.

180. CHARNES, ABRAHAM, WILLIAM W. COOPER, and R. O. FERGUSON. "Optimal estimation of executive compensation by linear programming" Management Sci. 1 (1954) 138-151.

181. CHARNES, ABRAHAM, WILLIAM W. COOPER, and A. HENDERSON. An Introduction to Linear Programming. New York, John Wiley and Sons, Inc., 1953, ix, 74.

182. CHARNES, ABRAHAM, WILLIAM W. COOPER, and B. MELLON. "Blending aviation gasolines - a study in programming interdependent activities." In reference 401, pp. 115-146.

183. CHARNES, ABRAHAM, WILLIAM W. COOPER, and B. MELLON. "Blending aviation gasolines - a study in programming interdependent activities in an integrated oil company" Econometrica 20 (1952) 135-159.

184. CHARNES, ABRAHAM, WILLIAM W. COOPER, and B. MELLON. "A model for optimizing production by reference to cost surrogates." In reference 17, pp. 117-150.

185. CHARNES, ABRAHAM, WILLIAM W. COOPER, and B. MELLON. "A model for programming and sensitivity analysis in an integrated oil company" Econometrica 22 (1954) 193-217.

186. *CHARNES, ABRAHAM and C. E. LEMKE. "Computational problems of linear programming" Proc. of the Assn. of Comp. Mach. Pittsburgh, 1952, pp. 97-82.

187. CHARNES, ABRAHAM and C. E. LEMKE. "Minimization of non-linear separable convex functionals" Naval Res. Log. Q. 1 (1954) 301-312.

188. CHARNES, A. and M. H. MILLER. "Mathematical programming and evaluation of freight shipment systems, Part II - Analysis" Naval Res. Log. Q. 4 (1957) 243-252.

189. CHARNES, ABRAHAM and M. H. MILLER. "A model for the optimal programming of railway freight train movements" Management Sci. 3 (1956-57) 74-92.

CHARNES, ABRAHAM. See Bartlett, T. E.

CHARNES, ABRAHAM. See Cooper, William.

190. CHENERY, HOLLIS B. "Il compito dell' industrializazione nello sviluppo Italia meridionale" L'industria (Milano) (1955) 378.

191. CHENERY, HOLLIS B. and KENNETH S. KRETSCHMER. "Resource allocation for economic development" Econometrica 24 (1956) 365-399.

192. CHENEY, L. K. "Linear program planning of refinery operations" Naval Res. Log. Q. 4 (1957) 9-16.

193. CHERNOFF, HERMAN. "Large sample theory - parametric case" Ann. Math. Stat. 27 (1956) 1-22.

194. CHERNOFF, HERMAN. "Rational selection of decision functions" Econometrica 22 (1954) 422-443.

CHÉRON, ANDRÉ. See Borel, Émile.

195. CHEVALLEY, CLAUDE. "Theory of games" View, Series 5, no. 1, March 1945, p. 43.

196. CHIANG, C. L., J. L. HODGES, and J. YERUSHALMY. "Statistical problems in medical diagnoses." In reference 728, pp. 121-133.

197. CHIPMAN, JOHN S. "Computational problems in linear programming" Rev. of Econ. and Stat. 35 (1953) 342-349.

198. CHIPMAN, JOHN. "Linear programming" Rev. of Econ. and Stat. 35 (1953) 101-117.

199. CHOQUET, GUSTAVE. "Sur le théorème des points-selle de la théorie des jeux" Bull. Sci. Math. 79 (1955) 48-53.

200. CHURCHMAN, C. WEST, RUSSELL L. ACKOFF, and LEONARD E. ARNOFF. Introduction to Operations Research. New York, John Wiley and Sons, Inc., 1957, x, 645 pp.

CLARK, CHARLES E. See Bellman, Richard.

201. CODEGONE, CESARE and PROSPERO NUVOLI. "Osservazioni sulla programmazione lineare" Rivista di Politica Economica 45 (1955) 794-799.

COOMBS, CLYDE H. See Thrall, Robert M.

202. COOPER, WILLIAM W. and ABRAHAM CHARNES. "Linear programming" Sci. Am. 191 (August 1954) 21-23.

COOPER, W. W. See Charnes, Abraham.

203. COPELAND, ARTHUR H. "John von Neumann and Oskar Morgenstern's theory of games and economic behavior" Bull. Amer. Math. Soc. 51 (1945) 498-504.

CRAFT, CLIFFORD J. See Bellman, Richard.

CRANDALL, H. W. See Garvin, W. W.

204. DALKEY, NORMAN. "Equivalence of information patterns and essentially determinate games." In reference 575, pp. 217-243.

205. DANIELSSON, S. "Spelteoretiska grundbegrepp" Artilleri Tidskr. 84 (1955) 104-111.

206. DANØ, SVEN. "Lineaer programmering" Nordisk Mat. Tidskr. 4 (1956) 121-138.

207. *DANØ, SVEN. "Lineaer programmering i produktionsteorien, I-III" Nationaløkonomisk Tidskr. 93 (1955) 94-117 og 205-233; 94 (1956) 47-61.

DANØ, SVEN. See Gale, David.

208. DANSKIN, J. M. "Fictitious play for continuous games" Naval Res. Log. Q. 1 (1954) 313-320.

209. DANSKIN, J. M. "Linear programming in the face of uncertainty: example of a failure." In reference 17, pp. 39-53.

210. DANSKIN, J. M. and L. GILLMAN. "A game over function space" Rivista Mat. Univ. Parma 4 (1953) 83-94.

211. van DANTZIG, D. "Utilité d'une distribution de probabilités ou distribution des probabilités des utilités" Econométrie, Colloques Internationaux du Centre National de la Recherche Scientifique, No. 40. Paris, Centre National de la Recherche Scientifique, 1953, pp. 235-241; discussion pp. 242-244.

van DANTZIG, D. See de Bruijn, N. G.

212. DANTZIG, GEORGE B. "Application of the simplex method to a transportation problem." In reference 552, pp. 359-373.

213. DANTZIG, GEORGE B. "Constructive proof of the min-max theorem" Pacific J. Math. 6 (1956) 25-33.

214. DANTZIG, GEORGE B. "Developments in linear programming." In reference 17, pp. 667-685.

215. DANTZIG, GEORGE B. "Discrete-variable extremum problems" J. Oper. Res. 5 (1957) 266-277.

216. *DANTZIG, GEORGE B. "Linear programming" Problems Numerical Analysis of Future (1951) 18-21.

217. DANTZIG, GEORGE B. "Linear programming under uncertainty" Management Sci. 1 (1955) 197-206.

218. DANTZIG, GEORGE B. "Maximization of a linear function of variables subject to linear inequalities." In reference 552, pp. 339-347.

219. DANTZIG, GEORGE B. "Note on Klein's 'Direct use of extremal principles in solving certain problems involving inequalities'" J. Oper. Res. 4 (1956) 247-249.

220. DANTZIG, GEORGE B. "Optimal solution of a dynamic Leontief model with substitution" Econometrica 23 (1955) 295-302.

221. DANTZIG, GEORGE B. "The programming of interdependent activities: mathematical model." In reference 552, pp. 19-32.

222. DANTZIG, GEORGE B. "Programming of interdependent activities. II. Mathematical model" Econometrica 17 (1949) 200-211.

223. DANTZIG, GEORGE B. "A proof of the equivalence of the programming problem and the game problem." In reference 552, pp. 330-335.

224. DANTZIG, GEORGE B. "Recent advances in linear programming" Management Sci. 2 (1956) 131-144.

225. DANTZIG, GEORGE B. "Thoughts on linear programming and automation" Management Sci. 3 (1956-57) 131-139.

226. DANTZIG, GEORGE B. "Upper bounds, secondary constraints, and block triangularity in linear programming" Econometrica 23 (1955) 174-183.

227. DANTZIG, GEORGE B., L. R. FORD Jr., and D. R. FULKERSON. "A primal-dual algorithm for linear programs." In reference 576, pp. 171-181.

228. DANTZIG, GEORGE B. and D. R. FULKERSON. "Computation of maximal flows in networks" Naval Res. Log. Q. 2 (1955) 277-283.

229. DANTZIG, GEORGE B. and D. R. FULKERSON. "Minimizing the number of tankers to meet a fixed schedule" Naval Res. Log. Q. 1 (1954) 217-222.

230. DANTZIG, GEORGE B. and D. R. FULKERSON. "On the max-flow min-cut theorem of networks." In reference 576, pp. 215-221.

231. DANTZIG, GEORGE B., D. R. FULKERSON, and SELMER JOHNSON. "Solution of a large-scale traveling salesman problem" J. Oper. Res. Soc. 2 (1954) 393-410.

232. DANTZIG, GEORGE B. and A. J. HOFFMAN. "Dilworth's theorem on partially ordered sets." In reference 576, pp. 207-214.

233. DANTZIG, GEORGE B. and SELMER JOHNSON. "A production smoothing problem." In reference 17, pp. 151-176.

234. DANTZIG, GEORGE B. and W. ORCHARD-HAYS. "The product form for the inverse in the simplex method" Math. Tables and other Aids to Computation, VIII, No. 45 (1954) 64-67.

235. DANTZIG, GEORGE B. and ALEX ORDEN. "A duality theorem based on the simplex method." In reference 401, pp. 51-55.

236. DANTZIG, GEORGE B., ALEX ORDEN, and PHILIP WOLFE. "The generalized simplex method for minimizing a linear form under linear inequality restraints" Pacific J. Math. 5 (1955) 183-195.

DANTZIG, GEORGE B. See Ferguson, Allen R.

DANTZIG, GEORGE B. See Johnson, Selmer.

DANTZIG, GEORGE B. See Wood, Marshall.

237. DAVIDSON, D., J. C. C. McKINSEY, and PATRICK SUPPES. "Outlines of a formal theory of value, I" Philos. Sci. 22 (1955) 140-160.

238. DAVIDSON, DONALD and PATRICK SUPPES (in collaboration with Sidney Siegel). Decision Making. Stanford, California, Stanford University Press, 1957, 121 pp.

239. DAVIDSON, DONALD and PATRICK SUPPES. "A finitistic axiomatization of subjective probability and utility" Econometrica 24 (1956) 264-275.

240. DAVIES, ROBERT. "The matrix for an industrial game" Indust. Math. 6 (1955) 1-5.

241. *DAVIS, CHANDLER. "Linear programming and computers. I, II" Computers and Automation 4 (1955) no. 7, 10-17; No. 8, 10-16.

DAVIS, ROBERT L. See Thrall, Robert M.

242. DEBEAU, DAVID E. "Linear programming isn't always the answer" J. Oper. Res. 5 (1957) 429-433.

243. DEBREU, GERARD. "The coefficient of resource utilization" Econometrica 19 (1951) 273-292.

244. DEBREU, GERARD. "Representation of a preference ordering by a numerical function." In reference 889, pp. 159-165.

245. DEBREU, GERARD. "A social equilibrium existence theorem" Proc. Nat. Acad. Sci. U.S.A. 38 (1952) 886-893.

DEBREU, GERARD. See Arrow, Kenneth J.

DEEMER, WALTER L. Jr. See Thomas, Clayton J.

246. DEEMS, PAUL S. "War gaming and exercises" Air Univ. Q. Rev. 9 (1956-57) 98-126.

247. *DEL PUNTA, VENIERO. "La programmazione lineare della produzione aziendale" Produttività (1955).

248. DEL PUNTA, VENIERO. "Sull' applicazione dell' input-output analysis e della programmazione lineare al problema dello sviluppo economico" Rivista di Politica Economica 46 (1956) 161-192.

DEL PUNTA, VENIERO. See Fei, John.

249. DEMARIA, GIOVANNI. "Prime applicazioni economiche alla teoria dei giochi" Tecnica ed Org. 5 (1954) 4-15.

250. DEMARIA, GIOVANNI. "Su una nuova logica economica" Giornale degli Econ. e. Ann. di Econ. 6 (1957) 661-671.

251. DEUTSCH, KARL W. "Game theory and politics: some problems of application" Canad. J. Econ. and Pol. Sci. 20 (1954) 76-83.

252. DILLON, J. L. and G. C. McFARLANE. "Using the simplex worksheet for linear programming" J. Farm Econ. 38 (1956) 622-623.

253. DINES, LLOYD L. "Convex extension and linear inequalities" Bull. Am. Math. Soc. 42 (1936) 353-365.

254. *DINES, LLOYD L. "On a theorem of von Neumann" Proc. Nat. Acad. Sci. U.S.A. 33 (1947) 329-331.

255. DINES, LLOYD L. "Systems of linear inequalities" Ann. of Math., 2nd Ser. 20 (1918-19) 191-199.

256. DORFMAN, ROBERT. Application of Linear Programming to the Theory of the Firm. Berkeley, Univ. of California Press, 1951, ix, 98 pp.

257. DORFMAN, ROBERT. "Application of the simplex method to a game theory problem." In reference 552, pp. 348-358.

258. DORFMAN, ROBERT. "Mathematical, or 'linear' programming: a non-mathematical exposition" Am. Econ. Rev. 43 (1953) 797-825.

259. DORFMAN, ROBERT, PAUL A. SAMUELSON, and ROBERT M. SOLOW. Linear Programming and Economic Analysis (A RAND Corporation Research Study) New York, McGraw-Hill, 1958, ix, 525 pp.

260. DORN, W. S. and H. J. GREENBERG. "Linear programming and plastic limit analysis of structures" Q. Appl. Math. 15 (1957-58) 155-167.

261. DRESHER, MELVIN. "Games of strategy" Math. Mag. 25 (1957) 93-99.

262. *DRESHER, MELVIN. "Methods of solution in game theory" Econometrica 18 (1950) 179-181.

263. DRESHER, MELVIN and SAMUEL KARLIN. "Solutions of convex games as fixed-points." In reference 575, pp. 75-86.

264. DRESHER, MELVIN, SAMUEL KARLIN, and LLOYD S. SHAPLEY. "Polynomial games." In reference 574, pp. 161-180.

265. DRESHER, MELVIN, ALBERT W. TUCKER, and PHILIP WOLFE, eds. Contributions to the Theory of Games, III. Annals of Mathematics Studies, No. 39, Princeton, N. J., Princeton University Press, 1957, vi, 435 pp.

266. DUBINS, L. E. "A discrete evasion game." In reference 265, pp. 231-255.

DUCKWORTH, ERIC. See Kay, Emil.

267. DUFFIN, R. J. "Infinite programs." In reference 576, pp. 157-170.

268. DULMAGE, A. L. and J. E. L. PECK. "Certain infinite zero-sum two-person games" Canad. J. Math. 8 (1956) 412-416.

DULMAGE, A. L. See Peck, J. E. L.

DUNBAR, JOHN O. See Puterbaugh, Horace L.

DUNNE, JOHN. See Otter, Richard.

269. DVORETZKY, A., J. KIEFER, and J. WOLFOWITZ. "Asymptotically minimax character of the sample distribution function and of the classical multinomial estimator" Ann. of Math. Stat. 27 (1956) 642-669.

270. DVORETZKY, A., J. KIEFER, and J. WOLFOWITZ. "The inventory problem. II. Case of unknown distributions of demand" Econometrica 20 (1952) 450-466.

271. DVORETZKY, A., J. KIEFER, and J. WOLFOWITZ. "Sequential decision problems for processes with continuous time parameter. Problems of estimation" Ann. Math. Stat. 24 (1953) 403-415.

272. DVORETZKY, A., J. KIEFER, and J. WOLFOWITZ. "Sequential decision problems for processes with continuous time parameter. Testing hypotheses" Ann. Math. Stat. 24 (1953) 254-264.

273. DVORETZKY, A., J. KIEFER, and J. WOLFOWITZ. "Asymptotically minimax character of the sample distribution function and of the classical multinomial estimator" Ann. Math. Stat. 27 (1956) 642-669.

274. DVORETZKY, A., A. WALD, and J. WOLFOWITZ. "Elimination of randomization in certain problems of statistics and of the theory of games" Proc. Nat. Acad. Sci. U.S.A. 36 (1950) 256-260.

275. DVORETZKY, A., A. WALD, and J. WOLFOWITZ. "Elimination of randomization in certain statistical decision procedures and zero-sum two-person games" Ann. Math. Stat. 22 (1951) 1-21.

276. DVORETZKY, A., A. WALD, and J. WOLFOWITZ. "Relations among certain ranges of vector measures" Pacific J. Math. 1 (1951) 59-74.

277. DWYER, PAUL S. "The detailed method of optimal regions" Psychometrika 22 (1957) 43-52.

278. DWYER, PAUL S. "Solution of the personnel classification problem with the method of optimal regions" Psychometrika 19 (1954) 11-26.

DWYER, PAUL S. See Galler, Bernard.

279. EDWARDS, WARD. "An attempt to predict gambling decisions" in Proceedings of a Symposium on Mathematical Models of Human Behavior, 1954. Dunlap and Associates, Inc. and The Commission on Accidental Trauma, Armed Forces Epidemiological Board, 1955, pp. 83-96.

280. EDWARDS, WARD. "The prediction of decisions among bets" J. of Expt'l. Psych. 50 (1955) 201-214.

281. EDWARDS, WARD. "Probability-preferences among bets with differing expected values" Am. J. Psych. 67 (1954) 56-67.

282. EDWARDS, WARD. "Probability-preferences in gambling" Am. J. Psych. 66 (1953) 349-364.

283. EDWARDS, WARD. "The reliability of probability-preferences" Am. J. Psych. 67 (1954) 68-95.

284. EDWARDS, WARD. "The theory of decision-making" Psych. Bull. 51 (1954) 380-417.

285. EDWARDS, WARD. "Variance preferences in gambling" Am. J. Psych. 67 (1954) 441-452.

286. EISENMANN, KURT. "Linear programming" Quart. Appl. Math. 13 (1955-56) 209-232.

287. EISENMANN, KURT. "Simplified treatment of degeneracy in transportation problems" Quart. Appl. Math. 14 (1956-57) 399-403.

288. EISENMANN, KURT. "The trim problem" Management Sci. 3 (1956-57) 279-284.

289. ELFVING, G. "Sufficiency and completeness in decision function theory" Ann. Acad. Sci. Fennicae, Ser. A I No. 135, 9S (1952).

290. ELLSBERG, D. "Classic and current notions of 'measurable utility'" Economic J. 64 (1954) 528-556.

291. ELLSBERG, D. "Theory of the reluctant duelist" Am. Econ. Rev. 46 (1956) 909-923.

292. ELMAGHRABY, SALAH. "A note on production scheduling by the use of the transportation method" J. Oper. Res. 5 (1957) 565-566.

293. EVERETT, H. "Recursive games." In reference 265, pp. 47-78.

294. *FABIAN, VÁCLAV. "A decision function" Czechoslovak Math. J. 6 (81) (1956) 31-45.

295. FAN, KY. "Fixed-point and minimax theorem in locally convex topological linear spaces" Proc. Nat. Acad. Sci. U.S.A. 38 (1952) 121-126.

296. FAN, KY. "Minimax theorems" Proc. Nat. Acad. Sci. 39 (1953) 42-47.

297. FAN, KY. "On systems of linear inequalities." In reference 576, pp. 99-156.

298. FAN, KY, IRVING GLICKSBERG, and A. J. HOFFMAN. "Systems of inequalities involving convex functions" Proc. Amer. Math. Soc. 8 (1957) 617-622.

299. FARKAS, JULIUS. "Theorie der einfachen Ungleichungen" J. Reine Angew. Math. 124 (1902) 1-27.

300. FARQUHARSON, ROBIN. "Sur une généralisation de la notion d'equilibrium" C. R. Acad. Sci. Paris 240 (1955) 46-48.

FARR, D. See Charnes, Abraham.

301. FARRELL, M. J. "An application of activity analysis to the theory of the firm" Econometrica 22 (1954) 291-302.

302. FAXEN, KARL-OLAF. "Spelteori, Förväntnings-analys och handelsavtal" Soertryk af Nationaløkonomisk Tidskr. (1949) 251-264.

303. FEI, JOHN and VENIERO DEL PUNTA. "L'uso della 'programmazione lineare' per lo studio della pianificazione dello sviluppo industriale con riferimento al problema del Mezzogiorno italiano" Rivista di Politica Economica 45 (1955) 901-926.

304. FELS, EBERHARD. "Einige Bemerkungen zu Burgers variante eines Spiel modells von Gillman" Mitteilungsbl. Math. Stat. 6 (1954) 53-55.

305. di FENIZIO, FERDINANDO. "La metodologia di Oskar Morgenstern" L'industria (1952) 36-61; 207-230.

306. FERGUSON, ALLEN R. and GEORGE B. DANTZIG. "The allocation of aircraft to routes - an example of linear programming under uncertain demand" Management Sci. 3 (1956-57) 45-73.

307. FERGUSON, ALLEN R. and GEORGE B. DANTZIG. "The problem of routing aircraft - a mathematical solution" Aeronaut. Engrg. Rev. 14 (1955) 51-55.

FERGUSON, R. O. See Charnes, Abraham.

308. de FINETTI, BRUNO. "Concetti sul 'compartamento induttivo' illustrati su di un esempio" Statistica, Bologna 14 (1954) 350-378.

309. de FINETTI, BRUNO. "Media di decisioni e media di opinioni" Bull. Inst. Internat. Statist. 23, part II (1951) 249-282.

310. de FINETTI, BRUNO. "Probability and the social sciences" International Soc. Sci. Bull. 6 (1954) 591-609.

311. de FINETTI, BRUNO. "Rôle de la théorie des jeux dans l'économie et rôle des probabilités personnelles dans la théorie des jeux" Économétrie, Colloques Internationaux du Centre National de la Recherche Scientifique Paris, Centre National de la Recherche Scientifique, 1953, pp. 49-61.

312. *de FINETTI, BRUNO. "Sulla preferibilità" Giornale degli Econ. e Ann. di Econ. (1952).

313. FISHER, R. A. "Randomisation and an old enigma of card play" Math. Gaz. 18 (1934) 294-297.

314. FISHER, WALTER D. "On a pooling problem from the statistical decision viewpoint" Econometrica 21 (1953) 567-585.

315. FISHER, WALTER D. and LEONARD W. SCHRUBEN. "Linear programming applied to feed-mixing under different price conditions" J. Farm Econ. 35 (1953) 471-483.

316. FLEMING, W. H. "A note on differential games of prescribed duration." In reference 265, pp. 407-412.

317. FLEMING, W. H. "On a class of games over function space and related variational problems" Ann. of Math. 60 (1954) 578-594.

FLEMING, W. H. See Berkovitz, L. D.

318. FLOOD, MERRILL M. "Application of transportation theory to scheduling a military tanker fleet" J. Oper. Res. Soc. 2 (1954) 150-162.

319. FLOOD, MERRILL M. "A group preference experiment" Proc. of a Symposium on Mathematical Models of Human Behavior, 1954, Dunlap and Associates, Inc. and The Commission on Accidental Trauma, Armed Forces Epidemiological Board, 1955, pp. 1-21.

320. FLOOD, MERRILL M. "On game-learning theory and some decision-making experiments." In reference 889, pp. 139-158.

321. FLOOD, MERRILL M. "On the Hitchcock distribution problem" Pacific J. Math. 3 (1953) 369-386.

322. FLOOD, MERRILL M. "On the Hitchcock distribution problem." In reference 401, pp. 74-99.

323. FLOOD, MERRILL M. "The traveling-salesman problem" J. Oper. Res. 4 (1956) 61-75.

324. FØLNER, ERLING. "Elementer af von Neumann's spilteori" Nordisk Mat. Tidskr. 1 (1953) 115-126; 144.

325. FORD, L. R. Jr. and D. R. FULKERSON. "Maximal flow through a network" Canad. J. Math. 8 (1956) 399-404.

326. FORD, L. R. Jr. and D. R. FULKERSON. "A primal-dual algorithm for the capacitated Hitchcock problem" Naval Res. Log. Q. 4 (1957) 47-54.

327. FORD, L. R. Jr. and D. R. FULKERSON. "A simple algorithm for finding maximal network flows and an application to the Hitchcock problem" Canad. J. Math. 9 (1957) 210-218.

328. FORD, L. R. Jr. and D. R. FULKERSON. "Solving the transportation problem" Management Sci. 3 (1956-57) 24-32.

FORD, L. R. Jr. See Dantzig, George B.

329. FOSTER, CAXTON and ANATOL RAPAPORT. "Parasitism and symbiosis in an n-person non-constant-sum continuous game" Bull. Math. Biophys. 18 (1956) 219-231.

330. FOULKES, J. "Linear programming and structural design." In reference 17, pp. 177-184.

331. FOURASTIÉ, JEAN. "La stratégie du jeu et la méthode dans les sciences économiques" Critique 53 (1957) 872-895.

332. FOX, KARL A. and RICHARD C. TAEUBER. "Spatial equilibrium models of the livestock-feed economy" Am. Econ. Rev. 45 (1955) 584-608.

FOX, KIRK. See Swanson, Earl R.

333. FRANK, MARGUERITE and PHILIP WOLFE. "An algorithm for quadratic programming" Naval Res. Log. Q. 3 (1956) 95-110.

334. FRANK, P. and J. KIEFER. "Almost subminimax and biased minimax procedures" Ann. Math. Stat. 22 (1951) 465-468.

335. FRASER, D. A. S. "A regression analysis using the invariance method" Ann. Math. Stat. 28 (1957) 517-520.

336. FRÉCHET, MAURICE. "Commentary on the three notes of Émile Borel" Econometrica 21 (1953) 118-124.

337. FRÉCHET, MAURICE. "Émile Borel, initiator of the theory of psychological games and its application" Econometrica 21 (1953) 95-96.

338. FRÉCHET, MAURICE. "Sur l'importance en économétrie de la distinction entre les probabilités rationelles et irrationelles" Econometrica 23 (1955) 303-306.

339. FRENCH, CHARLES E. "Activity analysis: an agricultural marketing tool" J. Farm Econ. 37 (1955) 1236-1248. Discussion by Kirk Fox 1261-1263.

FRENCH, CHARLES E. See Snodgrass, Milton.

340. FREUND, RUDOLF J. "The introduction of risk into a programming model" Econometrica 24 (1956) 253-263.

341. FRIEDMAN, LAWRENCE. "An analysis of stewardess requirements and scheduling for a major airline. Annex B - calculation of the school table" Naval Log. Res. Q. 4 (1957) 199-205.

342. FRIEDMAN, LAWRENCE and ARTHUR J. YASPAN. "An analysis of stewardess requirements and scheduling for a major airline. Annex A - the assignment problem" Naval Log. Res. Q. 4 (1957) 193-197.

343. FRIEDMAN, M. and L. J. SAVAGE. "The expected utility-hypothesis and the measurability of utility" J. Pol. Econ. 60 (1952) 463-474.

344. FRIEDMAN, M. and L. J. SAVAGE. "The utility analysis of choices involving risk" J. Pol. Econ. 56 (1948) 279-304. Reprinted in G. J. Stigler & K. E. Boulding Eds., Readings in Price Theory, Chicago: Richard D. Irwin, 1952, pp. 57-96.

345. *FRISCH, R. "Methods of solving linear programming problems" Seminario Internazionale sull' Analisi dell' Input-Output, Varenna, June-July, 1954.

346. FRISCH, M. RAGNAR. "La résolution des problèmes de programme linéaire par la méthode du potentiel logarithmique" Cahiers du seminaire d'économetrie, No. 4 Programme Linéaire, Aggregation et Nombre Indices. Paris, Centre National de la Recherche Scientifique, pp. 7-20.

347. FULKERSON, D. R. and S. M. JOHNSON. "A tactical air game" J. Oper. Res. 5 (1957) 704-712.

FULKERSON, D. R. See Dantzig, G. B.

FULKERSON, D. R. See Ford, L. R. Jr.

348. GADDUM, J. W., A. J. HOFFMAN and D. SOKOLOWSKY. "On the solution of the caterer problem" Naval Res. Log. Q. 1 (1954) 223-229.

349. GAINEN, LEON. "Linear programming in bid evaluations." In reference 17, pp. 28-38.

350. GAINEN, LEON, D. P. HONIG, and E. D. STANLEY. "Linear programming in bid evaluations" Naval Res. Log. Q. 1 (1954) 48-54.

GAINEN, LEON. See Stanley, Commander E. D.

351. GALE, DAVID. "The basic theorems of real linear equations, inequalities, linear programming, and game theory" Naval Res. Log. Q. 3 (1956) 193-200.

352. GALE, DAVID. "The closed linear model of production." In reference 576, pp. 285-303.

353. GALE, DAVID. "Convex polyhedral cones and linear inequalities." In reference 552, pp. 287-297.

354. GALE, DAVID. "Information in games with finite resources." In reference 265, pp. 141-145.

355. GALE, DAVID. "The law of supply and demand" Math. Scand. 3 (1955) 155-169.

356. GALE, DAVID. "Mathematics and economic models" American Scientist 44 (1956) 33-44.

357. GALE, DAVID. "Neighboring vertices on a convex polyhedron." In reference 576, pp. 255-263.

358. GALE, DAVID. "A theorem on flows in networks" Pacific J. Math. 7 (1957) 1073-1082.

359. GALE, DAVID. "A theory of n-person games with perfect information" Proc. Nat. Acad. Sci. U.S.A. 39 (1953) 496-501.

360. GALE, D. and S. DANØ. "Linear programming: an introduction to the problems and methods" Nordisk Tidskr. for Teknish Økonomi 40 (1954) 5-32.

361. GALE, DAVID, HAROLD W. KUHN, and ALBERT W. TUCKER. "Linear programming and the theory of games." In reference 552, pp. 317-329.

362. GALE, DAVID, HAROLD W. KUHN, and ALBERT W. TUCKER. "On symmetric games." In reference 574, pp. 81-87.

363. GALE, DAVID, HAROLD W. KUHN, and ALBERT W. TUCKER. "Reductions of game matrices." In reference 574, pp. 89-96.

364. GALE, DAVID and S. SHERMAN. "Solutions of finite two-person games." In reference 574, pp. 37-49.

365. GALE, DAVID and F. M. STEWART. "Infinite games with perfect information." In reference 575, pp. 245-266.

366. GALLER, BERNARD A. and PAUL S. DWYER. "Translating the method of reduced matrices to machines" Naval Res. Log. Q. 4 (1957) 55-71.

367. GARVIN, W. W., H. W. CRANDALL, J. B. JOHN, and R. A. SPELLMAN. "Applications of linear programming in the oil industry" Management Sci. 3 (1956-57) 407-430.

368. GASS, SAUL I. "A first feasible solution to the linear programming problem." In reference 17, pp. 495-508.

369. GASS, SAUL I. "On the distribution of manhours to meet scheduled requirements" Naval Res. Log. Q. 4 (1957) 17-25.

370. GASS, SAUL I. and THOMAS SAATY. "The computational algorithm for the parametric objective function" Naval Res. Log. Q. 2 (1955) 39-45.

371. GASS, SAUL I. and THOMAS L. SAATY. "Parametric objective function (Part 1)" J. Oper. Res. Soc. 2 (1954) 316-319.

372. GASS, SAUL I. and THOMAS L. SAATY. "Parametric objective function (Part 2) - generalization" J. Oper. Res. Soc. 3 (1955) 395-401.

GASS, SAUL I. See Saaty, Thomas.

373. GEISLER, M. A. and H. W. KARR. "The design of military supply tables for spare parts" J. Oper. Res. 4 (1956) 431-442.

GEISLER, M. A. See Wood, Marshall K.

374. GELBAUM, B. R. "Symmetric zero-sum n-person games." This Study.

375. GEORGESCU-ROEGEN, NICHOLAS. "The aggregate linear production function and its applications to von Neumann's economic model." In reference 552, pp. 98-115.

376. GEORGESCU-ROEGEN, NICHOLAS. "Choice, expectations and measurability" Q. J. Econ. 68 (1954) 503-534.

377. GEORGESCU-ROEGEN, NICHOLAS. "Limitationality, limitativeness, and economic equilibrium." In reference 17, pp. 295-330.

378. GEORGESCU-ROEGEN, NICHOLAS. "Note on the economic equilibrium for nonlinear models" Econometrica 22 (1954) 54-57.

379. GERSTENHABER, MURRAY. "Theory of convex polyhedral cones." In reference 552, pp. 298-316.

380. GHOSH, M. N. "An extension of Wald's decision theory to unbounded weight functions" Sankhyā 12 (1952) 8-26.

381. *GIARDINA, BASILIO. "Basi statistiche della 'programmazione lineare' della produzione" Commucazione presentata al Convegno di studi di statistica aziendale, Torino, Ottobre 1952, p. 5.

382. GIARDINA, BASILIO. "La programmazione lineare della produzione nell' impresa" L'industria (1953) 576-632.

GIBRAT, R. See Massé, P.

383. GILLETTE, DEAN. "Stochastic games with zero stop probabilities." In reference 265, pp. 179-187.

384. GILLIES, D. B. "Discriminatory and bargaining solutions to a class of symmetric n-person games." In reference 575, pp. 325-342.

385. GILLIES, D. B., J. P. MAYBERRY, and J. von NEUMANN. "Two variants of poker." In reference 575, pp. 13-50.

386. GILLIES, D. B. "Solutions to general non-zero-sum games." This Study.

387. GILLMAN, L. "Operations analysis and the theory of games: an advertising example" J. Am. Stat. Assn. 45 (1950) 541-545.

GILLMAN, L. See Danskin, J. M.

388. *GIRSHICK, MEYER A. "An elementary survey of statistical decision theory" Review of Ed. Res. 24 (1954) 448-466.

389. GIRSHICK, MEYER A., SAMUEL KARLIN, and H. L. ROYDEN. "Multistage statistical decision procedures" Ann. Math. Stat. 28 (1957) 111-125.

390. GIRSHICK, MEYER A. and HERMAN RUBIN. "A Bayes approach to a quality control model" Ann. Math. Stat. 23 (1952) 114-125.

391. GIRSHICK, MEYER A., HERMAN RUBIN, and ROSEDITH SITGREAVES. "Estimates of bounded relative error in particle counting" Ann. Math. Stat. 26 (1955) 276-285.

392. GIRSHICK, M. A. and L. J. SAVAGE. "Bayes and minimax estimates for quadratic loss functions." In reference 727, pp. 53-73.

GIRSHICK, M. A. See Arrow, K. J.

393. GLEAVES, V. B. "Cyclic scheduling and combinatorial topology; assignment and routing of motive power to meet scheduling and maintenance requirements. Part I - a statement of the operation problem of the Frisco railroad" Naval Res. Log. Q. 4 (1957) 203-205.

394. GLEYZAL, A. N. "An algorithm for solving the transportation problem" J. Res. Nat. Bur. Standards 54 (1955) 213-216.

395. GLICKSBERG, IRVING L. "A derivative test for finite solutions of games" Proc. Amer. Math. Soc. 4 (1953) 895-897.

396. GLICKSBERG, IRVING L. "A further generalization of the Kakutani fixed point theorem, with application to Nash equilibrium points" Proc. Amer. Math. Soc. 3 (1952) 170-174.

397. GLICKSBERG, IRVING L. and OLIVER GROSS. "Notes on games over the square." In reference 575, pp. 173-182.

GLICKSBERG, IRVING L. See Bellman, R.

GLICKSBERG, IRVING L. See Fan, Ky.

398. GOLDMAN, ALAN J. "Resolution and separation theorems for polyhedral convex sets." In reference 576, pp. 41-51.

399. GOLDMAN, ALAN J. and ALBERT W. TUCKER. "Polyhedral convex cones." In reference 576, pp. 19-40.

400. GOLDMAN, ALAN J. and ALBERT W. TUCKER. "Theory of linear programming." In reference 576, pp. 53-97.

401. *GOLDSTEIN, L. and A. ORDEN, eds. Project SCOOP, Symposium on Linear Inequalities and Programming. Washington, D. C. Planning Research Division, Director of Management and Analysis Service, Comptroller, Hq. USAF, April 1952.

402. GOLDSTEIN, LEON. "The problem of contract awards." In reference 401, pp. 147-154.

GOODMAN, LEO A. See Bahadur, Raghu Raj.

GREENBERG, H. J. See Dorn, W. S.

403. GREENWALD, DAKOTA U. Linear Programming. New York, The Ronald Press Co., 1957, vii, 75 pp.

404. GRIESMER, JAMES H. "Extreme games with three values." This Study.

405. GROSS, OLIVER. "A rational game on the square." In reference 265, pp. 307-311.

GROSS, OLIVER. See Bellman, Richard.

GROSS, OLIVER. See Glicksberg, Irving.

406. GRUNDY, P. M. and C. A. B. SMITH. "Disjunctive games with the last player losing" Proc. Cambridge Philos. Soc. 52 (1956) 527-533.

407. GUILBAUD, GEORGES TH. "Leçons sur les éléments principaux de la théorie mathématique des jeux. I. La regle des partis (26 pp.). II. Jeux de reflexion (29 pp.). III. La lutte (27 pp.). Iv. La ruse (28 pp.). V. Coalitions (15 pp.) in Stratégies et Décisions Économiques, Etudes Théoriques et Applications aux Enterprises. Paris, Centre National de la Recherche Scientifique, 1954.

408. GUILBAUD, GEORGES TH. "Les problemes de partage" Économie Appliquée 5 (1952) 93-137.

409. GUILBAUD, G. TH. "La théorie des jeux" Revue d'Économie Politique 25 (1955) 153-188.

410. GUILBAUD, GEORGES TH. "La théorie des jexu - contributions, critiques a la théorie de la valeur" Économie Appliquée 2 (1949) 275-319.

411. GUILBAUD, GEORGES TH. "The theory of games" in International Economic Papers No. 1 (1951) 37-65 (translated by A. L. Minkes from "La théorie des jexu" in Économie Appliquée (1949)).

412. *GUNTHER, PAUL. "Use of linear programming in capital budgeting" J. Oper. Res. Soc. 3 (1955) 219-224.

413. GURK, H. M. "Five-person, constant-sum, extreme games." This Study.

414. GURK, H. M. and J. R. ISBELL. "Simple solutions." This Study.

415. GUTTMAN, LOUIS. "Some inequalities between latent roots and minimax (maximin) elements of real matrices" Pacific J. Math. 7 (1957) 897-902.

416. GUY, RICHARD K. and CEDRIC A. B. SMITH. "The G-values of various games" Proc. Cambridge Philos. Soc. 52 (1956) 514-526.

417. HALE, J. K. and H. H. WICKE. "An application of game theory to special weapons evaluation" Naval Res. Log. Q. 4 (1957) 347-356.

418. HANNAN, JAMES. "Approximation to Bayes risk in repeated play." In reference 265, pp. 97-139.

419. HANNAN, JAMES and HERBERT ROBBINS. "Asymptotic solutions of the compound decision problem for two completely specified distributions" Ann. Math. Stat. 26 (1955) 37-51.

HANSSMAN, FRED. See McCloskey, Joseph F.

HARRIS, THEODORE. See Arrow, Kenneth J.

420. HARSANYI, JOHN C. "Approaches to the bargaining problem before and after the theory of games: a critical discussion of Zeuthen's, Hicks', and Nash's theories" Econometrica 24 (1956) 144-157.

421. HARSANYI, JOHN C. "A bargaining model for the cooperative n-person game." This Study.

422. HARSANYI, JOHN C. "Cardinal utility in welfare economics and in the theory of risk-taking" J. Pol. Econ. 61 (1953) 434-435.

423. HARTMAN, S. "Quelques remarques sur le jeu de Banach et Mazur" Colloq. Math. 4 (1957) 261.

424. HAUSNER, MELVIN. "Multidimensional utilities." In reference 889, pp. 167-186.

425. HAYWOOD, Col. O. G. Jr. "Military decision and game theory" J. Oper. Res. Soc. 2 (1954) 365-385.

426. *HAYWOOD, Col. O. G. Jr. "Military decision and the mathematical theory of games" Air University Q. Rev. 4 (1950) 67 ff.

427. HEADY, EARL O. "Economic concepts in directing and designing research for programming use of range resources" J. Farm Econ. 38 (1956) 1604-1616. Discussion by C. B. Baker: 1630-1633.

428. HEADY, EARL O. "Simplified presentation and logical aspects of linear programming technique" J. Farm. Econ. 36 (1954) 1035-1048. Discussion 1048-1050.

HEADY, EARL O. See Babbar, M. M.

429. HELLER, I. "Least ballast shipping required to meet a specified shipping problem." In reference 401, pp. 164-171.

430. HELLER, I. "On the travelling salesman's problem." In reference 17, pp. 643-665.

431. HELLER, I. and TOMPKINS, C. B. "An extension of a theorem of Dantzig's." In reference 576, pp. 247-254.

432. HENDERSON, A. "The theory of duopoly" Q. J. Econ. 68 (1954) 565-584.

HENDERSON, A. See Charnes, A.

433. HENDERSON, JAMES M. "A short run model for the coal industry" Rev. Econ. and Stat. 37 (1955) 336-346.

434. HERSTEIN, I. N. "Metodi et tecniche matematiche in economica" L'industria (Milano) (1956) 13-32.

435. HERSTEIN, I. N. "Some mathematical methods and techniques in economics" Q. Appl. Math. 11 (1953) 249-262.

436. HERSTEIN, I. N. and JOHN MILNOR. "An axiomatic approach to measurable utility" Econometrica 21 (1953) 291-297.

437. HESELDON, G. P. M. and D. S. VAJDA. "Linear programming of an air-life" Conference on Linear Programming, May 1954 (arranged by Ferranti Ltd.) pp. 5-15. Appendix by E. M. L. Beale, pp. 16-17. Discussion p. 19.

438. HILDRETH, CLIFFORD. "Alternative conditions for social orderings" Econometrica 21 (1953) 81-94.

439. HILDRETH, CLIFFORD. "Economic implications of some cotton fertilizer experiments" Econometrica 23 (1955) 88-98.

440. HILDRETH, CLIFFORD. "A quadratic programming procedure" Naval Res. Log. Q. 4 (1957) 79-85.

441. HILDRETH, CLIFFORD. "Point estimates of ordinates of concave functions" J. Am. Stat. Assn. 49 (1954) 598-619.

442. HITCHCOCK, FRANK L. "The distribution of a product from several sources to numerous localities" J. Math. and Physics 20 (1941) 224-230.

443. HODGES, J. L. Jr. and E. L. LEHMANN. "Some applications of the Cramér-Rao inequality." In reference 727, pp. 13-22.

444. HODGES, J. L. Jr. and E. L. LEHMANN. "Some problems in minimax point estimation" Ann. Math. Stat. 21 (1950) 182-197.

445. HODGES, J. L. Jr. and E. L. LEHMANN. "The use of previous experience in reaching statistical decisions" Ann. Math. Stat. 23 (1952) 396-407.

HODGES, J. L. Jr. See Blackwell, David.

HODGES, J. L. Jr. See Chiang, C. L.

446. HOEFFDING, WASSILY. "The role of assumptions in statistical decisions." In reference 728, pp. 105-114.

447. HOFFMAN, A. J. "How to solve a linear programming problem." In reference 17, pp. 397-424.

448. HOFFMAN, A. J. "On approximate solutions of systems of linear inequalities" J. Res. Nat. Bur. Standards 49 (1952) 263-265.

449. HOFFMAN, A. J. and WALTER JACOBS. "Smooth patterns of production" Management Sci. 1 (1954) 86-91.

450. HOFFMAN, A. J. and J. B. KRUSKAL. "Integral boundary points of convex polyhedral." In reference 576, pp. 223-246.

451. HOFFMAN, A. J. and HAROLD W. KUHN. "On systems of distinct representatives." In reference 576, pp. 199-206.

452. HOFFMAN, A. J. and HAROLD W. KUHN. "Systems of distinct representatives and linear programming" Amer. Math. Monthly 63 (1956) 455-460.

453. HOFFMAN, A., M. MANNOS, D. SOKOLOWSKY and N. WIEGMANN. "Computational experience in solving linear programs" J. Soc. Ind. Appl. Math. 1 (1953) 17-33.

HOFFMAN, A. J. See Antosiewicz, H.

HOFFMAN, A. J. See Dantzig, G. B.

HOFFMAN, A. J. See Fan, Ky.

HOFFMAN, A. J. See Gaddum, J. W.

454. HOLLADAY, JOHN C. "Cartesian products of termination games." In reference 265, pp. 189-200.

HONIG, Lieut. D. P. See Gainen, Leon.

HONIG, Lieut. D. P. See Stanley, Commander E. D.

455. HODD, WILLIAM C. "Linear programming and the firm" Canad. J. Econ. and Pol. Sci. 18 (1952) 208-212.

456. *HOSKIN, MICHAEL. "The theory of games" Eureka 19 (1957) 5-10.

457. HOUTHAKKER, H. S. "On the numerical solution of the transportation problem" J. Oper. Res. Soc. 3 (1955) 210-214.

458. HURWICZ, LEONID. "Game theory and decisions" Sci. Am. 192 (Feb. 1955) 78-83.

459. HURWICZ, LEONID. "La teoria dei giochi e le decisioni" L'industria (1955) 261-268.

460. HURWICZ, LEONID. "The theory of economic behaviour" Am. Econ. Rev. 35 (1945) 909-925. Reprinted in G. J. Stigler & K. E. Boulding eds. Readings in Price Theory, Chicago, R. D. Irwin, 1952.

461. HURWICZ, LEONID. "What has happened to the theory of games" Am. Econ. Rev. Papers and Proceedings 43 (1953) 398-405.

HURWICZ, LEONID. See Arrow, Kenneth J.

462. HUTTON, R. F. "A row sum check upon computational accuracy in mathematical programming problems" J. Farm Econ. 37 (1955) 734-736.

463. INADA, KEN-ICHI. "Alternative incompatible conditions for a social welfare function" Econometrica 23 (1955) 396.

464. INADA, KEN-ITI. "On a certain decision problem under some constraints" Ann. Inst. Stat. Math Tokyo 4 (1952-53) 65-82.

465. ISAACS, RUFUS. "A card game with bluffing" Amer. Math. Monthly 62 (1955) 99-108.

466. ISAACS, RUFUS. "The problem of aiming and evasion" Naval Res. Log. Q. 2 (1955) 47-67.

467. ISBELL, JOHN R. "Absolute games." This Study.

468. ISBELL, JOHN R. "A class of game solutions" Proc. Amer. Math. Soc. 6 (1955) 346-348.

469. ISBELL, J. R. "A class of majority games" Quart. J. Math. Oxford 7 (1956) 183-187.

470. ISBELL, J. R. "Finitary games." In reference 265, pp. 79-96.

471. ISBELL, J. R. "On a theorem of Richardson" Proc. Am. Math. Soc. 8 (1957) 928-929.

472. ISBELL, J. R. and W. H. MARLOW. "Attrition games" Naval Res. Log. Q. 3 (1956) 71-94.

473. ISBELL, J. R. and W. H. MARLOW. "Methods of mathematical tactics" The George Washington University Logistics Research Project, Logistics Papers, Issues No. 14, September 1956. Chap. I-42; II-52; III-93.

ISBELL, J. R. See Gurk, H. M.

ISODA, KAZUO. See Nikaidô, Hukukane.

474. JACKSON, JAMES R. "On the existence problem of linear programming" Pacific J. Math. 4 (1954) 29-36.

475. JACOBS, WALTER. "Air Force progress in logistics planning" Management Sci. 3 (1956-57) 213-224.

476. JACOBS, WALTER. "The caterer problem" Naval Res. Log. Q. 1 (1954) 154-165.

477. JACOBS, WALTER. "Loss of accuracy in simplex computations" Naval Res. Log. Q. 4 (1957) 89-94.

478. JACOBS, WALTER. "Military applications of linear programming." In reference 17, pp. 1-27.

JACOBS, WALTER. See Hoffman, Alan J.

479. JENSEN, ARNE. "The applicability of decision theory in the planning and operation of telephone plant" in Proceedings of the First International Congress on the Application of the Theory of Probability in Telephone Engineering and Administration, 1955. Copenhagen, 1957. (These Congress Proceedings are published simultaneously as Volume 1, 1957 of Teleteknik.)

480. JENSEN, ARNE. A Distribution Model Applicable to Economics. Kopenhagen, Munksgaard, 1954, 99 pp.

481. JEWELL, WILLIAM S. "Warehousing and distribution of a seasonal product" Naval Res. Log. Q. 4 (1957) 29-34.

JOHN, J. B. See Garvin, W. W.

482. JOHNS, M. V. Jr. "Non-parametric empirical Bayes procedures" Ann. Math. Stat. 28 (1957) 649-669.

483. JOHNSON, GLENN L. "The Friedman-Savage utility hypothesis in the Interstate Managerial Study" J. Farm Econ. 37 (1955) 1110-1114.

484. JOHNSON, SELMER and GEORGE DANTZIG. "A production smoothing problem." In reference 17, pp. 151-176.

JOHNSON, S. M. See Dantzig, George B.

JOHNSON, S. M. See Fulkerson, D. R.

485. *de JONGH, B. H. "An extension of the theory of games and economic behaviour" Verzekerings-Arch. Actuarieel Bijvoegsel 30 (1953) 54-61.

JUNCOSA, M. L. See Kalaba, R. E.

486. JUSTMAN, E. "La théorie des jeux. Une nouvelle théorie de l'équilibre économique" Revue d'Econ. Politique 59 (1949) 616-633.

487. KAKUTANI, SHIZUO. "A generalization of Brouwer's fixed point theorem" Duke Math. J. 8 (1941) 457-459.

488. KALABA, R. E. and M. L. JUNCOSA. "Communication-transportation networks. (Abstract)" Naval Res. Log. Q. 4 (1957) 221-222.

489. KALABA, R. E. and M. L. JUNCOSA. "Optimal design and utilization of communication networks" Management Sci. 3 (1956-57) 33-44.

490. KALISCH, G. K. "Generalized quota solutions of n-person games." This Study.

491. KALISCH, G. K., J. W. MILNOR, J. F. NASH, and E. D. NERING. "Some experimental n-person games." In reference 889, pp. 301-327.

492. KALISCH, G. K. and E. D. NERING. "Countably infinitely many person games." This Study.

493. KALMAR, L. "Zur Theorie der abstrakten Spiele" Acta Sci. Math. Szeged 4 (1928-29) 65-85.

494. KANTOROVITCH, L. "On the translocation of masses" Dokl. Akad. Nauk SSSR 37 (1942) 199-201.

495. KAPLAN, ABRAHAM. "Sociology learns the language of mathematics" Commentary (September 1952) 274, 282-284.

496. KAPLAN, MORTON. System and Process in International Politics. New York, John Wiley and Sons, Inc., 1957, xxiv, 283.

497. KAPLANSKY, IRVING. "A contribution to von Neumann's theory of games" Ann of Math. 46 (1945) 474-479.

498. KARLIN, SAMUEL. "Continuous games" Proc. Nat. Acad. Sci. U.S.A. 37 (1951) 220-223; errata 381.

499. KARLIN, SAMUEL. "An infinite move game with a lag." In reference 265, pp. 257-272.

500. KARLIN, SAMUEL. "On a class of games." In reference 575, pp. 159-171.

501. KARLIN, SAMUEL. "On games described by bell shaped kernels." In reference 265, pp. 365-391.

502. KARLIN, SAMUEL. "Operator treatment of minmax principle." In reference 574, pp. 133-154.

503. KARLIN, SAMUEL. "Pólya-type distributions, II" Ann. Math. Stat. 28 (1957) 281-308.

504. KARLIN, SAMUEL. "Pólya-type distributions, III: Admissibility for multi-action problems" Ann. Math. Stat. 28 (1957) 839-860.

505. KARLIN, SAMUEL. "Reduction of certain classes of games to integral equations." In reference 575, pp. 125-158.

506. KARLIN, SAMUEL. "The theory of infinite games" Ann. of Math. 58 (1953) 371-401.

507. KARLIN, SAMUEL and RODRIGO RESTREPO. "Multistage poker models." In reference 265, pp. 337-363.

508. KARLIN, SAMUEL and HERMAN RUBIN. "Distributions possessing a monotone likelihood ratio" J. Amer. Stat. Assoc. 51 (1956) 637-643.

509. KARLIN, SAMUEL and HERMAN RUBIN. "The theory of decision procedures for distributions with monotone likelihood ratio" Ann. Math. Stat. 27 (1956) 272-299.

510. KARLIN, SAMUEL and LLOYD S. SHAPLEY. "Geometry of reduced moment spaces" Proc. Nat. Acad. Sci. U.S.A. 35 (1949) 673-677.

KARLIN, SAMUEL. See Bohnenblust, H. F.

KARLIN, SAMUEL. See Bradt, Russell N.

KARLIN, SAMUEL. See Dresher, Melvin.

KARLIN, SAMUEL. See Girshick, Meyer A.

KARR, H. W. See Geisler, M. A.

511. KARUSH, W. and A. VAZSONYI. "Mathematical programming and service scheduling" Management Sci. 3 (1956-57) 140-148.

512. KARUSH, W. and A. VAZSONYI. "Mathematical programming and employment scheduling" Naval Res. Log. Q. 4 (1957) 297-320.

513. KATZMAN, I. "Solving feed problems through linear programming" J. Farm Econ. 38 (1956) 420-429.

514. KAY, EMIL and ERIC DUCKWORTH. "Linear programming in practice" Appl. Stat. 6 (1957) 26-39.

515. KAYSEN, CARL. "The minimax rule of the theory of games, and the choice of the strategy under conditions of uncertainty" Metroecon. 4 (1952) 5-14.

516. KAYSEN, CARL. "A revolution in economic theory?" Rev. Econ. Studies 14 (1946-47) 1-15.

517. KEEPING, E. S. "Statistical decisions" Amer. Math. Monthly 63 (1956) 147-159.

KEHRBERG, EARL W. See Puterbaugh, Horace L.

518. KELLEY, J. E. Jr. "A dynamic transportation model" Naval Res. Log. Q. 2 (1955) 175-180.

519. KELLEY, J. E. Jr. "A threshold method for linear programming" Naval Res. Log. Q. 4 (1957) 35-45.

520. KEMENY, J. G. "A new approach to n-person games." This Study.

521. KEMENY, JOHN G., OSKAR MORGENSTERN, and GERALD L. THOMPSON. "A generalization of the von Neumann model of an expanding economy" Econometrica 24 (1956) 115-135.

522. KEMENY, JOHN G. and LAURIE SNELL. "Game-theoretic solution of baccarat" Amer. Math. Monthly 64 (1957) 465-469.

523. KEMENY, JOHN G., LAURIE SNELL, and GERALD L. THOMPSON. Introduction to Finite Mathematics. Englewood Cliffs, N. J., Prentice-Hall, Inc., 1957, xi, 372 pp.

524. KEMENY, JOHN G. and GERALD L. THOMPSON. "The effect of psychological attitudes on the outcomes of games." In reference 265, pp. 273-298.

525. KIEFER, J. "Invariance, minimal sequential estimation, and continuous time processes" Ann. Math. Stat. 28 (1957) 573-601.

526. KIEFER, J. "On Wald's complete class theorems" Ann. Math. Stat. 24 (1953) 70-75.

527. KIEFER, J. "Optimum sequential search and approximation methods under minimum regularity assumptions" J. Soc. Indust. Appl. Math. 5 (1957) 105-136.

528. KIEFER, J. "Sequential minimax estimation for the rectangular distribution with unknown range" Ann. Math. Stat. 23 (1952) 586-593.

529. KIEFER, J. "Sequential minimax search for a maximum" Proc. Amer. Math. Soc. 4 (1953) 502-506.

530. KIEFER, J. and LIONEL WEISS. "Some properties of generalized sequential probability ratio tests" Ann. Math. Stat. 28 (1957) 57-74.

KIEFER, J. See Dvoretzky, A.

KIEFER, J. See Frank, P.

KIMBALL, GEORGE E. See Morse, Philip M.

531. KING, RICHARD A. "Some applications of activity analysis in agricultural economics" J. Farm Econ. 35 (1953) 823-833.

532. KING, R. A., C. E. BISHOP, and J. G. SUTHERLAND. "Programming resource use and capital investment in agriculture" Management Sci. 3 (1956-57) 173-184.

533. KITAGAWA, TOSIO. "Empirical functions and interpenetrating sampling procedures" Mem. Fac. Sci. Kyusyu Univ. 8 (1954) 109-152.

534. KITAGAWA, TOSIO. "Successive processes of statistical controls. I" Mem. Fac. Sci. Kyusyu Univ., Ser. A 7 (1952) 13-28.

535. KITAGAWA, TOSIO. "Successive process of statistical inferences, (1)" Mem. Fac. Sci. Kyusyu Univ. 5 (1950) 139-180.

536. KITAGAWA, TOSIO. "Successive process of statistical inferences, (2)" Mem. Fac. Sci. Kyusyu Univ. 6 (1951) 55-95.

537. KITAGAWA, TOSIO. "Successive process of statistical inferences, (3)" Mem. Fac. Sci. Kyusyu Univ. 6 (1952) 131-155.

538. KITAGAWA, TOSIO. "Successive process of statistical inferences. IV" Bull. Math. Stat. 5 (1952) 35-50.

539. KITAGAWA, TOSIO. "Successive process of statistical inferences, (5)" Mem. Fac. Sci. Kyusyu Univ. 7 (1953) 81-106.

540. KITAGAWA, TOSIO. "Successive process of statistical inferences, (6)" Mem. Fac. Sci. Kyusyu Univ. 8 (1953) 1-29.

541. KLEIN, BERTRAM. "Direct use of extremal principles in solving certain optimizing problems involving inequalities" J. Oper. Res. Soc. 3 (1955) 168-175.

542. KLEIN, LAWRENCE. A Textbook of Econometrics. Evanston, Illinois; White Plains, New York, Row, Peterson and Company, 1953.

543. KLEIN, MORTON. "Some production planning problems" Naval Res. Log. Q. 4 (1957) 269-286.

544. KNESER, HELLMUTH. "Soziologie und Wirtschaftswissenschaft in heutiger mathematischer Behandlung" Studium Gen. 6 (1953) 666-678.

545. KNESER, HELLMUTH. "Sur un théorème fondamental de la théorie des jeux" C. R. Acad. Sci. Paris 234 (1952) 2418-2420.

546. KOCH, WOLDEMAR. "Die Bedeuting der Theoretischen Ökonomik für die Allgemeine Soziologie" Vorlesung von Dr. Woldemar Koch. Recht und Staat in Geoschichte und Gegenwart. Tubingen, Verlag J. C. B. Mahr, 1955, pp. 3-21.

547. KONIJN, H. S. "A further remark on the characterization of minimax procedures" Ann. Inst. Statist. Math. Tokyo 6 (1954) 123.

548. KONIJN, H. S. "On certain classes of statistical decision procedures" Ann. Math. Stat. 24 (1953) 440-448; corrections 25 (1954) 170-171.

549. KONIJN, H. S. "A remark on the characterization of minimax procedures" Ann. Inst. Statist. Math. Tokyo 4 (1953) 103-105.

550. KONIJN, H. S. "Some estimates which minimize the least upper bound of a probability together with the cost of observation" Ann. Inst. Statist. Math. Tokyo 7 (1955-56) 143-158.

551. KOOPMANS, T. "Activity analysis and its applications" Amer. Econ. Rev. 43 (1953) 406-414.

552. KOOPMANS, TJALLING C. and OTHERS. Activity Analysis of Production and Allocation. Cowles Commission Monograph No. 13. New York, John Wiley and Sons, Inc., 1951, xiv, 404.

553. KOOPMANS, TJALLING C. "Analysis of production as an efficient combination of activities." In reference 552, pp. 33-97.

554. KOOPMANS, T. C. "Optimum utilization of the transportation system" Proc. of the International Statistical Conference, 1947, Washington, D. C., vol. 5, pp. 136-146.

555. KOOPMANS, T. C. Three Essays on the State of Economic Science. New York/Toronto, McGraw-Hill, 1957, 231 pp.

556. KOOPMANS, TJALLING C. and MARTIN BECKMANN. "Assingment problems and the location of economic activities" Econometrica 25 (1957) 53-76.

557. KOOPMANS, TJALLING C. and STANLEY REITER. "A model of transportation." In reference 552, pp. 222-259.

KOOPMANS, TJALLING C. See Brown, George W.

558. KOSE, T. "Solutions of saddle value problems by differential equations" Econometrica 24 (1956) 59-70.

559. *KRELLE, W. "Möglichkeiten und Grenzen der Oekonometrie" Z. Gesamte Staatswissenschaft (1957) 415.

560. KRENTEL, W. D., J. C. C. McKINSEY, and W. W. QUINE. "A simplification of games in extensive form" Duke Math. J. 18 (1951) 885-900.

KRETSCHMER, KENNETH S. See Chenery, Hollis B.

561. *KRIENS, J. "Doeltreffende beleidskeuze door lineaire programmering" in Keus en kans; voor drachten gehouden op de Statistische dag 1956 (22 maart) 86 p. Statistica Neerlandica (Gravenhage) no. 1 (1956).

562. *KROPA, E. L. "Game theory in development" Ind. and Eng. Chem. 48 (1956) 388-392.

KRUSKAL, J. B. See Hoffman, A. J.

563. *KUDO, AKIO. "On the invariant multiple decision procedures" Bull. Math. Statist. 6 (1956) 57-68.

564. *KUDO, HIROKICHI. "On minimax invariant estimates of the transformation parameter" Nat. Sci. Rep. Ochanomizu Univ. 6 (1955) 31-73.

565. KUHN, HAROLD W. "Extensive games" Proc. Nat. Acad. Sci. U.S.A. 36 (1950) 570-576.

566. KUHN, H. W. "Extensive games and the problem of information." In reference 575, pp. 193-216.

567. KUHN, H. W. "The Hungarian method for the assignment problem" Naval Res. Log. Q. 2 (1955) 83-97.

568. KUHN, H. W. "A note on Prager's transportation problem" J. Math. Phys. 36 (1957) 107-111.

569. KUHN, H. W. "A note on 'the law of supply and demand'" Math. Scand. 4 (1956) 143-146.

570. KUHN, H. W. "On a theorem of Wald." In reference 576, pp. 265-273.

571. KUHN, H. W. "A simplified two-person poker." In reference 574, pp. 97-103.

572. KUHN, H. W. "Solvability and consistency for linear equations and inequalities" Amer. Math. Monthly 63 (1956) 217-232.

573. KUHN, H. W. "Variants of the Hungarian method for assignment problems" Naval Res. Log. Q. 3 (1956) 253-258.

574. KUHN, HAROLD W. and ALBERT W. TUCKER, eds. Contributions to the Theory of Games, Volume I. Annals of Mathematics Studies, No. 24. Princeton, N. J., Princeton University Press, 1950.

575. KUHN, HAROLD W. and ALBERT W. TUCKER, eds. Contributions to the Theory of Games, Volume II. Annals of Mathematics Studies, No. 28. Princeton, N. J., Princeton University Press, 1953.

576. KUHN, HAROLD W. and ALBERT W. TUCKER, eds. Linear Inequalities and Related Systems. Annals of Mathematics Studies, No. 38. Princeton, N. J., Princeton University Press, 1956.

577. *KUHN, H. W. and A. W. TUCKER. "Nonlinear programming." In reference 727, pp. 481-492.

578. *KUHN, HAROLD W. and ALBERT W. TUCKER. "Theory of games" The Encyclopaedia Britannica (1956 printing) vol. 10.

579. KUHN, H. W. and A. W. TUCKER. "John von Neumann's work in the theory of games and mathematical economics" Bull. Amer. Math. Soc. 64 (1958) von Neumann memorial supplement (no. 3, part 2) 100-122.

KUHN, HAROLD W. See Gale, David.

KUHN, HAROLD W. See Hoffman, A. J.

580. LADERMAN, JACK. "On the asymptotic behavior of decision procedures" Ann. Math. Stat. 26 (1955) 551-575.

LADERMAN, JACK. See Beckmann, Martin J.

581. LAGACHE, MICHEL. "L'analyse structurale en économie; la théorie des jeux" Revue d'Économie Pol. 60 (1950) 399-418.

582. LAND, Mrs. A. H. "A problem in transportation" Conference on Linear Programming (arranged by Ferranti, Ltd.) May 1954, pp. 20-28.

LAND, Mrs. A. H. See Vajda, S.

583. *de LASALA, JESUS. "Study of a rapid method of calculation for application of the simplex method to the Hitchcock transportation problem" Trabajos de Estadistica 6 (1955) 209-236.

584. LeCAM, LUCIEN. "An extension of Wald's theory of statistical decision functions" Ann. Math. Stat. 26 (1955) 69-81.

585. LeCAM, LUCIEN. "Note on a theorem of Lionel Weiss" Ann. Math. Stat. 25 (1954) 791-794.

586. *LeCAM, LUCIEN. "On some asymptotic properties of maximum likelihood estimates and related Bayes' estimates" Univ. California Publ. Stat. 1 (1953) 277-329.

587. LEHMANN, E. L. "A general concept of unbiasedness" Ann. Math. Stat. 22 (1951) 587-592.

588. LEHMANN, E. L. "Multiple decision problems, I" Ann. Math. Stat. 28 (1957) 1-25.

589. LEHMANN, E. L. "On the existence of least favorable distributions" Ann. Math. Stat. 23 (1952) 408-416.

590. LEHMANN, E. L. "Some principles of the theory of testing hypotheses" Ann. Math. Stat. 21 (1950) 1-26.

591. LEHMANN, E. L. "Testing multiparameter hypotheses" Ann. Math. Stat. 23 (1952) 541-552.

592. LEHMANN, E. L. "A theory of some multiple decision problems, II" Ann. Math. Stat. 28 (1957) 547-572.

593. LEHMANN, E. L. and C. M. STEIN. "The admissibility of certain invariant statistical tests involving a translation parameter" Ann. Math. Stat. 24 (1953) 473-479.

LEHMANN, E. L. See Bahadur, Raghu Raj.

LEHMANN, E. L. See Hodges, J. L. Jr.

LEHMAN, SHERMAN. See Bellman, Richard.

594. LEMKE, C. E. "The dual method of solving the linear programming problem" Naval Res. Log. Q. 1 (1954) 36-47.

LEMKE, C. E. See Charnes, Abraham.

595. LEONTIEF, WASSILY. "Mathematics in economics" Bull. Amer. Math. Soc. 60 (1954) 215-233.

596. *LEUNBACH, G. "An illustration of the application of statistical decision functions in a telephone plant." Summary in Proc. of the First International Congress on the Application of the Theory of Probability in Telephone Engineering and Administration, 1955, Copenhagen, 1957, p. 30. (These Congress Proceedings published simultaneously as Vol. 1, 1957, No. 1 of Teleteknik. Published in Nordisk Tidskr. for Teknisk Økonomi (1955) 146.)

597. LEUNBACH, G. "The theory of games and economic behavior, Johann v. Neumann and Oscar Morgenstern" Nordisk Tidskr. for Teknisk Økonomi (1948) 175-178.

598. LINDLEY, D. V. "On a measure of the information provided by an experiment" Ann. Math. Stat. 27 (1956) 986-1005.

599. LINDLEY, D. V. "Statistical inference" J. Roy. Statist. Soc. Ser. B. 15 (1953) 30-65; discussion, 65-76.

600. LLOSÁ, RICARDO SAN JUAN. "El metodo del 'simplex' de la programación lineal" Trabajos de Estadistica 7 (1956) 199-219.

601. *LOCKE, E. L. "The Finan-Seer" Astounding Science Fiction 44 (1949).

602. LOMAX, K. S. "Allocation and programming in modern economics" Manchester School Econ. Soc. Stud. 21 (1953) 177-223.

603. LOOMIS, LYNN H. "On a theorem of von Neumann" Proc. Nat. Acad. Sci. U.S.A. 32 (1946) 213-215.

604. LOUWES, S. L. "Activiteits analyse, lineaire programmering en hun toepassingen, I" De Economist (Haarlem) 103 (1955) 817-853.

605. LOUWES, S. L. "Activiteits analyse, lineaire programmering en hun toepassingen, II" De Economist (Haarlem) 104 (1956) 40-56.

606. LUCE, R. DUNCAN. "A definition of stability for n-person games" Ann. of Math. 59 (1954) 357-366.

607. LUCE, R. DUNCAN. "K-stability of symmetric and of quota games" Ann. of Math. 62 (1955) 517-527.

608. LUCE, R. DUNCAN. "A note on the article 'Some experimental n-person games.'" This Study.

609. LUCE, R. DUNCAN. "ψ-stability: A new equilibrium concept for n-person game theory" Proc. of a Symposium on Mathematical Models of Human Behavior, 1954. Dunlap and Associates, Inc., and the Commission on Accidental Trauma. Armed Forces Epidemiological Board, 1955, pp. 1-21.

610. LUCE, R. DUNCAN. "Semiorders and a theory of utility discrimination" Econometrica 24 (1956) 178-191.

611. LUCE, R. DUNCAN and ERNEST W. ADAMS. "The determination of subjective characteristic functions in games with misperceived payoff functions" Econometrica 24 (1956) 158-171.

612. LUCE, R. DUNCAN and HOWARD RAIFFA. Games and Decisions. Introduction and Critical Survey. New York, John Wiley and Sons, Inc., 1957, xix, 509.

613. LUCE, R. DUNCAN and A. A. ROGOW. "A game theoretic analysis of congressional power distributions for a stable two-party system" Behavioral Science 1 (1956) 83-95.

614. LUCE, R. DUNCAN and ALBERT W. TUCKER, eds. Contributions to the Theory of Games, Volume IV. Annals of Mathematics Studies, No. 40. Princeton, N. J., Princeton University Press, 1958.

615. LUNDQUIST, N. H. "Luftförsvarets -- en spelteoretisk tillämpning" Artillen Tidskr. 84 (1955) 122-130.

616. McCLOSKEY, JOSEPH F. and FRED HANSSMAN. "An analysis of stewardess requirements and scheduling for a major domestic airline" Naval Res. Log. Q. 4 (1957) 183-191.

617. McCORKLE, CHESTER O. Jr. "Linear programming as a tool in farm management analysis" J. Farm. Econ. 37 (1955) 1222-1235. Discussion by James S. Plaxico 1258-1261.

618. *McCORKLE, CHESTER O. Jr. and JAMES BOLES. "Use of linear programming in cotton acreage adjustment research in California - 1954" Proc. of the Western Farm Economics Association, Twenty-seventh Annual Meeting, July 26-28, 1954, Estes Park, Colorado.

McDERMOTT, JAMES. See Baldwin, Roger R.

619. McDONALD, JOHN DENNIS. Strategy in Poker, Business and War. New York, Norton, 1950, 128 pp.

620. McDONALD, JOHN and FRANC RICCIARDI. "The business decision game" Fortune (1958) 140-142; 208; 213-214.

621. *McDONALD, JOHN. "Poker: an American game" Fortune 37 (1948) 128-131 and 181-187.

622. *McDONALD, JOHN and JOHN W. TUKEY. "Colonel Blotto: A problem of military strategy" Fortune (June 1949) 102.

623. *McDONALD, JOHN. "The theory of strategy" Fortune 38 (1949) 100-110.

McFARLANE, G. C. See Dillon, J. L.

624. McKENZIE, LIONEL W. "Competitive equilibrium with dependent consumer preferences." In reference 17, pp. 277-294.

625. McKENZIE, LIONEL W. "Equality of factor prices in world trade" Econometrica 23 (1955) 239-257.

626. McKINSEY, J. C. C. Introduction to the Theory of Games. New York, McGraw-Hill, 1952, x, 371.

627. McKINSEY, J. C. C. "Isomorphism of games, and strategic equivalence." In reference 574, pp. 117-130.

628. McKINSEY, J. C. C. "Some notions and problems of game theory" Bull. Amer. Math. Soc. 58 (1952) 591-611.

McKINSEY, J. C. C. See Krentel, W. D.

629. MACHOL, ROBERT E. "The mechanical blackboard" J. Oper. Res. 5 (1957) 422-428.

630. MAILLET, PIERRE. "Une nouvelle technique économique: les programmes linéaires" Revue d'Economie Pol. 63 (1953) 114-123.

MAISEL, HERBERT. See Baldwin, Roger R.

631. MAJUMDAR, TAPAS. "Choice and revealed preference" Econometrica 24 (1956) 71-73.

MALCOLM, DONALD G. See Bellman, Richard.

632. MALINVAUD, E. "Note on von Neumann-Morgenstern's strong independence axiom" Econometrica 20 (1952) 679.

633. MANDELBROT, BENOIT. "Contribution a la théorie mathématique des jeux de communication" Publ. Inst. Stat. Univ. Paris 2, Nos. 1 and 2 (1953) 1-124.

634. MANNE, ALAN S. "A note on the Modigliani-Hohn production smoothing model" Management Sci. 3 (1956-57) 371-379.

635. MANNE, ALAN S. Scheduling of Petroleum Refinery Operations. Cambridge, Mass., Harvard University Press, 1956, xii, 185.

636. MANNE, ALAN S. "The strong independence assumption-gasoline blends and probability mixtures" (with additional note by A. Charnes) Econometrica 20 (1952) 665-669.

MANNE, ALAN S. See Markowitz, Harry M.

MANNOS, M. See Hoffman, A.

637. MARKOWITZ, HARRY. "Concepts and computing procedures for certain X_{ij} programming problems." In reference 17, pp. 509-565.

638. MARKOWITZ, HARRY M. "The elimination form of the inverse and its application to linear programming" Management Sci. 3 (1956-57) 255-269.

639. MARKOWITZ, HARRY M. "The optimization of a quadratic function subject to linear constraints" Naval Res. Log. Q. 3 (1956) 111-133.

640. MARKOWITZ, HARRY M. "The utility of wealth" J. Pol. Econ. 60 (1952). Also: Proc. of a Symposium on Mathematical Models of Human Behavior, 1954. Dunlap and Associates, Inc. and The Commission on Accidental Trauma. Armed Forces Epidemiological Board, 1955, pp. 54-62.

641. MARKOWITZ, HARRY M. and ALAN MANNE. "On the solution of discrete programming problems" Econometrica 25 (1957) 84-110.

MARLOW, W. H. See Isbell, J. R.

642. MARSCHAK, J. "Neumann's and Morgenstern's new approach to static economics" J. Pol. Econ. 54 (1946) 97-115. Also circulated as "Cowles Commission Paper" N.S. 13b.

643. MARSCHAK, J. "Norms and habits of decision making under certainty" Proc. of a Symposium on Mathematical Models of Human Behavior, 1954. Dunlap and Associates, Inc. and The Commission on Accidental Trauma. Armed Forces Epidemiological Board, 1955, pp. 45-53.

644. MARSCHAK, JACOB. "Rational behavior, uncertain prospects, and measurable utility" Econometrica 18 (1950) 111-141.

645. MARSCHAK, JACOB. "The role of liquidity under complete and incomplete information" Amer. Econ. Rev. 39 (1949) 182-195. With discussion 196-210. Reprinted as Cowles Commission Paper, N.S. 37.

646. MARSCHAK, JACOB. "Verso una scala di preferenza per prendere decisioni" L'industria (1955) 319-331.

647. MARSCHAK, JACOB. "Why 'should' statisticians and business men maximize 'moral expectation'?" In reference 727, pp. 493-506. Reprinted as Cowles Commission Paper, N.S. 53.

648. MARSCHAK, JACOB and M. R. MICKEY. "Optimal weapon systems" Naval Res. Log. Q. 1 (1954) 116-140.

MARSCHAK, JACOB. See Arrow, Kenneth.

MARSCHAK, JACOB. See Radner, Roy.

MARSCHAK, T. See Beckmann, Martin.

649. MARTIN, A. D. Jr. "Mathematical programming of portfolio selections" Management Sci. 1 (1955) 152-166.

650. MASSÉ, P. and R. GIBRAT. "Application of linear programming to investments in the electric power industry" Management Sci. 3 (1956-57) 149-166.

651. MASSÉ, P. and P. MORLAT. "Sur le classement économique des perspective aléatories" Économetrie. Colloques Internationaux du Centre National de la Recherche Scientifique, Paris, Centre National de la Recherche Scientifique, 1953, pp. 165-193; discussion 194-199.

652. MATUSITA, KAMEO. "Decision rule by probability ratio" Ann. Inst. Stat. Math. Tokyo 6 (1954) 143-151.

653. MATUSITA, KAMEO. "Decision rules, based on the distance, for problems of fit, two samples, and estimation" Ann. Math. Stat. 26 (1955) 631-640.

654. MATUSITA, KAMEO. "On the theory of statistical decision functions" Ann. Inst. Stat. Math. Tokyo 3 (1951-52) 17-35; correction 4 (1952) 51-53.

655. MATUSITA, KAMEO. "A remark to 'on the estimation by the minimum distance method'" Ann. Inst. Stat. Tokyo 6 (1954) 124.

656. MATUSITA, KAMEO. "A remark to the Wald's theory of statistical inference" Ann. Inst. Stat. Math. Tokyo 1 (1949) 141-148.

657. MATUSITA, KAMEO and HIROTUGU AKAIKE. "Decision rules, based on the distance, for the problems of independence, invariance and two samples" Ann. Inst. Stat. Math. Tokyo 7 (1955-56) 67-80.

658. MATUSITA, KAMEO and HIROTUGU AKAIKE. "Note on the decision problem" Ann. Inst. Stat. Math. Tokyo 4 (1952-53) 11-14.

659. MATUSITA, KAMEO and MINORU MOTOO. "On the fundamental theorem for the decision rule based on distance" Ann. Inst. Stat. Math. Tokyo 7 (1955-56) 137-142.

660. MAURICE, RITA J. "A minimax procedure for choosing between two populations using sequential sampling" J. Roy. Statist. Soc. Ser. B 19 (1957) 255-261.

661. MAYBERRY, J. P. "A geometric interpretation of the simplex method." In reference 401, pp. 56-67.

662. MAYBERRY, J. P., J. F. NASH, and M. SHUBIK. "A comparison of treatments of a duopoly situation" Econometrica 21 (1953) 141-154.

MAYBERRY, J. P. See Gilles, D. B.

MELLON, B. See Charnes, A.

van METER, DAVID. See Middleton, David.

MICKEY, M. R. See Marschak, Jacob.

663. MIDDLETON, DAVID and DAVID van METER. "Detection and extraction of signals in noise from the point of view of statistical decision theory, I and II" J. Soc. Indust. Appl. Math. 3 (1955) 192-253; 4 (1956) 86-119.

664. MIDUTANI, KAZUO. "The maximum expansion of bank credit and the theory of games" Osaka Economic Papers 4 (1955) 53-56.

665. MIHALEVIC, V. S. "Bayes solutions and optimal methods of statistical acceptance control" Ukrain. Mat. Z. 8 (1956) 454-459.

MILLER, M. H. See Charnes, A.

666. MILLS, EDWIN S. "The theory of inventory decision" Econometrica 25 (1957) 222-238.

667. MILLS, HARLAN D. "Marginal values of matrix games and linear programs." In reference 576, pp. 183-193.

668. MILLS, W. H. "The four person game - edge of the cube" Ann. of Math. 59 (1954) 367-378.

669. MILLS, W. H. "The four person game - finite solutions on the face of the cube." This Study.

670. MILNOR, JOHN. "Games against nature." In reference 889, pp. 49-59.

671. MILNOR, JOHN. "Sums of positional games." In reference 575, pp. 291-301.

672. MILNOR, JOHN and LLOYD S. SHAPLEY. "On games of survival." In reference 265, pp. 15-45.

MILNOR, JOHN. See Herstein, I. N.

MILNOR, J. W. See Kalisch, G. K.

673. MINTY, GEORGE. "A comment on the shortest-route problem" J. Oper. Res. 5 (1957) 724.

674. von MISES, RICHARD. "Uber die J. von Neumannsche Theorie der Spiele" Math. Nachr. 9 (1953) 363-378.

675. *MIYASAWA, KÔICHI. "Minimax estimations" Bull. Math. Stat. 5 (1952) 59-76.

676. *MIYASAWA, KÔICHI. "On the minimax point estimations" Bull. Math. Stat. 5 (1953) 1-17.

677. *MIYASAWA, KÔICHI. "On the statistical decision function, I" Bull. Math. Stat. 4 (1951) 22-32.

678. MORGENSTERN, OSKAR. "Demand theory reconsidered" Q. J. Econ. 62 (1948) 165-201.

679. MORGENSTERN, OSKAR. "Economics and the theory of games" Kyklos 3 (1949) 294-308.

680. MORGENSTERN, OSKAR. "Oligopoly, monopolistic competition, and the theory of games" Amer. Econ. Rev. 38 (suppl.) (1948) 10-18; discussion 19-32.

681. MORGENSTERN, OSKAR. "La teoria dei giochi e del comportamento economico" L'industria (1951) 315-346. Trans. from J. für Sozialwissenschaft.

682. MORGENSTERN, OSKAR. "Spieltheorie." In Handwörterbuch der Sozialwissenschaften. Hrsg. von E. von Beckerath, C. Brinkmann. Stuttgart, G. Fischer (1952-56) (Volume 13) 706-713.

683. MORGENSTERN, OSKAR. "Die Theorie der Spiele und des wirtschlaftlichen Verhaltens" Jahrbuch für Sozialwissenschaft 1 (1950) 113-139.

684. MORGENSTERN, OSKAR. "Theorie des Spiels" Die Amerikanische Rundschau 5 (1949) 76-87.

685. MORGENSTERN, OSKAR. "The theory of games" Sci. Amer. 180 (1949) 22-25.

MORGENSTERN, OSKAR. See Kemeny, John J.

MORGENSTERN, OSKAR. See von Neumann, John.

686. MORIGUTI, SIGEITI. "Notes on sampling inspection plans" Rep. Stat. Appl. Res. Un. Jap. Sci. Engrs. 3 (1955) 99-121.

MORLAT, P. See Massé, P.

687. MORSE, P. M. "Mathematical problems in operations research" Bull. Amer. Math. Soc. 54 (1948) 602-621.

688. MORSE, PHILIP M. "Operations research" Comm. Pure Appl. Math. 8 (1955) 1-12.

689. MORSE, PHILIP M. "Statistics and operations research" J. Oper. Res. 4 (1956) 1-18.

690. MORSE, PHILIP M. and GEORGE E. KIMBALL. Methods of Operations Research. New York, John Wiley and Sons, Inc., 1951, vii, 158.

691. MORTON, D. G. "An application of dynamic programming." In Conference on Linear Programming (arranged by Ferranti and Co., Ltd.) pp. 32-38. Discussion 39-40.

692. MORTON, GEORGE. "Economic theory in a new light" Z. für Ökonometrie 1 (1950) 128-132.

693. MORTON, GEORGE. "Notes on linear programming" Economica N.S. 18 (1951) 397-411.

MORTON, G. See Vajda, S.

694. MOSTELLER, FREDERICK and PHILIP NOGEE. "An experimental measurement of utility" J. Pol. Econ. 59 (1951) 371-404.

MOTOO, MINORU. See Matusita, Kameo.

695. MOTZKIN, T. S. "The assignment problem" Proc. of the Sixth Symposium on Applied Mathematics: Numerical Analysis, vol. 6, ed. by J. H. Curtiss. New York, McGraw-Hill Book Company, Inc., 1956, pp. 109-125.

696. MOTZKIN, T. S. "New techniques for linear inequalities and optimization." In reference 401, pp. 15-27.

697. MOTZKIN, T. S. "The probability of solvability of linear inequalities." In reference 17, pp. 607-611.

698. MOTZKIN, T. S. "Remarks on the history of linear inequalities." In reference 401, pp. 179-181.

699. MOTZKIN, T. S. "Two consequences of the transposition theorem on linear inequalities" Econometrica 19 (1951) 184-185.

700. MOTZKIN, T. S., H. RAIFFA, G. L. THOMPSON, and R. M. THRALL. "The double description method." In reference 575, pp. 51-73.

701. MOTZKIN, T. S. and I. J. SCHOENBERG. "The relaxation method for linear inequalities" Canad. J. Math. 6 (1954) 393-404.

702. MUNKRES, JAMES. "Algorithms for the assignment and transportation problems" J. Soc. Indust. Appl. Math. 5 (1957) 32-38.

703. MYCIELSKI, JAN, S. SWIERCZKOWSKI, and A. ZIĘBA. "On infinite positional games" Bull. Acad. Polon. Sci. Cl. III, Vol. 4 (1956) 485-488.

704. MYCIELSKI, JAN and A. ZIĘBA. "On infinite games" Bull. Acad. Polon. Sci. Cl. III, Vol. 3 (1955) 133-136.

705. NASH, JOHN F. "The bargaining problem" Econometrica 18 (1950) 155-162.

706. NASH, JOHN F. Jr. "Equilibrium points in n-person games" Proc. Nat. Acad. Sci. U.S.A. 36 (1950) 48-49.

707. NASH, JOHN. "Non-cooperative games" Ann. of Math. 54 (1951) 286-295.

708. NASH, JOHN Jr. "Two-person cooperative games" Econometrica 21 (1953) 128-140.

709. NASH, JOHN Jr. and L. S. SHAPLEY. "A simple three-person poker game." In reference 574, pp. 105-116.

NASH, JOHN Jr. See Kalisch, G. K.

NASH, JOHN Jr. See Mayberry, John P.

710. NEISSER, HANS. "L'oligopole, les anticipations et la théorie des jeux" Économie Appl. 5 (1952) 225-259. Traduit de l'anglais par Pierre Pujade.

711. NEISSER, HANS. "Oligopoly as a non-zero-sum game" Rev. Econ. Studies (1957) 1-22.

712. NEISSER, HANS. "The strategy of expecting the worst" Social Research 19 (1952) 346-363.

713. *NELSON, MICHAEL, E. N. CASTLE, and W. G. BROWN. "The use of the production function and linear programming in valuation of intermediate products" Land Econ. 33 (1957) 257-261.

714. NERING, EVAR D. "Symmetric solutions for general-sum symmetric 4-person games." This Study.

NERING, E. D. See Kalisch, G. K.

715. von NEUMANN, JOHN. "A certain zero-sum two-person game equivalent to the optimal assignment problem." In reference 575, pp. 5-12.

716. von NEUMANN, JOHN. "Communication on the Borel notes" Econometrica 21 (1953) 124-127.

717. von NEUMANN. JOHN. "A model of general economic equilibrium" Rev. Econ. Studies 13 (1945-46) 1-9. Translated by G. Morgenstern from reference 721.

718. von NEUMANN, JOHN. "Un modello di equilibrio economico generale" L'industria (Milano) (1952) N. 1, 1-13. Translated by Flaminio Mentasti from Rev. Econ. Studies 1945.

719. von NEUMANN, JOHN. "A numerical method to determine optimum strategy" Naval Res. Log. Q. 1 (1954) 109-115.

720. von NEUMANN, JOHN. "On the theory of games of strategy." Translated by Mrs. Sonja Bargmann. This Study.

721. von NEUMANN, JOHN. "Über ein ökonomisches Gleichungssystem und eine Verallgemeinerung des Brouwerschen Fixpunktsatzes" Ergebnisse eines Math. Kolloq. 8 (1937) 73-83.

722. von NEUMANN, JOHN. "Zur Theorie der Gesellschaftsspiele" Math. Ann. 100 (1928) 295-320.

723. von NEUMANN, JOHN and OSKAR MORGENSTERN. Theory of Games and Economic Behavior, Princeton, Princeton University Press, 1944, xviii, 625 pp.

724. von NEUMANN, JOHN and OSKAR MORGENSTERN. Theory of Games and Economic Behavior, 2nd ed., Princeton, Princeton University Press, 1947, xviii, 641 pp.

725. von NEUMANN, JOHN and OSKAR MORGENSTERN. Theory of Games and Economic Behavior, 3rd ed., Princeton, Princeton University Press, 1953, xx, 641 pp.

von NEUMANN, JOHN. See Brown, G. W.

von NEUMANN, JOHN. See Gillies, D. B.

726. NEYMAN, JERZY, ed. Proceedings of the First Berkeley Symposium on Mathematical Statistics and Probability, Berkeley, University of California Press, 1949.

727. NEYMAN, JERZY, ed. Proceedings of the Second Berkeley Symposium on Mathematical Statistics and Probability, Berkeley, University of California Press, 1951, xi, 666 pp.

728. NEYMAN, JERZY, ed. Proceedings of the Third Berkeley Symposium on Mathematical Statistics and Probability, Berkeley, University of California Press, 1956. I - ix, 208; II - ix, 246; III - ix, 252; IV - ix, 179; V - ix, 184 pp.

729. *NIELSEN, N. "Lineaer programmering" Nordisk Tidskr. for Teknisk Økonomi 42 (1955) 107-138.

730. NIKAIDÔ, HUKUKANE. "New aspects of von Neumann's model with special regard to computational problems" Ann. Inst. Stat. Math. Tokyo 6 (1954) 223-230.

731. NIKAIDÔ, HUKUKANE. "Note on the general economic equilibrium for nonlinear production functions" Econometrica 22 (1954) 49-53.

732. NIKAIDÔ, HUKUKANE. "On a minimax theorem and its applications to functional analysis" J. Math. Soc. of Japan 5 (1953) 86-94.

733. NIKAIDÔ, HUKUKANE. "On von Neumann's minimax theorem" Pacific J. Math. 4 (1954) 65-72.

734. NIKAIDÔ, HUKUKANE. "Zum Beweis der Verallgemeinerung des Fixpunktsatzes" Kodai Math. Sem. Rep. 5 (1953) 13-16.

735. NIKAIDÔ, HUKUKANE. "Zusatz und Berichtigung für meine Mitteilung 'Zum Beweis der Verallgemeinerung des Fixpunktsatzes in diesen Reports, Bd. 5, Nr. 1, 1953'" Kodai Math. Sem. Rep. 6 (1954) 11-12.

736. NIKAIDÔ, HUKUKANE and KAZUO ISODA. "Note on non-cooperative convex games" Pacific J. Math. 5 (1955) 807-815.

737. NORDBOTTEN, SVEIN. "Allocation in stratified sampling by means of linear programming" Skand. Aktuarietidskr. 39 (1956) 1-6.

NUVOLI, PROSPERO. See Codegone, Cesare.

738. NYBLÉN, GOREN. "Quelques réflexions sur le vieux problème de l'avantage collectif à la lumière des développements récents" Écon. Appliquée 5 (1952) 485-499.

739. NYBLÉN, GOREN. The Problem of Summation in Economic Science. Lund, C. W. K. Gleerup, 1951, xii, 289 pp.

740. ORCHARD-HAYS, WILLIAM. "Computational experience in solving linear programming problems" Proc. of a Symposium on Operations Research in Business and Industry, April 8 and 9, 1954, Kansas City, Mo., Midwest Research Institute, 1954, 92-103, discussion 104.

ORCHARD-HAYS, W. See Dantzig, G. B.

741. ORDEN, ALEX. "Application of the simplex method to a variety of matrix problems." In reference 401, pp. 28-50.

742. *ORDEN, ALEX. "Solutions of systems of linear inequalities on a digital computer" Proc. of the Association of Computing Machinery, Pittsburgh, 1952, 91-95.

743. ORDEN, ALEX. "The transhipment problem" Management Sci. 2 (1956) 276-285.

ORDEN, ALEX. See Dantzig, G. B.

ORDEN, ALEX. See Goldstein, L.

744. OTTER, RICHARD and JOHN DUNNE. "Games with equilibrium points" Proc. Nat. Acad. Sci. U.S.A. 39 (1953) 310-314.

745. OXTOBY, JOHN C. "The Banach-Mazur game and Banach category theorem." In reference 265, pp. 159-163.

746. OZGA, S. A. "Measurable utility and probability — a simplified rendering" Econ. J. 66 (1956) 419-430.

747. *PAULL, A. E. "Linear programming; a key to optimum newsprint production" Pulp and Paper of Canada 57 (1956) 145-150.

748. PEARSON, CARL E. "Note on linear programming" Q. Appl. Math. 14 (1956) 205-206.

749. PECK, J. E. L. and A. L. DULMAGE. "Games on a compact set" Canad. J. Math. 9 (1957) 450-458.

750. PERCUS, JEROME and LEON QUINTO. "The application of linear programming to competitive bond bidding" Econometrica 24 (1956) 413-428.

PESTON, M. H. See Beresford, R. S.

751. PETERSON, G. A. "Selection of maximum profit combinations of live-stock enterprises and crop rotations" J. Farm Econ. 37 (1955) 546-554.

752. PETERSON, RAYMOND P. "Constant risk minimax point estimates" Univ. Washington Publ. Math. 3 (1952) 77-84.

753. PETERSON, RAYMOND P. "The determination of classes of constant risk estimates" Univ. Washington Publ. Math. 3 (1952) 85-87.

754. PETERSON, RAYMOND P. "Uniformly best constant risk and minimax point estimates" J. Res. Nat. Bur. Standards 48 (1952) 49-53.

PETERSON, R. P. See Beaumont, R. A.

755. de POSSEL, RENÉ. "Sur la théorie mathématique des jeux des hasard et de réflexion" Conférence du Centre Universitaire Méditerraneen de Nice Actualites Sci. et Ind., No. 436, Paris, Hermann and Cie., 1936, 44 pp.

756. PRAGER, WILLIAM. "A generalization of Hitchcock's transportation problem" J. Math. Phys. 36 (1957) 99-106.

757. *PRAGER, WILLIAM. "Management science: a new field for applied mathematicians" Pi Mu Epsilon J. 2 (1956) 204-219.

758. PRAGER, WILLIAM. "Numerical solution of the generalized transportation problem" Naval Res. Log. Q. 4 (1957) 253-261.

759. PRAGER, WILLIAM. "On the caterer problem" Management Sci. 3 (1956-57) 15-23.

760. PRAGER, WILLIAM. "On the role of congestion in transportation problems" Z. Angew. Math. Mech. 35 (1955) 264-268.

761. PRAGER, WILLIAM. "On warehousing problems" J. Oper. Res. 5 (1957) 504-512.

762. PRAGER, WILLIAM. "Problems of traffic and transportation" Proc. of a Symposium on Operations Research in Business and Industry, April 8 and 9, 1954, Kansas City, Mo., Midwest Research Institute, 1954, pp. 105-113. Abstract.

763. PREDETTI, ALDO. "Come si risolve un problema di programmazione lineare mediante il metodo del potenziale logaritmico" Statistica 16 (1956) 189-211.

764. *PREDETTI, ALDO. "Un problema di programmazione lineare risolto col metodo del simplesso a variabili base" Riv. Industriale (1955).

765. PREDETTI, ALDO. "Sull' apporto di R. Frisch alla soluzione di problemi economici di programmazione lineare: il metodo del simplesso a variabili base" L'industria (Milano) 4 (1955) 526.

766. PRINZ, Dr. D. G. "Some experiences on the Manchester computer with the simplex method" Conference on Linear Programming (arranged by Ferranti and Co., Ltd.) May 1954.

767. PUTERBAUGH, HORACE L. "Computation of the R column of the simplex linear programming worksheet" J. Farm Econ. 38 (1956) 1058.

768. PUTERBAUGH, HORACE L, EARL W. KEHRBERG, and JOHN O. DUNBAR. "Analyzing the solution tableau of a simplex linear programming problem in farm organization" J. Farm Econ. 39 (1957) 478-489.

QUINE, W. V. See Krentel, W. D.

QUINTO, LEON. See Percus, Jerome.

769. RABIN, MICHAEL O. "Effective computability of winning strategies." In reference 265, pp. 147-157.

770. RADNER, ROY. "The linear team: an example of linear programming under uncertainty." In reference 17, pp. 381-396. Reprinted as Cowles Commission Paper N.S. No. 100.

771. RADNER, ROY and JACOB MARSCHAK. "Note on some proposed decision criteria." In reference 889, pp. 61-68. Reprinted as Cowles Commission Paper No. 96.

772. RAIFFA, HOWARD. "Arbitration schemes for generalized two-person games." In reference 575, pp. 361-387.

773. RAIFFA, HOWARD, G. L. THOMPSON, and R. M. THRALL. "An algorithm for the determination of all solutions of a two-person zero-sum game." In reference 401, pp. 100-114.

RAIFFA, HOWARD. See Luce, R. Duncan.

RAIFFA, HOWARD. See Motzkin, T. S.

774. RAPAPORT, ANATOL. "Some game-theoretic aspects of parasitism and symbosis" Bull. Math. Biophys. 18 (1956) 15-30.

RAPAPORT, ANATOL. See Foster, Caxton.

REITER, STANLEY. See Koopmans, Tjalling C.

775. RESTREPO, RODRIGO. "Tactical problems involving several actions." In reference 265, pp. 313-335.

RESTREPO, RODRIGO. See Karlin, Samuel.

776. *RICCIARDI, FRANC M. "Business war games for executives. A new concept in management training" Management Rev. (May 1957) 45-56.

RICCIARDI, FRANC. See Bellman, Richard.

RICCIARDI, FRANC. See McDonald, John.

777. RICHARDSON, MOSES. "Extension theorems for solutions of irreflexive relations" Proc. Nat. Acad. Sci. U.S.A. 39 (1953) 649-655.

778. RICHARDSON, M. Fundamentals of Mathematics, rev. ed., New York, The Macmillan Company, 1958, 387-401; also 262-263.

779. RICHARDSON, MOSES. "On finite projective games" Proc. Amer. Math. Soc. 7 (1956) 458-465.

780. RICHARDSON, MOSES. "On weakly ordered systems" Bull. Amer. Math. Soc. 52 (1946) 113-116.

781. RICHARDSON, MOSES. "Relativization and extension of solutions of irreflexive relations" Pacific J. Math. 5 (1955) 551-584.

782. RICHARDSON, MOSES. "Solutions of irreflexive relations" Ann. of Math. 58 (1953) 573-590. Errata 60 (1954) 595.

783. RICOSSA, SERGIO. "Un esempio di applicazione del metodo del simplesso nella teoria della produzione" Giornale degli Econ. e Ann. di Econ. 14 (1955) 129-140.

784. RICOSSA, SERGIO. "Grandi mezzi per le grandi decisioni di politica economica" Giornale degli Econ. e Ann. di Econ. 14 (1955) 455-467.

785. RICOSSA, SERGIO. "La programmazione lineare ed il metodo del simplesso" L'industria (1953) 401-435.

786. RICOSSA, SERGIO. "La programmazione lineare: un esempio di 'Operations Research'" Tecnica ed Organ 5 (1954) 16-19 (no. 13).

787. RICOSSA, SERGIO. "Note comparative su alcuni modelli della produzione" L'industria (1954) 238-243.

788. RICOSSA, SERGIO. "Nuove applicazione della teoria economica della produzione" Giornale degli Econ. e Ann. di Econ. 14 (1955) 33-51.

789. RIOS, SIXTO. "Metodos y problemas de la investigación operativa" Trabajos de Estadestica 7 (1956) 187-198.

790. *ROBBINS, HERBERT. "Asymptotically subminimax solutions of compound statistical decision problems." In reference 727, pp. 131-148.

ROBBINS, HERBERT. See Bahadur, Raghu Raj.

ROBBINS, HERBERT. See Hannan, James F.

791. ROBBINS, JAMES. "Military applications of operations research in Sweden" J. Oper. Res. 4 (1956) 347-352.

792. ROBINSON, JULIA. "An iterative method of solving a game" Ann. of Math. 54 (1951) 296-301.

ROGOW, A. A. See Luce, R. Duncan.

793. RHODE, VIRGINIA F. "Bibliography on linear programming" J. Oper. Res. 5 (1957) 45-62.

794. ROSENBLETH, WALTER A. "A review of J. von Neumann and O. Morgenstern, 'Theory of Games and Economic Behavior'" Psychometrika 16 (1951) 141-146.

ROYDEN, H. L. See Girshick, Meyer A.

RUBIN, HERMAN. See Karlin, Samuel.

795. RUIST, ERIK. "Comparison of tests for non-parametric hypotheses" Ark. Mat. 3 (1955) 133-163.

796. RUIST, ERIK. "Speltieri och ekonomoska problem" Ekonomisk Tidskr. 5 (1949) 112-117.

797. SAATY, T. L. "The number of vertices of a polyhedron" Amer. Math. Monthly 62 (1955) 326-331.

798. SAATY, THOMAS and SAUL GASS. "Parametric objective function (part I)" J. Oper. Res. Soc. 2 (1954) 316-319.

SAATY, THOMAS. See Gass, Saul.

799. SAKAGUCHI, MINORU. "On minimax tests of hypotheses" Rep. Stat. Appl. Res. Un. Jap. Sci. Engrs. 3 (1955) 130-139.

800. *SALINE, L. E. "Quadratic programming of interdependent activities for optimum performance" Am. Soc. Me. Engr. Trans. 78 (1956) 37-46.

801. SALVESON, M. E. "The assembly line balancing problem." In reference 17, pp. 55-101.

802. SAMUELSON, PAUL A. "Comparative statics and the logic of economic maximizing" Rev. Econ. Studies 14 (1946-47) 41-43.

803. SAMUELSON, PAUL A. "Linear programming and economic theory." In reference 17, pp. 251-272.

804. SAMUELSON, PAUL A. "Probability, utility, and the independence axiom" Econometrica 20 (1952) 670-678.

805. SAMUELSON, PAUL A. "Spatial price equilibrium and linear programming" Amer. Econ. Rev. (1952) 283-303.

SAMUELSON, PAUL A. See Dorfman, R.

806. SARDÁ, JUAN. "Nuevos aspectos del análisis de la conducta ecónomica" Moneda y Credito 58 (1956) 3-11.

807. SAVAGE, L. S. "Une axiomatisation de comportement raisonnable face à l'incertitude" Économetrie, Colloques Internationaux du Centre National de la Recherche Scientifique, 1953, Paris, Centre National de la Recherche Scientifique, 1953, 29-33.

808. SAVAGE, LEONARD J. The Foundations of Statistics. New York, John Wiley and Sons, Inc., 1954, xv, 294.

809. SAVAGE, LEONARD J. "The theory of statistical decision" J. Amer. Statist. Assoc. 46 (1951) 55-67.

SAVAGE, LEONARD J. See Bahadur, Raghu Raj.

SAVAGE, LEONARD J. See Friedman, M.

810. SCARF, HERBERT E. "On differential games with survival payoff." In reference 265, pp. 393-405.

811. SCARF, HERBERT E. and LLOYD S. SHAPLEY. "Games with partial information." In reference 265, pp. 213-229.

812. SCHELL, EMIL D. "Distribution of a product by several properties." In reference 17, pp. 616-642.

813. SCHELLING, THOMAS C. "An essay on bargaining" Amer. Econ. Rev. 46 (1956) 281-306.

SCHLAIFER, R. See Henderson, A.

SCHOENBERG, I. J. See Motzkin, T. S.

814. SCHUMANN, JOCHEM and HORST TODT. "Classical theory, Graham's theory and linear programming in international trade: Comment" Quart. J. Econ. 71 (1957) 464-469.

SCHÜTZENBERGER, MARCEL P. See Ville, Jean.

815. SCHWAN, H. T. and J. J. WILKINSON. "Linear programming; what can it do?" Chem. Eng. 63 (1956) 211-214.

816. SHACKLE, G. L. S. "Expectation and cardinality" Economic J. 66 (1956) 211-219.

817. SHANNON, CLAUDE E. "Programming a computer for playing chess" Phil. Mag. 41 (1950) 256-275.

818. SHAPLEY, LLOYD S. "Quota solutions of n-person games." In reference 575, pp. 343-359.

819. SHAPLEY, LLOYD S. "A solution containing an arbitrary closed component." This Study.

820. SHAPLEY, LLOYD S. "The solutions of a symmetric market game." This Study.

821. SHAPLEY, LLOYD S. "Stochastic games" Proc. Nat. Acad. Sci. U.S.A. 39 (1953) 1095-1100.

822. SHAPLEY, LLOYD S. "A value for n-person games." In reference 575, pp. 307-317.

823. SHAPLEY, LLOYD S. and MARTIN SHUBIK. "Method for evaluating the distribution of power in a committee system" Amer. Pol. Sci. Rev. 48 (1954) 787-792.

824. SHAPLEY, LLOYD S. and MARTIN SHUBIK. "Solutions of n-person games with ordinal utilities" (Abstract) Econometrica 21 (1953) 348-349.

825. SHAPLEY, LLOYD S. and R. N. SNOW. "Basic solutions of discrete games." In reference 574, pp. 27-35.

SHAPLEY, LLOYD S. See Bohnenblust, H. F.

SHAPLEY, LLOYD S. See Dresher, Melvin.

SHAPLEY, LLOYD S. See Karlin, Samuel.

SHAPLEY, LLOYD S. See Milnor, John.

SHAPLEY, LLOYD S. See Nash, J. F. Jr.

826. SHERMAN, SEYMOUR. "Comment on 'smooth patterns of production'" Management Sci. 1 (1955) 271.

827. SHERMAN, SEYMOUR. "Games and sub-games" Proc. Amer. Math. Soc. 2 (1951) 186-187.

SHERMAN, SEYMOUR. See Gale, David.

828. SHIFFMAN, MAX. "Games of timing." In reference 575, pp. 97-123.

829. SHOJI, URA. "Minimax approach to a single sampling inspection by attributes" Rep. Statist. Appl. Res. Un. Jap. Sci. Engrs. 3 (1955) 140-148.

SHRUBEN, LEONARD W. See Fisher, Walter D.

830. SHUBIK, MARTIN. "A business cycle model with organized labor considered" Econometrica 20 (1952) 284-294.

831. SHUBIK, MARTIN. "A comparison of treatments of a duopoly problem (Part II)" Econometrica 23 (1955) 417-431.

832. SHUBIK, MARTIN. "Does the fittest necessarily survive?" In reference 837, pp. 43-46.

833. SHUBIK, MARTIN. "Edgeworth market games." This Study.

834. SHUBIK, MARTIN. "A game theorist looks at the antitrust laws and the automobile industry" Stanford Law Review 8 (1956) 594-630.

835. SHUBIK, MARTIN. "Information, risk, ignorance, and indeterminacy" Quart. J. Econ. 68 (1954) 629-640.

836. *SHUBIK, MARTIN. "Information, theories of competition and the theory of games" J. Pol. Econ. 60 (1952) 142-150.

837. SHUBIK, MARTIN, ed. Readings in Game Theory and Political Behavior. Doubleday Short Studies in Political Science, Garden City, New York: Doubleday and Co., 1954.

838. SHUBIK, MARTIN. "The role of game theory in economics" Kyklos 6 (1953-54) 21-34.

839. SHUBIK, MARTIN. "The uses of game theory in management science" Management Sci. 2 (1955) 40-54.

SHUBIK, MARTIN. See Mayberry, J. P.

SHUBIK, MARTIN. See Shapley, Lloyd S.

840. SIMON, HERBERT A. "A comparison of game theory and learning theory" Psychometrika 21 (1956) 267-272.

841. SION, MAURICE. "On general minimax theorems" Pacific J. Math. 8 (1958) 171-176.

842. SION, MAURICE and PHILIP WOLFE. "On a game without a value." In reference 265, pp. 299-306.

SITGREAVES, ROSEDITH. See Girshick, Meyer A.

843. SJOBERG, GIDEON. "Strategy and social power: some preliminary formulations" The Southwestern Social Science Quarterly 33 (1953) 297-308.

844. SLATER, MORTON L. "A note on Motzkin's transportation theorem" Econometrica 19 (1951) 185-187.

SMITH, CEDRIC A. B. See Grundy, P. M.

SMITH, CEDRIC A. B. See Guy, Richard K.

845. SMITH, JACK W. "A plan to allocate and procure electronic sets by the use of linear programming techniques and analytical methods of assigning values to qualitative factors" Naval Res. Log. Q. 3 (1956) 151-162.

846. SMITH, VICTOR E. "Continuous and discontinuous factor substitution" So. Econ. J. 22 (1955) 189-201.

847. SMITH, VICTOR. "Perfect vs. discontinuous input markets: a linear programming analysis" J. Farm Econ. 37 (1955) 538-545.

848. SMITH, WALTER. "A note on truncation and sufficient statistics" Ann. Math. Stat. 28 (1957) 247-252.

849. SMITH, L. WHEATON Jr. "Current status of the industrial use of linear programming" Management Sci. 2 (1956) 156-158.

SNELL, J. LAURIE. See Kemeny, John J.

850. SNODGRASS, MILTON M. "Linear programming — a new approach to interregional competition in dairying" J. Farm Econ. 38 (1956) 1501-1510.

851. SNODGRASS, MILTON M. and CHARLES E. FRENCH. "Simplified presentation of 'transportation problem procedure' in linear programming" J. Farm Econ. 39 (1957) 40-51.

SNOW, R. N. See Shapley, Lloyd S.

852. SNYDER, RICHARD C. "Game theory and the analysis of political behavior." Stephen Bailey, et al. Research Frontiers in Politics and Government, Washington, D. C., The Brookings Institution, 1955, pp. 70-103 (Ch. 4).

853. SOBEL, MILTON. "An essentially complete class of decision functions for certain standard sequential problems" Ann. Math. Stat. 24 (1953) 319-337.

SOKOLOWSKY, D. See Gaddum, J. W.

SOKOLOWSKY, D. See Hoffman, A.

SOLOW, ROBERT M. See Dorfman, Robert.

854. SOMERVILLE, PAUL N. "Optimum sampling in binomial populations" J. Amer. Stat. Assoc. 52 (1957) 494-502.

855. SOYSTER, H. R. "Mathematical programming and evaluation of freight shipment systems, Part I applications" Naval Res. Log. Q. 4 (1957) 237-242.

856. ŠPAČEK, A. "Note on minimax solutions of statistical decision problems" Colloq. Math. 2 (1951) 275-281.

SPELLMAN, R. A. See Garvin, W. W.

857. STANLEY, Commander E. D., Lieutenant D. P. HONIG, and LEON GAINEN. "Linear programming in bid evaluation" Naval Res. Log. Q. 1 (1954) 48-54.

STANLEY, Commander E. D. See Gainen, Leon.

858. STEIN, CHARLES. "The admissibility of Hotelling's T^2 test" Ann. Math. Stat. 27 (1956) 616-623.

859. STEIN, CHARLES. "A necessary and sufficient condition for admissibility" Ann. Math. Stat. 26 (1955) 518-522.

STEIN, C. M. See Lehmann, E. L.

860. STEIN, P. and S. ULAM. "Experiments in chess on electronic computing machines" Chess Review (1957) 13-17.

861. *STEINHAUS, H. "Definicze potrzehne de teorii gier i puscige" Zlotz. Mysl. Akad. (Lwow) 1929.

862. STEINHAUS, H. "The problem of estimation" Ann. Math. Stat. 28 (1957) 633-648.

863. STEINHAUS, H. "Sur la division pragmatique" Econometrica 17 (Supplement) (1949) 315-319.

STEWART, F. M. See Gale, David.

864. STIEMKE, E. "Über positive Lösungen homogener linearer Gleichungen" Math. Ann. 76 (1915) 340-342.

865. STIGLER, GEORGE J. "The cost of subsistence" J. Farm. Econ. 27 (1945) 303-314.

866. STONE, JEREMY J. "An experiment in bargaining games" Econometrica 26 (1958) 286-296.

867. STONE, RICHARD. "The theory of games" Economic J. 58 (1948) 185-201.

868. STROTZ, ROBERT H. "Cardinal utility" Papers and Proc. Amer. Econ. Rev. 43 (1953) 384-397. W. J. Baumol, 415-416.

869. SUPPES, PATRICK and MURIEL WINET. "An axiomatization of utility based on the notion of utility differences" Management Sci. 1 (1955) 259-270.

SUPPES, PATRICK. See Davidson, Donald.

SUTHERLAND, J. G. See King, R. A.

870. SUZUKI, GEORGE. "Procurement and allocation of naval electronic equipments" Naval Res. Log. Q. 4 (1957) 1-7.

871. SVERDRUP, ERLING. "Similarity, unbiassedness, minimaxibility and admissibility of statistical test procedures" Skand. Aktuarietidskr. 36 (1953) 64-86.

872. *SVERDRUP, ERLING. "Weight functions and minimax procedures in the theory of statistical inference" Arch. Math. Naturvid. 51 (1952) 117-192.

873. SWANSON, EARL R. "Application of programming analysis to corn belt farms" J. Farm Econ. 38 (1956) 408-419.

874. SWANSON, EARL R. "Integrating crop and livestock activities in farm management activity analysis" J. Farm Econ. 37 (1955) 1249-1258. Discussion by R. J. Hildreth 1263-1265.

875. SWANSON, EARL R. and KIRK FOX. "The selection of livestock enterprises by activity analysis" J. Farm Econ. 36 (1954) 78-86.

SWIERCZKOWSKI, S. See Mycielski, Jan.

876. SYMONDS, GIFFORD H. "How to use linear programming" Petroleum Processing 10 (1955) 674-680.

877. *SYMONDS, GIFFORD H. "Linear programming solves gasoline refining and blending problems" Ind. and Eng. Chem. 48 (1956) 394-401.

878. SYMONDS, GIFFORD. Linear Programming: The Solution of Refinery Problems. New York, Esso Standard Oil Company, 1955, 74 pp.

879. *SYMPOSIUM ON LINEAR PROGRAMMING. J. Roy. Statist. Soc. Ser. B (1955) 165-208. S. Vajda, E. M. L. Beale, G. Morton, and A. H. Land.

TAEUBER, RICHARD C. See Fox, Karl A.

880. THOMAS, CLAYTON J. and WALTER L. DEEMER Jr. "The role of operational gaming in operations research" J. Oper. Res. 5 (1957) 1-27.

THOMAS, C. J. See also Caywood, T. E.

881. THOMPSON, DOROTHEA M. and GERALD L. THOMPSON. "A bibliography of game theory." This Study.

882. THOMPSON, GERALD L. "Bridge and signaling." In reference 575, pp. 279-289.

883. THOMPSON, GERALD L. "Decision making and new mathematics" Naval Res. Log. Q. 3 (1956) 141-149.

884. THOMPSON, GERALD L. "On the solution of a game-theoretic problem." In reference 576, pp. 275-284.

885. THOMPSON, GERALD L. "Signaling strategies in n-person games." In reference 575, pp. 267-277.

THOMPSON, GERALD L. See Kemeny, John J.

THOMPSON, GERALD L. See Motzkin, T. S.

THOMPSON, GERALD L. See Raiffa, Howard.

THOMPSON, GERALD L. See Thompson, Dorothea M.

886. THOMPSON, P. M. "Editing large linear programming matrices" Naval Res. Log. Q. 4 (1957) 97-100.

887. THOMSEN, PAUL. "The mathematical treatment of a well known two-person game" Mat. Tidskr. Ser. A (1952) 63-72.

888. THRALL, ROBERT M. "Some results in non-linear programming." In reference 17, pp. 471-493.

889. THRALL, ROBERT M., CLYDE H. COOMBS, and ROBERT L. DAVIS, eds. Decision Processes. New York, John Wiley and Sons, Inc., 1954, viii, 332.

890. THRALL, ROBERT M. and LEONARD TORNHEIM. Vector Spaces and Matrices. New York, John Wiley and Sons, Inc., 1957, xii, 318, Chap. 11.

THRALL, ROBERT M. See Motzkin, T. S.

THRALL, ROBERT M. See Raiffa, Howard.

891. TINTNER, GERHARD. "Game theory, linear programming and input-output analysis" Z. für Nationalökonomie 17 (1957) 1-38.

892. TINTNER, GERHARD. "Programmazione lineare stocastica con applicazioni a problemi di economia agraria" Giornali d' Econ. 14 (1955) 299-317.

893. TINTNER, GERHARD. "Stochastic linear programming with applications to agricultural economics." In reference 17, pp. 197-228.

894. TINTNER, GERHARD. "Strategische Spieltheorie und ihre Anwendung in den Sozialwissenschaften" Sonderdruck aus Allgemeines Statistisches Archiv 3 (1957) 242-251.

895. TINTNER, GERHARD. "La teoria dei giochi, la programmazione lineare e l'analisi della interdependenze strutturali" L'industria (1957). Translated by Dr. Adelberto Predetti from Z. für Nationalökonomie 17 (1957) 1-38.

TINTNER, GERHARD. See Babbar, M. M.

896. TODD, JOHN. "Motivation for working in numerical analysis" Comm. Pure Appl. Math. 8 (1955) 97-116.

897. TODD, JOHN. "Begründung für die Beschäftigung mit numerischer Analysis" Jber. Deutsch. Math. Verein. 58 (1955) 11-38.

TODT, HORST. See Schumann, Jochen.

898. TOMPKINS, C. "Machine attacks on problems whose variables are permutations" Proc. of the Sixth Symposium on Applied Mathematics, vol. VI Numerical Analysis, ed. by J. H. Curtiss. New York, McGraw-Hill Book Co., Inc., 1956, pp. 195-211.

899. TOMPKINS, C. "Projection methods in calculation." In reference 17, pp. 425-448.

TOMPKINS, C. B. See Heller, I.

TORNHEIM, LEONARD. See Thrall, Robert M.

900. TUCKER, ALBERT W. "Dual systems of homogeneous linear relations." In reference 576, pp. 3-18.

901. TUCKER, ALBERT W. Game Theory and Programming. National Science Foundation Summer Mathematics Institute Notes, Stillwater, The Oklahoma Agricultural and Mechanical College, 1955, 86 pp.

902. TUCKER, ALBERT W. "Linear and nonlinear programming" J. Oper. Res. 5 (1957) 244-257.

903. TUCKER, ALBERT W. "Linear inequalities and convex polyhedral sets." In reference 17, pp. 569-602.

904. TUCKER, ALBERT W. "Theorems of alternatives for pairs of matrices." In reference 401, pp. 180-181.

TUCKER, ALBERT W. See Dresher, M.

TUCKER, ALBERT W. See Gale, David.

TUCKER, ALBERT W. See Kuhn, H. W.

TUCKER, ALBERT W. See Luce, R. Duncan.

905. UZAWA, HIROFUMI. "A generalization of Laplace criterion for decision problems" Ann. Inst. Stat. Math. Tokyo 7 (1955-56) 123-129.

906. UZAWA, HIROFUMI. "Note on the rational selection of decision functions" Econometrica 25 (1957) 166-174.

907. VADJA, STEFAN. "Spieltheorie und statistische Entscheidungsverfahren" Mitteilungsbl. Math. Statist. 5 (1953) 52-61.

908. VAJDA, S. "Theory of games" Science News 28 (May 1953) 29-40. London, Baltimore, Melbourne. Penguin Books.

909. VAJDA, S., E. M. L. BEALE, G. MORTON, A. H. LAND, and DISCUSSANTS. "Symposium on linear programming" J. Roy Statist. Soc. Ser. B 17 (1955) 165-203.

910. VAJDA, S. The Theory of Games and Linear Programming, London, Methuen and Co., New York, John Wiley and Sons, Inc., 1956 , 106 pp.

911. VASWANI, R. "Aircraft assignment problem; an analytic linear programming solution" Aeronaut. Eng. Rev. 16 (1957) 64-68.

912. VAZSONYI, ANDREW. "Optimizing a function of additively separated variables subject to a simple restriction." In reference 17, pp. 453-469.

VAZSONYI, A. See Karush, W.

913. VERHULST, MICHEL. "The concept of a 'mission'" Naval Res. Log. Q. 3 (1956) 45-57.

914. VICKREY, WILLIAM. "Self-policing properties of certain imputation sets." This Study.

915. VIDALE, MARCELLO L. "A graphical solution of the transportation problem" J. Oper. Res. 4 (1956) 193-203.

916. VILLE, JEAN A. "Leçons sur quelques aspects nouveaux de la théorie des probabilités" Ann. Inst. H. Poincaré 14 (1954) 61-143.

917. VILLE, JEAN A. "Sur la théorie génerale des jeux ou intervient l'habilité des jouers" Applications aux Jeux de Hasard by Émile Borel and Jean Ville, Tome IV, Fascicule II, in the Traité du Calcul des Probabilités et de ses Applications par Émile Borel (1938) 105-113.

918. VILLE, JEAN et MARCEL PAUL SCHÜTZENBERGER. "Les problèmes de diagnostic séquentiel" C. R. Acad. Sci. Paris 232 (1951) 206-207.

919. VOGEL, WALTER. "Eine allgemeine Klasse von Zwei-Personen-Spielen" Bericht über die Tagung Wahrscheinlichkeitsrechnung und matematische Statistik in Berlin, Oktober 1954. Berlin, Duetscher Verlag der Wissenschaften (1956) pp. 73-79.

920. VOROBIEV, N. N. "Reduced strategies for games in the generalized form" Dokl. Akad. Nauk. 115 (1957) 855-857.

921. VOTAW, D. F. Jr. "Methods of solving some personnel-classification problems" Psychometrika 17 (1952) 255-266.

922. VOTAW, D. F. Jr. "Programming under conditions of uncertainty." In reference 17, pp. 187-195.

923. VOTAW, D. F. Jr. and A. ORDEN. "The personnel assignment problem." In reference 401, pp. 155-163.

924. WAGNER, HARVEY M. "A comparison of the original and revised simplex methods" J. Oper. Res. 5 (1957) 361-369.

925. WAGNER, HARVEY M. "A linear programming solution to dynamic Leontief type models" Management Sci. 3 (1956-57) 234-254.

926. WAGNER, HARVEY M. "A supplementary bibliography on linear programming" J. Oper. Res. 5 (1957) 555-563.

927. WAGNER, HARVEY. "A two-phase method for the simplex tableau" J. Oper. Res. 4 (1956) 443-447.

928. WAGNER, HARVEY M. "Unified treatment of bargaining theory; theory of games" So. Econ. J. 23 (1956-57) 380-397.

929. *WALD, ABRAHAM. "Asymptotic minimax solutions of sequential point estimation problems." In reference 727, pp. 1-11.

930. WALD, ABRAHAM. "Basic ideas of a general theory of statistical decision rules" Proc. of the International Congress of Mathematicians, Cambridge, Mass. Amer. Math. Soc. 1952, pp. 231-243.

931. WALD, ABRAHAM. "An essentially complete class of admissible decision functions" Ann. Math. Stat. 18 (1947) 549-555.

932. WALD, ABRAHAM. "Estimation of a parameter when the number of unknown parameters increases indefinitely with the number of observations" Ann. Math. Stat. 19 (1948) 220-227.

933. WALD, ABRAHAM. "Foundations of a general theory of sequential decision functions" Econometrica 15 (1947) 279-313.

934. WALD, ABRAHAM. "Generalization of a theorem by von Neumann concerning zero sum two person games" Ann. of Math. 46 (1945) 281-286.

935. WALD, ABRAHAM. "Note on zero sum two person games" Ann. of Math. 52 (1950) 739-742.

936. WALD, ABRAHAM. "On some systems of equations of mathematical economics" Econometrica 19 (1951) 368-403.

937. WALD, ABRAHAM. Statistical Decision Functions. New York, John Wiley and Sons, Inc., ix, 179 pp.

938. WALD, ABRAHAM. "Statistical decision functions" Ann. Math. Stat. 20 (1949) 165-205.

939. WALD, ABRAHAM. "Statistical decision functions which minimize the maximum risk" Ann. of Math. 46 (1945) 265-280.

940. WALD, ABRAHAM. "'Theory of Games and Economic Behavior,' by John von Neumann and Oskar Morgenstern" Rev. of Econ. Stat. 29 (1947) 47-52.

941. WALD, ABRAHAM. "Über einige Gleichungssysteme der mathematischen Ökonomie" Z. für Nationalökonomie 7 (1936) 637-670.

942. WALD, ABRAHAM and J. WOLFOWITZ. "Bayes solutions of sequential decision problems" Proc. Nat. Acad. Sci. U.S.A. 35 (1949) 99-102.

943. WALD, ABRAHAM and J. WOLFOWITZ. "Bayes solutions of sequential decision problems" Ann. Math. Stat. 21 (1950) 82-99.

944. *WALD, ABRAHAM and J. WOLFOWITZ. "Characterizations of the minimal complete class of decision functions when the number of distributions and decisions is finite." In reference 727, pp. 149-157.

945. WALD, ABRAHAM and J. WOLFOWITZ. "Optimum character of the sequential probability ratio test" Ann. Math. Stat. 19 (1948) 326-339.

946. WALD, ABRAHAM and J. WOLFOWITZ. "Two methods of randomization in statistics and the theory of games" Ann. of Math. 53 (1951) 581-586.

WALD, ABRAHAM. See Dvoretzky, A.

947. WALDEN, W. "A study of simple games through experiments on computing machines." In reference 265, pp. 201-211.

948. WALTER, EDWARD. "Bemerkungen zur Theorie der Entscheidungsfunktionen" Mitteilungsbl. Math. Statist. 5 (1953) 62-63.

949. WARD, L. E. Jr. "On the optimal allocation of limited resources" J. Oper. Res. 5 (1957) 815-819.

950. WAUGH, FREDERICK V. "The minimum cost dairy feed (an application of 'linear programming')" J. Farm. Econ. 33 (1951) 299-310.

951. WAUGH, FREDERICK V. and GLENN L. BURROWS. "A short cut to linear programming" Econometrica 23 (1955) 18-29.

952. WEINBERGER, OTTO. "Wirtschaftshandlungen und Spielstrategie" Son. aus der Statistichen Vierteljahresschrift 2 (1949) 24-31.

953. WEISS, LIONEL. "A higher order complete class theorem" Ann. Math. Stat. 24 (1953) 677-680.

954. WEISS, LIONEL. "Sequential procedures that control the individual probabilities of coming to the various decisions" Ann. Math. Stat. 25 (1954) 779-784.

955. WEISS, LIONEL. "Testing one simple hypothesis against another" Ann. Math. Stat. 24 (1953) 273-281.

WEISS, LIONEL. See Kiefer, J.

956. WELINDER, CARSTEN and GUNNAR BERGENDAL. "Något om spelteoriens användning inom nationalekonomien" Ekonomisk Tidskr. 54 (1952) 203-223.

957. WELTER, C. P. "The theory of a class of games on a sequence of squares, in terms of the advancing operation in a special group" Nederl. Akad. Wetensch. Proc. Ser. A 57 (1954) 194-200.

958. WEYL, H. "Elementare theorie der konvexen polyeder" Comment. Math. Helv. 7 (1935) 290-306.

959. WEYL, HERMANN. "Elementary proof of a minimax theorem due to von Neumann." In reference 574, pp. 19-25.

960. WEYL, HERMANN. "The elementary theory of convex polyhedra." In reference 574, pp. 3-18. (Translation of reference 958 by Harold W. Kuhn.)

961. WETZEL, WOLFGANG G. "Comments on 'production scheduling by the transportation method of linear programming'" J. Oper. Res. 5 (1957) 440-442.

962. WHEELER, HARVEY. "Problems of Stalinism" The Western Pol. Q. 10 (1957) 634-674.

963. WHITIN, T. M. "Classical theory, Graham's theory and linear programming in international trade" Q. J. of Econ. 67 (1953) 520-544.

WICKE, H. H. See Hale, J. K.

WIEGMANN, N. See Hoffman, A.

WILKINSON, J. J. See Schwan, H. T.

964. WILLIAMS, JOHN. The Compleat Strategyst. New York, McGraw-Hill, 1954, xiii, 234 pp.

965. *WILLIAMS, JOHN. La Stratégie dans les Actions Humaines, les Affaires, la Guerre, les Jeux. Traduit par M. Mesnage. Préface de G. Th. Guilbaud. Paris, Dunod, 1956, xvii, 278 pp.

966. WILLIAMS, NOEL. "The allocation of scarce materials between products" Appl. Statist. 5 (1956) 166-176.

967. WILLIAMS, NOEL. "An application of linear programming to the selection of raw materials" Appl. Statist. 4 (1955) 22-31.

WINET, MURIEL. See Suppes, Patrick.

968. WOLD, H. "Ordinal preferences or cardinal utility?" (with additional notes by G. L. S. Shackle, L. J. Savage, and H. Wold) Econometrica 20 (1952) 661-664.

969. WOLFE, PHILIP. "Determinateness of polyhedral games." In reference 576, pp. 195-198.

970. WOLFE, PHILIP. "The strict determinateness of certain infinite games" Pacific J. Math. 5 (1955) 841-847.

WOLFE, PHILIP. See Dantzig, G. B.

WOLFE, PHILIP. See Dresher, M.

WOLFE, PHILIP. See Frank, Marguerite.

WOLFE, PHILIP. See Sion, Maurice.

971. WOLFOWITZ, J. "Minimax estimates of the mean of a normal distribution with known variance" Ann. Math. Stat. 21 (1950) 218-230.

972. WOLFOWITZ, J. "On ε-complete classes of decision functions" Ann. Math. Stat. 22 (1951) 461-465.

973. WOLFOWITZ, J. "The method of maximum likelihood and the Wald theory of decision functions" Nederl. Akad. Wetensch. Proc. Ser. A 56 (1953) 114-119.

WOLFOWITZ, J. See Dvoretzky, A.

WOLFOWITZ, J. See Wald, Abraham.

974. WOOD, MARSHALL K. and GEORGE B. DANTZIG. "Programming of interdependent activities. I. General discussion" Econometrica 17 (1949) 193-199.

975. WOOD, MARSHALL K. and GEORGE B. DANTZIG. "The programming of interdependent activities: general discussion." In reference 552, pp. 15-18.

976. WOOD, MARSHALL K. and MURRAY A. GEISLER. "Development of dynamic models for program planning." In reference 552, pp. 189-215.

977. WOODBURY, MAX A. "The design of an optimal battery of tests." In reference 401, pp. 172-175.

978. WOODBURY, MAX A. "Linear-convex games" J. Soc. Appl. Math. 1 (1953) 137-142.

YASPAN, ARTHUR J. See Friedman, Lawrence.

YERUSHALMY, J. See Chiang, C. L.

979. YOUNG, GAIL S. Jr. "A footnote to 'statistical decision functions'" Michigan Math. J. 1 (1952) 186-188.

980. ZACHRISSON, L. E. "En stridsvagnsduell med spelteoretiska konsekvenser" Artilleri Tidskr. 84 (1955) 112-121.

981. ZERMELO, E. "Über eine Anwendung der Mengenlehre auf ide theorie des schachspiels" Proc. Fifth International Congress of Mathematicians, Cambridge, England, 1913, vol. II, pp. 501-504.

982. ZIĘBA, A. "Continuous games with perfect information" Bull. Acad. Polon. Sci. Cl. III 3 (1955) 515-518.

983. ZIĘBA, A. "An elementary proof of von Neumann's minimax theorem" Colloq. Math. 4 (1957) 224-226.

984. ZIĘBA, A. "An example of the game of Banach and Mazur" Colloq. Math. 4 (1957) 230-231.

ZIĘBA, A. See Mycielski, Jan.

985. ZUBRZYCKI, S. "On the game of Banach and Mazur" Colloq. Math. 4 (1957) 227-229.

ADDED IN PROOF

A1. ACRIVOS, A. "Linear programming: how does it work?" Chem. Eng. 63 (August 1956) 215-216.

A2. BECKMANN, MARTIN. "A continuous model of transportation" Econometrica 20 (1952) 643-660.

A3. *BITTER, F. "The mathematical formulation of strategic problems." In reference 726, pp. 223-228.

A4. DEBREU, GERARD. "Valuation equilibrium and Pareto optimum" Proc. Nat. Acad. Sci. U.S.A. 40 (1954) 588-592.

A5. de FINETTI, BRUNO. "La notion de 'horizon bayesien'." In Colloque sur l'Analyse Statistique, Bruxelles, 1954. Centre Belge de Recherches Mathématiques, Bruxelles, Georges Thone Liège, 1955, pp. 57-71.

A6. de FINETTI, BRUNO. "Media di decisioni e media di opinioni" Bulletin de l'Institut International de Statistique 34, 2 ième livraison (1954) 144-157.

A7. FRANCKX, E. "Sur les jeux stratégiques finis." In Colloque sur l'Analyse Statistique, Bruxelles, 1954. Centre Belge de Recherches Mathématiques, Bruxelles, Georges Thone Liège, 1955, pp. 143-158.

A8. GILLIS, PAUL P. and SIMONE HUYBERECHTS. "Théorie des jeux sur le carré-unité." In Colloque sur l'Analyse Statistique, Bruxelles, 1954. Centre Belges de Recherches Mathématiques, Bruxelles, George Thone Liège, 1955, pp. 159-176.

A9. *GOODMAN, L. A. and HARRY MARKOWITZ. "Social welfare functions based on individual rankings" Amer. J. of Soc. 58 (1952) 257-262.

A10. HUTTON, R. F. and J. R. ALLISON. "A linear programming model for development of feed formulas under mill-operating conditions" J. Farm Econ. 39 (1957) 94-111.

A11. *INADA, KEN-ICHI. "Elementary proofs of some theorems about the social welfare function" Ann. Inst. Stat. Math. 6 (1954) 115-122.

A12. KARLIN, S. and L. S. SHAPLEY. Geometry of Moment Spaces. Memoirs of the American Mathematical Society, No. 12. Providence, R. I., American Mathematical Society, 1953, 93 pp.

A13. KRELLE, WILHELM. "Unbestimmtheitsbereiche in der Preisbildung als ein Erklärungsgrund für Anderungen der Einkommensverteilung" Schiften des Vereins für Sozialpolitik Gesellschaft für Wirtschafts und Sozialwissenschaften. 12 (1956) 61-96.

A14. *McDONALD, JOHN. "Strategy of the seller — or what businessmen won't tell" Fortune 46 (1952) 124-127, 197-198.

A15. MACKENZIE, H. C. "The linear programming approach to production planning" Scottish J. of Pol. Econ. 4 (1957) 29-45.

A16. MARSCHAK, J. "Towards an economic theory of organization and information." In reference 889, pp. 187-220.

A17. PRINZ, D. G. Programming the Transportation Problem for Digital Computers. Manchester, Ferranti, Ltd. Sept. 1955, 32 pp.

A18. *SELIGMAN, B. B. "Game theory and collective bargaining" Labor and Nation 8 (1952) 50-52.

A19. SHANNON, LYLE W. "An experimental approach to the development of a socio-economic model" Social Forces 28 (1950) 410-418.

A20. *STEINHAUS, HUGO. "Quality control by sampling (a plea for Bayes rule)" Colloq. Math. 2 (1951) 98-108.

A21. TUCKER, A. W. "Linear programming" Indust. Qual. Control 12 (1956) 1-4.

A22. *WELDON, J. C. "On the problem of social welfare functions" Canad. J. of Econ. and Pol. Sci. 18 (1952) 452-463.

A23. YAMADA, YUZO and TSUNEYOSHI SEKI. "Price analysis in the 'Theory of Games,' with mathematical appending" The Hitotsubashi Review 27 (June 1952) 30-56 (in Japanese).

A24. *ZEUTHEN, FREDERIK. Problems of Monopoly and Economic Warfare. G. Routledge and Sons, London, 1930.